Restoration of Aquatic Ecosystems

Science, Technology, and Public Policy

Committee on Restoration of Aquatic Ecosystems:
Science, Technology, and Public Policy

Water Science and Technology Board

Commission on Geosciences, Environment,
and Resources

National Research Council

NATIONAL ACADEMY PRESS
Washington, D.C. 1992

NOTICE: The project that is the subject of this report was approved by the Governing Board of the National Research Council, whose members are drawn from the councils of the National Academy of Sciences, the National Academy of Engineering, and the Institute of Medicine. The members of the committee responsible for the report were chosen for their special competences and with regard for appropriate balance.

This report has been reviewed by a group other than the authors according to procedures approved by a Report Review Committee consisting of members of the National Academy of Sciences, the National Academy of Engineering, and the Institute of Medicine.

Support for this project was provided by Living Lakes, Inc.; Chevron USA, Inc.; U.S. Department of Agriculture, Soil Conservation Service under Agreement No. 69-3A75-9-152/R; U.S. Environmental Protection Agency under Agreement No. X-816435-01-0/R; U.S. Department of the Interior, Bureau of Reclamation under Agreement No. 9-FG-81-16650/R; and the National Research Council. Although the results described in this document have been funded in part by the U.S. Environmental Protection Agency under Assistance Agreement X-816435-01-0 to the National Academy of Sciences, it has not been subjected to the Agency's peer and administrative review, and therefore may not necessarily reflect the views of the Agency, and no official endorsement should be inferred.

Library of Congress Cataloging-in-Publication Data

NationalResearch Council (U.S.). Committee on Restoration of Aquatic
    Ecosystems—Science, Technology, and Public Policy.
        Restoration of aquatic ecosystems : science, technology, and
    public policy / Committee on Restoration of Aquatic Ecosystems-
    Science, Technology, and Public Policy, Water Science and
    Technology Board, Commission on Geosciences, Environment, and
    Resources.
            p.   cm.
        "November 1991."
        Includes bibliographical references and index.
        ISBN (invalid) 0-309-04534-7
        1. Aquatic ecology. 2. Aquatic ecology—Government policy—United
    States. I. National Research Council (U.S.). Water Science and
    Technology Board. II. National Research Council (U.S.). Commission
    on Geosciences, Environment, and Resources. III. Title.
    QH541.5.W3N38   1992   91-43324
    333.91'153—dc20   CIP

This book is printed on acid-free recycled stock that is made from 70% de-inked fiber of which 10% is postconsumer waste. ♻

Printed in the United States of America

Committee on Restoration of Aquatic Ecosystems: Science, Technology, and Public Policy

Staff

STEPHEN D. PARKER, Director
SHEILA D. DAVID, Senior Staff Officer
CHRIS ELFRING, Senior Staff Officer
SARAH CONNICK, Staff Officer
JACQUELINE MACDONALD, Research Associate
JEANNE AQUILINO, Administrative Specialist
ANITA A. HALL, Administrative Secretary
PATRICIA CICERO, Senior Secretary
JOYCE SPARROW, Secretary

The National Academy of Sciences is a private, nonprofit, self-perpetuating society of distinguished scholars engaged in scientific and engineering research, dedicated to the furtherance of science and technology and to their use for the general welfare. Upon the authority of the charter granted to it by the Congress in 1863, the Academy has a mandate that requires it to advise the federal government on scientific and technical matters. Dr. Frank Press is president of the National Academy of Sciences.

The National Academy of Engineering was established in 1964, under the charter of the National Academy of Sciences, as a parallel organization of outstanding engineers. It is autonomous in its administration and in the selection of its members, sharing with the National Academy of Sciences the responsibility for advising the federal government. The National Academy of Engineering also sponsors engineering programs aimed at meeting national needs, encourages education and research, and recognizes the superior achievements of engineers. Dr. Robert M. White is president of the National Academy of Engineering.

The Institute of Medicine was established in 1970 by the National Academy of Sciences to secure the services of eminent members of appropriate professions in the examination of policy matters pertaining to the health of the public. The Institute acts under the responsibility given to the National Academy of Sciences by its congressional charter to be an adviser to the federal government and, upon its own initiative, to identify issues of medical care, research, and education. Dr. Stuart Bondurant is acting president of the Institute of Medicine.

The National Research Council was organized by the National Academy of Sciences in 1916 to associate the broad community of science and technology with the Academy's purposes of furthering knowledge and advising the federal government. Functioning in accordance with general policies determined by the Academy, the Council has become the principal operating agency of both the National Academy of Sciences and the National Academy of Engineering in providing services to the government, the public, and the scientific and engineering communities. The Council is administered jointly by both Academies and the Institute of Medicine. Dr. Frank Press and Dr. Robert M. White are chairman and vice chairman, respectively, of the National Research Council.

[E]very generation receives a natural and cultural legacy in trust from its ancestors and holds it in trust for its descendants. This trust imposes upon each generation the obligation to conserve the environment and natural and cultural resources for future generations. The human species faces a grave obligation: conserve this fragile planet Earth and its human cultural legacy for future generations. We now recognize that humans have the power to alter the planet irreversibly, on a global scale. Humans must be concerned with the condition of the planet that is passed to future generations.

E. BROWN-WEISS
Environment
April 1990

Preface

This report is the result of recognition by the Water Science and Technology Board of the National Research Council's (NRC) Commission on Geosciences, Environment, and Resources that it should be concerned with the emerging science of restoration ecology in relation to aquatic ecosystems.

During its deliberations, the Committee on Restoration of Aquatic Ecosystems found that almost every restoration effort it reviewed focused on some component of a larger hydrologic system. The components fit into one of four categories: lakes, rivers, streams, and wetlands. However, the committee was also acutely aware that each of these entities functions in a larger ecological landscape greatly influenced by other components of the hydrologic cycle, including adjacent terrestrial systems. Regrettably, the case histories of restoration attempts that involved this larger ecological landscape were exceedingly rare.

After much discussion, the committee finally decided to review restoration case studies in the components of lakes, river and streams, and wetlands because the available literature tended to be compartmentalized in this way and because it was a convenient and easily understood means of communicating a large body of information. At the same time, the committee believed very strongly that the spatial and temporal scope of most restoration efforts was far too small. Moreover, the committee felt that all too many environmental decisions, including those involving restoration, had been made in a fragmented fashion unlikely to produce a self-maintaining aquatic ecosystem integrated into the larger ecological landscape. As a result, there is a special chapter on integrated aquatic ecosystem restoration (Chapter 7) that discusses the failings of a fragmented approach and speculates on the advantages to be derived from a more integrated approach to restoring aquatic ecosystems.

The committee recognized the difficulty of producing a report of acceptable length while also providing a useful level of detail on the large number of restoration efforts that have been completed or are in progress. It was decided that a limited number of illustrative case studies would be selected for review and that the level of detail would include only the information necessary to communicate the unique attributes of each restoration effort.

This report describes the status and functions of surface water ecosystems; the effectiveness of aquatic restoration efforts; the technology associated with those efforts; and the research, policy, and institutional reorganization required to begin a national strategy for aquatic ecosystem restoration. Although ground water is an important natural resource in the United States and degradation of its quality has an effect on surface water supplies, the committee chose not to review restoration of ground water. Despite increasing awareness that some of the ground water in the United States is contaminated, public policy toward ground water protection is still in the formative stages. Increased technology and expanded monitoring activities probably will detect the effects of past contamination and land uses on water quality. Conclusive answers to questions about the location, extent, and severity of ground water contamination, and about trends in ground water quality, must await further collection and analysis of data from the nation's aquifers.[1] The Water Science and Technology Board has in progress at this time a separate, special, detailed assessment of ground water remediation.

The committee was much influenced by the strategy of the former NRC Committee on Applications of Ecological Theory to Environmental Problems.[2] Our committee shares the 1986 NRC committee's perception that, whereas much about the functioning of ecological systems remains poorly understood, it is common to fail to use even available information when attempting to solve environmental problems. Finally, our committee also decided to provide examples of the

---

[1]U.S. Geological Survey. 1987. National Water Summary 1987. U.S. Geological Survey Water-Supply Paper 2350. U.S. Geological Survey, Federal Center, Denver, Colorado.

[2]National Research Council. 1986. Ecological Knowledge and Environmental Problem-Solving. Commission on Life Sciences, National Academy Press, Washington, D.C.

creative use of ecological information, believing that a good example is more instructive than a bad one. In following this strategy, we also recognized that powerful analytical systems are not substitutes for biological insights or imaginative questioning and hypothesizing. Therefore, we joined the Committee on Applications of Ecological Theory to Environmental Problems in focusing on some important issues concerning restoration techniques.

This report does not address the need for reintroducing species in restoration attempts, except to note the need for source pools of species in each ecoregion. The 1981 National Research Council report *Testing for Effects of Chemicals on Ecosystems*[3] advocated the establishment of ecological preserves, although for a different purpose (test species for ecotoxicological procedures including the establishment of microcosms and mesocosms). The need for such ecological preserves as a source of recolonizing species will increase dramatically if the "no-net-loss" policy for wetlands and other aquatic ecosystems is not implemented expeditiously.

The committee carried out its tasks through a series of meetings in which the format of the report was decided. Subgroups were formed to draft the various chapters. Restoration case studies were selected by these groups to illustrate points made in each chapter. The committee made four field trips to sites where restoration of aquatic systems had taken place or was going on. Subcommittees made two other site visits.

An assignment of this complexity, especially in a newly developing field, requires an exceptional effort on the part of committee members. The linkages among various components of the aquatic ecosystems and the terrestrial system that so strongly affects them are numerous and complex, as are the economic and policy questions related to the restoration process. Committee members worked diligently to sort through an enormous amount of information pertaining to a variety of aquatic ecosystems involving an even wider variety of methods to identify and analyze components critical to restoration efforts. I am much indebted to the subcommittee chairs Patrick Brezonik, Donald Hey, Leonard Shabman, Richard Sparks, James Tripp, Dan Willard, and Joy Zedler, who facilitated the flow of information and the meeting of deadlines. Most importantly, their summaries at each committee meeting ensured that the entire committee was aware of the working of these subunits.

---

[3]National Research Council. 1981. Testing for Effects of Chemicals on Ecosystems. National Academy Press, Washington, D.C.

Many thanks are also due to the committee members, NRC staff, and NRC consultant, who prepared the case histories without which the quality of this report would be seriously diminished. This report has benefited greatly from the skilled and creative efforts of Sheila D. David, Senior Staff Officer for the NRC, in contributing to the conceptual development of this study. I am personally indebted to Ms. David for alerting me to situations that required immediate attention and for her thoughtful discussions on how this report might best fill the charge of the Water Science and Technology Board.

The committee's consultant, John J. Berger, has been exceedingly helpful in a variety of ways including major contributions to the case studies and several chapters of the report. The committee is deeply indebted to Jeanne Aquilino, Administrative Specialist, for the systematic and orderly distribution of materials, draft report production, and professional assistance during the scheduled meetings.

I also wish to thank those who made presentations and provided background material to the committee during visits to restoration sites. Special thanks to David Rosgen, hydrologist, Pagosa Springs, Colorado; committee member Donald Hey, Director, Des Plaines River Wetlands Demonstration Project; Louis Toth of the South Florida Water Management District (SFWMD); Kent Loftin (former project manager) of the SFWMD; Anne Galli, Carol Ceberio, Don Smith, and Anthony Scardino, Jr., of the Hackensack Meadowlands Development Commission; Tom Muir of the U.S. Fish and Wildlife Service; and Steve Cordle and Bill Sipple of the U.S. Environmental Protection Agency. In addition, this study could not have been accomplished without the financial support of the National Research Council Fund; Chevron, USA, Inc.; Living Lakes, Inc.; the Soil Conservation Service; the U.S. Bureau of Reclamation; and the U.S. Environmental Protection Agency.

All committee members commented that the boundary conditions for this topic were much more difficult to establish than for other study projects in which they had participated. Part of the reason for this is that the varied disciplinary information necessary for an informed decision is daunting. Nevertheless, despite these difficulties, no one on the committee had any reservations about the potential for improving damaged aquatic ecosystems appreciably through restoration efforts. Even if a major national effort to restore aquatic ecosystems is forthcoming, their protection and management will require continued advances in point and nonpoint pollution abatement. The management and restoration of aquatic ecosystems will require intensive monitoring, as well as increased interaction and cooperation among federal, state, and local agencies concerned with air, water, wildlife, soil, agriculture, forestry, and urban planning and development. We hope this report contributes to the knowledge base and advancement of restoration ecology.

JOHN CAIRNS, JR., *Chairman*
Committee on Restoration of
Aquatic Ecosystems: Science,
Technology, and Public Policy

Contents

Boxes

Summary

The acid test of our understanding is not whether we can take ecosystems to bits on pieces of paper, however scientifically, but whether we can put them together in practice and make them work.

A. D. Bradshaw, 1983

INTRODUCTION

Aquatic ecosystems perform numerous valuable environmental functions. They recycle nutrients, purify water, attenuate floods, augment and maintain streamflow, recharge ground water, and provide habitat for wildlife and recreation for people. Rapid population increases in many parts of the United States—accompanied by intensified industrial, commercial, and residential development—have led to the pollution of surface waters by fertilizers, insecticides, motor oil, toxic landfill leachates, and feedlot waste. At the same time that water pollution and releases of nutrient-laden municipal sewage effluents have increased, water consumption has also increased, thus reducing the flows available for the dilution of wastes.

Increased sediment delivery resulting from urban construction, agriculture, and forestry also has resulted in greater turbidity and sedimentation in downstream channels, lakes, and reservoirs, with attendant losses of water storage and conveyance capacity, recreational and aesthetic values, and quantity and quality of habitat for fish and wildlife. Increased demands for drainage of wetlands have been ac-

1

commodated by channelization, resulting in further loss of stream habitat. This has led to aquatic organisms becoming extinct or imperiled in increasing numbers and to the impairment of many beneficial water uses, including drinking, swimming, and fishing.

Although public and private decisions to manage aquatic ecosystems have enhanced water transportation, developed sources of hydroelectric power, reduced flood hazards, and provided water for municipal, industrial, and agricultural purposes, these activities have also altered the physical, chemical, and biological processes within aquatic ecosystems. This committee is convinced that U.S. public opinion strongly supports an increased level of attention to environmental protection. The nation's investment in different types of environmental programs has been considerable but piecemeal and has not always been effective. An accelerated effort toward environmental restoration and preservation is needed. The committee believes that a comprehensive and aggressive restoration component should be the centerpiece of such an effort.

The premise of this report is that *ecological restoration* of aquatic ecosystems is possible. Restoration means returning an ecosystem to a close approximation of its condition prior to disturbance. Accomplishing restoration means ensuring that ecosystem structure and function are recreated or repaired, and that natural dynamic ecosystem processes are operating effectively again. At times, however, restoration may be impractical or undesirable, as when a body of water that is naturally without fish is successfully transformed through stocking into a valuable trout fishery or when important urban developments have been situated on wetlands. In such cases, the committee recognizes that the economic value of these developments may preclude any attempt to restore preexisting natural systems at these locations. The committee also recognizes that preventive measures to protect aquatic ecosystems are important and that priority should be given to preventive measures that benefit more than one portion of the hydrologic cycle. Had environmental protection been adequate in the past, many expensive restoration projects would not be necessary today.

Naturally, restoration of aquatic ecosystems may be accomplished in stages, and particular ecosystem functions and characteristics—such as potable water—may be restored even when other ecosystem characteristics deviate from natural conditions. Thus, in certain situations, partial ecological restoration may be the operant management goal and may provide significant ecological benefits even though full restoration is not attained.

Therefore, since the loss and impairment of aquatic ecosystems is accompanied by loss and impairment of valuable environmental functions and amenities important to humans, and since restoration of aquatic ecosystems is possible, the committee concludes that a large-scale aquatic ecosystem restoration program in the United States should be implemented to regain and protect the physical, chemical, and biological integrity of surface water. Such a program should seek to:

- correct nonpoint source pollution problems;
- arrest the decline of wildlife populations; and
- restore all types of wildlife habitats with priority to endangered species habitat.

Failure to restore aquatic ecosystems promptly will result in sharply increased environmental costs later, in the extinction of species or ecosystem types, and in permanent ecological damage.

NATIONAL STRATEGY

The committee recommends that a national aquatic ecosystem restoration strategy be developed for the United States. This comprehensive program should set specific national restoration goals for wetlands, rivers, streams, and lakes, and it should provide a national assessment process to monitor achievement of those goals. The following recommendations are proposed as building blocks for the program and its guiding strategy. Details of the program design should be developed by federal and state agencies in collaboration with nongovernmental experts. A national strategy would include four elements:

1. National restoration goals and assessment strategies for each ecoregion (regions that have broad similarities of soil, relief, and dominant vegetation).
2. Principles for priority setting and decision making.
3. Policy and program redesign for federal and state agencies to emphasize restoration.
4. Innovation in financing and use of land and water markets.

Achieving these restoration goals will require planning, federal leadership, and federal funding, combined with financial resources and active involvement from all levels of government, as well as the involvement of nongovernmental organizations and businesses. Therefore, the federal government should initiate an interagency and intergovernmental process to develop the national aquatic ecosystem restoration strategy. The program should be developed and maintained under the firm leadership of a single responsible organization with the characteristics stipulated in Chapter 8. Implementation of the program should include reliance on local and regional environmental restoration boards for program planning, synthesis, and leadership. Current appropriate federal programs should be reviewed to identify available opportunities for aquatic ecosystem restoration.

CONGRESS

In light of existing budgetary constraints, innovative ways to finance restoration efforts are necessary. Thus, Congress should establish a National Aquatic Ecosystem Restoration Trust Fund. Private landowners and corporations should be given powerful federal and state incentives to restore their aquatic ecosystems. Every effort should be made to use federal and other governmental funding to encourage citizen participation in restoration. Citizen participation (either through private citizen groups or public interest groups) has been instrumental in initiating and continuing restoration activities. In addition, Congress should allow states and local governments to trade in federal water development construction, maintenance, and major repair funds to finance aquatic ecosystem restoration programs.

The Food, Agriculture, Conservation, and Trade Act of 1990 (P.L. 101-624) authorized the U.S. Department of Agriculture (USDA) to enter into long-term contracts with farmers to take former wetlands in agricultural use out of production and allow them to be restored as wetlands. However, the act limits the number of acres eligible for the program to 200,000 per year, with a maximum of 1 million acres. Each acre of cropland taken out of production and restored as wetland is no longer eligible for USDA program benefits. Thus, Congress should request that USDA investigate where and how an expansion of the Agricultural Wetland Reserve Program would result in a savings of USDA farm program expenditures; and saved funds could then be reallocated to expand the wetland reserve program beyond 1 million acres.

Any redirection of federal policies and programs for aquatic ecosystem restoration should take into consideration the following:

- use of a landscape perspective in restoration efforts;
- use of adaptive planning and management (this refers to analysis of alternative strategies, reviewing new scientific data, and reanalyzing management decisions);
- evaluating and ranking restoration alternatives based on an assessment of opportunity cost rather than on traditional benefit-cost analysis;

• incorporating the definition of restoration as the return of an ecosystem to a close approximation of its condition prior to disturbance, in the mandates of all appropriate federal agencies;

• reliance on nonfederal and federal units of government to coordinate restoration programs in local areas; and

• initiating an interagency and intergovernmental process to develop a unified national strategy for aquatic ecosystem restoration.

LONG-TERM, LARGE-SCALE, COORDINATED RESTORATION— PLANNING, EVALUATING, AND MONITORING

Although restoration ecology applied to aquatic ecosystems is in a very early stage of development, the prospect for substantive improvements in damaged aquatic ecosystems is excellent. However, current federal and state environmental programs and policies are fragmented and do not adequately emphasize restoration based on management of large, interconnected aquatic ecosystems. The diverse responsibilities of all layers of government affecting aquatic resources need to be better coordinated if large-scale restoration is to be accomplished efficiently and effectively. Because aquatic ecosystems are interconnected and interactive, effective restoration efforts should usually be conducted on a large enough scale to include all significant components of the watershed.

In addition, aquatic restoration efforts also need to be long-term to ensure that restoration project goals have been achieved and that restored ecosystems can endure stressful episodic natural events such as floods, droughts, storms, pestilence, freezing, heavy cyclical predation, invasion by exotics, and other perturbations. Because of limited resources, it is impossible in the short term to undertake all worthy aquatic ecosystem restoration projects. Criteria are thus needed to set priorities, select projects, and evaluate project designs. It is important to give priority to the repair of those systems that will be lost without intervention. A "triage" framework needs to be applied as a minimum initial step. In this approach, threatened systems would be divided into three categories: (1) those that will recover without intervention, (2) those that cannot be restored to a meaningful degree even with extensive intervention, and (3) those that can be significantly restored with appropriate action. Systems in the third group require further consideration. Selections from that group should be based on criteria such as the likelihood of success, opportunity cost, and technical review of the restoration plan. It is imperative that these criteria be applied to the selection of projects because many restoration projects will not coincide with political boundaries.

Planning a restoration project must start with specifying the project mission, goals, and objectives. Goals should be prioritized so that project designers and evaluators have a clear understanding of their relative importance. In addition to specifying goals, objectives, and performance indicators, project managers and designers need to propose a monitoring and assessment program that is appropriate in scale as well as in sampling frequency and intensity to measure the performance indicators accurately and reliably, and thereby assess progress toward the project's objectives, goals, and mission. Postproject evaluation will enable scientists to determine when and to what degree the system has become self-maintaining and whether or not the restoration attempt was effective.

Monitoring of a restoration effort should include both structural (state) and functional (process) attributes, and should not be restricted to one level of biological organization. Monitoring of attributes at population, community, ecosystem, and landscape levels is appropriate in a restoration effort.

LAKES

By far the most widespread problem facing lakes and reservoirs is agricultural nonpoint runoff of silt and associated nutrients and pesticides. Lakes often do not cleanse or restore themselves. They are sinks for incoming contaminants that recycle and maintain the impaired conditions. Federal drinking water standards, for example, cannot be met, except with great difficulty and expense, unless degraded lakes and reservoirs are improved and then protected from further contamination.

A net gain over the next 20 years of 2 million acres of restored lakes, out of the current 4.3 million acres of degraded lakes, is an achievable goal. By the year 2000, it is recommended that a minimum of 1 million acres of lakes be restored. The costs for research, development, and technical guidance are federal responsibilities. The costs for actual restorations should be borne by federal and nonfederal sources, working through individual state lake programs. The committee realizes that the goals for the restoration of lakes should be realistic and tailored to individual regions of the country. Further development of project selection, goal setting, and evaluation techniques based on the concept of "ecoregions" as explained in Chapter 4 should be encouraged and supported by the U.S. Environmental Protection Agency (EPA).

All states have degraded lakes, and each state should develop restoration plans and programs. States should consider establishing trust funds for environmental restoration and protection. The Clean Lakes Program (CLP) administered by EPA has been the most reliable source of grant support for lake restoration efforts. This program should receive stable administrative support and increased funding from Congress. The 1991 appropriation for the CLP was $8 million. Although this amount will help to maintain or initiate a few lake restoration programs, it is inadequate for the large task of lake restoration facing the country. This program's mandate should be broadened to include all aspects of lake ecosystems, including habitat restoration, elimination of undesirable species, and restoration of native species.

Knowledge of the current ecological condition of the nation's lakes is grossly inadequate, and a national assessment of lakes is necessary to determine the severity and extent of damage and to measure changes in their status. The CLP should increase support of research and development of effective tools for restoration, and should continue guiding states in developing lake restoration programs.

The federal government should support research and development for demonstration watershed-scale restorations that integrate lake, stream, and wetland components. Research could be coordinated under an interagency program, such as the Federal Coordinating Council for Science, Engineering, and Technology, to coordinate the selection, planning, and evaluation of demonstration projects. Although many techniques are available to restore lakes, further development is required to improve their efficiency and effectiveness. The research and development programs in lake restoration should take an experimental approach, emphasizing controlled manipulation of whole-lake ecosystems or large in-lake enclosures.

Research and development programs in applied limnology are needed to study

- improved techniques for littoral zone and aquatic macrophyte management;
- biomanipulation (food web management);
- contaminant cleanup in lakes, especially for mercury and polychlorinated biphenyls (PCBs);
- the relationships between loadings of stress-causing substances and responses of lakes;
- paleolimnological approaches to restoration; and
- prediction of lake trophic state from nutrient loading relationships.

RIVERS AND STREAMS

Given that healthy, vegetated riparian habitat and bottomlands are essential to the natural ecological functioning of associated streams and rivers—and are among the nation's rarest habitats due to prior devastation—riparian habitat and bottomland restoration should be made a high national priority along with the restoration of the stream or river channel itself.

Because a river and its floodplain are intimately linked, they should be managed and restored as integral parts of an ecosystem. Remnant and undisturbed large river and floodplain ecosystems are rare and ecologically valuable. Therefore, reaches of certain large rivers and their floodplain ecosystems (such as portions of the Atchafalaya River and the Upper Mississippi River Fish and Wildlife Refuge) and at least 50 other large rivers (greater than approximately 120 miles in length) should be designated as "reference reaches" for use as restoration templates and should be protected as quickly as possible. Reference reaches should be designated and protected on representatives of all orders of streams and rivers in each of the nation's ecoregions. Highest priority should be given to protecting representative orders of rivers and streams not already protected as national wild and scenic rivers, or by being located in national or state parks.

Stream and river restoration should begin with improved land management practices that will allow natural restoration of the stream or river to occur. Therefore, the committee recommends the following:

• Erosion control programs in watersheds should be accelerated, not just to conserve soil, but also for the purpose of restoring streams and rivers.

• Grazing practices on federal lands should be reviewed and then changed to minimize damages to river-riparian ecosystems and to restore damaged rivers and streams.

• Erosion control by "soft engineering" approaches, such as bio-engineering techniques for bank stabilization and repair, should be considered first, in preference to "hard engineering" approaches, such as dams, levees, channelization, and riprap.

• Dikes or levees no longer needed or cost-effective should be razed to reestablish hydrological connections between riparian and floodplain habitats and associated rivers and streams.

• Classification systems for land use and wetlands should explicitly designate riparian environments and floodplains that retain their periodic connections to rivers.

The committee could not find a recent national assessment of the number of stream and river miles affected by channelization or leveeing, but the total is probably much greater than the number of miles of river dammed. Although water resources agencies track their own development projects, the only nationwide inventory of rivers and streams was conducted in the 1970s (DOI, 1982) in response to passage of the Wild and Scenic Rivers Act of 1968.

Therefore, the committee believes there is a need for a comprehensive up-to-date nationwide assessment of rivers, comparable to the National Wetland Inventory. It would be very useful to know how many miles of free-flowing, unchannelized rivers remain in the United States, and where these reaches are located.

The Food, Agriculture, Conservation, and Trade Act of 1990 and Section 404 of the Clean Water Act of 1977 (P.L. 95-217) now encourage the restoration and protection of wetlands. These laws should be expanded to provide for the protection and restoration of large active floodplains and riparian zones that are key components of riverine ecosystems. In addition, the Conservation Reserve Program, the Environmental Easement Program, and short-term agricultural set-aside programs should be amended to ensure that riparian zones and floodplains of all kinds are eligible for inclusion along with wetlands.

Opportunities to allocate water to in-stream uses arise (1) when land with water rights is sold or transferred, (2) when municipalities and irrigators decrease water withdrawals through conservation, and (3) when operating permits for dams are scheduled for renewal. Although the prior appropriations system (the basis of water law in the West) initially did not permit in-stream flow rights, many western states now recognize in-stream flow water rights. Therefore, states that have not established a water right for in-stream uses should do so. Flow that becomes available as the result of water conservation or lapse of permits should not automatically be reassigned to a consumptive use or withdrawal. Instead, consideration should be given to assigning the flow to in-stream uses. In addition, operating plans for dams should consider the annual water regime required by riverine fish and wildlife.

Federal agencies should be requested to update channelization estimates and to estimate miles of bank stabilization work already performed. The agencies should provide average and mean costs per mile for construction and maintenance of these conventional river management strategies, so that unit costs are available for comparison of different strategies. Government agencies should also conduct post-project evaluations of fluvial modifications, enhancement, improvement, channelization, and restoration projects to determine whether these projects actually achieve the benefits (e.g., flood protection, fish and wildlife enhancement) for which they were designed at costs that were projected.

The committee also recommends that a national river and stream restoration target of 400,000 miles of river-riparian ecosystems be restored within the next 20 years. This target represents only about 12 percent of the total 3.2 million miles of U.S. rivers and streams, and is recommended because it is comparable to the miles of streams and rivers affected by point source and urban runoff (EPA, 1990).

WETLANDS

Historically, the most destructive alterations to wetlands have been physical, often eliminating the topographic and hydrologic characteristics that support the wetland ecosystem. Their position in the landscape, whether as isolated wetlands or floodplains contiguous with rivers and streams, gives wetlands a major role in storage of floodwater and abatement of flooding. When wetlands are converted to systems that are intolerant of flooding (drained agricultural lands, filled developed lands), their storage capacity decreases and downstream flooding occurs. Wetlands have properties of both aquatic and terrestrial ecosystems. Their most widely valued function is providing habitat for fish, birds, and other wildlife, which contributes to the maintenance of biodiversity.

Controversy exists as to whether or not certain wetland systems can be restored. The arguments are particularly important when wetland restoration is undertaken with the promise that because full restoration of a degraded site is possible, other natural wetlands can be destroyed without any net loss of wetland habitat. Wetland restoration should not be used to mitigate avoidable destruction of other wetlands until it can be scientifically demonstrated that the replacement ecosystems are of equal or better functioning. Funding priority should be given to programs for restoration of damaged wetlands over wetlands creation because of the superior chances of success. An exception would be cases in which restoration is part of a mitigation agreement that would result in a net loss of acreage.

Wetlands restored in regulatory contexts often receive little management after initial restoration because private and public landowners, who are not motivated to provide such management, may move on or have no legal obligation for such management. Similarly, the responsible federal agencies do not have staff to assess the adequacy of restoration projects and do not monitor or require monitoring of permit mitigation conditions for sufficient time periods (10 years or longer). As a result, such wetlands may be overrun by exotic species, quickly filled by sediment, polluted, or otherwise misused.

The practice of wetland restoration needs to move from a trial-and-error process to a predictive science. The following recommended practices should be applied by resource managers to wetland restorations:

• Strive to restore wetlands to self-sustaining ecosystems requiring minimal maintenance.

• Provide buffers to protect restored wetlands, ensuring that restored coastal wetlands have room to migrate inland as long-term increases in sea level occur.

• Develop innovative methods of accelerating the restoration process (e.g., better propagation techniques for native plant species and protocols for obtaining adequate genetic diversity in the transplant material), and establish regional and national data bases to provide comparisons of the natural functioning of different wetland ecosystem types in different regions.

• Design and conduct experimental research programs to examine wetland restoration techniques and functional development over time in different system types.

• Use wetland restoration sites for scientific experiments that are designed to accelerate the restoration process.

• Support baseline studies of wetland ecosystem functioning to provide comparisons of different wetland types among regions and at different stages of development.

Traditional research on wetlands and ecosystem development should also be continued, using both natural and restored wetlands. Examples of this traditional research include the following topics adapted from Kusler and Kentula (1989):

• The hydrologic needs and requirements of wetland plants and animals, including minimum water depths, hydroperiod, velocity, dissolved nutrients, the role of large-scale but infrequent events, such as floods, and the effects of long-term fluctuations in water levels.

• The importance and functional significance of substrate to wetland plants and animals and to chemical and biological functions.

• Characteristics of development rates for natural successional vegetation.

• Recolonization of restored sites by invertebrate and vertebrate fauna.

• Functions of wetlands, with special emphasis on habitat values for a broad range of species, food chain support, and water quality enhancement.

• Evaluation of the stability and persistence of wetland ecosystems.

• Evaluation of the impact of sediment deposition or erosion, nutrient loading or removal, toxic runoff, pedestrian and off-road vehicle use, grazing, and other impacts on wetland structure and function.

• The ability of microbes, which are important to global carbon, sulfur, and nitrogen cycles, to perform these roles in restored wetlands.

The committee recommends that inland and coastal wetlands be restored at a rate that offsets any further loss of wetlands and contributes to an overall gain of 10 million wetland acres by the year 2010, largely through reconverting crop and pastureland and modifying or removing existing water-control structures. This represents a tenfold increase in the wetlands restoration target included in the Agricultural Wetland Reserve Program of the Food, Agriculture, Conservation, and Trade Act of 1990. This number also represents less than 10 percent of the total number of acres of wetlands lost in the last 200 years. The committee further recommends that, in the long term, this acreage be expanded to restore more of the approximately 117 million acres of the wetlands that have been lost in the United States over the past 200 years.

EDUCATION AND TRAINING

To accomplish the preceding tasks, the nation will require resource management professionals with multidisciplinary training. Restoration of aquatic ecosystems requires an integrated, broad-based approach; those trained to help restore these systems must have an interdisciplinary education. Although specialization will still be necessary, professionals will need the ability to coordinate work that draws on aquatic biology and fisheries, chemistry, hydrology, ecology, fluvial geomorphology, hydraulic engineering, social sciences, and wildlife management.

Some well-intentioned restoration projects have failed because fluvial and biological processes were not adequately taken into account in their design and implementation. The public has become increasingly aware of the need for restoration of river-riparian ecosystems (as several case studies in Appendix A indicate), and numerous public and private agencies and citizen groups are likely to initiate further stream and river restoration projects. These organizations, if properly guided and supported, can be a valuable impetus for effective aquatic ecosystem restoration and, in some cases, a valuable source of volunteer labor to accomplish restoration.

A new emphasis on resource stewardship and restoration cannot succeed without public understanding and support. Thus, educational programs aimed at raising the level of public knowledge and comprehension of aquatic ecosystem restoration rationales, goals, and methods should receive adequate government funding.

The committee believes that hydrological advisory services should be operated by states or federal agencies to provide technical assistance to groups interested in stream and river restoration. Universities with experts in natural resources or hydrology and water resources institutes, based at universities in every state, also should contribute technical assistance required for the restoration of aquatic ecosystems through free or at-cost expert hydrological and biological advisory services.

CONCLUSION

Without an active and ambitious restoration program in the United States, our swelling population and its increasing stresses on aquatic ecosystems will certainly reduce the quality of human life for present and future generations. By embarking now on a major national aquatic ecosystem restoration program, the United States can set an example of aquatic resource stewardship that ultimately will also improve the management of other resource types and will set an international example of environmental leadership.

REFERENCES

Benke, A. C. 1990. A perspective on America's vanishing streams. J. Am. Benthol. Soc. 9(1):77-78.

Bradshaw, A. D. 1983. The reconstruction of ecosystems. J. Appl. Ecol. 20:1-17.

Clean Water Act of 1977. P.L. 95-217, Dec. 27, 1977, 91 Stat. 1566.

Food, Agriculture, Conservation, and Trade Act of 1990. P.L. 101-624.

Kusler, J. A., and M. E. Kentula, eds. 1989. Wetland Creation and Restoration: The Status of the Science. Vol. I-II. U.S. EPA/7600/3-89/038. U.S. EPA Environmental Research Laboratory, Corvallis, Ore.

U.S. Department of the Interior (DOI), National Park Service. 1982. The Nationwide Rivers Inventory. U.S. Government Printing Office, Washington, D.C.

U.S. Environmental Protection Agency (EPA). 1990. The Quality of Our Nation's Water. EPA 440/4-90-005, Washington, D.C.

1

Overview

[A]ny nation concerned about the quality of life, now and forever, must be concerned about conservation. It will not be enough to merely halt the damage we've done. Our natural heritage must be recovered and restored. . . . It's time to renew the environmental ethic in America—and to renew U.S. leadership on environmental issues around the world. Renewal is the way of nature, and it must now become the way of man.

Vice President George Bush, 1988

Aquatic ecosystems worldwide are being severely altered or destroyed at a rate greater than that at any other time in human history and far faster than they are being restored. Some of these losses occur through intentional exploitation of resources. Other losses occur cumulatively and unobtrusively through lack of knowledge or careless resource management. Maintenance and enhancement of economically valuable aquatic ecosystem functions—especially floodwater storage and conveyance, pollution control, ground water recharge, and fisheries and wildlife support—have all too often been largely ignored in aquatic resource management. Even when management has been directed to these ends, it has often been fragmentary in its emphasis on lakes, rivers and streams, or wetlands in isolation from their regional watershed contexts—despite clear hydrological and ecological linkages. Contemporary restoration work is often too narrow in emphasis, focusing in lakes, for example, on correcting nutrient

14

overenrichment in the water column but giving little consideration to sedimentation or loss of aquatic habitat. Similarly, stream restoration efforts often concentrate on fisheries without regard for the wildlife values of riparian zone vegetation. Wetland restoration efforts often focus on revegetation while paying little attention to deep-water zones.

The purpose of this report is to suggest and analyze strategies for repairing past and ongoing damage to aquatic ecosystems from all types of anthropogenic activities. The loss or alteration of a large percentage of lakes, rivers, streams, and wetlands and of their associated vital ecological functions has a major effect both on the quality of life and on carrying capacities for human societies. These ecosystems provide a variety of ecological services of value to society. To ensure their viability for sustained, long-term use, freshwater ecosystems require not only protection from pollutants but also restoration and informed management.

The thesis of this report is that restoring altered, damaged, or destroyed lakes, rivers, and wetlands is a high-priority task at least as urgent as protecting water quality through abatement of pollution from point and nonpoint sources. Indeed these two activities are not dissociated, but rather are part of a continuum that includes both protection from pollution, and restoration and management. Restoration is essential if per capita ecosystem service levels are to remain constant while the global human population increases.

This report describes the status and functions of surface water ecosystems; the effectiveness of aquatic restoration efforts; the technology associated with those efforts; and the kinds of research, policy, management, and institutional changes required for successful restoration. Even if a major national effort is made to restore aquatic ecosystems, their protection and management will require continued advances in point and nonpoint pollution abatement. In short, the first objective should be to ensure no net loss of the quality of aquatic ecosystems, followed by efforts to increase the number of robust, self-maintaining aquatic ecosystems. Management of aquatic ecosystems will require intensive monitoring, as well as increased interaction and cooperation among national agencies concerned with air, water, wildlife, soil, agriculture, forestry, and urban planning and development.

STUDY BACKGROUND

Restoration is increasingly becoming an integral part of a national effort to improve water quality and the ecology of aquatic ecosystems. In 1988, the Water Science and Technology Board (WSTB) discussed the possibility of the National Research Council (NRC) contributing to the literature on restoration science and technology by conducting a review of both successful and failed attempts to restore aquatic ecosystems—specifically lakes, rivers, and wetlands.

A planning session was organized in the summer of 1988 to see if an NRC study of aquatic restoration efforts was appropriate. The planning committee decided that the science developing to support the emerging techniques of aquatic ecosystem restoration could benefit from an NRC assessment and report that would bring together significant and useful information on aquatic restoration efforts.

In 1989, the NRC appointed the Committee on Restoration of Aquatic Ecosystems: Science, Technology, and Public Policy under the WSTB to conduct an evaluation of the status of the restoration of aquatic ecosystems. The committee was requested to identify restoration projects and attempt to ascertain if they had succeeded or failed. Scientific, technological, political, and regulatory aspects were to be considered, as well as other factors that aid or hinder restoration efforts.

The committee's task has been to

1. develop a scientifically useful definition of restoration that could be considered as a standard for the science of restoration as it develops;

2. formulate criteria by which to choose the restoration projects to be reviewed as case studies;

3. evaluate restoration attempts with respect to their scientific basis, their performance over time, the technologies used, the monitoring effort, the costs, the objectives of the effort, the degree to which these objectives have been fulfilled, and why the efforts were successes or failures, while taking political and regulatory factors into consideration;

4. identify common factors of successful restoration projects and, based on this review, provide a recommended list of criteria for successful restoration that could serve as a model for future efforts to restore aquatic ecosystems;

5. identify federal policies and policy conflicts and those agencies that have programs resulting in negative impacts on aquatic ecosystems; and

6. make general recommendations regarding data needs, the science required to better understand each system, and the necessary regulations and policies.

The committee was composed of 15 restoration experts from the fields of limnology, geomorphology, surface water hydrology, aquatic and terrestrial ecology, water chemistry, environmental engineering, environmental law and policy, wetlands science, agricultural economics, and land use planning.

During the study the committee visited several restoration sites to determine firsthand how restoration efforts are accomplished. In a 2-year period, various committee members visited the Des Plaines River Wetlands Demonstration Project in Illinois; the Blanco River restoration in Pagosa Springs, Colorado; the Hackensack Meadowlands in New Jersey; prairie pothole wetlands in Minnesota; bottomland and hardwood forests in Louisiana; and the Kissimmee River restoration project in Florida. Writing assignments were made to several subcommittees concentrating on restoration of rivers, lakes, wetlands, and large integrated systems. Another subgroup concentrated on the development of a national aquatic ecosystem restoration strategy and the changes in policy and institutions necessary to begin this process. Brief case studies were prepared by the committee, NRC staff, and an NRC consultant.

This report is intended for a broad audience, including:

• scientists and engineers restoring aquatic ecosystems;
• legislators and regulators concerned with bringing the nation's aquatic ecosystems back to ecological health;
• state departments of environmental protection;
• industrial environmental protection departments;
• public interest and other citizen groups interested in restoring lakes, rivers, and wetlands; and
• teachers and students in the natural and environmental sciences.

WHAT IS RESTORATION?

As used in this report, the term *restoration* (see Box 1.1) means the reestablishment of predisturbance aquatic functions and related physical, chemical, and biological characteristics (Cairns, 1988; Magnuson et al., 1980; Lewis, 1989). Restoration is different from habitat creation, reclamation, and rehabilitation—it is a holistic process not achieved through the isolated manipulation of individual elements. The holistic nature of restoration, including the reintroduction of animals, needs to be emphasized. The installation of a few grasses and forbs does not constitute restoration. The long-term maintenance of biodiversity depends on the survival of appropriate plant assemblages, which may require, for example, grazing by muskrat and beaver. Without critical faunal elements, an ecosystem may not survive long.

Merely recreating a form without the functions, or the functions in

BOX 1.1
THE MEANING OF RESTORATION

In this report, *restoration* is defined as the return of an ecosystem to a close approximation of its condition prior to disturbance. In restoration, ecological damage to the resource is repaired. Both the structure and the functions of the ecosystem are recreated. Merely recreating the form without the functions, or the functions in an artificial configuration bearing little resemblance to a natural resource, does not constitute restoration. The goal is to emulate a natural, functioning, self-regulating system that is integrated with the ecological landscape in which it occurs. Often, natural resource restoration requires one or more of the following processes: reconstruction of antecedent physical hydrologic and morphologic conditions; chemical cleanup or adjustment of the environment; and biological manipulation, including revegetation and the reintroduction of absent or currently nonviable native species.

It is axiomatic that no restoration can ever be perfect; it is impossible to replicate the biogeochemical and climatological sequence of events over geological time that led to the creation and placement of even one particle of soil, much less to exactly reproduce an entire ecosystem. Therefore, all restorations are exercises in approximation and in the reconstruction of naturalistic rather than natural assemblages of plants and animals with their physical environments.

Berger, 1990

an artificial configuration bearing little resemblance to a natural form, does not constitute restoration. The objective is to emulate a natural, self-regulating system that is integrated ecologically with the landscape in which it occurs. Often, restoration requires one or more of the following processes: reconstruction of antecedent physical conditions; chemical adjustment of the soil and water; and biological manipulation, including the reintroduction of absent native flora and fauna or of those made nonviable by ecological disturbances. An aquatic ecosystem that was disturbed at some earlier time (e.g., 2, 20, or 200 years ago) is a candidate for restoration. An approximate point in time must be selected to develop criteria for restoration. Restoring an aquatic ecosystem to its predisturbance condition may be a difficult problem. For some ecosystems, the fossil record (fossil plants, pollen) can be helpful. For lakes, paleoecological methods can be used. For prairies, soil core analysis is used. Sometimes what is required is some "historical investigative ecology."

Whereas *restoration* aims to return an ecosystem to a former natural condition, the terms *creation, reclamation,* and *rehabilitation* imply putting a landscape to a new or altered use to serve a particular human purpose (creation or reclamation) (see Glossary, Appendix B, for definitions).

The term *restoration* is used in numerous regulations and public laws when what is meant is reclamation, rehabilitation, or mitigation. In 1937, Congress enacted the Federal Aid in Wildlife Restoration Act (P.L. 75-415), which was intended to aid wildlife restoration projects. In the statement of purpose, however, the terms *restoration* and *rehabilitation* are used interchangeably. Further, the bill deals only with ". . . improvement of areas of land or water adaptable as feeding, resting, or breeding places for wildlife . . . ". In a similar vein, a memorandum of agreement between the U.S. Army Corps of Engineers and the U.S. Environmental Protection Agency (1990) defines restoration as "measures undertaken to return the existing fish and wildlife habitat resources to a modern historic condition. Restoration then includes mitigation as well as some increments of enhancement." Mitigation is simply the alleviating of any or all detrimental effects arising from a given action (although this may not truly occur). Mitigation for filling a wetland in order to build a shopping center may involve restoring a nearby wetland that had been filled for some other reason, or it could involve creating a wetland on an adjacent area that was formerly upland. Mitigation need not, and often does not, involve in-kind restoration or creation. For example, the loss of floodwater storage due to filling a wetland might be mitigated by creating a detention basin. Although the functional attributes of flood control are rehabilitated, the chemical and biological characteristics or other functional values of the wetland are not. Mitigation of frequently and rapidly fluctuating water levels in a flood control reservoir may be achieved simply by altering the release schedule from the reservoir. In this case, mitigation is achieved by reclamation, not by restoration or creation.

Preservation is the maintenance of an aquatic ecosystem. Preservation involves more than preventing explicit alterations, such as removing timber from a bottomland hardwood wetland or preventing the construction of levees and tide gates on a coastal marsh. Preservation also implies management (e.g., weed and pest control) of the aquatic ecosystem to maintain its natural functions and characteristics. Preservation is sometimes mistakenly linked to mitigation via the assumption that a preserved aquatic ecosystem at one location will offset or mitigate the losses of displaced aquatic functions at another. Although such preservation may prevent further losses, it cannot compensate for losses already incurred. Preservation is distinct from restoration and creation in that the functions and characteristics of the preserved ecosystem are presumed to exist, more or less, in their desired states. This is not to say that the aquatic ecosystem has not been subject to changes over the years but that the ecosystem is performing in an acceptable manner not requiring reclamation or rehabilitation.

Whether restored, created, rehabilitated, mitigated, or preserved, most, if not all, aquatic ecosystems subject to the pressures of large human populations need to be managed. Management is the manipulation of an ecosystem to ensure the maintenance of one or more functions or conditions. In the case of preserved, created, or restored aquatic ecosystems, management activities should be directed toward maintaining all functions and characteristics. This is distinct from the management of an aquatic ecosystem for more limited objectives. Controlling water levels in a wetland for duck production is a limited management objective. Another limited objective is releasing water from a reservoir to maintain in-stream flows for trout fishing. These activities generally ignore the needs of other organisms and bias an ecosystem's characteristics in support of a desired single function. However, management of an aquatic ecosystem need not be limited in scope. Controlled burns of mesic prairies will prevent the introduction of weedy plant species and increase plant and habitat diversity. The management strategy of using beaver to build dams to prevent stream-bank erosion (Spencer, 1985) may also aid the restoration process when, for example, the beavers graze on woody vegetation and the beaver ponds trap nutrients and sediments (Seton, 1929; Naiman, 1988).

Selectively restoring a river meander or a chemical characteristic of a lake is not restoring the aquatic ecosystem unless that is the only significant aspect that has been degraded. To restore the aquatic ecosystem, all functions and characteristics must be considered, an approach that may in practice be difficult to achieve. However, the term *restoration* should be applied only to those activities directed to rebuilding an entire ecosystem: reconstructing topography without using the appropriate soils or plant materials is unlikely to lead to recreating the plethora of functional values of the natural or predisturbed aquatic ecosystem. Although it may seem appropriate to describe as restoration the building of wetlands in backwater areas of a flood control or water supply reservoir, this application distorts the meaning and masks the true purpose of such a created aquatic ecosystem. These ecosystems may be desired in backwater areas for duck habitat and hunting, water quality management, or even additional flood control. However, such created ecosystems will not possess the full range of physical, chemical, and biological characteristics of their natural counterparts. For example, their hydrologic characteristics will differ markedly from the prototype.

The distinctions among the terms *restoration, creation, rehabilitation,* and *reclamation* are important, and it is necessary to understand also how these terms relate to mitigation and preservation. Using consistent definitions, scientists and engineers will be better able to communicate their intentions and activities among themselves, policymakers, and the general public. This should facilitate setting clear goals and establishing effective programs for improving our environment.

STATUS OF AQUATIC RESOURCES IN THE UNITED STATES

This report on the status of our aquatic ecosystems must start with an assessment of the conditions of the land surface. Ninety-seven percent of this country's surface area is land; consequently, most of the water moving into and through aquatic ecosystems interacts with the surface of the land. Of the land surface in the 50 states, comprising 2.3 billion acres, 54 percent is managed for agricultural purposes (Bureau of the Census, 1990). Excluding Alaska, agricultural lands account for 65 percent of the land surface. Of the agricultural lands, 39 percent are grazed and 37 percent are cropped (Frey and Hexem, 1985). Regardless of the activity, the 1.2 billion acres of agricultural land have been substantially altered. Grazing, plowing, chemical applications, and drainage have changed the vegetative cover and soil conditions to such an extent that they no longer exhibit the characteristics of preagricultural conditions. These activities are necessary to support our highly productive agricultural industry, but one of the side effects is the degradation of aquatic ecosystems on a continental scale.

Smaller in scale but more extreme in effect is the alteration of the land surface to accommodate urban development. In building cities, wetlands and floodplains have been filled and made impervious by asphalt and concrete. Although only 3 percent of the nation's land surface is designated as urban, within an urban area, the hydrological and biological changes are extreme. In Chicago, a city of 228 square miles, 45 percent of the land is now covered by impervious surfaces. The once verdant wet prairies and marshes that dominated the landscape before this great city was built are gone. The roofs, streets, and roads have greatly changed the quantity and quality of water flowing into Lake Michigan and into the Des Plaines and Illinois Rivers. The change in flow was accompanied by a dramatic change in water quality due to the large waste loads conveyed by storm water runoff and by domestic and industrial wastewater. Both the hydrologic and the water quality effects extend miles beyond the limits of the city.

The U.S. agricultural industry and urban systems have had to rely, to a great extent, on the diverse functions of aquatic ecosystems. Uplands, wetlands, and floodplains have been drained to build houses, factories, and farms. Approximately 117 million acres of wetlands alone have been lost in the United States since the 1780s (Dahl, 1990). This represents 5 percent of the total land surface in the 50 states but about 30 percent of the presettlement wetlands (excluding Alaska, the wetland loss is approximately 53 percent; Dahl, 1990). The effects of increased losses have been harmful, if for no other reason than increased flooding. The dispersive capabilities of streams and rivers were and are inadequate to handle the large amounts of runoff generated and diverted to them from uplands and former wetlands, which once acted as flood control reservoirs. In 1912, the state engineer for Illinois observed that floods on the Des Plaines River were increasing in severity and frequency (Horton, 1914). He ascribed this hydrologic phenomenon to the clearing of land and draining of wetlands in the watershed.

The widespread loss of U.S. wetlands is illustrated in Figure 1.1. When one considers the losses from 1780 to 1980 in the central United States, it is no wonder that floods ravaged the river valleys of the Ohio, Wabash, Illinois, Missouri, and Mississippi. Unfortunately, wetlands continue to be drained by ditching, and storage areas continue to be blocked by levees, so that flood damage continues to increase.

Whereas more than 60 percent of the U.S. land surface is manipulated for human needs (urban development, forests, and agricultural areas), more than 85 percent of the inland water surface area in the United States is artificially controlled (Bureau of Census, 1990). Surface water controls range from very simple fixed weirs to very complex multigated dams and extend from small farm ponds and streams to our largest rivers and the Great Lakes. They benefit us in numer-

Wetland Distribution Circa 1780's

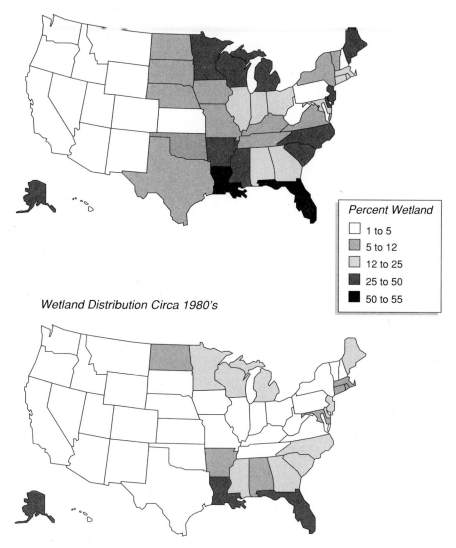

Wetland Distribution Circa 1980's

FIGURE 1.1 Comparison of wetland acreage in the United States in the 1780s and the 1980s. Source: Dahl, 1990.

TABLE 1.1 U.S. Water Budget for 1980 (billion gallons per day)

Region[a]	Supply[b]	Ground Water Depletion	Consumptive Use	Reservoir Evaporation	Yield[c]
1	77.3	0	0.4	0.2	76.7
2	96.5	0	1.7	0.2	94.6
3	212.6	0	5.1	0.5	207
4	76.8	0	1.3	0.3	75.2
5	140.1	0	1.7	0.4	138
6	43.3	0	0.4	0	42.9
7	79.7	0	1.5	0.6	77.6
8	75.4	0.04	7.14	0.30	68
9	7.7	0	0.1	0.4	7.2
10	67.3	2.2	16	3.3	50.2
11	63.7	3.6	9.6	1.4	56.3
12	35.9	3.1	6.5	1.8	30.7
13	5	0	2.4	0.8	1.8
14	12.3	0	2.3	1.7	8.3
15	−1.1	2.1	4.9	1.9	-5.8
16	17.1	12	3.9	0.2	25
17	290.6	0	12	0.6	278
18	86.9	1.4	25	0.5	62.8
19	921.04	0	0.04	0	921
20	14.3	0	0.7	0	13.6
Total	2,322.54	24.44	103	15.1	2,229.1

[a]Regions relate to the hydrologic units assigned by the U.S. Geological Survey (see Figure 1.2).

[b]Surface runoff before adding ground water and subtracting consumptive use and evaporation.

[c]Surface water discharge from the region.

SOURCE: Solley et al., 1988.

ous ways. They stabilize lakes at levels that afford reliable access for recreational boating, and they maintain navigational conditions for commercial barges and ships. Manipulation of water levels offers optimal flood protection and water supply for drinking and irrigation. However, the controls also may have detrimental effects on wildlife and other functions of aquatic ecosystems, and wetlands in the littoral zone suffer from either too much or too little water. Dynamic hydrologic cycles are all but eliminated, causing the degradation of plant and animal communities.

Of the 2,200 billion gallons of water available per day in the United States, approximately 4.7 percent is consumed (Table 1.1 and Figure 1.2). This total assumes, however, that the availability of water is uniformly distributed over time throughout the year. On a sustained basis, perhaps only 25 percent of the water is available on average, so that the consumption rate is thus quadrupled to 18.7 percent relative to the sustained yield—still a small percentage of the total available resource. A much higher percentage is extracted and recycled. The U.S. Geological Survey (Solly et al., 1988) estimated that in 1985 a total of 338 billion gallons of water per day was used for off-stream purposes (Table 1.2). This represented approximately 15 percent of the total resource, or 61 percent of the sustained yield. In-stream uses were an order of magnitude larger. The production of hydropower utilizes more than 3,000 billion gallons per day, an amount that exceeds the available supply but includes the repetitive use of water as

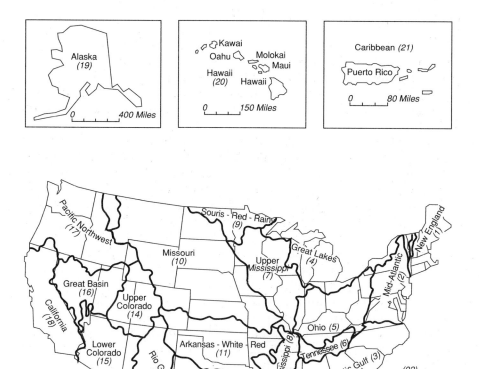

FIGURE 1.2 Hydrologic units of the United States. Source: U.S. Geological Survey, 1987.

TABLE 1.2   Water Use in the 50 States for 1985 (million gallons per day)

| Sector | Off-stream | | | | | On-stream |
	Ground	Surface	Total	Con-sumed	Re-turned	
Domestic-commercial	3,989	31,311	35,300	6,884	28,417	
Industrial-mining	5,267	25,533	30,800	4,928	25,872	
Thermoelectric	655	130,345	131,000	4,323	126,677	
Irrigation-livestock	48,504	92,496	141,000	75,999	65,001	
Hydropower						3,050,000
Total	58,415	279,685	338,100	92,134	245,967	3,050,000

SOURCE: Solley et al., 1988.

it moves through river systems.   Given that there are well over 2.5 million dams in the United States (Johnston Associates, 1989), only a small probability exists that a drop of water could make its way from its cloud of origin, over the land surface, through the drainage system, and back into an ocean without passing through a man-made structure.

Both off-stream and on-stream uses change the physical and chemical characteristics of the water. Reservoirs alter the thermal properties of the waters in rivers and streams by changing the surface area and depth characteristics. During the winter the larger surface areas created by a reservoir release more heat than an undammed stream would have, whereas during the summer they absorb more heat; consequently, the downstream thermal regime is changed. Thermal electric plants discharge heat to streams, rivers, and lakes via the dispersal of cooling waters. Domestic and industrial (including thermal electric) uses alter the hydrology at the point of both withdrawal and discharge. The return flows introduce elevated concentrations of nutrients and toxic substances despite modern wastewater treatment technology. Relative to the sustained yield, industrial and domestic wastewaters represent about 32 percent of the water treated. Dissolved solids are added to the stream from irrigation return flows and agricultural drainage in general. These flows account for 12 percent of the sustained yield. The high concentrations of dissolved solids result, in part, from the evaporation of irrigation water. Evaporative losses account for 14 percent of the sustained yield. Other sources such as runoff from roads, parking lots, and farm fields contribute substantial amounts of solids and nutrients to our rivers, lakes, and streams.

Despite the investment of more than $260 billion (1990 costs) from 1970 to 1984 in the construction and operation of public and private wastewater treatment facilities, the chemistry of our streams seems to have improved only slightly (U.S. EPA, 1984; Smith et al., 1987). Based on an analysis of 380 sampling stations distributed throughout the country, the concentrations of chloride, sulfate, nitrate, magnesium, sodium, and potassium (Smith et al., 1987) have increased. Suspended solids and pH have also increased at most stations, as have the concentrations of heavy metals, including arsenic, cadmium, iron, and manganese. Although most stations reported that dissolved oxygen increased, a beneficial change, the ratio was only about 3 to 2. Decreases were reported in the concentrations of calcium and phosphorus. Based on analyses undertaken by state personnel, the U.S. Environmental Protection Agency has concluded that progress has been made but that much remains to be done (U.S. EPA, 1990). However, only 758,000 miles of stream were surveyed, 23 percent of the total streams in the United States.

The apparent lack of concern for the physical structure of our nation's streams perhaps stems from the fact that no one seems to have a very clear idea of how many stream miles there are in the country, let alone their physical, chemical, and biological state of repair. Although basic documentation is lacking, one estimate is that there are more than 3.25 million miles of U.S. stream channels (Leopold et al., 1964) and, based on EPA's estimate, 758,000 of these miles are affected by effluents from municipal and industrial treatment plants. An additional 155,000 miles are constructed agricultural drains (Wooten and Jones, 1955). Incorporated into our major river systems are close to 12,000 miles of inland waterways. For these waterways, navigational channels are maintained at depths of 8 to 16 ft. Along our streams, levees and flood walls traverse an estimated 25,000 miles (Johnston Associates, 1989) and enclose more than 30,000 square miles of floodplain. The floodplain estimate is extrapolated from the ratio of length of levees to enclosed area for the Upper Mississippi River. Channelization, for navigation or drainage, and levees have drastically reduced the flow area of streams. At the same time, increased runoff from the draining of uplands and wetlands has been forced into the drainage system. The hydrological effects of this loss of storage are enormous.

The environmental stress and altered characteristics and functions of our aquatic ecosystems caused by dispersive and extractive uses and stream modifications are reflected in the status of our fisheries, as reported by the U.S. Fish and Wildlife Service (Judy et al., 1984). Of 666,000 miles of perennial U.S. streams surveyed, more than 40

TABLE 1.3   Water Quality Limitations
on Fisheries of Perennial Streams

Limit	Miles Affected	Percent[a]
Turbidity	277,000	41.6
Elevated temperature	215,000	32.3
Excess nutrients	144,000	21.6
Toxic substances	90,900	13.6
Dissolved oxygen	75,400	11.3
pH	26,000	3.9
Salinity	14,600	2.2
Gas supersaturation	5,500	0.8

NOTE: Streams surveyed in 1982.

[a]Percent of the 666,000 miles surveyed.

SOURCE: Judy et al., 1984.

percent of the stream miles were adversely affected by turbidity, 32 percent by elevated temperature, and 21 percent by excess nutrients (Table 1.3). Water quantity problems resulting from diversions and dams affected approximately 18 percent of the reaches (Table 1.4). The physical limitations most frequently cited were siltation, bank erosion, and channel modifications. Of these, siltation was cited most often and was identified as impairing 40 percent of the miles surveyed (Table 1.5). This survey was conducted once in 1977 and again in 1982.

TABLE 1.4   Water Quantity Limitations
on Fisheries of Perennial Streams

Limit	Miles Affected	Percent[a]
Diversions		
Agricultural	105,000	15.8
Municipal	10,700	1.6
Industrial	3,290	0.5
Dams		
Water supply	30,800	4.6
Flood control	26,900	4.0
Power	24,800	3.7

NOTE: Streams surveyed in 1982.

[a]Percent of the 666,000 miles surveyed.

SOURCE: Judy et al., 1984.

TABLE 1.5   Physical Limitations on
Fisheries of Perennial Streams

	Miles	Percent[a]
Siltation	265,000	39.8
Bank erosion	152,000	22.8
Channel modifications	143,500	21.5
Migratory blockages	39,700	6.0
Bank encroachment	9,000	1.4

NOTE: Streams surveyed in 1982.

[a]Percent of the 666,000 miles surveyed.

SOURCE: Judy et al., 1984.

Little change seemed to occur over the intervening 5-year period (Table
1.6). Regardless of when the survey was conducted, only 5 or 6 percent
of the miles surveyed supported high-quality sport fisheries or exotic
species. Minimal or lower-quality species of fish were found in more
than one-third of the streams. Approximately three-quarters of the streams
would support only a low-quality sport fishery.

TABLE 1.6   Level of Aquatic Sport Species Supported by Fisheries of
Perennial Streams Surveyed in 1977 and 1982

Class	Level Supported[a]	1977		1982	
		Miles	Percent	Miles	Percent
0	No species	29,000	4	29,000	4
1	Nonsport species	48,000	7	49,000	7
2	Minimal sport species	170,000	26	166,000	25
3	Low sport species	224,000	34	228,000	34
4	Moderate sport species	155,000	23	156,000	23
5	High sport and special species	38,000	6	35,000	5
Surveyed		666,000	100	666,000	100

NOTE: Streams surveyed in 1982.

[a]The fish are classed according to nongame (e.g., carp), game (e.g., bass), and special species (e.g., cutthroat trout). The descriptors of abundance (minimal, low, moderate, and high) were subjectively determined, the assessment being made by personnel of the U.S. Fish and Wildlife Service and state fish management agencies.

SOURCE: Judy et al., 1984.

Restoration Initiative

Because of the highly modified and disturbed state of many of our aquatic ecosystems, particularly those closely associated with large population centers or located in agricultural areas, there is considerable potential for the use of restoration to solve water quality, wildlife, and flooding problems. A restoration initiative must be broad and also must encompass large tracts of land; yet these areas need not impinge on the economic viability of agricultural or urban centers. For example, restoration of about 50 percent (approximately 59 million acres) of the nation's lost wetlands (117 million acres in the past 200 years) would affect less than 3 percent (Table 1.7) of the land used for agriculture, forestry, and urban settlement. Of course, most wetland restoration would take place on floodprone land that is uneconomical for farming or other activities. Given the 162 million acres of flood-prone land (Table 1.8) and if the nation restored 59 million acres of wetlands in the long term, only 36.4 percent of the flood-prone areas would have to be given over to wetland restoration. The restoration could take place in littoral zones around lakes and reservoirs and along the floodplain, creating circular greenways and along the floodplain creating green corridors.

TABLE 1.7 Allocation of Wetland Areas (in millions of square acres) by Land Category[a]

| | | | Current State of Wetlands | |
Category	Total Area	Presettlement Wetlands Area	Existing	Destroyed
Agriculture	1,233	134	40	94
Forest	497	54	41	13
Parks	211	23	21	2
Tundra	189	170	170	0
Urban	74	8	2	6
Defense	24	3	1	2
Desert	21	0	0	0
Other	16	0	0	0
Total	2,265	392	275	117

[a]Presettlement wetlands represent 11 percent of the relevant land use category except for tundra, which was taken from Dahl (1990).

SOURCES: McGinnies et al., 1968; Joint Federal-State Land Use Planning Commission for Alaska, 1973; Frey and Hexem, 1985; Bureau of the Census, 1990; Dahl, 1990.

TABLE 1.8 Allocation of Flood-Prone Areas and Wetlands
(in millions of square acres) by Land Category

Category	Total	Flood-plain	Existing Wetland	Restored Wetland	Total Wetland	Restored Wetland as Percentage of Land Category	
						Flood-plain	Total
Agriculture	1,233	98	40	35	75	36	3
Forest	497	39	41	14	55	36	3
Parks	211	17	21	6	27	35	3
Tundra	189	0	170	0	170	0	0
Urban	74	6	2	3	4	50	3
Defense	24	2	1	1	1	50	3
Desert	21	0	0	0	0	0	0
Other	16	0	0	0	0	0	0
Total	2,265	162	274	59	332		

SOURCE: Johnston Associates, 1989.

The restoration of river corridors would directly address the recommendations made by the President's Commission on Americans
Outdoors (1986). The riverways called for in its recommendations
fully embrace the concept of riverine floodplain restoration. If 2,000
river and stream segments are protected and revitalized as the commission recommended, the 59 million acres of restored wetland could
be distributed along these corridors. Given that the average river
segment length is 200 miles, the total length of restored river corridors would be 400,000 miles. This would be only 2.6 times the length
of outlet drains, equivalent to half of the streams surveyed by the
U.S. Environmental Protection Agency (EPA, 1990), and less than 1.3
percent of the total length of streams in the United States. Distribut-
ing the 59 million acres of land along the stream and river segments
would create a corridor with an average width of 1,000 ft.

Conditions of Lakes

Lakes provide many examples of why abatement of pollutant loading
is a necessary but often insufficient step toward improving and restoring freshwater quality and quantity, and ecosystem functions.
Many lakes have lost significant storage capacity through siltation,
which reduces their recreational and water supply usefulness, im-

pairs their capacity to control flooding, and constitutes a severe economic loss. Siltation also remains a serious problem in the United States; 1.7 billion tons of topsoil are lost to erosion every year (U.S. Department of Agriculture, 1982).

Pollution abatement alone will not return many lakes and reservoirs to their former condition because nutrients and toxic materials are recycled from lake sediments. These processes maintain eutrophic conditions or continue to contaminate food webs and associated fisheries, even though loading has been reduced or eliminated. Invasions and planned introductions of nonnative species have become serious problems, impairing fisheries or recreational use (see Chapter 4 for further details).

The extent of lake damage in the United States is substantial. A recent survey by the U.S. Environmental Protection Agency (1990) indicates that about 2.6 million acres of lakes are impaired (relative to suitability for intended uses), and this most likely is a significant underestimate of the acreage that is ecologically degraded and potentially restorable. By far the most common source of stress leading to impairment is agricultural activity (almost 60 percent of impaired acreage is attributed to this source); nutrient and organic enrichment and siltation problems are the most common causes of impairment. It must be noted, however, that survey information regarding some problems such as exotic species and toxic metals is grossly inadequate. These lakes and reservoirs, and others like them, require active restoration and subsequent protection and management, in part because sites for new reservoirs are rare or absent in most areas of the United States (Brown and Wolfe, 1984). Acidification of lakes by acid rain is widespread in the northeastern United States and Canada, and in Norway, Sweden, and the United Kingdom (NAPAP, 1990). Acidified lakes will recover only slowly after cessation of sulfur deposition and may require significant restorative efforts (Schindler, 1988; Schindler et al., 1989).

Conditions of Rivers and Streams

Streams and rivers perform numerous ecological and economic functions. They are conveyances; diluents; sources of power generation; sources of potable water, water for industrial uses, and water for irrigation; and recreation sites. Unfortunately, multiple problems afflict many U.S. rivers today. Our rivers have been diverted, dammed for navigation and hydropower (FERC, 1988; Benke, 1990), channelized, polluted, their wetlands removed, their basins silted in from soil and bank erosion, and their sediments contaminated with toxins. In places such as the Grand Canyon, dams have prevented or slowed sediment transport downstream, causing erosion of beaches in the canyon (NRC, 1987). The combination of dams on the upper Mississippi River and levees along the lower Mississippi has reduced replenishment of the Mississippi delta by sedimentation during the annual floods and thereby contributed to the problem of land subsidence, shoreline erosion, and loss of coastal marshes (Keown et al., 1981; Penland, 1982; Penland and Boyd, 1985). More than half of the nation's rivers have fish communities adversely affected by turbidity, high temperature, toxins, and low levels of dissolved oxygen. Almost 40 percent of perennial streams in the United States are affected by low flows, and 41 percent by siltation, bank erosion, and channelization (Council on Environmental Quality, 1989).

The problems affecting aquatic resources cannot be solved without examining the deleterious land management practices that contribute to those problems. For example, failure to control wind and water erosion and destruction of forested riparian areas has produced heavy silt loads. Increased sediment delivery resulting from forestry practices has also increased sedimentation and turbidity in downstream channels, lakes, and reservoirs, with attendant loss of capacity for water storage and conveyance, recreational and aesthetic values, and quantity and quality of habitat for fish and wildlife. Low or nonexistent dry season flows are one result, leading to water shortages, elimination of river biota, and the increased potential for flash floods. Annual sediment loads in major rivers range from 111 million to 1.6 trillion metric tons, three-fourths of which is deposited in riverbeds, on floodplains, or in reservoirs. One of the major items in the budget of the U.S. Army Corps of Engineers is the cost of dredging, particularly of the lower Mississippi River (Brown and Wolfe, 1984).

Although there have been measurable improvements in stream quality over the last 20 years in the United States, these are associated primarily with improvements in municipal wastewater discharges (Smith et al., 1987). River sediments remain contaminated with toxic substances in many areas, flash floods are common and occasionally lethal, costs to treat water prior to its use have increased, and streambeds remain covered with silt. Vast stretches of rivers and streams have been channelized, a practice that destroys wetlands; increases sediment, nutrient loss, and bank erosion; and often eliminates streamside vegetation that is essential to maintain cool stream temperatures and to stabilize banks. Thousands of miles of rivers and streams are affected by acid mine drainage. Eight percent of the samples of 59,000 stream segments (21,000 km) examined in the National Surface Water Inventory between 1984 and 1986 were acidic (NAPAP, 1990). A systematic restoration of U.S. streams and rivers, along with continued pollution controls, is essential.

Conditions of Wetlands

Wetlands provide essential functions, including flood control, soil and nutrient retention, and wildlife habitat. In some agricultural areas such as the state of California, more than 90 percent of the natural wetlands have been drained or filled. Many riverine wetlands, so essential to water storage, aquifer recharge, and wildlife, have been converted to agricultural areas or destroyed by channelization and urban sprawl. The average rate of wetland loss in the conterminous United States from the mid-1950s to the mid-1970s was nearly 460,000 acres per year, leading to an aggregate loss over all time of about half the wetlands believed to have been here before settlement began—an area greater than Massachusetts, Connecticut, and Rhode Island combined (The Conservation Foundation, 1988; Council on Environmental Quality, 1989). The rate of wetland loss declined to approximately 290,000 acres per year from 1975 to 1984 (Dahl and Johnson, 1991).

Although a "no-net-loss" policy for U.S. wetlands was advocated by President George Bush as a presidential candidate in 1988, the policy's implementation strategy is still being developed at this writing (fall, 1991). During his campaign, then-Vice President Bush declared that all existing wetland should be preserved. His stand was an endorsement of a no-net-loss policy recommendation made by the National Wetlands Policy Forum, a broadly based group including representatives of both industry and environmental groups. In 1989, the U.S. Environmental Protection Agency and three other federal agencies implementing wetlands protection provisions of the Clean Water Act of 1977 (P.L. 95-217), as amended in 1980, produced a wetland delineation manual to help decision makers identify wetlands. This federal manual confirmed a 1983 U.S. Fish and Wildlife Service estimate that 100 million acres of the nation are wetlands. Since the appearance of the manual, however, a number of interest groups, lawmakers, and several federal agencies urged the administration to make the definition of wetlands less encompassing, thereby reducing the amount of land designated as wetlands. These groups have contended that the federal definition of wetlands contained in the wetland delineation manual was so broad as to include areas that are not truly wetlands and that have long been regarded as dry. It is essential that this matter be resolved in order to develop a workable restoration policy.

In response to the criticism, the Bush administration has now developed a new definition of wetlands that would permit construction and farming on up to 10 million acres of land previously classified as wetlands and off limits to development (Schneider, 1991); representatives of the Environmental Defense Fund, an environmental group, have asserted that the new definition would allow the development of up to 30 million acres—one-third of the nation's remaining wetlands. The new definition has had strong backing from the administration's Council on Competitiveness, chaired by Vice President Quayle.

At best, even the original no-net-loss policy meant only no further loss in the aggregate of wetland function or area. Hence, it meant no net return of lost ecological functions and no increase in the nation's wetland area. To recover some of the lost area and functions (e.g., control of soil and nutrient loss, aquifer recharge, control of floods, and provision of nutrient subsidies to fisheries), a major wetland restoration and protection program, particularly in agricultural and coastal regions, is needed. In view of the tremendous losses that have been sustained by the wetland resource base, our national goal should in fact be a *net gain* in wetlands, rather than no additional loss. A similar line of reasoning leads us to believe that, at a minimum, a no-net-loss policy for all other aquatic resources should be implemented as well. Detailed national studies should be conducted of wetlands and of each major aquatic resource type to set national goals for achieving net gains in all aquatic resources through resource restoration.

NEED FOR NATIONAL AQUATIC ECOSYSTEM RESTORATION

This report presents major elements of an agenda for restoration of aquatic resources. Although the details of this agenda will have to be articulated by scientists, public officials, and citizens working together, some characteristics of a national restoration strategy are already discernible. In the broadest terms, aquatic ecosystem restoration objectives must be a high priority in a national restoration agenda: such an agenda must provide for restoration of as much of the damaged aquatic resource base as possible, if not to its predisturbance condition then to a superior ecological condition that far surpasses the degraded one, so that valuable ecosystem services will not be lost.

Despite a continuing national pattern of loss of aquatic resources in area, quality, and function, comparatively little is being invested today on a national scale to restore aquatic ecosystems. Although no reliable estimate of current national spending on aquatic ecosystem restoration is available, the total is most likely to be only in the tens of millions of dollars for the entire nation. This sum is tiny relative to the multibillion-dollar scale of investments made in water development and pollution abatement. Numerous restoration projects at all levels of government and by the private sector are significant and promising, but unfortunately, the vast majority are small in scale and uncoordinated on a regional or a national basis. Much more restoration of aquatic ecosystems is needed to slow and reduce the loss of national aquatic resources, ecosystem services, and wildlife.

Concurrent with the overall decline of aquatic resources, demographic and climatological trends are threatening to exacerbate the underlying ecological problems that make aquatic ecosystem restoration necessary. The world's population is now increasing at a rate of 90 million people per year, adding the equivalent of more than the entire U.S. population to the earth every 3 years. If the United Nations has projected correctly that the world population will be 9 billion people within 40 years, global demand for water, as for other resources, will increase greatly, causing water shortages and further damage to aquatic ecosystems (Postel, 1985). Coupled with the likelihood of significant global climate change (Abrahamson, 1989; Cairns and Zweifel, 1989; Schneider, 1989a; Ehrlich et al., 1990), this increased demand could disrupt not only agricultural systems, but also rivers, lakes, streams, estuaries, and ground water sources at the very time when the human population is at a peak. Already there is worldwide evidence of excessive ground water removal coupled with dramatic drops in ground water tables (Postel, 1985). This means not only water shortages but also land subsidence and saltwater intrusion into aquifers—currently major concerns in Texas, Florida, the Middle East, and China (Postel, 1985). Climate change and population expansion may well be the most serious ecological problems now confronting the world and threatening aquatic ecosystems. Even if this nation embarks on a large-scale aquatic resource restoration and protection program, the impacts of climate change will have to be carefully factored into those plans to avoid expending precious restoration efforts on aquatic resources that are likely to parch from the combined effects of global warming and increased water diversion for human use.

Many prominent atmospheric scientists are now warning that within the next century, the planet may be warmer than it has been in 100,000 years if present trends continue (Schneider, 1989a,b). Most climate models predict a drier United States with less runoff from the Rockies in the arid West. The Midwest and Great Plains are also expected to become drier. (See Chapter 6 for further discussion of climate changes and effects of sea level rise on wetlands.) Lashof (1989) examined an array of biotic and abiotic feedback processes that might affect both the magnitude and the rate of greenhouse warming. These are not routinely included in many general atmospheric models. All but one of the feedback loops Lashof considered were positive; that is, they enhance, rather than reduce, the magnitude and rate of global warming. For this reason alone, the prospects for our already stressed aquatic resources are extremely precarious.

Negative trends in the quality of aquatic resources have been apparent for decades. We continue to find examples of the decline in some functions of major U.S. aquatic ecosystems—for example, San Francisco Bay, Long Island Sound, the coastal marshes and bottomland hardwood forests of the Mississippi delta, the Great Lakes, and the Everglades, to name but a few. If the damage to these ecosystems is not reversed, they will most likely undergo further significant, and in some cases irreversible, ecological deterioration (Wilson, 1988; Woodwell, 1990). To withstand the possible compound stresses from increasing population, and increased demands for aquatic ecosystem services, prudence requires that the nation adopt a national aquatic ecosystem restoration agenda.

REFERENCES AND RECOMMENDED READING

Abrahamson, D. E., ed. 1989. The Challenge of Global Warming. Island Press, Washington, D.C. and Covelo, Calif.

Benke, A. C. 1990. A perspective on America's vanishing streams. J. Am. Benthol. Soc. 9(1):77-78.

Berger, J. J. 1990. Evaluating Ecological Protection and Restoration Projects: A Holistic Approach to the Assessment of Complex, Multi-Attribute Resource Management Problems. Doctoral dissertation. University of California, Davis.

Brown, L. R., and E. C. Wolfe. 1984. Soil erosion: Quiet crisis in the world economy. Worldwatch Paper 60. World Watch Institute, Washington, D.C.

Bureau of the Census. 1990. Statistical Abstract of the United States, 1990: The National Data Book. U.S. Department of Commerce. U.S. Government Printing Office, Washington, D.C.

Bush, G. 1988. Vice President George Bush, Remarks to the Ducks Unlimited Sixth International Waterfowl Symposium, June 8. Crystal Gateway Marriott, Crystal City, Va.

Cairns, J., Jr. 1988. Increasing diversity by restoring damaged ecosystems. Pp. 333-343 in E. O. Wilson, ed., Biodiversity. National Academy Press, Washington, D.C.

Cairns, J., Jr., and P. F. Zweifel, eds. 1989. On Global Warming. Virginia Polytechnic Institute and State University, Blacksburg, Va.

Clean Water Act of 1977. P.L. 95-217, Dec. 27, 1977, 91 Stat. 1566.

Council on Environmental Quality. 1989. Environmental Trends—Chapter 2. Water.

Interagency Advisory Committee on Environmental Trends. Executive Office of the President, Council on Environmental Quality, Washington, D.C. 152 pp.

Dahl, T. E. 1990. Wetland Losses in the United States 1780s to 1980s. U.S. Department of the Interior, U.S. Fish and Wildlife Service, Washington, D.C.

Dahl, T. E., and C. E. Johnson. 1991. Status and Trends of Wetlands in the Contermi- nous United States, mid-1970's to mid-1980's, U. S Department of the Interior, Fish and Wildlife Service, Washington, D.C.

Duda, A. M., and R. J. Johnson. 1984. Lakes are losing the battle in clean water pro- grams. J. Water Pollut. Control Fed. 56:815-822.

Duda, A. M., M. L. Iwanski, R. J. Johnson, and J. A. Joksch. 1987. Numerical standards for managing lake and reservoir water quality. Lake Reservoir Manage. 3:1-16.

Ehrlich, P. R., G. Daily, A. H. Ehrlich, P. Matson, and P. Vitowsek, eds. 1989. Global Change and Carrying Capacity: Implications for Life on Earth. Global Change and Our Common Future. National Academy Press, Washington, D.C.

El-Swaify, S. A., and E. W. Dangler. 1982. Rainfall erosion in the tropics: A state of the art. In Soil Erosion and Conservation in the Tropics. American Society of Agronomy. Madison, Wis.

Federal Aid in Wildlife Restoration Act. P.L. 75-415, Sept. 2, 1937, Pitman-Robertson Wildlife Restoration Act.

Federal Energy Regulatory Commission (FERC). 1988. Hydroelectric power resources of the United States—Developed and undeveloped. Superintendent of Documents. U.S. Government Printing Office, Washington, D.C.

Frey, H. T., and R. W. Hexem. 1985. Major Uses of Land in the United States: 1982. Economic Research Service. Agricultural Economic Report No. 535. U.S. Depart- ment of Agriculture, Washington, D.C.

Horton, A. H. 1914. Water Resources of Illinois. State of Illinois River and Lake Commission, Springfield, Ill.

Johnston, L. R., Associates. 1989. A Status Report on the Nation's Floodplain Manage- ment Activity. An Interim Report. Prepared for the Interagency Task Force on Flood- plain Management. Contract No. TV-72105A. Knoxville, Tenn.

Joint Federal-State Land Use Planning Commission for Alaska. 1973. Major Ecosys- tems of Alaska, Juneau, Alaska.

Jordan, W. R., III, M. E. Gilpin, and J. D. Aber, eds. 1987. Restoration Ecology: Eco- logical Restoration as a Technique for Basic Research. Cambridge University Press, New York. 342 pp.

Judy, R. D., Jr., and P.N. Seeley, T.M. Murray, S.C. Svirsky, M. R. Whitworth, L. S. Ischinger. 1984. 1982 National Fisheries Survey, Vol. 1. Technical Report: Initial Findings. U.S. Department of the Interior, Washington, D.C.

Keown, M. P., E. A. Dardeau, Jr., and E. M. Causey. 1981. Characterization of the Suspended-Sediment Regime and Bed-Material Gradations of the Mississippi River Basin. Potamology Program (P-I). Report 1, Volume II. U.S. Army Corps of Engineers District, New Orleans, La. 375 pp.

Lashof, D. 1989. The dynamic greenhouse: Feedback processes that may influence future concentrations of trace atmospheric trace gases and climate change. Climatic Change 14(3):213-242.

Leopold, L. B., M. G. Wolman, and J. P. Miller. 1964. Fluvial Processes in Geomorphol- ogy. W. H. Freeman, San Francisco, Calif. 522 pp.

Lewis, R. R., III. 1989. Wetlands restoration, creation, and enhancement technology: Suggestions for standardization. Wetland Creation and Restoration: The Status of the Science, Vol. II. EPA 600/3/89/038B. U.S. Environmental Protection Agency, Washington, D.C.

Magnuson, J. J., H. A. Regier, W. J. Christie, and W. C. Sonzogni. 1980. To Reinhabitat and Restore Great Lakes Ecosystems. The Recovery Process in Damaged Ecosystems. Ann Arbor Science Publishers, Ann Arbor, Mich.

McGinnies, W. G., B. J. Goldman, and P. Paylor. 1968. Deserts of the World. University of Arizona Press, Tucson, Ariz.

Naiman, R. J. 1988. Animal influence on ecosystem dynamics. BioScience 38(11).

National Acid Precipitation Assessment Program (NAPAP). 1990. Integrated Assessment: Questions 1 and 2. NAPAP Interagency Program, Washington, D.C.

National Research Council (NRC). 1987. River and Dam Management. Water Science and Technology Board. National Academy Press, Washington, D.C.

Penland, S. 1982. Assessment of geological and human factors responsible for Louisiana coastal barrier erosion. Pp. 14-38 in D. F. Boesch, ed. Proceedings of the Conference on Coastal Erosion and Wetland Modification in Louisiana: Causes, Consequences, and Options. Louisiana Universities Marine Consortium/U.S. Fish and Wildlife Service. FWS/OBS-82/59.

Penland, S. and R. Boyd, eds. 1985. Mississippi Delta barrier shoreline development. Pp. 53-121 in Transgressive Depositional Environments of the Mississippi River Delta Plain: a Guide to the Barrier Islands, Beaches, and Shoals in Louisiana. Louisiana Geological Survey Guidebook Series No. 3, Baton Rouge, La. 233 pp.

Postel, S. 1984. Water: Rethinking management in an age of scarcity. Worldwatch Paper 62. Worldwatch Institute, Washington, D.C.

Postel, S. 1985. Managing freshwater supplies. Pp. 42-72 in State of the World 1985, Chapter 2. L. R. Brown World Watch Institute Report. W. W. Norton Publishers, New York.

President's Commission on Americans Outdoors. 1986. Report and Recommendations to the President of the United States. U.S. Government Printing Office, Washington, D.C.

Schindler, D. W. 1988. Effects of acid rain on freshwater ecosystems. Science 239:149-157.

Schindler, D. W., S. E. M. Kasian, and R. H. Hesslein. 1989. Biological impoverishment in lakes of the midwestern and northeastern United States from acid rain. Environ. Sci. Technol. 23:573-580.

Schneider, K. 1991. Plan may open some wetlands for developers. The New York Times, May 15, 1991.

Schneider, S. 1989a. Global Warming. Sierra Club Books, San Francisco, Calif.

Schneider, S. 1989b. Global warming: Causes, effects, and implications. Pp. 33-52 in J. Cairns, Jr., and P. F. Zweifel, eds., On Global Warming. Virginia Polytechnic Institute and State University Press, Blacksburg, Va.

Seton, E. T. 1929. Lives of Game Animals. Volume IV, Part 2. Doubleday, Doran & Company, Garden City, N.Y.

Smith, R. A., R. B. Alexander, and M. G. Wolman. 1987. Water-quality trends in the nation's rivers. Science 235:1607-1615.

Solley, W. B., C. F. Merk, and R. R. Pierce. 1988. Estimated Use of Water in the United States in 1985. Circular 1004. U.S. Geological Survey. U.S. Government Printing Office, Washington, D.C.

Spencer, J. 1985. A Plague of Beavers. Am. For. 91(5):22-27, 62-63.

The Conservation Foundation. 1988. Protecting America's Wetlands: An Action Agenda. National Wetlands Policy Forum. The Conservation Foundation, Washington, D.C.

U.S. Army Corps of Engineers. 1981. Final Report to Congress: The Streambank Erosion Control Evaluation and Demonstration Act of 1974, Section 32, P.L. 93-251. Main Report. Washington, D.C.

U.S. Army Corps of Engineers and U.S. Environmental Protection Agency. 1990. Memorandum of Agreement Regarding Mitigation for Dredged or Fill Material Disposal in Wetlands. February 6.

U.S. Department of Agriculture (USDA). 1979. Farm Drainage in the United States. History, Status and Prospects. Miscellaneous Publication No. 1455. Economic Research Service, Washington, D.C.

U.S. Department of Agriculture (USDA). 1982. Basic Statistics. Soil Conservation Service, Washington, D.C.

U.S. Environmental Protection Agency (EPA). Office of Policy and Analysis. 1984. The Cost of Clean Air and Water Report to Congress. Washington, D.C.

U.S. Environmental Protection Agency (EPA). 1990. The Quality of Our Nation's Water. EPA 440/4-90-005. Washington, D.C.

U.S. Fish and Wildlife Service. 1990. Wetlands: Meeting the President's Challenge. Washington, D.C.

U. S. Geological Survey. 1987. Hydrologic unit maps. USGS Water Supply Paper 2294. Reston, Virginia

Wilson, E. O., ed. 1988. Biodiversity. National Academy Press, Washington, D.C.

Woodwell, G. M., ed. 1990. The Earth in Transition: Patterns and Processes of Biotic Impoverishment. Cambridge University Press, Cambridge, Mass.

Wooten, H. H., and L. A. Jones. 1955. The history of our drainage enterprises. In Drainage of Fields. Water—The Yearbook of Agriculture 1955. U.S. Department of Agriculture, Washington, D.C.

2

A Selective History of Changing Goals and Authority for Aquatic Ecosystem Management

Throughout the nation's history, public policy often has supported modification of aquatic ecosystems. Now, a new focus is needed for these aquatic ecosystem management policies. However, to appreciate the challenges ahead in both science and public policy, a historical perspective on the changing goals of aquatic ecosystem management will set the stage for the rest of this report.

Early lock and dam systems were put in place to facilitate the primary means of transportation for bulk goods—the inland waterways. The Swamplands Acts of the mid-1800s granted vast tracts of wetlands to the states in the lower Mississippi River Valley, on the condition that the lands be drained and used in agricultural production. By the turn of the twentieth century, the leaders of the progressive conservation movement argued for a more comprehensive program of water resource development, including storage projects, channels, and levees, for the purpose of assuring the long-term material prosperity of the nation. Thus, the Reclamation Act of 1902 was an effort to develop water projects in the west to create small communities and farms by providing low-cost and reliable irrigation water. Gifford Pinchot, a leader of the progressive conservation movement declared, "Conservation stands emphatically for the development and use of water power now, without delay . . . [and] for the immediate construction of navigable waterways . . ." (Nash, 1968). In the 1920s a series of congressional actions began to increase federal flood control project construction, primarily in the lower Mississippi River

Valley, to protect existing properties and open up new lands to agricultural production.

By the mid-1930s the progressive vision for water development had become national policy. Initial federal efforts to engage in river basin water management began with the Lower Mississippi Valley Commission during the presidency of Franklin Roosevelt. The 1934 National Resources Planning Board (NRPB), which undertook the task of defining how the natural resources of the nation could direct that era's weak economy to economic health, argued that water control structures were a part of the nation's economic relief and recovery effort; it stated (NRPB, 1934, p. 255):

> [I]n the interests of the national welfare there must be national control of all running waters of the United States, from the desert trickle that might make an acre or two productive to the rushing flood waters of the Mississippi.

The NRPB's comprehensive watershed management program also included permanently converting steeply sloped lands that were in agricultural use to forest cover. The purpose served by reforested land was limited: these restored lands would reduce the intensity of runoff in order to reduce flooding. Deep percolation would store rainfall in ground water that would later be available for economic uses.

In 1950, President Truman's Water Policy Commission stated that integrated river basin planning could lead to the development of the nation's economy:

> . . . the American people are awakening to the new concept that the river basins are economic units; that many problems center around the use and control of the water resources. . . .

In summarizing the thinking of this era, Gilbert White articulated three elements to what Wengert (1981) later called the "pure doctrine" of river basin development: the multiple-purpose water storage project, an integrated system of projects within river basins, and the goal of water resources management being regional economic development. Plans for water development projects were expected to be defined through rational analysis by water management scientists, who would foresee the opportunities for water development and formulate the optimal sequence of projects to be put in place over time. This faith in scientific planning could be traced to the progressive era. For example, President Theodore Roosevelt, in a 1908 letter transmitting the report of the Inland Waterways Commission to the Congress (Morell, 1956), stated,

[T]he decision to undertake any project should rest on actual need ascertained by investigation and judgment of experts and on its relation to the great river system and the general plan, never on mere clamor.

The scientific expertise to direct watershed development rationally was said to reside in the federal government. In addition, the commerce clause of the constitution, and the fact that basin boundaries crossed state lines, demanded the exercise of federal authority. In the 1930s, the NRPB proposed developing plans for 17 separate river basins that would culminate in ". . . detailed engineering, social, financial and legal studies of water projects . . ." (NRPB, 1934). It was only then "[t]hat further studies . . . [were] needed concerning the division of responsibility and costs among federal, state, and local authorities" (NRPB, 1934). The execution of the rationally determined water development plan assumed that new organizations would be created to implement the actions dictated by the technical analysis, but only the Tennessee Valley Authority was established.

On the matter of cost distribution, the federal government was believed to have the greater financial capacity for basin-scale development, and this belief resulted in limitations on cost sharing and repayment obligations for the beneficiaries of federal expenditures on water development. This repayment philosophy encouraged the political demand for water development projects. Largely because of incentives created by federal cost-sharing policy, projects identified within plans were not executed as expected. With the federal government paying most project costs, the choice of projects for funding priority was made in Congress as part of a logrolling process. The politics of project funding meant that individual federal water project plans of the U.S. Army Corps of Engineers (COE), the Bureau of Reclamation, the Tennessee Valley Authority, and the Soil Conservation Service in fact defined the content of watershed plans. These projects were designed by separate agencies to meet each agency's internal engineering design and project performance criteria.

The regional planning approach to water management failed for another reason. Water management projects and programs are ultimately limited tools for directing economic change in a region. Water planning tends to be oriented toward the solution of perceived problems, such as controlling flooding or providing transportation where it is needed. Water planning reacts to larger technical, social, and economic forces, rather than directing those forces. Because water management is a reactive process, inflexible plans to implement projects that are based on conditions at some point in time are rapidly made obsolete by dynamic technical, social, and economic change.

A renewed effort to prioritize projects according to federal river basin plans came with the passage of the Water Resources Planning Act of 1965 (P.L. 89-80). That act created a federal Water Resources Council and authorized a national system of river basin commissions. The council, with members from several cabinet departments, was expected to integrate federal water resource management efforts, recognizing that environmental protection concerns were receiving increasing public attention. The river basin commissions were expected to integrate federal and nonfederal activities in river basin planning and project implementation, reducing federal dominance of that process. However, the Water Resources Council, with leadership dominated by the federal water project construction agencies, emphasized traditional federal water development projects. New project evaluation guidelines were developed, and project cost-sharing reforms were studied. River basin planning continued to be development project oriented: national water assessments focused exclusively on hydrology and water supply, and the regional water management plans developed became lists of federal water projects.

As much as any other factor, the intent of the Water Resources Planning Act was undone by the council's failure to incorporate many of the new social concerns for the quality of the environment in its program development. Through the 1960s, the nation had built a large capital stock of dams and water delivery systems, but as the nation moved into the 1970s, the concept of a "capital stock" in water resources expanded beyond engineering works to include the remaining free-flowing rivers and, of more importance, the associated riparian lands, uplands, wetlands, and environmental attributes associated with them. The result was a steady reduction of public support for federal water project investment. By the 1980s, both the council and the federal commitment to the basin commission were gone as change in social demands on aquatic ecosystems shifted from traditional economic development to environmental concerns.

WATER QUALITY MANAGEMENT

By the 1970s, the U.S. Environmental Protection Agency (EPA) had superseded federal water project construction agencies as the focal point for federal water resource management. The EPA mission, defined under the Clean Water Act of 1977 (P.L. 95-217), was to lead the nation in ". . . restoration of the physical, chemical and biological integrity of the nation's waters. . . ." Under the Clean Water Act of

1977, EPA emphasized programs addressing the chemical (more specifically, nutrients, bacteria, organic enrichment, toxics, heavy metals, pesticides, and salts) contamination of the nation's waters. In addition to water quality management, EPA was given responsibility for implementing a suite of programs created by environmental legislation in the 1970s and 1980s, much of which was motivated by concerns about the nation's aquatic ecosystems.

Historically, wastewater management emerged as a national concern at the same time that federal water development programs were being established. However, wastewater management was treated as a public health concern, unlike the emphasis in water project development, which was to promote the material welfare of the nation. With the recognition of the germ theory of disease in the second half of the nineteenth century, wastewater was expected to be collected and removed from areas of population concentration through sewers that discharged to nearby rivers, lakes, and estuaries downstream of the waste discharge. Communities were expected to treat waters that were taken from the rivers for public water supply. Construction of facilities for wastewater handling and water supply was the financial responsibility of local governments and affected industries.

The broad concept of "water quality" management for environmental purposes developed slowly as the aquatic ecosystem effects of concentrated point source discharges became evident. Even as the national concern shifted from wastewater management for public health to broadly defined water quality, the federal government did not take a strong management role. The 1948 Federal Water Pollution Control Act was an initial effort that limited the federal government's role to research and monitoring. A series of amendments to the 1948 act, beginning in 1956, culminated in the Federal Water Pollution Control Act Amendments of 1972 (P.L. 92-500). These were more than simple amendments, because they gradually altered federal-state relationships for point source pollution control. With the Federal Water Pollution Control Act Amendments of 1972 (subsequently renamed the Clean Water Act of 1977), the states became the implementers of federally established standards and regulations for municipal and industrial discharges. Because the federal government, represented by EPA, needed to overcome the inertia of state historical dominance, the cost of the new federal presence was eased by massive federal aid to municipalities for pollution control equipment and by special income tax treatment for private costs of pollution control equipment.

With this history and policy environment, it is not surprising that as EPA sought to establish a strong and dominant federal role (vis-à-vis local and state governments) the agency focused almost exclusively on setting and enforcing standards for discharges of specific contaminants to the nation's waters. Nonetheless, there was some limited attention to watershed-scale "water quality" planning and management. However, efforts at areawide planning under Section 208 of P.L. 92-500 fell short for a number of institutional reasons. Also, the Clean Water Act of 1977 did recognize the need to consider in-stream flows, nonpoint source pollution, riparian habitat, and wetlands as part of a watershed-scale program for improvement of the nation's waters. Still, the EPA focus remained on reduction of chemical inputs and resulting concentrations in the waters; EPA's attention was not elevated to aquatic ecosystem restoration as conceived of in this report. Thus, with its focus on chemical contamination, land use was a concern solely in relation to sediment and chemical delivery to the waters. Stream flow patterns and levels were considered only as they determined the assimilative capacity of receiving waters. To illustrate, the establishment of waste discharge standards in relation to the 7-day low-flow regime in streams represented a recognition that the pattern and volume of hydrologic flows determined the assimilative capacity of the waters and hence set waste discharge control requirements. The Willamette River case study (Appendix A) in this report argues that stream flow regulation through water release from dams has been essential to chemical water quality in that river, but other elements of the aquatic ecosystem have not been restored.

Permits for alteration of wetlands, under the Section 404 permit authority of the Clean Water Act of 1977 were reviewed primarily in terms of the effects on water quality; however, other aquatic ecosystem functions of wetlands were slowly introduced as part of this Section 404 review process. Starting in the mid-1970s, EPA recognized that continued destruction of wetlands, rivers, and other shallow aquatic habitat was inconsistent with its efforts to improve the chemical quality of water. At the same time, disagreements involving EPA, COE, permit applicants, and environmental groups concerning the geographic scope of federal authority to protect wetlands and riparian habitats, as well as the kinds of development and agricultural activities subject to the program, were occasionally resolved in the courts.

Until the mid-1980s (the following section), COE was often put in the position of using its Section 404 authority to protect aquatic ecosystems at the same time that federal water development pressures supported the physical destruction of those systems. However, although COE's administration of the Section 404 program has changed significantly in recent years, the program at best has become a tool for retarding the loss of and protecting aquatic ecosystems, not for restoring them, with the incidental exception of some restoration mitigation projects. Further, the Clean Water Act of 1977, despite its objective including physical restoration, established no programs for EPA or COE to use to pursue actual physical restoration of aquatic ecosystems.

Over the last 20 years, the nation has made considerable progress in controlling and reducing certain kinds of chemical pollution of its rivers, lakes, and wetlands. Biological oxygen demand loadings from sewage treatment plants have been reduced significantly. Direct industrial discharges have been controlled. The use of certain agricultural pesticides, such as dichlorodiphenyltrichloroethane (DDT) and other chlorinated hydrocarbons, has been restricted or banned. As a result, the chemical water quality, including dissolved oxygen (DO) levels, in many lakes and rivers has improved, and loadings of some toxic contaminants have decreased.

At the same time, it is well recognized that the nation's water quality programs have not been effective in controlling and reducing loadings of nutrients, sediments, and some toxicants associated with "nonpoint source" pollution from agricultural, urban storm water discharge, mining, and oil and gas extraction activities. This kind of pollution results typically from material changes in the landscape or watershed of affected aquatic ecosystems—removal of forests or other native vegetation, diversion and replacement by exposed soil or impervious material—coupled with dispersed addition of agricultural or lawn fertilizers, animal manure, and other chemicals. Airborne contaminants, including sulfate, nitrate, and metals, add to these nonpoint source loadings. The 1990 amendments to the Clean Air Act should contribute to reducing these atmospheric inputs. The causes of nonpoint source pollution suggest that restoration of the land surface within aquatic ecosystem watersheds on a landscape basis may be one strategy to reduce loadings of sediments, nutrients, and toxicants. However, limited attention has been paid to aquatic ecosystem restoration as defined in this report; the Clean Water Act's mandate "to restore . . ." was confined to a particular and, from an aquatic ecosystem perspective, narrow focus on chemical pollution, concerns in Section 404 about the evolution of wetlands notwithstanding. Within this limited domain there has been success, but challenges remain.

NONSTRUCTURAL APPROACHES TO
FLOODPLAIN MANAGEMENT

Prior to 1966, the predominant federal approach to flood loss reduction was the construction of flood control works to reduce the height and velocity of floodwaters. These works often have severe impacts on aquatic environments. In 1966, a prestigious federal task force on federal flood control policy recommended that nonstructural measures such as flood warning systems and zoning to control building in the floodplain be placed on a par with structural measures such as dams, dikes, channelization, and levees. The goal of nonstructural measures is to adjust the use of the floodplain to the flood threat, rather than modify flooding. This 1966 report was a turning point in federal floodplain management policy. It led, initially, to adoption by Congress in 1968 of a National Flood Insurance Program. This program authorized the mapping of the nation's floodplains and offered federally subsidized flood insurance for existing structures in the floodplain, providing communities and states agreed to regulate new construction.

Since the program's adoption in 1968, federal agencies have prepared floodplain maps for more than 20,000 communities. Almost 18,000 communities have enrolled in the program and adopted floodplain regulations meeting federal standards for new development. These regulations tightly control development or fills within "floodway" areas, near the channels of rivers and streams. This helps to protect natural vegetation and floodplain contours. Regulations also require that development in outer flood-fringe areas be elevated or floodproofed to the 100-year flood elevation.

Floodplain mapping, flood insurance, and regulations are the best-known nonstructural floodplain management measures in use across the nation. However, relocation efforts, flood warning systems, floodproofing of structures, and evacuation from the floodplain are other nonstructural measures applied in both pre- and postflood disaster contexts.

Gradually, federal agencies and Congress have, over a period of years, placed greater emphasis on protection and maintenance of natural values such as wetlands in both structural and nonstructural floodplain management. A "Unified National Program for Floodplain Management," prepared for Congress by a task force of federal agencies in 1976, called for protection of floodplain "beneficial" natural values. Congress has gradually required greater cost sharing for structural flood control measures. The trend toward nonstructural approaches and protection of natural values culminated in the 1990

Omnibus Water Bill (P.L. 101-640). This act (Water Resources Development Act of 1990) requires that COE achieve the wetland no-net loss goal based on both acreage and function for new water projects (Section 13). The COE is also directed to enhance existing environmental values of projects and is required to carry out wetland restoration and creation demonstration projects. The act excludes from the benefit base for justifying new water projects any new or substantially reconstructed structure built in the floodplain after July 2, 1991 (Section 14).

FEDERALLY OWNED LANDS AND PUBLIC INCENTIVES FOR PRIVATE DECISIONS

Federal influence on the nation's aquatic resources has not been limited to water project construction and water quality regulation. Land use decisions on vast acreages of federally owned lands have affected watersheds. The Bureau of Land Management's (BLM) grazing right allocations and grazing fees determined the number of animal units on the land over much of the West's watersheds. The Forest Service's management of national forestland has affected aquatic ecosystems. Federal wildlife refuges have been managed in ways that altered habitats, favoring one species over another. Planning processes for all these agencies have been undertaken, frequently with the purpose of balancing varying management goals, including aquatic ecosystem goals. Nonetheless, much as the water development agencies had a mission for flood control, navigation, or irrigation, these land management agencies were driven by their own missions. The BLM "produced" animal grazing units; the Forest Service produced timber; the Fish and Wildlife Service produced deer and ducks. Water and related lands under the control of these agencies were managed to achieve these resource goals.

Perhaps the most powerful federal force for change in aquatic ecosystems has been the national agricultural policy. Early encouragement of wetlands drainage through the Swamplands Acts was followed years later by federal financial and technical assistance for private drainage of wetlands for agricultural purposes. The drainage decisions of landowners were also indirectly influenced by federal flood control and drainage projects (Stavins and Jaffee, 1991), by provisions of the federal tax code, by agricultural price and income support programs, and by other public efforts to encourage agricultural production (Kramer and Shabman, 1988).

Federal programs likewise influence crop choice and tillage decisions that can affect aquatic ecosystems. The sugar program supports continued production of that crop in South Florida, with the resulting possibility of adverse effects on water distribution and quality in Lake Okeechobee and the Everglades. The structure of the price support programs for all crops encourages the production of crops in places and at times in which soil erosion may be aggravated. Export policy can create boom periods when erodible lands are brought into cultivation. Although chemical use patterns on farms may be a product of agricultural policy and of the price and production incentives it creates, the link between agricultural policy and farm decisions on land, water, and chemical use is not always clear cut. It is not certain that in the absence of agricultural policy, different land, water, and chemical use decisions would be made. What is clear is that agricultural producers' decisions currently are made in response to government policy constraints and incentives as much as to competitive market prices.

CHANGE AT CENTURY'S END

In the last few years, federal policies toward management of aquatic ecosystems have undergone a dramatic change. In general, through a series of congressional actions in the mid-1980s, federal incentives to destroy or alter aquatic ecosystems were significantly reduced. These actions have, therefore, contributed to protection of these systems but not to their physical restoration. Whatever progress has been made in reducing certain kinds and sources of chemical pollution and in reducing physical loss, restoration of the nation's aquatic ecosystems has not been high on the nation's agenda. It was only in 1990 that Congress took a number of initial steps to put the federal government behind actual physical restoration measures.

Congressional Actions in the Mid-1980s

In the mid-1980s, Congress took a number of steps that reduced federal financial support for aquatic ecosystem degradation. The 1986 Water Resources Development Act (P.L. 99-662) placed major new cost burdens on the beneficiaries of water project construction, often states and their political subdivisions. The Swampbuster program of the 1985 Food Security Act eliminated U.S. Department of Agriculture (USDA) benefits in many circumstances where farmers cleared and drained wetlands for crop production. The Internal Revenue Service Tax Code was amended to alter the tax treatment of agricultural drainage expenses in wetlands.

A number of factors explained these actions. Due to budgetary constraints, the federal government reduced its willingness to finance water projects, just as it phased down the EPA construction grants program for wastewater treatment facilities. The federal government has shifted the burden for a variety of programs to the states. In addition, the demand for environmental protection services has grown.

Changes in these economic incentive programs have contributed to a substantial retardation in the rate of wetland loss per year and a substantial slow-down in the rate of loss of free-flowing rivers and natural lakes due to construction of dams, levees, and water diversions. Beaumont (1978) indicated that the building of large dams in North America peaked in 1968 and has since declined. Building of nonfederal dams decreased from 2,000 per year in the 1960s to about 1,240 per year in the 1970s (Johnston Associates, 1989). As worthwhile as they are, these programs were not designed to accomplish actual physical restoration of aquatic ecosystems. What physical restoration has occurred has been largely incidental. For example, although the Conservation Reserve Program (CRP) was designed primarily to give farmers incentives to take highly erodible lands out of production, the Fish and Wildlife Service (FWS) and the Soil Conservation Service (SCS) have used the CRP to restore some small wetlands, such as prairie pothole wetlands in Minnesota (see Chapter 6 and Appendix A). Likewise, the FWS has been able to put easements on and restore some former wetlands subject to forfeiture by the Farmers Home Administration or the Resolution Trust Corporation.

Congressional Initiatives in the 1989-1990 Session

In the 1989-1990 session, Congress and the Bush administration took several actions designed to achieve real physical restoration of certain aquatic ecosystems. Some of these actions were in part a response to the initiatives of states and nonprofit organizations. Indeed, the states are developing the necessary planning expertise and research capacity to execute water management programs. We mention some examples.

In the mid-1980s, Florida initiated its Restore the Everglades program. This included the Kissimmee River demonstration project, designed largely by the South Florida Water Management District. In 1990, Congress appropriated some $6 million for COE to pursue further Kissimmee River restoration work and design. In addition, Congress appropriated $550,000 to have the National Park Service conduct a detailed hydrologic study of the Everglades to increase understanding of the current distribution of water in South Florida and of the flows needed to restore the Everglades ecosystem.

In 1988, in part in response to recommendations made by the newly formed Coalition to Restore Coastal Louisiana, the governor of Louisiana established an office to coordinate all of the state's coastal management and restoration efforts, and in 1989, by a 2-to-1 margin, the voters of the state approved a referendum setting up a coastal wetlands restoration fund, financed by new oil and gas taxes, with revenues up to $25 million per year. In 1990, Congress enacted the Coastal Wetlands Planning, Protection, and Restoration Act (United States Senate, 1990) that establishes a joint federal-state task force to identify and implement wetland restoration projects in Louisiana and a joint planning group to devise an overall plan for the restoration of coastal Louisiana, with the source of funding being a portion of the federal tax surcharge on nonmotor vehicular oil and gas consumption.

A third example concerns the Stillwater wetlands in Nevada. A plan to acquire water rights to restore Stillwater wetlands was initiated largely by an environmental group, the Environmental Defense Fund, following a proposal put together by a task force in the mid-1970s. What helped to make the proposal a reality was the appropriation of about $1 million in federal funds. The proposal calls for the acquisition of 50,000 acre-feet of water to be purchased from willing sellers who are farmers. Nevada funding has come from a State Parks and Wildlife Bond. The Nature Conservancy has advanced funds as well.

The largest commitment to wetland restoration made by Congress in 1990 was the adoption of the Agricultural Wetland Reserve Program as part of the 1990 Food, Agriculture, Conservation, and Trade Act of 1990 (P.L. 101-624). This program could help to reconvert one million acres of cropland to wetlands, and it may be funded largely, albeit indirectly, through reduction of USDA subsidies that would otherwise apply to these croplands.

EMERGING ROLE OF THE STATES

The history of aquatic ecosystem management in the twentieth century in the United States has been one of federal domination. The federal river basin planning process came to be identified with the missions and priorities of federal water project construction agencies; the states tended to retreat from an active role in this process and take a "let the federal government do it—and pay for it" attitude (Allee et al., 1982). The original Kissimmee River project, which made major alterations in that aquatic ecosystem, was designed by a federal agency (with state acceptance of the plan) in accord with the long-standing federal priorities of flood reduction and land enhancement for economic development.

Federal domination of water quality management for point source control was a response to a perceived failure of the states to take adequate action. The nature of point source control made such federal leadership institutionally possible. The recognition that point source wastewater treatment will not yield water quality goals has now directed water quality management to land use-dominated issues. The decline in federal water development activities, which sought to increase the reliable "supply" of water, meant that reallocation of waters among aquatic ecosystem values has become paramount. Water allocation and land use decisions fall logically to the states and their political subdivisions, because it is there that the tools for water management (land use controls, water laws, etc.) exist. The National Governors Association made this point in a 1973 water policy statement that "[t]he states have primary authority and responsibility for water management" (Wilson, 1981). Now, nearly 20 years later, substantial responsibilities for water management are returning to the states and the states are acting on those responsibilities.

The federal government has reduced its willingness to finance traditional water development projects. The Water Resources Development Act of 1986 placed major new cost burdens on the beneficiaries of water project construction, often states and their political subdivisions. The construction grants program for wastewater treatment facilities has been phased down. In this environment the states are doing more. Many states have established and capitalized revolving fund accounts to assist in wastewater treatment plant construction. In Minnesota, state financial resources have been put behind a "Reinvest in Minnesota" program to restore wetlands.

As important as funding is leadership. The states are developing the necessary planning expertise and research capacity to execute water management programs. The Kissimmee River restoration project has been designed largely by the South Florida Water Management District. States in the Chesapeake Bay region have developed effective management programs that now dwarf federal efforts in that region in treatment facility planning and land use management. Also, where states have felt a need to act in concert they have often done so without federal prompting.

Congress got its feet wet in the 1989-1990 session in promoting certain wetland restoration programs. Many states have shown a strong interest in taking the initiative in developing programs for aquatic ecosystem restoration. The stage is therefore set for a major national aquatic ecosystem restoration program both to build on and to stimulate grassroots, local and state government restoration efforts.

REFERENCES

Allee, D., L. Dworsky, and R. North, eds. 1982. United States water planning and management. Pp. 1-42 in United River Basin Management—Stage II. American Water Resources Association, Minneapolis, Minn.

Beaumont, P. 1978. Man's impact on river systems: A world-wide view. Area 10:38-41.

Clean Water Act of 1977. P.L. 95-217, Dec. 27, 1977, 91 Stat. 1566.

Coastal Wetlands Planning, Protection and Restoration Act. P.L. 101-646, Nov. 29, 1990.

Federal Water Pollution Control Act Amendments of 1972. P.L. 92-500, Oct. 18, 1972. 86 Stat. 816.

Food, Agriculture, Conservation, and Trade Act of 1990. P.L. 101-624.

Johnston, L. R., Associates. 1989. A status report on the nation's floodplain management activity. An Interim Report. Contract No. TV-72105A. Prepared for the Interagency Task Force on Floodplain Management. Knoxville, Tenn. 465 pp.

Kramer, R. A., and L. A. Shabman. 1988. Incentives for agricultural development of U.S. wetlands: A case study of the bottomland hardwoods of the lower Mississippi River valley, Chapter 6. Pp. 175-201 in T. Phipps, P. Crosson, and K. Price, eds., Agriculture and the Environment. Resources for the Future, Washington, D.C.

Morrell, B. 1956. Our Nation's Water Resources—Policies and Politics. University of Chicago Press, Chicago, Ill.

Nash, R. 1968. P. 9 in The American Environment: Readings in the History of Conservation. Addison Wesley, Reading, Mass.

National Resources Planning Board (NRPB). 1934. Report on National Planning and Public Works in Relation to Natural Resources and Including Land Use and Water Resources. U.S. Government Printing Office, Washington, D.C.

President's Water Policy Commission. 1950. Ten Rivers in America's Future. U.S. Government Printing Office, Washington, D.C.

Stavins, R. N., and A. D. Jaffee. 1991. Unintended impacts of public investments on private decisions: The depletion of forested wetlands, June 1980. Am. Econ. Rev. 80(30):337-352.

United States Senate. 1990. Congressional Record—Senate October 26, S. 17360.

Water Resources Development Act of 1986. P.L. 99-662, Nov. 17, 1985, 100 Stat. 4082.

Water Resources Development Act of 1990. P.L. 101-640, 1990 Omnibus Water Bill, Nov. 28, 1990, 104 Stat. 4604.

Water Resources Planning Act of 1965. P.L. 89-80.

Wengert, N. 1981. A critical review of the river basin as a focus for resources planning, development, and management. Pp. 9-27 in R. North, L. Dworsky, and D. Allee, eds., Unified River Basin Management. American Water Resources Association, Minneapolis, Minn.

Wilson, L. U. 1981. State water management capacity. Pp. 399-404 in D. Allee, L. Dworsky, and R. North, eds., Unified River Basin Management—Stage II. American Water Resources Association, Minneapolis, Minn.

3

Planning and Evaluating
Aquatic Ecosystem Restoration

INTRODUCTION

The fundamental goal of aquatic ecosystem restoration is to return it to a condition that resembles its natural predisturbance state as closely as possible. Achievement of this goal entails restoration of the target ecosystem's structure and function both locally and within its broader landscape or watershed context. To measure the degree of success in achieving restoration goals, physical, chemical, and biological evaluation data are necessary to verify that an ecosystem is performing as it should.

To achieve long-term success, aquatic ecosystem restoration should address the causes and not just the symptoms of ecological disturbance. Sometimes these causes are obvious; sometimes they are subtle and far removed in space and time from the ecological damage, as in the case of Grove Lake in Pope County, Minnesota. In the 1800s, small prairie potholes were ditched and drained there in the headwaters of the Crow River (see Prairie Potholes case study, Appendix A). Runoff quantities and velocities were increased by the straightened, more efficient drainage system. This increased the movement of nutrients and sediments downstream. These materials entered Grove Lake and several downstream lakes, causing water quality problems that resulted in accelerated eutrophication and other changes in plant composition. The lakes also became progressively shallower and less attractive to wildlife. Dredging the lakes or altering the water chemistry produced temporary restoration of certain lake functions, but once these symptom-oriented treatments had been completed, the symptoms began to reappear. Restoration of a river or other aquatic system requires replacing not only the predisturbance morphology but the hydrologic conditions as well. To accomplish this, land uses may have to be altered, vegetation may need to be reestablished, and interrelated ecosystems—tributaries or adjacent wetlands—may have to be given fundamental corrective ecological attention as well.

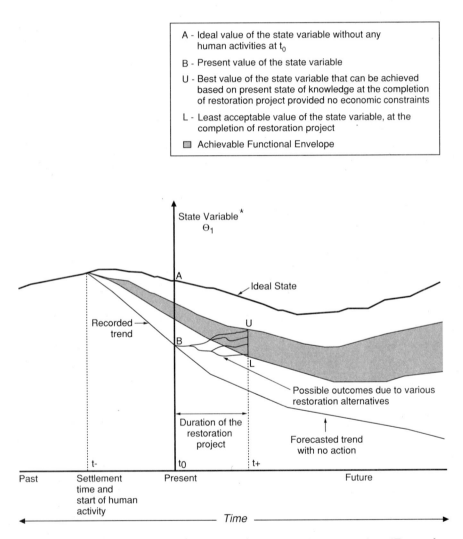

FIGURE 3.1 Schematic representation of a restoration scenario. *Examples of state variables include river stage, water temperature, and fish species.

In the development of restoration strategies, restoration of an ecosystem to an approximation of its natural predisturbance condition should be pursued as the first goal. However, in some situations, this ideal may be impractical, as illustrated in Figure 3.1. The shaded area represents an "envelope" in which the morphology and function of the ecosystem are considered to be acceptable and achievable under existing social, political, economic, and engineering constraints. The goals in this restoration scenario would be to move the ecosystem by the time the project is complete from its present state to some point within the achievable envelope.

RESTORATION PROJECT PLANNING

Planning a restoration project starts with specifying the project mission, goals, and objectives (Table 3.1). The goals and objectives then become the basis for the evaluation assessment criteria. The

TABLE 3.1 Restoration Checklist

Project Planning and Design

1. Has the problem requiring treatment been clearly understood and defined?
2. Is there a consensus on the restoration program's mission?
3. Have the goals and objectives been identified?
4. Has the restoration been planned with adequate scope and expertise?
5. Does the restoration management design have an annual or midcourse correction point in line with adaptive management procedures?
6. Are the performance indicators—the measurable biological, physical, and chemical attributes—directly and appropriately linked to the objectives?
7. Have adequate monitoring, surveillance, management, and maintenance programs been developed along with the project, so that monitoring costs and operational details are anticipated and monitoring results will be available to serve as input in improving restoration techniques used as the project matures?
8. Has an appropriate reference system (or systems) been selected from which to extract target values of performance indicators for comparison in conducting the project evaluation?
9. Have sufficient baseline data been collected over a suitable period of time on the project ecosystem to facilitate before-and-after treatment comparisons?
10. Have critical project procedures been tested on a small experimental scale in part of the project area to minimize the risks of failure?
11. Has the project been designed to make the restored ecosystem as self-sustaining as possible to minimize maintenance requirements?
12. Has thought been given to how long monitoring will have to be continued before the project can be declared effective?
13. Have risk and uncertainty been adequately considered in project planning?

continued

TABLE 3.1 (*Continued*)

___

During Restoration

1. Based on the monitoring results, are the anticipated intermediate objectives being achieved? If not, are appropriate steps being taken to correct the problem(s)?
2. Do the objectives or performance indicators need to be modified? If so, what changes may be required in the monitoring program?
3. Is the monitoring program adequate?

Post-Restoration

1. To what extent were project goals and objectives achieved?
2. How similar in structure and function is the restored ecosystem to the target ecosystem?
3. To what extent is the restored ecosystem self-sustaining, and what are the maintenance requirements?
4. If all natural ecosystem functions were not restored, have critical ecosystem functions been restored?
5. If all natural components of the ecosystem were not restored, have critical components been restored?
6. How long did the project take?
7. What lessons have been learned from this effort?
8. Have those lessons been shared with interested parties to maximize the potential for technology transfer?
9. What was the final cost, in net present value terms, of the restoration project?
10. What were the ecological, economic, and social benefits realized by the project?
11. How cost-effective was the project?
12. Would another approach to restoration have produced desirable results at lower cost?

___

project mission is the overall general purpose, such as the restoration of a particular stream and perhaps a fringe of adjoining riparian wetlands. The goals might include restoring water quality, benthic substrate, hydrology, channel stability, riverine flora and fauna, and wetland flora and fauna. Goals should be prioritized so that project designers and evaluators have a clear understanding of their relative importance. Objectives are then derived from the goals, giving, for example, the specific characteristics of water quality to be achieved, the particle size and condition of the benthic substrate, the species composition and population sizes of the various communities of aquatic biota expected, and so on. Finally, the evaluator must construct specific "performance indicators" linked to each objective. These performance indicators are specific measurable quantities that reveal to what extent the objectives are being achieved. In the case of water quality, they might include indicators such as pH, amount of chlorophyll in a water sample, or Secchi disk visibility.

In addition to specifying goals, objectives, and performance indicators, project managers and designers should propose a monitoring and assessment program that is appropriate in scale (areal extent), as well as in sampling frequency and intensity, to measure the performance indicators accurately and reliably, and thereby assess progress toward the project's objectives, goals, and mission.

Project Schedule

A realistic restoration schedule needs to be set to avoid inappropriate expenditures of effort and money. Enough time must be allowed for pre- and postproject monitoring so that the estimates of baseline and reference conditions used are representative and reliable. Monitoring must be maintained long enough for resource managers to confirm that the restoration can withstand unusual environmental events, such as floods, droughts, and frosts. In regions where environmental conditions are highly variable from year to year, the time frame will probably have to be long. For example, at least one wetland restoration project in San Francisco Bay has a 20-year monitoring requirement. Adequate financing must be provided to guarantee long-term maintenance and surveillance of the project. However, detailed 20-year monitoring and assessment programs will not be needed for every restoration project, particularly simple projects for which there is a large experience base. As restoration technology improves in reliability, selective monitoring using cost-effective indicators should become possible.

Project Scale

The areal extent of a restoration project is important for four reasons. First, the project area needs to be large enough to limit deleterious effects that boundary conditions may impose on interior aquatic functions. For example, a prairie slough restored too close to a highway may be stunted in its development by de-icing agents in road runoff. Second, project managers must be able to exert influence over zones in which major causes of ecological disturbance to the project are occurring, so that the disturbance can be controlled or eliminated. Third, the area needs to be large enough so that important effects of the project can be monitored for project assessment purposes. Finally, the project should be of an affordable size.

The restoration of a site should be considered in a regional context, and the area that will be available to reestablished wildlife communities should be considered in relation to the size and longevity of the organisms that will occupy the restoration site. Restoration of a vernal pool may be viewed in the context of its local watershed or with respect to the distribution of its main species or subspecies; however, plans for the restoration of migratory waterfowl or wading bird habitat must include continental or intercontinental considerations. In many cases, the restoration planner must review land uses adjacent to the project for potential disturbances or other effects of interactions, including hydrological connections.

In developing specific plans for ecosystem reconstruction from the landscape perspective, it may be necessary to look quite far for undisturbed systems to serve as reference systems. For southern California coastal wetlands, the nearest low-disturbance reference site is 300 km south of the border in Mexico. Biogeographic information (i.e., distributional limits of species known to inhabit local, more disturbed sites) is clearly required to estimate whether the species lists and relative abundances of organisms at a distant site are appropriate for the restoration site. Knowledge of the migratory routes of birds and fish, and of dispersal patterns for invertebrate larvae and seeds, is critical in determining what scale to use in planning aquatic restorations.

Ideally, an effective restoration will have a positive ecological influence beyond the immediate project site. For example, an isolated wetland may have been restored partly in the hope that migratory birds would use it. Its ability to provide the desired migratory bird habitat function, however, depends in part on processes operating on a continental scale; thus, assessment may require a much broader evaluation of waterfowl behavior and production than merely at the restoration site itself. For example, a restored wetland that fails to attract birds for a year or more need not be considered a failure if migratory patterns have shifted for reasons other than the quality of the restored habitat. The contrast between the temporal and spatial scales of existing restoration assessment practices and assessment needs as proposed in this chapter is depicted in Figure 3.2.

Genetic Issues

The scale of genetic variation is an important but little-known factor affecting restoration efforts. Until genetic inventories are available for species to be planted or transplanted to restoration sites, and until we understand how great the genetic variability must be in the

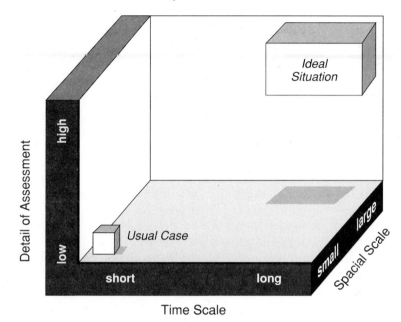

FIGURE 3.2 Time scale of the assessment. The diagram compares the usual (and very limited) protocol for assessing the "success" of a restoration site with an idealized framework, involving long-term, detailed, and large-scale evaluation.

transplanted stock, it is advisable to select material from local sources by sampling broadly within those populations. Transplantation of *Spartina foliosa* from San Francisco Bay, where it is abundant, to San Diego Bay (more than 800 km away), where it is uncommon, may provide plant cover, but there are two risks if these populations are genetically distinct. The northern population may not tolerate the higher soil salinities that develop in some years; alternatively, a foreign population may grow better and outcompete the local ecotype. Only the use of local genotypes can preserve and maintain local biodiversity.

The local range of genetic variation is also of concern. Because many of the favorite transplant species for marshes reproduce vegetatively (and are propagated vegetatively by suppliers), there is a risk that sites will be established from single clones, whose descendants may someday die en masse if a rare environmental event occurs or an unusual parasite infests the restoration site. Thus, care must be taken to specify local, diverse plant material, and suppliers must keep permanent records of the sources of their materials.

Human Influences

It would be impractical to develop or implement a restoration program in most areas of the United States without explicitly considering the effects of humans (see Chapter 6). Consequently, project design criteria should reflect both human behavior and needs, and the biological needs of project species. For example, humans may demand that restoration work address acute water quality problems, reduce the threat of flooding, increase biodiversity, or simply create a more aesthetically pleasing landscape. In addition, humans may themselves damage the experimental ecosystem before it has a chance to develop. Adequate provision should therefore be made for project fencing or other access control, when necessary to safeguard against vandalism or depredations by domestic animals, or to provide wildlife seclusion and protection.

The integration of human values and ecological performance is illustrated by the project assessment matrix shown in Figure 3.3. The unacceptable position in the lower left corner is identified by a solid black box. In this case, neither human nor ecological values are provided by the project. This position might be represented in the case of a dam constructed to provide irrigation where the stream flow was inadequate to fill the reservoir. Therefore the dam provided no agricultural benefits and destroyed wildlife habitat.

The progression from the lower left corner (cell A0) to the upper right corner (cell C2) of the matrix can be represented by a project using wetlands for wastewater treatment. No ecological or human value is achieved if, when wastewater is discharged into the wetland, the wetland is destroyed and no nutrient removal occurs. However, some ecologic and human values can be achieved if the wetlands are able to survive wastewater discharge but are then converted to a low-density or a monotypic plant community. To achieve ultimate success (cell C2), the wastewater effluent would be treated to the desired standards while the wetland simultaneously supported a high density of plants and animals. This latter project might be termed *restoration* if wastewater flows emulated historic hydrologic conditions and if the plants, animals, and landscape adequately represented predisturbance conditions.

A restoration often cannot follow a vertical path from low to high ecological values (cell A0 to C0). Some economic or social benefit often must be produced, tilting the line to the right toward C2. This need not be undesirable because ecological and human values can often be served simultaneously.

Restoration failures may occur for several different kinds of reasons. In the first case, restoration projects can be carelessly implemented, as when a contractor disregards engineering or horticultural specifications. Examples include installation of wetlands using nonlocal biota that may not be adapted to local soil salinity or temperature, or planting saltwater wetlands at the wrong tidal elevation. In these cases, the project might have succeeded had engineering design criteria and restoration protocols been observed. The failure is thus not a fault of ecological science or engineering knowledge but of implementation.

Another kind of failure occurs when design criteria are scrupulously followed and the project designer's knowledge proves inad-

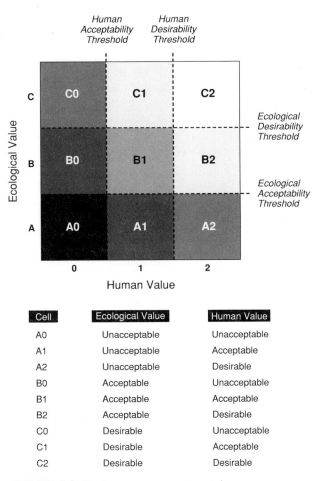

Cell	Ecological Value	Human Value
A0	Unacceptable	Unacceptable
A1	Unacceptable	Acceptable
A2	Unacceptable	Desirable
B0	Acceptable	Unacceptable
B1	Acceptable	Acceptable
B2	Acceptable	Desirable
C0	Desirable	Unacceptable
C1	Desirable	Acceptable
C2	Desirable	Desirable

FIGURE 3.3 Project assessment matrix.

equate to produce a functional restoration design. A third class of failure occurs when objectives and criteria are not established prior to the project (Kusler and Kentula, 1989). These projects lack milestones to judge progress, and in the absence of assessment criteria for use in monitoring, it is hard to obtain early warnings that the restoration is not "on track."

PURPOSE OF EVALUATION

The general purpose of evaluating an aquatic ecosystem restoration project is either to determine how effective the restoration attempt was in replicating the target ecosystem or to select from among competing restoration projects the one most likely to prove effective. Evaluation of a completed restoration project (postproject evaluation) is vital to learning whether the permit requirements of a mandatory restoration project (e.g., one performed for mitigation purposes) have been achieved and is also essential for people conducting discretionary restoration projects to know how effective their efforts have been. Evaluation before the fact (preproject evaluation) can help a decision maker identify the project most likely to provide the greatest ecological benefits at the lowest cost, an especially important consideration in an era of budgetary constraints and enormous environmental challenges.

Naturally, 100 percent similarity of a restored system to predisturbance conditions is impossible to achieve; even two parts of any single pristine aquatic system are never 100 percent similar in either structure or function. Therefore, perfection should not be expected in restoration, and restoration planners must recognize that restoration is an exercise in approximating prior conditions.

SELECTING ASSESSMENT CRITERIA AND
SYNTHESIZING DATA

An evaluation of aquatic restoration must include procedures for synthesizing data to be produced by monitoring the restoration project or by analyzing the restoration proposals. The evaluation framework should define the problem, specify what data are to be collected when, and explain how the data are to be used once collected. The latter seemingly self-evident point is actually a critical issue in ecosystem evaluation. Because restoration strives to alter an existing ecosystem so that it becomes more similar to a predisturbance model, the evaluator needs to gather a comprehensive data set relating the restored system to the antecedent one in biological, physical, and chemical terms. The data gathered on these diverse aspects of project performance can be extensive and conflicting, because achievements may differ greatly from one aspect of the restoration to another. (For example, a restoration may be effective in producing good water quality, but poor in reproducing some of the ecosystem's floral or faunal characteristics, or vice versa.) Therefore, if a decision maker is to use the results of evaluation for a policy decision, rules may have to be established for synthesizing large quantities of observational data into a form in which comparisons between projects become possible, clear, and meaningful.

Because in completely evaluating a restoration, one is in effect evaluating an entire ecosystem, a broadly representative range of assessment criteria must be used to reflect the major dimensions of the ecosystem, including its complex food webs, habitat heterogeneity, and dynamic physical, chemical, and biological processes. Thus, thorough evaluation of a restoration may become a complex, multidisciplinary process involving a great deal of data collection and necessitating that the resulting body of basically incomparable or unrelated data be reduced to manageable terms by using multiattribute decision techniques. (For a discussion of multiattribute decision techniques to compare complex restoration projects, see MacCrimmon, 1968; Raiffa, 1969; Stokey and Zeckhauser, 1978; Tecle et al., 1988; Berger, 1991). Interpretation of the results of evaluation is always facilitated by a skillfully written narrative explanation of project outcomes. Often this documentation will be all that is required, especially for relatively simple, straightforward projects.

One solution to submerging the decision maker in a sea of data is to strategically select assessment criteria that suggest the presence of a host of other complex desired ecological states. For example, use of a measure such as the biomass of key indicator plants in a wetland species assemblage may provide a great deal of information about the reestablished vegetation. Use of the wetland by wading birds, waterfowl, and fish also provides "an integrated measure (i.e., [one] dependent on an array of structural features) . . . of floodplain integrity" (Toth, 1991).

Assessment Criteria

Assessment criteria should include both structural and functional attributes of the ecosystem, and should be based on known antecedent conditions of the target or reference ecosystem. These criteria should be established well before the assessment takes place and should be linked, as described above, to specific project objectives.

Assessment measurements should take into account both temporal variation and spatial heterogeneity. Thus, attributes that are patchy in time or space need widespread and long-term characterization. Multiple criteria should be employed to minimize the risk of overlooking important ecological effects, and a range of reference systems and long-term data sets should be compared with the project's attributes if possible.

As noted in this chapter, performance indicators that are implicitly or explicitly derived from project objectives are the assessment criteria that are actually used in the evaluation process. However, selecting an appropriate subset of indicators from the universe of possible evaluation factors is a skill and an art—in essence, a separate decision problem that is of great importance to the feasibility, cost, and validity of the evaluation. To assist the decision maker in developing appropriate indicators, the next three sections contain annotated lists of possible structural, functional, and holistic ecological assessment criteria, adapted from Berger (1990). (Additional evaluation criteria for aquatic restoration are provided in Chapter 6 and Zedler et al., 1988; Berger, 1990, 1991; PERL, 1990; Southland, 1991.)

STRUCTURAL CHARACTERISTICS

The following are examples of structural characteristics:

1. *Water quality* both on and off the project site, wherever affected by the restoration. Measures include dissolved oxygen, dissolved salts, dissolved toxics and other contaminants, floating or suspended matter, pH, odor, opacity, temperature profiles, and other indicators.

2. *Soil condition* as revealed by soil chemistry; erodibility; permeability; organic content; soil stability; physical composition, including particle sizes and microfauna; and other factors.

3. *Geological condition* as indicated by surface and subsurface rock and other strata, including aquifers (see hydrology).

4. *Hydrology*, including quantity of discharge on annual, seasonal, and episodic basis; timing of discharge; surface flow processes, including velocities, turbulence, shear stress, bank/stream storage, and exchange processes; ground water flow and exchange processes; retention times; particle size distribution and quantities of bed load and suspended sediment; and sediment flux (aggradational or degradational tendencies) (Rosgen, 1988).

5. *Topography* as indicated by surface contours; the relief (elevations and gradients) and configuration of site surface features; and project size and location in the watershed, including position relative

to similar or interdependent ecosystems. Riverine floodplain wetlands, for example, require a river to provide periodic inundation.

6. *Morphology* (may be subsumed by topography) as indicated by the shape and form of the ecosystem, including subsurface features. For a lake, morphology includes shoreline circumference-to-area ratio, mean depth, and mean-depth-to-maximum-depth ratio. For rivers and streams, it includes channel patterns (braided, meandering, or straight); bank width-to-depth ratios; meander geometry (amplitude, length, radius of curvature); cross-sectional depth profiles; and riffle-to-pool ratio (river and stream descriptions). For wetlands, morphology includes inlets and outlets, channels, islands, adjacent uplands-to-wetlands ratio, fetch and exposure, and vegetation-water interspersion (Adamus et al., 1987).

7. *Flora and fauna*, including density, diversity, growth rates, longevity, species integrity (presence of full complement of indigenous species found on the site prior to disturbance), productivity, stability, reproductive vigor, size- and age-class distribution, impacts on endangered species, incidence of disease, genetic defects, genetic dilution (by nonnative germ plasm), elevated body burdens of toxic substances, and evidence of biotic stress.

8. *Carrying capacity, food web support, and nutrient availability* as determined for specific indicator species. Ultimately, these will be a function of nutrient availability in conjunction with other site-specific factors. Nutrient availability and nutrient flux patterns are therefore subsumed under "carrying capacity." However, an understanding of nutrient dynamics will give the resource manager more predictive capability than simply knowing current carrying capacity. Two questions of interest are whether the ecosystem is gaining or losing nutrients, and whether the nutrient flux is comparable to that in the antecedent system.

FUNCTIONAL CHARACTERISTICS

The following are examples of functional characteristics:

1. Surface and ground water storage, recharge, and supply.
2. Floodwater and sediment retention.
3. Transport of organisms, nutrients, and sediments.
4. Humidification of atmosphere (by transpiration and evaporation).
5. Oxygen production.
6. Nutrient cycling.
7. Biomass production, food web support, and species maintenance.

8. Provision of shelter for ecosystem users (e.g., from sun, wind, rain, or noise).
9. Detoxification of waste and purification of water.
10. Reduction of erosion and mass wastage.
11. Energy flow.

See Chapter 6, Tables 6.1 and 6.2, for a more complete list and discussion of functional characteristics.

EMERGENT PROPERTIES

The following are examples of emergent properties (i.e., those exhibited by the ecosystem as a whole):

1. *Resilience*, the ability of the ecosystem to recover from perturbation.
2. *Persistence*, the ability of the ecosystem to undergo natural successional processes or persist in a climax sere (a stage in ecological succession), all without active human management. Persistence incorporates the notion of self-sufficiency, the ability of the ecosystem to survive as a dynamic system, evolving in a manner and at a rate regarded as normal for that type of ecosystem at its particular stage of development. To measure the persistence or degree to which reestablished biota can sustain themselves within the context of succession, measurement units may include time between needed management intervention or units of management effort required. Examples of typical postproject modifications or maintenance include grading, replanting, and controlling weeds and pests.
3. *Verisimilitude*, a broad, summative, characteristic of the restored ecosystem reflecting the overall similarity of the restored ecosystem to the standard of comparison, be it prior conditions of the ecosystem or of a reference system.

See Chapter 6, Table 6.4, for additional emergent properties.

CONCLUSIONS AND RECOMMENDATIONS

The aim of restoration is to return ecosystems to a close approximation of their natural, self-sustaining, and predisturbance condition. The function of evaluating a restoration effort is to determine in a reliable scientific manner how effective a particular restoration has been, i.e., how similar the restored ecosystem is to the target ecosystem.

For comprehensive preproject evaluation of prospective restoration alternatives, economic and social impacts must be considered along with ecological effects, in addition to factors such as risk and social equity (the incidence of benefits and costs for different classes of people affected by the project). For dependable evaluations, assessment criteria must include both structural and functional attributes of the ecosystem. The scale of the restoration must be adequate to account for spatial heterogeneity of habitat and for interactions between the target system and its surrounding landscape. The duration of the project and its monitoring must be sufficient to encompass unusual environmental events that periodically stress the ecosystem. Assessment criteria, evaluation methodology, restoration techniques, and project implementation must all be able to stand up to the scrutiny of peer review.

If, because of budgetary or other problems, a comprehensive restoration project cannot be completed, efforts should be made to conserve valuable and unique plants and animals so that they or their gene pools will be available when restoration becomes feasible.

REFERENCES AND RECOMMENDED READING

Adamus, P. R., E. J. Clairain, Jr., R. D. Smith, and R. E. Young. 1987. Wetland Evaluation Technique (WET). Vol. II. Methodology Operational Draft. U.S. Army Corps of Engineers Waterways Experiment Station, Vicksburg, Miss.

Berger, J. J. 1990. Evaluating Ecological Protection and Restoration Projects: A Holistic Approach to the Assessment of Complex, Multi-Attribute Resource Management Problems. Ph.D. dissertation. University of California, Davis.

Berger, J. J. 1991. A generic framework for evaluating complex restoration and conservation projects. Environ. Prof. 13(3):254-262.

Cairns, J., Jr. 1991. The status of the theoretical and applied science of restoration ecology. Environ. Prof. 13(3):1-9.

Kusler, J. A., and M. E. Kentula, eds. 1989. Executive Summary. Wetland Creation and Restoration: The Status of the Science. Vol. I. EPA 600/3-89/038A. U.S. Environmental Protection Agency, Washington, D.C.

MacCrimmon, K. R. 1968. Decisionmaking Among Multiple-Attribute Alternatives: A Survey and Consolidated Approach. Memorandum RM-4823-ARPA. The Rand Corporation, Santa Monica, Calif.

Pacific Estuarine Research Laboratory (PERL). 1990. A Manual for Assessing Restored and Natural Coastal Wetlands with Examples from Southern California. California Sea Grant Report No. T-CSGCP-021. La Jolla, Calif.

Raiffa, H. 1969. Preferences for Multi-Attributed Alternatives. Memorandum RM-5868-DOT/RC. Prepared for U.S. Department of Transportation, Federal Railroad Administration. Office of High Speed Ground Transportation, The Rand Corporation, Santa Monica, Calif. April.

Rosgen, D. L. 1988. The conversion of a braided river pattern to meandering—A landmark restoration project. Paper presented at the California Riparian Systems Conference, September 22-24. Davis, Calif.

Southland, M. T. 1991. Ecosystem Restoration Criteria: A Review and Conceptual Framework. Dynamac Corporation, Environmental Services, Rockville, Md.

Stokey, E., and R. Zeckhauser. 1978. A Primer for Policy Analysis. W. W. Norton, New York.

Tecle, A., M. M. Fogel, and L. Duckstein. 1988. Multicriterion analysis of forest watershed management alternatives. Water Resour. Bull. 24(6):1169-1178.

Toth, L. 1991. Environmental Responses to the Kissimmee River Demonstration Project. Technical Publication 91-02. Environmental Sciences Division, Research and Evaluation Department, South Florida Water Management District, West Palm Beach, Fla. March.

Water Resources Development Act of 1990. P.L. 101-640, Nov. 28, 1990, 104 Stat. 4604.

Zedler, J. B., R. Langis, J. Cantilli, M. Zalejko, K. Swift, and S. Rutherford. 1988. Assessing the functions of mitigation marshes in Southern California. Pp. 323-330 in J. Kusler, S. Daly, and G. Brooks, eds., Urban Wetlands: Proceedings of the National Wetland Symposium, Oakland, California. Association of Wetland Managers, Berne, N.Y.

4

Lakes

OVERVIEW

The fact that lakes occupy such a small fraction of the landscape belies their importance as environmental systems and resources for human use. They are major recreational attractions for Americans. Sport fishing, swimming, and boating are highly popular pastimes, and lake-front property has a high economic value. Large lakes and reservoirs are used as drinking water supplies; the Great Lakes alone serve as the domestic water supply for approximately 24 million Americans, and many more Americans rely on man-made reservoirs and smaller lakes for their source of drinking water. Lakes are used by humans for many commercial purposes, including fishing, transportation, irrigation, industrial water supplies, and receiving waters for wastewater effluents. Aside from their importance for human use, lakes have intrinsic ecological and environmental values. They moderate temperatures and affect the climate of the surrounding land. They store water, thereby helping to regulate stream flow; recharge ground water aquifers; and moderate droughts. They provide habitat to aquatic and semiaquatic plants and animals, which in turn provide food for many terrestrial animals; and they add to the diversity of the landscape.

The myriad ways in which humans use lakes, along with the numerous pollutant-generating activities of society, have stressed lake ecosystems in diverse ways, frequently causing impairment of lake quality for other human uses. Stresses to lakes arise from easily identifiable point sources such as municipal and industrial wastewater, from nonpoint degradation, from urban and agricultural runoff within a lake's watershed, and from more insidious long-range atmospheric transport of contaminants. Major categories of stresses include excessive eutrophication from nutrient and organic matter loadings; siltation from inadequate erosion control in agricultural, construction, logging, and mining activities; introduction of exotic species; acidification from atmospheric sources and acid mine drainage; and contamination by toxic (or potentially toxic) metals such as mercury and organic compounds such as polychlorinated biphenyls (PCBs) and pesticides. In addition, physical changes at the land-lake interface (e.g., draining of riparian wetlands) and hydrologic manipulations (e.g., damming outlets to stabilize water levels) also have major impacts on the structure and functioning of lake ecosystems.

No lake in the United States is entirely free from such stresses, but the stresses are not always severe enough to impair lake ecosystems or their usefulness for human activities. Nonetheless, thousands of U.S. lakes (and reservoirs) covering several million acres of water surface have become degraded to the extent that some type of activity is necessary to make them more usable resources and ecosystems.

Lake restoration is a relatively recent activity. Historically, the term *restoration* has been applied broadly in lake management to an array of actions aimed at improving lake conditions for designated human uses (e.g., contact recreation, fishing, water supply). Return of a lake to its pristine condition has not been an explicit goal of most lake restoration projects, although these actions often improve some aspects of a lake's ecological attributes. As such, most so-called lake restoration projects are actually rehabilitation efforts (in the sense of the definitions in Chapter 1), and many are merely designed to manage (mitigate) undesirable consequences of human perturbations. For reasons of historical precedence, a broader definition of the term restoration is used in this chapter, but a distinction is made between methods that improve ecosystem structure and function (restoration in the broad sense) and methods that merely manage the symptoms of stress. Lake restoration began in the United States about 20 years ago, primarily in response to problems of nutrient overenrichment. A lake improvement program, the Clean Lakes Program was established in 1975 within the U.S. Environmental Protection Agency by Section 314 of the 1972 Federal Water Pollution Control Act Amendments (P.L. 92-500). Between 1975 and 1985, federal funds were provided for Clean Lakes projects on 313 lakes in 47 states and Puerto Rico; 87 percent of the Clean Lakes funds have been used for lake improvement projects (U.S. EPA, 1985). Matching state and/or local funds typically are involved in these projects, and several states with large numbers of lakes have developed their own programs. As problems of lake acidification became more widely recognized during the past decade, restoration of acidified lakes by addition of limestone has become a relatively common practice in some northeastern states, as well as in Scandinavia.

For long-term restoration, it is essential to control the source of the problem. In the case of eutrophication, this means decreasing the loading of nutrients, particularly phosphorus, from various watershed sources. In some cases, this also means that loadings of silt and organic matter must be decreased. Control of external sources is sufficient to return some lakes to their former conditions, but in many cases the changes in the lake have been so dramatic—major shifts in biota, loss of habitat, physical changes in bottom sediments, and lake hydrology—that merely turning off the loadings is not sufficient to improve water quality and ecosystem structure, at least in a reasonable time frame. In-lake restoration techniques must be employed.

Numerous methods have been developed to restore lakes or improve their condition; this chapter describes more than 25 such methods. Available methods range widely in effectiveness, cost, frequency of use, and range of applicability. For example, methods that require addition of chemical agents to lake water are limited to small- and medium-sized lakes for economic reasons. Methods that use biological agents are potentially effective at low cost even in large systems because of low initial costs and the absence of labor and maintenance expenses. Many methods are applicable only to a single type of problem (e.g., liming to mitigate acidification). Others are potentially useful in restoring lakes degraded by a range of stresses; for example, dredging may be used for siltation, nutrient buildup, and toxic contaminant problems. Because eutrophication is the most widespread and longest-studied lake problem, more methods have been developed to restore eutrophic lakes than to address all other problems put together. Aside from removing contaminated sediments by dredging or covering them with uncontaminated sediment, few methods are available to restore lakes degraded by toxic substances.

Our ability to assess the effectiveness of past lake restoration projects and to compare the effectiveness of different restoration methods is severely limited by three factors. First, and perhaps most important, surveillance of lake conditions for an adequate period of time before and after a restoration attempt has been done on relatively few lakes. In some cases, sufficient surveillance probably was done, but rigorous analysis and interpretation of the data were not a part of the surveillance effort. All too often the data are not readily available for others to assess. Second, lake restoration projects usually are considered to be operational activities rather than research and development projects, and as a result they are designed to produce the desired effect—a restored lake—by whatever combination of methods seems likely to succeed. It usually is not possible to determine which of several techniques used simultaneously on a lake actually produced the measured improvements, even if detailed monitoring is done. Third, the goals of restoration projects are not always clearly defined, and it is difficult to judge the degree of success when clear objectives have not been set.

The above comments notwithstanding, many successful lake "restorations" have been documented, starting perhaps with the widely publicized case of Lake Washington, a large, deep lake in Seattle that was becoming increasingly eutrophic from municipal sewage effluent and was restored in the mid-1960s by diverting the effluent from the lake. Success in this and other cases generally has been defined in terms of restoring an aquatic resource for some human activities rather than restoring an ecosystem to its original condition. It is often assumed that improvements that benefit human uses of lakes lead to an improvement in the lake's ecology. There is no basis to assume, however, that water quality enhancements such as improved water clarity actually restore lake ecosystems to their original (presettlement) conditions. Restoration failures are less widely publicized, of course, but several cases have been described in which a project produced fewer improvements than anticipated in lake quality (see Appendix A). Analysis of these failures is important because we can learn as much about the factors leading to successful restoration from such projects as we can from success stories.

Lake restoration projects typically focus on restoring only one part (the lake) of a connected stream-wetland-lake system within a watershed. When wetlands are considered at all in lake restoration projects, it is typically for diversion of nutrient-laden storm water runoff or sewage effluent into the wetland in an effort to obtain nutrient uptake by wetland vegetation. Such diversions may provide a temporary lowering of nutrient loadings to lakes, but wetland flushing during high flow periods may result in little net annual retention of nutrients by the wetlands. The impacts of diversion on wetland ecology generally are not taken into account in deciding whether to proceed with such projects.

Although many techniques are potentially available to restore degraded lakes, the science of lake restoration is inexact, and the outcome of applying a given technique to a particular lake is difficult to predict accurately. Lake restoration technology can be advanced by ensuring that projects are monitored adequately so that the effects of various manipulations can be assessed properly. In this context, a lake restoration *project* should be considered as part of a long-term, ongoing management *program* rather than a one-time, permanent solution to a lake's problems.

INTRODUCTION—IMPORTANCE OF LAKES

Humans have always been attracted to lakes. Human settlement on lakeshores can be explained by practical reasons—lakes provide food and drinking water and a convenient means for personal transport and conveyance of goods—but can there be any doubt that even the Neolithic Swiss lake dwellers enjoyed their homes partly because of the beauty of their surroundings? Today, we prize small inland lakes especially for their recreational assets, including their visual appeal and the feeling of being close to nature that a "day at the lake" provides. Fishing, swimming, and boating are highly popular pastimes throughout the United States. Recreational fishing on inland lakes is estimated to generate more than $1.3 billion (1985 dollars) in economic activity annually in the state of Minnesota alone (Minnesota Department of Natural Resources, Office of Public Information, unpublished data, 1990), and comparable figures can be cited for many other states. In urban areas, lakefront homes are in high demand and command premium price tags; lakefront property in rural areas has a high commercial value for development of vacation homes. All too often, the attributes that give rise to a lake's recreational value—clear, high-quality water; scenic shorelines; prized game fish—are impaired by developments that were stimulated by the presence of these values.

There are about 100,000 lakes with areas greater than 40 hectares (1 ha = 2.47 acres) in the conterminous United States (Duda et al., 1987). Although natural lakes are found in most of the 50 states, they are especially common in several regions, owing to specific geological conditions: in the Upper Midwest, New England, New York, and Alaska, as a result of glacial activity; in Florida, where most lake basins are the result of chemical dissolution of underlying limestone; along major rivers like the Mississippi, where channel meandering has formed lake basins; and in mountainous areas of the Far West, where glaciers and volcanic activity have produced most of the lakes. In regions where natural lakes are rare or absent, artificial lakes (reservoirs) have been developed by damming rivers and streams to provide the benefits (e.g., recreation, water supply, water storage capacity) that natural lakes provide elsewhere.

Large lakes and reservoirs are used as public water supplies; the American Water Works Association (Achtermann, 1989) estimates that 68 percent of the water used for domestic purposes by the 600 largest utilities (>50,000 customers) comes from impounded surface waters (natural lakes and man-made reservoirs). For simplicity, in this chapter the term *lake* refers both to natural impoundments and to man-made reservoirs. The five Great Lakes alone supply domestic water to some 24 million Americans. Lakes provide many other economic benefits to society and are used for such diverse purposes as commercial fishing, transportation, irrigation, and dilution of wastewater effluents. Not all of these uses are compatible. The use of lakes as receptacles for wastewater obviously is likely to impair their usefulness as water supplies and recreational resources, but more subtle incompatibilities also exist. For example, the production of warmwater game fish is enhanced by increasing nutrient levels, at least up to a point, but swimmers prefer water to be as clear (hence, unproductive) as possible.

STRESSES ON LAKES

Classes of Stresses and Their Effects

Lake ecosystems are subject to stress from a wide range of human activities within their watersheds and along their shorelines and from the variety of ways that humans use them. These stresses often have caused significant impairment of lake quality. Six major classes of stresses have been important in degrading the quality of U.S. lakes in recent decades:

1. excessive inputs of nutrient and organic matter, leading to eutrophication;
2. hydrologic and physical changes such as water-level stabilization;
3. siltation from inadequate erosion control in agricultural and mining activities;
4. introduction of exotic species;
5. acidification from atmospheric sources and acid mine drainage; and
6. contamination by toxic (or potentially toxic) metals such as mercury and organic compounds such as PCBs and pesticides.

In addition, chemical stresses to lakes can be categorized according to source as (1) point sources (such as municipal wastewater), which generally are the easiest to identify and control; (2) nonpoint or diffuse sources such as urban and agricultural runoff from a lake's watershed; and (3) long-range atmospheric transport of contaminants (the most difficult to measure and control). These stresses result in a variety of impacts on lake quality relative to human use and ecological integrity.

The specific impacts of stresses on lake ecosystems depend on the nature of the stress and the characteristics of the lake, but some responses are common to several categories of stress. For example, stress-impacted lakes tend to lose sensitive native species. Their replacement by stress-tolerant native or exotic species often does not fully compensate for the loss and leads to lower biodiversity and simplified food webs. Many types of stress result in loss of habitat; often this is the proximate cause of species losses. Many kinds of stress produce "nuisance conditions," that is, proliferation of a native or exotic organism or deterioration in a physical-chemical property (such as water clarity) to the extent that beneficial uses of the lake are impaired. Finally, the development of toxic levels of contaminants in biota results not only from direct loading of toxic materials to lakes but also from indirect effects of other stresses (e.g., solubilization of aluminum as pH is decreased by acid deposition).

EUTROPHICATION

Of the six categories of stress, problems related to nutrient overenrichment and excessive plant production are probably the most common and have received public and scientific attention for the longest time. Concern about lake eutrophication from municipal wastewater extends back at least to the 1940s and the classic studies of Sawyer (1947) on the relationship between springtime concentrations of inorganic phosphorus and nitrogen and the occurrence of algal blooms in summer. By the 1960s, widespread concern existed about increasing eutrophication of the Great Lakes, and nutrient enrichment problems were recognized in numerous inland lakes. A large-scale research program funded primarily by federal agencies was undertaken on eutrophication in the 1960s and 1970s. This program led to improved understanding of the extent of the problem in U.S. lakes, delineated specific causes of the problem in some lakes, generated quantitative relationships between rates of nutrient loadings (especially of phosphorus) to lakes and water column responses in the lakes, and developed techniques to restore lakes degraded by eutrophication.

Eutrophication results in numerous ecological and water quality changes in lakes. The chain of events leading to use impairment is roughly as follows. Increased input of nutrients, especially phosphorus, leads to an increased incidence of nuisance blooms of algae (especially blue-green algae), leading to a loss of water clarity, a buildup of organic and nutrient-rich sediments, loss of oxygen from the bottom waters of the lake (which in turn, accelerates nutrient recycling processes), and changes in the lake's food web structure. Secondary nutrient limitation by silica or nitrogen that results when phosphorus levels are elevated also leads to changes in the phytoplankton community and to the development of nuisance species of algae (e.g., blue-green forms). Proliferation of macrophytes is also associated with eutrophication, especially in shallow lakes, but these problems are not tied *directly* to excessive rates of nutrient loading (see "Exotic Species," below). Although increases in nutrient levels enhance fish production, the loss of habitat (e.g., by sediment buildup, deoxygenation, undesirable proliferation of macrophytes) and food sources (by food web simplification) causes a shift from more desirable game fish to less desirable species, especially in more extreme cases of eutrophication. Stocking of exotics and overfishing exacerbate this problem. From a human use perspective these changes create numerous problems, including the following: fouling of boats and structures (by algal growths), loss of aesthetic appeal, accessibility problems for swimmers and boaters (because of macrophyte proliferation), economic damage to resort and property owners, and increased costs and technical difficulties of treating water for drinking purposes (because of taste and odor problems and increased potential for trihalomethane production).

The causes of eutrophication resulting from human activity are reasonably well understood. Once an oligotrophic lake has been made eutrophic, processes develop that may delay recovery after nutrient loadings have been decreased. If the hypolimnion becomes anoxic, recycling of phosphorus from the sediments is enhanced, in effect increasing the efficiency of use of the phosphorus input. During the eutrophic phase many changes may occur that will not be automatically reversed by a reduction in nutrient supply, such as loss of desirable macrophyte, invertebrate, and fish species. Nutrient reduction is a necessary, but not always a sufficient, condition for reversal of eutrophy.

Point sources of nutrients are the primary cause of excessive loadings in some lakes, but nonpoint sources (urban and agricultural runoff) contribute most of the nutrient input to the majority of U.S. lakes. Based on a modeling exercise with loading data on phosphorus for 255 lakes in the eastern United States, Gakstatter et al. (1978) concluded that only 18 to 22 percent of the lakes would show a measurable improvement in trophic conditions (which they assumed would require at least a 25 percent reduction in phosphorus inputs) if an effluent standard of 1 mg of phosphorus per liter were imposed on municipal wastewater treatment plants. Only 28 percent of the lakes would show measurable improvement if all their point sources of phosphorus were removed. Thus, most of the lakes (72 to 82 percent) in this analysis would require control of nonpoint sources of nutrients to achieve measurable improvements in trophic conditions.

HYDROLOGIC AND RELATED PHYSICAL CHANGES

The watersheds of lakes in urban and agricultural areas clearly are no longer ecologically the same as they were in presettlement days, and such land use changes are a primary cause of the stresses described in this section. What is not so widely recognized is the fact that important physical properties of lakes themselves, such as water residence time, water level, and basin morphology, are often modified significantly in developed areas. In turn, these changes can have untoward effects on water quality and ecological conditions. The importance of morphology in determining a lake's basic level of productivity is a fundamental concept in limnology.

Diversion of stream flow into lakes to provide water for urban or agricultural uses outside the watershed has occurred in some western states; Mono Lake, California is probably the best known example. The resulting decline in water supply to the lake has caused long-term lowering of the lake level, an increase in the lake's salinity, and ecological damage to tributary streams and to the lake itself (NRC, 1987). A much more widespread practice nationwide is the stabilization of lake levels by regulating outflows with a control structure (dam) at the lake outlet. This practice minimizes flooding of shoreline developments during wet periods and prevents loss of access to the lake due to receding shorelines during dry periods. However, long-term water-level stabilization also leads to loss of ephemeral wetlands in nearshore areas, converting them either to permanently dry upland areas or to lake littoral area. Fluctuating water levels are thought to have a cleansing effect on littoral sediments (oxidizing organic deposits); accumulation of such deposits in nearshore areas of lakes with stabilized water levels contributes to poor water quality and loss of fish spawning areas.

Changes in water level also affect fish reproduction directly by regulating access to spawning areas in the littoral zone, streams, or surrounding wetlands. Consequently, coordination between agencies that regulate water level and agencies that manage fisheries can have significant benefits. For example, the level of Lake Mendota, Wisconsin, had generally been lowered in winter to protect shoreline structures from ice damage. As a consequence, northern pike were prevented from spawning in the marshes around the lake. This problem was recognized in 1987. Beginning in 1988, the water level was raised about 15 cm during the spawning season (late March to early April). Numbers of spawning northern pike increased about twofold in 1988 and about eightfold in 1989 (Johnson et al., 1992). There has been no increase in the incidence of ice damage to shoreline structures.

Water residence times of lakes in developed areas are affected by water-level stabilization, as well as by diversion of streams into or out of a lake's drainage basin (thus also affecting watershed size and loading rates of nutrients and pollutants). Lake Okeechobee, Florida, is an extreme case of human-induced changes in lake morphometry, watershed area, water level, and other hydrologic characteristics that resulted in a variety of water quality problems (see Kissimmee River case study, Appendix A).

SEDIMENTATION

Problems of excessive sediment loading occur in lakes with large drainage basins where agricultural practices result in excessive soil erosion. Such problems are common in the central and southeastern parts of the United States, where row crop farming and erosive soils coexist, but some large reservoirs in the arid West also suffer from excessive sediment buildup. Siltation problems are significant in urban lakes as well. In extreme cases, excessive sedimentation leads to significant loss of reservoir storage capacity, diminishing the usefulness of lakes for regulating water availability (i.e., supplying water during droughts and controlling floods). Excessive sediment buildup renders large areas of lakes unusable for recreational purposes, as well as for fish spawning and habitat. Because nutrients (especially phosphorus) tend to adsorb onto sediments and because suspended sediments prevent penetration of light, lakes with very high loadings of sediment may not have sufficient plant productivity to support a good sport fishery; Lake Chicot, Arkansas, is an example (Stefan et al., 1990).

EXOTIC SPECIES

Lakes are island habitats. Like islands, they are highly susceptible to invasion by exotic species that lead to extirpation of native species

(Magnuson, 1976). In some cases, invasions by exotic species have had severe environmental and economic consequences. The most notorious species invasions have widespread effects that reverberate throughout an ecosystem. The seemingly random nature and explosive development of biological invasions have fascinated ecologists for many years (Elton, 1958); the status of basic research on this topic was reviewed by Mooney and Drake (1986).

Many thousands of acres of inland lakes suffer from problems of excessive macrophyte growths, and in most cases the problem plants are exotic (nonnative) species. Some of these plants were introduced to this country by the aquarium industry; others, such as water hyacinth (*Eichhornia crassipes*), were imported because they were regarded as visually attractive. The natural predators and pathogens that tend to keep the plant populations in check in their native lands usually are not present in this country. The resulting uncontrolled growth causes a variety of problems: clogging of irrigation canals, hydroelectric systems, and navigational waterways; flooding due to obstructed drainage systems; and impairment of boating and contact recreational activities (Barrett, 1989). Cases have been reported of swimmers becoming entangled in excessive growths of macrophytes and drowning. Dense beds of plants alter water chemistry and habitat structure, leading to changes in invertebrate and fish communities, and they are a major source of organic matter to the water column and sediments. Some exotic plants (e.g., purple loosestrife and water hyacinth) have low nutritive value to aquatic animals and provide a poor base for the food chain. Aquatic weed invasions contributing to major management problems include water hyacinth in 50 countries on five continents, kariba weed (*Salvinia molesta*) in tropical regions worldwide, hydrilla (*Hydrilla verticillata*) and Eurasian water milfoil (*Myriophyllum spicatum*) in North America, and *Elodea canadensis* in Europe (Hutchinson, 1975; Barrett, 1989).

Exotic species problems are by no means limited to plants. Benthic invertebrate invaders also have created problems. An example is the invasion of lakes throughout northern Wisconsin and Minnesota by the rusty crayfish, *Orconectes rusticus* (Lodge et al., 1985). This species displaces native species from their burrows, exposing them to predation. Rusty crayfish are voracious consumers of game fish eggs and obliterate macrophyte beds, essential habitat for recruitment of game fish (Lodge et al., 1985). Thus, the crayfish tend to eliminate their main predators, smallmouth bass. Ironically, the invasion originated with releases from anglers' bait buckets. Spread of the crayfish is now perpetuated by the development of commercial harvesting of the rusty crayfish (primarily for export to Scandinavia). Crayfishers have transplanted rusty crayfish to new lakes to increase the harvestable resource.

Exotic fish have displaced native species, contributed to the collapse of fisheries, and even led to water quality problems (Magnuson, 1976; see Lake Michigan case study, Appendix A). The common carp, *Cyprinus carpio*, is not native to this country but was introduced to many northern lakes and rivers in the late 1870s by the U.S. Fish Commission in response to requests from European immigrants. Carp are widely eaten in European countries but are rarely consumed in this country and are not a sought-after game fish. Because carp are benthivorous (bottom feeders) and stir up bottom sediments, they accelerate nutrient recycling from sediments, destroy spawning areas for other fish, and cause turbidity problems in lakes and rivers.

The Great Lakes have a long and unfortunate history of invasions by exotic species. The sea lamprey (*Petromyzon marinus*), a large parasite of game fish, is a native of the Atlantic Ocean that made its way into Lake Erie through the Welland Canal in 1921. It gradually worked its way as far as Lake Superior, where it remains a significant cause of fish mortality (especially for lake trout). The lamprey has been controlled (but not eliminated) by applying a "lampricide," 3-trifluoromethyl-4 nitrophenol (TFM), to tributary streams where adult lamprey spawn. The TFM selectively kills young lamprey. The alewife, a small forage fish, was also introduced into the Great Lakes inadvertently, as a result of development of the St. Lawrence Seaway. The fish grew to great abundance in the 1960s, and episodes of massive mortality in alewife populations caused problems along urban beaches. The fish was controlled in the Great Lakes primarily by stocking the lakes with other exotic fish, coho, and Chinook salmon.

The latest in a series of exotic species to invade the Great Lakes, and potentially the most devastating, is the zebra mussel (*Dreissena polymorpha*). First found in Lake St. Clair in 1988, this rapidly spreading species was found throughout the western basin of Lake Erie in 1989 and as far as the Duluth-Superior harbor in western Lake Superior in 1990. The organism was most likely introduced to the Great Lakes by discharge of ballast water from oceangoing vessels. A native of Asia, the zebra mussel has been a problem in European waters for more than 100 years. It is already causing obstruction problems with water intake for power plants and municipal and industrial water treatment plants in Lake Erie. Because fouling organisms historically have not been a problem in inland waters of the United States and Canada, most facilities have not been designed to control or compensate for these problems, and the potential costs are enormous (Mackie et al., 1989). The zebra mussel has become abundant enough that it may already have had an impact on the food web in Lake Erie. A filter feeder, it is thought to be responsible for an increase in water clarity in the lake during 1989 and 1990. Fishery scientists are concerned that the organism will divert enough primary and secondary production from pathways that support fish growth to affect the lake's economically important walleye fishery. No control techniques are currently available to address a problem of this magnitude. Although it is not yet found in U.S. waters outside the Great Lakes, the zebra mussel is expected to spread widely throughout the surface waters of the eastern United States over the next several years.

ACIDIFICATION

Acidification of poorly buffered lakes (and other surface waters) by acidic precipitation has been a major environmental issue in the United States and Canada (as well as parts of western Europe) for the past two decades. The ecological changes caused by acidification are fairly well understood (e.g., Schindler, 1988), but the severity of the problem is still controversial, despite more than a decade of extensive research. Acidification tends to simplify the biotic structure of lakes, as acid-sensitive species are lost and relatively fewer acid-tolerant species remain. However, ecological impacts generally are greater at the population level than at the community level, and effects on some integrative measures of ecosystem performance, such as total primary production and community respiration, have not been demonstrated conclusively, especially for mild levels of acidification.

In contrast, rates of decomposition of organic matter, especially leaves and other terrestrially produced materials, are slowed in acidic lakes (Perry et al., 1987; Brezonik et al., 1991a), and certain pathways in the biogeochemical cycles of major elements such as nitrogen and sulfur may be altered or inhibited under acidic conditions (e.g., Rudd et al., 1988). Water column concentrations of several minor metals (manganese, iron, and especially aluminum) and trace metals (cadmium, lead, zinc, and mercury) are higher in acidic lakes because of increased solubility and decreased tendency to adsorb onto particles, and the free (uncomplexed) chemical forms of the metal ions tend to predominate in acidic waters. Other factors being equal, this trend should increase metal bioaccumulation and toxicity to aquatic biota. Indeed, increased aluminum toxicity is thought to be a major factor in the loss of fish species in many acidic lakes, but the situation is less certain for other potentially toxic trace metals (Campbell and Stokes, 1985; Brezonik et al., 1991a). Increased competition for metal-binding sites on organisms by the higher H^+ concentrations in acidic waters may actually *decrease* biological uptake of trace metals and reduce their toxicity to aquatic biota.

From a perspective of water use, lake acidification has three major effects: loss of fish populations; increased water clarity, caused primarily by loss of colored organic matter (so-called humic material) from the water column; and increased abundance of acid-tolerant, filamentous algae (primarily *Mougeotia*), huge, unsightly masses of which may cover the bottom in littoral areas. Fish species differ widely in their sensitivity to acidity (Table 4.1). Smallmouth bass are much more sensitive than largemouth bass. Rainbow trout are impacted in the pH range 5.5 to 6.0; brook trout are much less sensitive. Perch survive and reproduce at pH 5, but survival of young-of-the-year perch is strongly affected at pH 4.7 (Brezonik et al., 1991b). Some Florida lakes with a pH as low as 4.5 have apparently healthy fish communities, although fish production is low because acidic lakes tend to be very oligotrophic. In general, fish production is much more closely related to a lake's nutritional status than to its pH. In a given species, adults are more tolerant than immature forms; lack of

TABLE 4.1 Approximate pH Range in Which Various Fish Species Suffer Reproductive Failure or Mortality

pH	Species
6.0 to 5.5	Smallmouth bass (*Micropterus dolomieui*) Walleye (*Stizostedion vitreum*) Rainbow trout (*Oncorhynchus mykiss*) Common shiner (*Notropis comutus*) Burbot (*Lota lota*)
5.5 to 5.2	Lake trout (*Salvelinus namaycush*) Trout perch (*Percopsis omiscomaycus*) Fathead minnow (*Pimephales promelas*)
5.2 to 4.7	Brook trout (*Salvelinus fontinalis*) Brown bullhead (*Ictalurus nebulosus*) White sucker (*Catostomus commersoni*) Largemouth bass (*Micropterus salmoides*) Rock bass (*Ambloplites rupestris*)
4.7 to 4.5	Cisco (*Coregonus artedii*) Yellow perch (*Perca flavescens*) Lake chub (*Couesius plumbeus*)

NOTE: Compiled by the committee from various sources.

spawning success and year-class recruitment failures occur before the condition of adults or their mortality is affected. Without question, potential damage to fishing has caused the greatest public concern about lake acidification, but the actual extent of losses has been very difficult to quantify. Several early studies purporting to show that acidification caused a significant loss of game fish (e.g., trout) in Adirondack lakes over the past 50 to 60 years were shown later to be flawed. For example, trout populations disappeared from some lakes because fishery management practices changed (i.e., stocking of young fish was stopped for unknown reasons). Nonetheless, the recently completed integrated assessment of the National Acid Precipitation Program (NAPAP, 1990a) concluded "with reasonable confidence" that acidification had resulted in a loss of one or more fish populations in about 16 percent of the Adirondack lakes.

On the one hand, the lakes most sensitive to acidification tend to be small and relatively unproductive (oligotrophic). On the other hand, these lakes tend to occur in relatively unspoiled forested areas and are valued for their pristine nature. It is difficult to compare the value of the total experience of catching a trout in such a lake (on a dollar-per-fish or dollar-per-pound basis) with that of catching a perch in a more highly developed lake. Moreover, small lakes do act as sensitive indicators of environmental damage and may be viewed as early warning indicators of environmental stress.

CONTAMINATION BY TOXIC SUBSTANCES

Lakes are sinks for many materials (i.e., inputs from their drainage basins exceed losses through outlet streams). Such materials tend to accumulate in certain compartments of lakes—ultimately in bottom sediments, but also (and more importantly) in biotic components. In several well-documented cases, toxic substances (metals or synthetic organic compounds) have accumulated to problem proportions in the food web of a lake (particularly in game fish) because of industrial accidents or inadequate disposal practices, but in other cases, the source of the toxic material is more diffuse—nonpoint source runoff or deposition from the atmosphere.

The list of metals that have been identified with use impairment in lakes is lengthy and includes silver, arsenic, cadmium, copper, mercury, manganese, lead, selenium, and zinc. Excessive levels of selenium in two North Carolina reservoirs resulted from discharges from coal-fired power plants (U.S. EPA, 1989) and caused drastic declines in fish populations and reproduction. Mining and mineral processing activities caused accumulations of toxic metals in biota of Lake Coeur d'Alene, Idaho.

Long-range atmospheric transport from widespread sources is blamed for high body burdens of mercury (Hg) in the fish of many otherwise pristine lakes in forested regions of the Upper Midwest (Henning, 1989; Swain and Helwig, 1989). The problem in these states is more pronounced in low-alkalinity (acid-sensitive) lakes, but levels of bioaccumulation are not closely correlated with water pH. Mercury contamination of fish is at least indirectly related to acidic deposition in that fossil fuel burning by power plants contributes to both problems. The accumulation of mercury varies widely among different species of fish; biomagnification proceeds as mercury moves through the food web, and top carnivores such as walleye have the highest body burdens. Within a given species, body burdens increase with size (and age) of the fish.

Several states routinely issue consumption advisories related to mercury contamination of fish in lakes, and there is much concern about the economic impacts of these advisories on sport fishing in the affected regions. The nature of the advisories varies from state to state, and depending on the level of contamination, the advisories may recommend that a certain size range and species of fish not be eaten at all or that consumption be limited to one meal per week or per month. Problems caused by mercury in lakes are not limited to human consumption of contaminated fish; wildlife whose diet includes fish are also at risk. Body burdens of mercury in piscivorous loons in northern Minnesota are high enough to cause acute toxicity and may explain some incidents of loon mortality (Swain and Helwig, 1989).

Contamination problems involving organochlorine compounds such as pesticides and PCBs have been induced in lakes by all three types of sources for chemical stress (point sources, nonpoint watershed sources, and long-range atmospheric transport). High levels of PCBs in fish of the lower Great Lakes are attributed to general, widespread use of these chemicals from the 1920s to the 1970s, but localized cases of sediment contamination can usually be traced to one or a few specific industrial operations. For example, severe contamination of sediments in Waukegan harbor (Lake Michigan) occurred as the result of disposal practices by one manufacturer. At the other extreme, high body burdens of PCBs are found in some large lake trout in Lake Superior (at levels sufficient to cause a consumption advisory), in spite of the fact that the lake has only minor point sources and nonpoint watershed sources of PCBs. Atmospheric transport (on scales of hundreds or even thousands of miles) is the principal source of PCBs in Lake Superior (Eisenreich, 1987) and the major source of toxaphene for all the Great Lakes. A chlorinated insecticide, toxaphene was used principally on cotton fields in southern states to control the boll weevil until it was banned in the 1970s. Nonetheless, residues of toxaphene are commonly found in water and fish of the Great Lakes (Camanzo et al., 1987). Another illustration of the importance of long-range atmospheric transport is provided by Siskiwit Lake, on Isle Royale, more than 90 km from the nearest shore in western Lake Superior. The island is a wilderness area (and a national park), and Siskiwit Lake has no watershed sources of contamination (past or present). Nonetheless, elevated levels of polychlorinated dibenzodioxins (PCDDs) and PCBs are found in fish from the lake (Swain, 1978; Czuczwa et al., 1984, 1985), a fact that can be attributed only to atmospheric inputs.

Responses to Stresses—Status of U.S. Lakes

OVERVIEW

Several assessments of conditions in U.S. lakes have been made in the past two decades (Ketelle and Uttormark, 1971; Duda and Johnson, 1984; ASIWPCA, 1984, 1985; U.S. EPA, 1989, 1990b). All were based on responses to questionnaires to administrators of state water agencies. The earliest surveys focused on trophic conditions (eutrophication was considered the major lake problem in the 1960s and 1970s), but more recent surveys also considered other types of degradation. The responses are largely qualitative and vary widely among the states in accuracy and completeness. States use differing criteria for classifying lakes and defining problem conditions, and all have incomplete data. Some states reported on only a small fraction of their lakes in a given assessment. For example, Florida, which has about 7,700 lakes, assessed trophic conditions in only 91 lakes for EPA's 1988 water quality survey (U.S. Environmental Protection Agency, 1989, 1990). Only one Florida lake was listed as hypereutrophic and thirteen as eutrophic; this grossly understates the seriousness of eutrophication problems in that state. States that lack active lake programs did not participate in some of the surveys. Omitted from the data are thousands of private lakes and small water bodies. All U.S. Army Corps of Engineers projects, which include 783 reservoirs with a total of 27,000 km^2 (66.7 million acres; Kennedy and Gaugush, 1988), have also been excluded from the surveys.

The most recent survey (U.S. EPA, 1989, 1990), which was conducted in 1988, compiled data from 40 responding states or territories. All 40 respondents provided some assessment of surface acreage of lakes supporting designated uses versus acreage of lakes with impaired or partially impaired uses, and threatened lakes (Table 4.2), but the percentage of total acreage that was assessed in a given state ranged from about 25 to 100 percent, and only 32 states specified the basis of their assessment decisions. Only 26 states provided information on the degree of impairment (minor, moderate, or major), 33 provided data on the nature (causes) of impairment (e.g., nutrients, siltation, and toxic substances), and 28 provided information on the sources of pollution (agriculture, storm sewers, municipal wastewater, and so on). The survey concluded that 26 percent of the assessed lake acreage suffered from some kind of use impairment (Table 4.2); of the 16.3 million acres assessed, almost 4.3 million acres were impaired (defined here as acreage not supporting or only partially supporting the designated uses). An additional 18 percent of the assessed acreage (2.9 million acres) was reported to be threatened. Four states (Florida, North Dakota, South Dakota, and Wisconsin) each had more than 600,000 acres of threatened or impaired waters (Table 4.2). Eight states had 120,000 acres of lakes with threatened or impaired waters, and an additional nine states had at least 160 km^2 (40,000 acres) of lake waters in those categories. These 21 states accounted for most of the threatened or impaired waters in this survey.

The EPA's 1988 survey identified 12 causes of impairment and estimated the percentage of total use-impaired lake acreage affected by each of these, as well as by identified sources of pollution for 33 responding states (Table 4.3A,B). Because of the nature of the survey, the numbers reported in Tables 4.2 and 4.3A and B cannot be extrapolated to the total population of lakes in the country, and probably should not be used even to estimate total numbers or acreage of impaired or threatened lakes in a given state. It is clear even from these limited statistics, however, that many of the nation's lakes are degraded to the extent that their use is impaired and that a wide variety of problems and causes are responsible for this situation. Moreover, the data in Table 4.2 probably represent highly conservative estimates of the surface area of impounded water that could be improved by proper restoration and management measures.

According to Duda and Johnson (1984), EPA Regions IV (Southeast), V (North Central), and VIII (Missouri Basin) have the highest fractions of impaired lakes (accounting for >80 percent of the impaired acreage described in their report). Duda et al. (1987) reported that 22 of the 32 major Tennessee Valley Authority (TVA) reservoirs have some form of use impairment, and 16 of the 21 non-TVA reservoirs in the region are impaired or threatened. Aside from the Great

TABLE 4.2   Designated Use Support in Lakes and Reservoirs

State	Number of Lakes	Total Acres of Lakes	Total Acres Surveyed	Acres Fully Supporting	Acres Partially Supporting	Acres Not Supporting
AL	43	504,336	491,566	405,486	0	86,080
CA	4,955	1,417,540	1,076,891	568,739	95,505	412,647
CO	4,069	265,982	124,973	123,300	1,673	0
CT	6,000	82,900	21,701	9,312	12,389	0
DC	8	377	136	0	0	136
FL	7,712	2,085,120	947,200	309,760	536,320	101,120
GA	175	417,730	417,730	412,357	5,347	26
IL	2,940	247,188	183,572	22,931	100,591	60,050
IN	560	104,540	104,540	104,361	63	116
IA	282	81,400	80,249	26,801	52,058	1,390
KS	232	175,189	173,911	116,655	48,141	9,115
KY	92	228,385	214,483	179,335	31,471	3,677
LA	101	713,719	517,476	376,335	141,141	0
ME	5,779	994,560	994,560	958,080	36,480	0
MD	59	17,448	17,448	14,838	2,603	7
MI	35,000	840,960	424,021	304,185	62,834	57,002
MN	12,034	3,411,200	1,435,554	1,198,709	67,622	169,223
MS	—	500,000	500,000	481,740	18,260	0
MO	362	288,012	288,012	285,701	2,311	0
MT	4,018	756,450	663,363	345,367	305,396	12,600
NB	412	145,300	85,518	82,304	2,779	435
NH	1,300	151,000	149,854	130,708	18,756	390
NM	—	126,500	119,666	72,358	47,308	0
NY	7,500	750,000	750,000	454,668	267,343	27,989
NC	1,500	305,367	305,367	293,470	2,075	9,822
ND	216	625,503	619,333	571,208	48,125	0
OH	2,500	117,323	90,771	30,936	50,988	8,847
OR	6,095	610,808	504,928	374,303	58,918	71,707
PR	38	11,146	11,146	3,801	4,240	3,105
RI	113	16,520	16,089	14,688	787	614
SC	1,418	525,000	410,407	409,242	840	325
SD	789	1,598,285	662,532	567,812	17,984	76,736
TN	117	538,657	538,657	452,009	50,830	35,818
TX	5,700	1,410,240	1,410,240	1,225,629	0	184,611
VT	719	229,146	227,121	177,915	37,713	11,493
VA	248	161,562	161,089	147,352	13,737	0
WA	808	613,582	156,518	122,834	33,104	580
WV	94	19,171	19,171	0	17,441	1,730
WI	14,998	971,000	971,000	249,000	478,000	244,000
WY	2,629	427,219	427,219	396,815	30,404	0
Total	131,615	22,486,365	16,314,012	12,021,044	2,701,577	1,591,391

SOURCE: Reprinted from U.S. Environmental Protection Agency, 1990.

TABLE 4.3A Lake Acres Affected by Causes of Impairment

Total Impaired Waters[a]	Nutrients	Siltation	Organic Enrichment	Salinity	Habitat Mod	Pathogens	Priority Organics	Suspended Solids	Metals	Pesticides	pH	Flow Alteration
2,658,839	1,297,044	676,664	671,923	380,831	301,354	228,246	217,258	200,239	197,803	141,136	136,723	86,737
	48.8%	25.4%	25.3%	14.3%	11.3%	8.6%	8.2%	7.5%	7.4%	5.3%	5.1%	3.3%

[a]The sum of partially and nonsupporting lake acres.

SOURCE: U.S. Environmental Protection Agency, 1990.

TABLE 4.3B Impaired Lake Acres Affected by Sources of Pollution

Total Impaired Waters[a]	Agriculture	Hydro/ Habitat Mod	Storm Sewers/ Runoff	Land Disposal	Municipal	Industrial	Resource Extraction	Con- struction	Silvi- culture	Combined Sewers
2,686,889	1,564,382	889,760	744,214	710,998	404,846	207,591	112,977	87,879	25,034	7,981
	58.2%	33.1%	27.7%	26.5%	15.1%	7.7%	4.2%	3.3%	0.9%	0.3%

[a]The sum of partially and nonsupporting lake acres.

SOURCE: U.S. Environmental Protection Agency, 1990.

Lakes, many lakes of national or regional significance are impaired (e.g., Lakes Apopka and Okeechobee, Florida; Ocean Lake, Wyoming; and Reelfoot Lake, Tennessee). Reelfoot Lake, a natural lake in the south-central United States, is a classic example. Silt and nutrients from agriculture and channelization of inflowing streams have increased sedimentation rates in the lake, and associated weed and algal growths have reduced its area from 208 km^2 (51,400 acres), to 52 km^2 (12,800 acres). The habitat of two endangered species of birds is threatened, and changes in the lake itself have affected the economy of the area (Duda and Johnson, 1984). Based on current rates of sedimentation, McIntyre and Naney (1990) predicted that the lake will become too shallow for recreational purposes in as little as 60 years (for the shallowest of its three basins). Changes in land management are needed to alter this situation.

The condition of the nation's lakes appears to be deteriorating. The 1984 ASIWPCA survey assessed changes from 1972 to 1982 and concluded that the acreage of lakes that had degraded was four times that of the acreage that had improved during the decade. Similarly, a 1983 survey of state lake administrators by the North American Lake Management Society (NALMS; Duda and Johnson, 1984) showed an alarming increase in problem lakes since the survey of Ketelle and Uttormark (1971). The NALMS survey was marred by the lack of lake programs in many states or the inability of some states to respond, but good documentation appears to be available in nine states located in six EPA regions. The number of problem lakes *reported* in these states increased by a factor of 20 between 1971 and 1983 (Duda and Johnson, 1984).

TROPHIC STATE

The National Eutrophication Survey (NES), conducted by EPA in 1973 to 1976, sampled several hundred lakes throughout the continental United States and constructed nutrient budgets on many of the lakes. Results of the survey showed that the great majority of surveyed lakes had eutrophic conditions and experienced some form of water quality degradation. Lakes were not selected for the NES based on a random sampling of U.S. lakes. The survey was designed to assess the severity of eutrophication in lakes with municipal sewage treatment plants in their drainage basin, and it would not be appropriate to extrapolate NES statistics to estimate the trophic status of the nation's lakes.

The surveys listed at the beginning of this section did attempt nationwide trophic state assessments. In the most recent (1988) assessment by EPA, 39 states provided trophic classification on a total of 15,514 lakes (U.S. EPA, 1990b). About 30 percent of the surveyed lakes were classified as eutrophic or hypereutrophic, and 23 percent were mesotrophic. Trophic conditions were unknown in about 30 percent of the lakes included in the survey. In some cases, a lake is eutrophic simply as a result of natural circumstances (e.g., ecoregional characteristics), but nonpoint pollution from agricultural and urban run-off is the cause of use impairment from excess nutrients in most lakes.

The trophic status of the North American Great Lakes, including Great Bear Lake and Great Slave Lake, was summarized by Robertson and Scavia (1984). They concluded that Lakes Ontario and Erie are eutrophic and that Green Bay (Lake Michigan), Saginaw Bay (Lake Huron), and the Lake Erie western basin are highly eutrophic. The other lakes are mesotrophic or oligotrophic.

Canada has the largest acreage of lakes in the world, and a complete inventory, much less an assessment of their trophic states, is not available at this time. Most of them are thought to be oligotrophic, and in terms of raw numbers, the great majority of Canadian lakes lie in wilderness or undeveloped forests. Nonetheless, many lakes in agricultural areas of southern Canada have water quality problems resulting from excessive nutrients, and recreational developments have led to impaired water quality in some lakes located within driving distance of major urban areas such as Toronto. A small sample of 130 Canadian lakes found 16 of them to be eutrophic (Janus and Vollenweider, 1981).

Summary reports (e.g., Vollenweider and Kerekes, 1981; Forsberg, 1987) show that eutrophication problems are widespread throughout Europe. Reports of this nature do not exist for other continents, but accounts of extensive soil erosion and massive siltation of reservoirs everywhere, coupled with the absence of wastewater treatment in many areas (Brown and Wolf, 1984; Postel, 1985), suggest that water bodies worldwide are affected by excessive biological production and its consequences. Rapid in-filling of major impoundments in Third World nations is particularly troubling in view of their needs for irrigation water, potable supplies, and flood control. Deforestation and cultivation of marginal lands are causing soil losses at rates that will fill some impoundments in these countries in 5 to 20 years (Brown and Wolf, 1984).

ACIDIFICATION

The National Surface Water Survey (NSWS), a major survey of lakes and streams in acid-sensitive regions of the United States, was conducted by EPA in the mid-1980s. Because the survey design was based on stratified-random sampling procedures, the results can be extrapolated to the *population* of surface waters in acid-sensitive regions of the United States. According to the survey (NAPAP, 1990a,b), 4.2 percent of the NSWS lakes (representing about 1,180 lakes in the total population of lakes in the sampled regions) were acidic, defined as having an alkalinity less than 0 (pH < ca. 5.0 - 5.5). The acidic lakes are about equally divided among three regions: (1) the Northeast (primarily the Adirondacks), (2) the Upper Midwest (primarily northeastern Wisconsin and the Upper Peninsula of Michigan, and (3) interior Florida. Because the total number of lakes occurring in each region is different, the *percentage* of acidic lakes varies regionally (Figure 4.1), with Florida having the highest percentage (23) and the Upper Midwest having the lowest (3). About three-fourths of the acidic lakes were attributable to acidic deposition; most of the remainder were colored and were thought to be acidic from the presence of natural organic acids.

Although only a small percentage of lakes in the NSWS were found to be acidic, several cautions must be considered before the results are accepted as an accurate portrayal of the impact of acidic precipitation on U.S. lakes. First, the NSWS sampled lakes only one time—in the fall—and this is not the season during which the pH is lowest in lakes or the most critical season for biological impacts. Second, the NSWS did not sample small lakes (those <4 ha in area and >1 ha in the western lakes survey, where acidification is of less concern; no acidic lakes were found in western regions), and survey results indicate that the frequency of acidic conditions increases as lake size decreases. Third, the definition of acidic conditions used in the NSWS is arbitrary; ecological damage may occur at higher pH and alkalinity values than the NSWS used for its criteria. The criterion (acid-neutralizing capacity) was selected because it is considered to be a fairly unambiguous indicator of anthropogenic acidification, at least for lakes not affected by natural organic acids. Finally, the survey data indicate the status of lakes at a particular point in time and do not indicate the extent to which any lake has become more acidic as a result of acidic deposition.

The NSWS concluded that about 8 percent of the streams sampled were chronically acidic (acid-neutralizing capacity). On a length basis, approximately 7,900 km of streams were acidic; this represents about 4 percent of the total length of streams (211,000 km) in the NSWS regions. The acidic streams occurred mainly in the Mid-Atlantic Highlands and Mid-Atlantic Coastal Plain regions (Figure 4.1). The NSWS also concluded that 26,400 km (13 percent) of streams

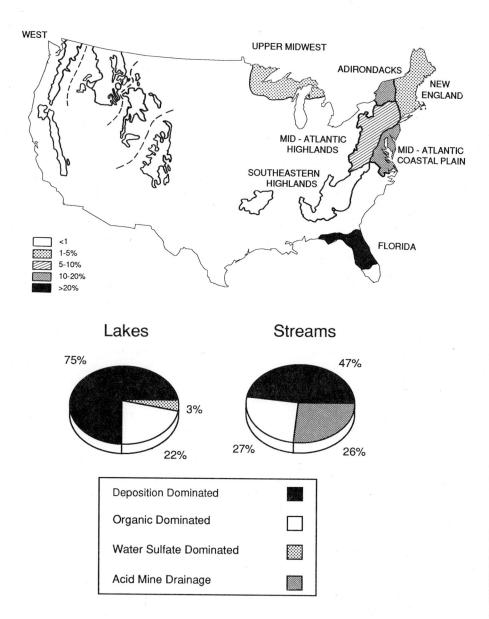

FIGURE 4.1 Percentage of acidic surface water (acid-neutralizing capacity) in the regions of EPA's National Surface Water Survey (upper) and sources of dominant acid anions in the acidic lakes and streams of the NSWS (lower). Source: National Acid Precipitation Assessment Program, 1990a.

had very low alkalinity values ($\leq$50 µeq per liter). These streams are especially susceptible to episodic acidification (e.g., during spring snowmelt or large rain events), and this is considered to be a significant problem in the Mid-Atlantic Highlands. Many miles of streams in Appalachia are acidic from acid mine drainage; the NSWS concluded that about 60 percent of the acidic stream length in the Mid-Atlantic Highlands is from this source and about 40 percent is caused by atmospheric deposition. Acidic mine drainage results from oxidation of pyritic minerals exposed to the atmosphere during mining activity. Oxidation of the minerals produces sulfuric acid. Regions in which acid mine drainage is a problem have few natural lakes, and this source accounts for only a few acidic lakes nationwide.

TOXIC SUBSTANCES

The importance of toxic substances in lake degradation can be illustrated with data from the Great Lakes. The International Joint Commission has identified 42 areas of concern in the Great Lakes, and 41 of these have problems associated with toxic substances (Hartig and Thomas, 1988). As mentioned earlier, all Great Lakes have fish contaminated by PCBs and organochlorine pesticides. Toxic substances reported from the analysis of sediments include metals (arsenic, cadmium, chromium, copper, lead, mercury, manganese, and zinc), cyanide, grease and oil, and a variety of chlorinated organic compounds: polychlorinated biphenyls (PCBs), polynuclear aromatic hydrocarbons (PAHs), hexachlorobenzene (HCB), dioxins, phthalates, and dibenzofurans. A much longer list has been identified from the analysis of water samples.

No large-scale field surveys have been conducted to determine the status of U.S. lakes with regard to the problem of toxic chemicals, but EPA's 1988 water quality assessment (based on questionnaires) shows that such problems are widespread (U.S. EPA, 1990b). A total of 556,000 acres of lakes in 18 states was reported to be impaired by priority organic pollutants, metals, or pesticides; this represents almost 21 percent of the total impaired acreage for which the cause of impairment is known (Table 4.3A). Some inland lakes and many rivers have fish consumption advisories because of contamination by these compounds. Elevated chlordane levels have been found in fish from Kansas lakes; PCB contamination is a common problem in New York lakes and has also led to consumption advisories for some Minnesota and Wisconsin lakes. Reliable statistics are lacking on the pervasiveness and seriousness of such problems because adequate surveys have not been done.

Accurate data are also lacking on the number of lakes degraded by toxic metals such as mercury, but several lines of evidence suggest the number could be very large. For example, 21 states currently issue fish consumption advisories because of mercury contamination problems. Almost 90 percent of the Minnesota lakes from which fish have been analyzed for mercury (233 out of 261 lakes) had at least one species with burdens high enough to issue a consumption advisory (>0.16 µg/g for a one-meal-per-week advisory), and 98 of the lakes (38 percent) had fish with mercury levels higher than 0.65 µg/g (the trigger level for a one-meal-per-month advisory) (D. Helwig, Minnesota Pollution Control Agency, personal communication, 1991). Most of these lakes are in undeveloped forested areas of northeastern Minnesota. The 1990 consumption advisory of the Wisconsin Department of Natural Resources includes 157 lakes and 11 rivers with mercury-contaminated fish, as well as parts of 11 rivers, Green Bay, Lake Michigan, and Lake Superior for PCBs. The cited lakes and rivers are found throughout the state. The number of lakes with consumption advisories increases as more lakes are sampled, and problems are not limited to low-alkalinity lakes or to the Midwest. Mercury-contaminated fish have been reported in several western and many East Coast states, including Florida and New York.

OTHER STRESSES

According to the 1988 EPA survey, almost 700,000 lake acres are impaired by siltation in the United States (Table 4.3A; U.S. EPA, 1990b). Given the incompleteness of the data on which this number is based, the actual area impacted by excess sediment is probably significantly greater. No national statistics are available on the extent of lakes impaired by exotic species. Nonetheless, it is common knowledge that problems with exotic macrophytes are pervasive, especially in southern states, and that many thousands of acres are affected. The Great Lakes all suffer from a variety of exotic species problems, and with the recent zebra mussel invasion, problems caused by exotic species appear to be getting worse. No national statistics are available on the extent of damage caused by physical and hydrologic changes to lakes; in many cases, these manipulations are not even recognized as a factor in lake degradation.

LAKE RESTORATION AND MANAGEMENT

Definitions

The definitions of restoration, rehabilitation, mitigation, and management discussed in Chapter 1 apply to lakes as well as to other aquatic systems, but as noted earlier in this chapter, limnologists have applied the term *restoration* rather broadly to actions designed to alleviate degraded conditions in lakes. There are some important differences between lakes and other surface waters relative to ease of restoration, and many of the methods used to restore lakes are not applicable to the restoration of wetlands and running waters. For example, rivers and streams degraded by chemical contaminants can be restored in many cases simply by eliminating the source of contamination and relying on their self-cleansing properties, but this approach seldom is sufficient for lakes, which tend to have long water and substance residence times and behave more as closed systems. In-lake manipulations are usually necessary (in addition to source controls) to restore lakes. Loss of habitat by *physical* alterations (channelization, installation of flow-regulating structures) is probably the most common reason that rivers need restoration, but most degraded lakes suffer from some sort of chemical contamination by excess nutrients, organic matter, toxic substances, or acidity. (Loss of littoral habitat (macrophyte beds) is a common condition in recreational lakes but often is not recognized as a problem. Instead, such losses are viewed by swimmers and boating enthusiasts as an "improvement.") Differences in the source of degradation lead to differing approaches in restoring lakes versus restoring rivers or wetlands, as well as to different approaches to managing the three types of aquatic systems to prevent further degradation or minimize the impacts of stress.

RESTORATION USING THE CONCEPT OF ECOREGIONS

A major determinant of lake and reservoir productivity is the steady-state, long-term average concentration of nutrients, especially those that can be growth limiting, such as phosphorus, nitrogen, and silica. Increased nutrient and organic matter loading, usually from cultural sources such as wastewater treatment plants and runoff from urban or agricultural land, often leads to sharply increased nutrient concentrations in the water column and ultimately to algal blooms, dis-

solved oxygen depletion, and other symptoms of cultural eutrophication. Elimination or significant reduction of these cultural sources of
stress is essential if a lake or reservoir is to be restored to its previous
condition.

The nutrient concentration attainable in a lake following significant reduction or elimination of cultural loading will depend on several factors, including basin morphometry, hydrologic conditions, land
use, and the geographic region in which the lake is located. Lake
morphometry plays a major role in determining the amount of "internal loading" of nutrients from the sediments to the water column.
Shallow lakes, particularly those exposed to wind-induced mixing,
are likely to have high internal loading rates. Water residence time
also plays a role in determining lake water column nutrient concentration. As water residence time decreases, the concentration of nutrients approaches the concentration in incoming streams or rivers,
and sedimentation of nutrients becomes less of a factor.

Morphometric features and hydrologic factors can vary widely from
lake to lake even within a small region, but nonetheless the earth can
be characterized as containing ecological regions (or "ecoregions")
that have broad similarities of soil, relief, and dominant vegetation.
Omernik (1987) divided the conterminous United States into 76 ecoregions, or areas of regional similarity in soil, land use, land surface
form, and potential natural vegetation (Figure 4.2). The water quality of streams within an ecoregion would be expected to be more
similar (in terms of nutrients, silt, organic matter, and major ions)
than would the water quality of streams of different ecoregions
(Hughes et al., 1986). It follows that trophic conditions of lakes in an
ecoregion characterized by highly erodible, nutrient-rich soils would
differ, even without any cultural nutrient loading, from those of lakes
in an area of sandy soils and low relief, simply because of differences
in loading from their drainage basins.

These expectations have been verified through studies of phosphorus concentrations, fish, and invertebrates in streams of Arkansas,
Kansas, Minnesota, Ohio, and Oregon, and lakes of Michigan, Minnesota, Ohio, and Wisconsin (Hawkes et al., 1986; Hughes and Larsen,
1988; Omernik et al., 1988; Wilson and Walker, 1989; Fulmer and
Cooke, 1990; and others). For example, Larsen et al. (1988) described
the patterns of water quality in streams of the five ecoregions that
extend into Ohio. Strong differences were found between ecoregions
with regard to nutrients and major ion variables, and with regard to
the complexity and health of fish assemblages.

Heiskary et al. (1987) and Wilson and Walker (1989) used the ecoregion
concept to develop lake restoration priorities and strategies for Min-

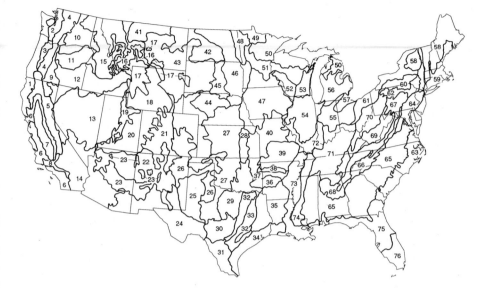

1. Coastal Range
2. Puget Lowland
3. Willamette Valley
4. Cascades
5. Sierra Nevada
6. Southern and Central California Plains and Hills
7. Central California Valley
8. Southern California Mountains
9. Eastern Cascades Slopes and Foothills
10. Columbian Basin
11. Blue Mountains
12. Snake River Basin/High Desert
13. Northern Basin and Range
14. Southern Basin and Range
15. Northern Rockies
16. Montana Valley and Foothill Prairies
17. Middle Rockies
18. Wyoming Basin
19. Wasatch and Uinta Mountains
20. Colorado Plateaus
21. Southern Rockies
22. Arizona/New Mexico Plateau
23. Arizona/New Mexico Mountains
24. Southern Deserts
25. Western High Plains
26. Southwestern Tablelands
27. Central Great Plains
28. Flint Hills
29. Central Oklahoma - Texas Plains
30. Central Texas Plateau
31. Southern Texas Plains
32. Texas Blackland Prairies
33. East Central Texas Plains
34. Western Gulf Coastal Plain
35. South Central Plain
36. Cuachila Mountains
37. Arkansas Valley
38. Boston Mountains
39. Ozark Highlands
40. Central Irregular Plains
41. Northern Montana Glaciated Plains
42. Northwestern Glaciated Plains
43. Northwestern Great Plains
44. Nebraska Sand Hills
45. Northeastern Great Plains
46. Northern Glaciated Plains
47. Western Corn Belt PLains
48. Red River Valley
49. Northern Minnesota Wetlands
50. Northern Lakes and Forest
51. North Central Hardwood Forests
52. Driftless Areas
53. Southeastern Wisconsin Till Plains
54. Central Corn Belt Plains
55. Eastern Corn Belt Plains
56. S. Michigan/N. Indiana Till Plains
57. Huron/Erie Lake Plain
58. Northeastern Highlands
59. Northeastern Coastal Zone
60. Northern Appalachian Plateau and Uplands
61. Erie/Ontario Lake Plain
62. North Central Appalachians
63. Middle Atlantic Coastal Plain
64. Northern Piedmont
65. Southeastern Plains
66. Blue Ridge Mountains
67. Central Appalachian Ridges and Valleys
68. Southwestern Appalachians
69. Central Appalachians
70. Western Allegheny Plateau
71. Interior Plateau
72. Interior River Lowland
73. Mississippi Alluvial Plain
74. Mississippi Valley Loess Plains
75. Southern Coastal Plain
76. Southern Florida Coastal Plain

FIGURE 4.2 Ecoregions of the United States. Source: Omernik, 1987.

nesota. Although seven ecoregions extend into Minnesota, 98 percent of the state's 12,500 lakes with surface areas greater than 10 ha occur in four of them. It is apparent from Table 4.4 and Figure 4.3 that lakes in the North Central Hardwood Forest (NCHF) and Northern Lakes and Forests (NLF) ecoregions differ substantially from lakes in the Western Corn Belt Plains (WCBP) and Northern Glaciated Plains (NGP) ecoregions. Lakes in the latter two ecoregions are unlikely to have water with few algal blooms, regardless of the amount of lake management activity. However, lakes with high algal biomass, low transparency, and severe dissolved oxygen depletion in the NCHF or NLF ecoregions are likely to have deviated significantly from their

TABLE 4.4 Summary of Land Use and Water Quality Data for Four Ecoregions in Minnesota

Variable	Units	Ecoregion			
		NCHF	NLF	NGP	WCBP
Number of lakes		36	30	8	11
Land uses					
Cultivated	%	34.8	1.8	73.0	60.6
Pasture	%	18.0	3.9	9.2	5.9
Urban	%	0.7	0.0	2.0	1.5
Residential	%	6.4	4.8	0.4	9.9
Forested	%	16.4	66.2	0.0	7.0
Marsh	%	2.5	2.1	0.6	1.2
Water	%	20.9	20.9	14.4	13.6
Watershed area	ha	4,670	2,140	2,464	756
Lake area	ha	364	318	218	107
Mean depth	m	6.6	6.3	1.6	2.5
Total phosphorus (P)	µg/liter	33	21	156	98
Chlorophyll *a*	µg/liter	14	6	61	67
Secchi disk	m	2.5	3.5	0.6	0.9
Total P load	kg/yr	1,004	305	1,943	590
Inflow P	µg/liter	183	58	5,666	564
Areal P load	kg km^{-2} yr^{-1}	276	96	891	551
Outflow	km^3/yr	6.2	5.3	0.9	1.0
Water residence time	yr	9.3	5.0	36.2	4.8
Stream total P	µg/liter	148	52	1,500	570

NOTE: Data are listed as averages. Ecoregions: NCHF, Northern Central Hardwood Forests; NLF, Northern Lakes and Forests; NGP, Northern Glaciated Plains; WCBP, Western Corn Belt Plains.

SOURCE: Reprinted by permission from Wilson and Walker, 1989. Copyright © 1989 by North America Lake Management Society.

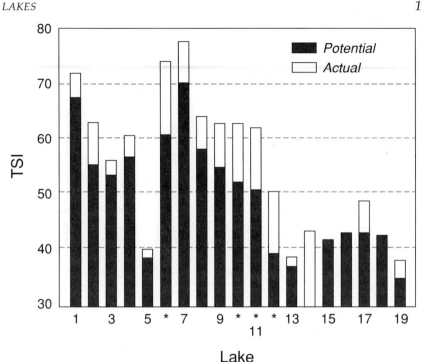

FIGURE 4.3 Actual and attainable trophic state (as indicated by Carlson trophic state index values) in 19 Ohio reservoirs (TSI < 40 = oligotrophy; TSI > 50 = eutrophy).

attainable conditions, probably through cultural nutrient loading. Lake restoration is possible for a eutrophic lake in any of the ecoregions, but the attainable trophic state for lakes can vary significantly between adjacent ecoregions.

The ecoregion idea has been used to predict the restoration potential of Ohio reservoirs subjected to varying degrees of nutrient, silt, and organic matter loading (Fulmer and Cooke, 1990). These authors used the 25th percentile values of stream phosphorus concentrations in the least affected streams of the ecoregion for each reservoir (from Larsen et al., 1988), along with hydrologic and morphometric data for the reservoirs and Canfield and Bachmann's (1981) loading model, to predict the steady-state phosphorus concentration attainable in the deep water zone of each reservoir. The 25th percentile concentration was chosen for purposes of illustration as a stream concentration that probably can be reached through technologically feasible changes in the watershed, such as advanced wastewater treatment, feedlot runoff detention systems, and other land management practices. Other concentrations appropriate to a specific stream or ecoregion could be chosen. The predicted phosphorus concentrations in the reservoirs were compared with measured values, and the data were transformed into an index number, a Carlson trophic state index (TSI) value (see "Water Quality and Human Use Criteria," below) to describe the lake conditions expected for that concentration.

Four reservoirs were identified that have much higher phosphorus concentrations and trophic states than those predicted by the model (Figure 4.4). These eutrophic reservoirs have trophic conditions in the mesotrophic range (TSI of 40–50; i.e., they can be shifted from conditions of prolonged and severe algal blooms to conditions of higher transparency and fewer problems with nuisance algae). Additional studies are needed to ascertain causes of the deviations from attainable quality, but the four reservoirs represent the best opportunities among the 19 studied reservoirs for obtaining significant lake improvements.

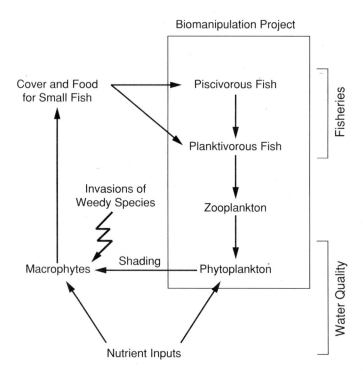

FIGURE 4.4   Linkages of algae, macrophytes, and fisheries in lakes.

The above approach to determining the best candidates for lake restoration differs significantly from the traditional one of simply selecting the lakes with the worst trophic states and then spending the available money in an attempt to restore them. The desired trophic state may not be attainable or may require large and continued expenditures of energy and money to be maintained. For example, lake 1 in Figure 4.3 is hypereutrophic, and its water quality is among Ohio's poorest. It is located in an ecoregion with rich humic soils in which the principal land use is agricultural. Its attainable trophic state (Figure 4.3) does not differ significantly from its current state. Although management activities (e.g., aeration, weed harvesting, dredging) could improve the lake for recreation, continued loading will refill it with silt and maintain its current trophic state. If this lake were assigned a top priority for restoration without considering its attainable condition, scarce restoration funds could be wasted. Lakes 6, 10, 11, and 12 (Figure 4.3) have much better attainable quality, have deviated significantly from this condition, and are thus better candidates for restoration.

One of the values of the ecoregion concept in lake restoration and management is that it provides a rational basis for setting regional rather than national lake water quality standards. The approach can take into account regional factors related to *attainable* water quality and thus can be used to designate lakes for protection and to establish lake restoration goals appropriate for each ecoregion.

Stream water quality in some watersheds of an ecoregion, and ultimately lake trophic state, can be greatly improved through changes in land use (e.g., wetland restoration, improved agricultural practices) and through stream restoration itself. In these cases, the additional use of in-lake procedures, such as enhancement of biological controls on algal populations or application of chemicals to control sediment phosphorus release, may improve a lake beyond expectations based on original ecoregion characteristics. Although no lake in the Eastern Corn Belt Plains (ECBP) ecoregion of Indiana and Ohio will look like the oligotrophic lakes of the Northern Lakes and Forests ecoregion of Minnesota, Wisconsin, and Michigan, it is important to recognize that the various ecoregions were defined based on existing land use conditions and that intensive row crop farming is not the native condition of land in the ECBP. If farming became less intensive or less prevalent in the ECBP, or if best management practices became effective in reducing the export of soil and nutrients to streams in the region, land use would become a reduced factor in determining stream quality, and lakes could improve to some degree beyond the conditions currently defined as *attainable*.

There are few case histories of the uses of in-lake procedures on lakes that also have had wastewater inflows eliminated and major improvements in land use practices. It is not yet known whether and how far lakes can be restored beyond the attainable condition associated with their ecoregion.

Criteria for Successful Restoration and Measures of Success

Most lake restoration projects undertaken in the United States over the past 20 years have focused on improving the fitness of a degraded lake for human uses such as swimming, other water contact sports, fishing, or drinking water supply. Undoing ecological damage from past human perturbations and restoring the lake's ecosystem to its pristine state are not the primary goals of typical restoration efforts. Nonetheless, restoration proponents generally assume (at least implicitly) that improving a lake's usability for human activities will improve functions of the lake's ecosystem, and indeed there is merit in the assumption. To the extent that a project stops excessive (human-induced) inputs of nutrients, sediments, or acids; controls or eliminates nonnative species; or removes unnatural toxic substances, it will promote return of the aquatic ecosystem to a less-stressed (more "natural") condition. The structural and functional properties of ecosystems change over time, however, because of natural ecological succession, as well as long-term shifts in climate and hydrology. Consequently, it is unrealistic to think that we can restore a lake precisely to the ecological state it was in before a stress occurred or to the unknown (and unknowable) state to which it would have evolved in the absence of the stress.

ECOLOGICAL CRITERIA

Improving the ecology of a lake is a laudable goal, and the success of lake restoration projects should be measured, at least in part, on the basis of ecological criteria, not just on economic or human use criteria. Although ecologists may disagree on the detailed characteristics of a functioning ecosystem, some general principles can be stated. Restoration should promote a self-sustaining, stable system. The system should have the ability to resist stress and the resiliency to rebound from stresses once they have been removed. Production and respiration should be roughly in balance, and the food web should be complicated enough (involving many checks and balances and negative feedback loops) to minimize uncontrolled growth and wild population swings by one or a few species. Often this is best achieved by restoring native species. Although ecologists no longer equate diversity directly with ecosystem stability, biotic diversity, often stated in terms of species or genetic diversity, is still an important measure of ecosystem quality.

Many so-called lake restoration projects really are only mitigation and management efforts to rid a lake, by whatever means, of some nuisance. Criteria for selection of the procedure are based primarily on cost and effectiveness relative to the specific target (e.g., elimination of a nuisance organism). Some procedures that rank highly based on those criteria fare poorly when evaluated more broadly in terms of total ecosystem restoration. For example, chemical herbicides are commonly used to control rampant macrophyte growths in littoral zones of lakes and, if applied properly, can achieve the goal of removing the nuisance, at least temporarily. However, the dead plant material may release a substantial load of nutrients to the water column, inducing excessive algal growth (substituting one problem for another), or the herbicide may adversely affect nontarget organisms. Most important, herbicide treatments and similar palliatives do not get at the underlying causes of a problem. In the example given, once the herbicide disappears, the macrophyte problem probably will return.

WATER QUALITY AND HUMAN USE CRITERIA

As stated above, the goal of most lake restoration projects is to remove a specific problem—a nuisance organism, excess chemical(s), or unwanted physical condition—and make the lake more desirable (based on human-centered criteria) and more usable for specific human purposes. The success of a restoration project is (and must be) evaluated according to the extent to which these human-oriented goals are met (e.g., Was the fishery restored? Was swimming improved?). Chemical measures of water quality (and associated numerical criteria for specific chemical species) may be used to measure the success of restoration when the problem has a simple cause (e.g., presence of a toxic chemical), but more typically the causes and symptoms of degradation are much more complicated. Quantitative measurements of improvements in recreational and aesthetic attributes are notoriously difficult to obtain, however, and therefore success commonly is measured in terms of quantifiable water quality characteristics such as Secchi disk transparency (a measure of water clarity) and concentration of chlorophyll a (a measure of algal biomass) that are loosely related to recreational and aesthetic conditions.

Quantitative evaluation of trophic state conditions has been aided by use of simple trophic state indices. The most widely used TSIs are those developed by Carlson (1977), based on Secchi disk transparency and on concentrations of total phosphate and chlorophyll *a*. These strongly intercorrelated parameters are, respectively, the best quantified physical, chemical, and biological measures of trophic conditions, and Carlson developed a simple index based on each parameter (Table 4.5). The approximate range of each index is 0 to 100, and values greater than about 50 denote eutrophic conditions. An increase of 10 units in an index represents a doubling of algal biomass. Carlson recommended that the indices be considered separately in evaluating trophic state, but others (e.g., Kratzer and Brezonik, 1981) recommended averaging the three values to obtain a single number integrating the components contributing to trophic state.

More complicated, multidimensional indices have been proposed to express the concepts of trophic state (Shannon and Brezonik, 1972; Uttormark and Wall, 1975) or water quality (Brown et al., 1972; Harkins, 1974; Walski and Parker, 1974) in a single number, but none of these indices has been used routinely in lake management and restoration programs. Dierberg et al. (1988b) used modified versions of Carlson's

TABLE 4.5   Values of Secchi Disk Transparency, Total Phosphorus Concentration, and Chlorophyll *a* Concentration Corresponding to Carlson Trophic State Index (TSI) Values

TSI	Secchi Disk Transparency (m)	Total Phosphorus Concentration (μg/liter)	Chlorophyll *a* Concentration (μg/liter)
0	64	1	0.04
10	32	2	0.12
20	16	4	0.34
30	8	8	0.94
40	4	16	2.6
50	2	32	6.4
60	1	65	20
70	0.5	130	56
80	0.25	260	154
90	0.12	519	427
100	0.06	1,032	1,183

SOURCE: Reprinted by permission from Carlson, 1977. Copyright © 1977 by the American Society of Limnology and Oceanography, Inc.

TSIs to evaluate water quality changes in Florida lakes that had undergone restoration.

The cost-effectiveness of a restoration effort must be measured in terms of the economic benefits obtained relative to expenses incurred. The difficulties in assigning a dollar amount to such elusive attributes as ecological health and wilderness values are well known, however, and cost-effectiveness as measured by conventional economic procedures should not be the sole measure of success. Restoration projects should also be evaluated in terms of ecological criteria (i.e., the extent to which a project improves lake ecosystems as measured by the criteria described above). In most cases, these criteria do not conflict with those related to human-centered goals, but success in attaining the latter goals does not necessarily translate to success relative to the former. As a minimum, lake restoration professionals should strive to ensure that human-centered goals are not achieved at the expense of ecological goals, for that would be shortsighted. Finally, because human and financial resources are limited, success should be measured in terms of the longevity of effects and the extent to which a restored lake is self-sustaining.

FEDERAL AND STATE PROGRAMS FOR LAKE RESTORATION AND MANAGEMENT

Federal Programs

The principal federal program dealing with restoration of degraded lakes is EPA's Clean Lakes Program (CLP), which was established by P.L. 92-500, the Federal Water Pollution Control Act Amendments of 1972. The CLP began in 1975 through a congressional appropriation pursuant to Section 314 of P.L. 92-500. The purpose of Section 314 was to develop a national program to clean up publicly owned freshwater lakes. The CLP requires that all point sources of pollution be treated or have treatment planned under Sections 201 and 402 of the 1972 Clean Water Act before a grant is awarded for in-lake restoration activities (Duda et al., 1987).

From 1975 to 1978, $35 million in research and development grants was issued to identify restoration techniques and restore specific lakes. In 1980, a four-part program was established that included (1) a classification survey, wherein states were to identify and rank their lakes according to trophic state; (2) "Phase I" projects, which were awarded for diagnosis and feasibility studies on lakes ranked by the states as having the greatest need (for restoration); (3) "Phase II" projects, in which funds were awarded to implement Phase I recommendations;

and (4) "Phase III" projects to assess the responses of restored lakes (U.S. EPA, 1985). Some Phase III funds are now being awarded. The Water Quality Act of 1987 (P.L. 100-4) reauthorized the CLP and mandated some new initiatives and requirements. For example, to remain eligible for CLP grant funds, each state is required to submit a biennial report that includes a revised lake classification list, an assessment of status and trends in lake water quality, and a restoration plan for degraded lakes. In addition, EPA was authorized to establish a Clean Lakes demonstration program to enhance understanding of the effectiveness of various lake restoration techniques, and the Water Quality Act directs that specific attention be paid to mitigation of acidified lakes.

A key feature of the CLP is its emphasis on assisting states in setting up their own programs. Federal funds have been limited to 70 percent of the cost of the classification studies and Phase I projects (up to a maximum of $100,000 in each category). The federal share of Phase II projects is 50 percent. States administer their own programs, and there is considerable emphasis on local involvement in raising the matching funds.

In fiscal years 1976 to 1980, more than $60 million was spent on classification, Phase I, and the initial funding of Phase II projects. In fiscal years 1981 to 1985, no funds for lake restoration and protection were requested in the President's budgets, but a total of $32.64 million was added to the EPA's budget by Congress for Phase II lake programs (Duda and Johnson, 1984; U.S. EPA, 1985).

The CLP has had some success in stimulating states to develop lake programs and in encouraging citizen involvement. During the period 1975 to 1985, 313 CLP studies and projects were funded. Four percent of the total funds ($93 million) were spent on classification, 9 percent on diagnosis and feasibility studies (Phase I), and 87 percent on restoration (Phase II). Projects were distributed among 47 states; only Alabama, West Virginia, and Hawaii did not participate. Through 1985, 67 Phase II projects had been completed, and 92 Phase II projects were in progress in 29 states (U.S. EPA, 1985). Half of the projects completed by 1985 were in four states: New York, Wisconsin, South Dakota, and Minnesota.

Many other federal programs are concerned, at least indirectly, with restoration of lakes. For example, EPA's Section 201 program deals with improvement of municipal wastewater treatment facilities in urban areas, and its Section 208 program (the numbers refer to sections of P.L. 92-500) is concerned with areawide planning for water quality management (with emphasis on nonpoint sources of pollution in storm water runoff). The U.S. Army Corps of Engineers has responsibility for managing numerous impoundments and for authorizing permits for dredge and fill operations on surface waters. The U.S. Fish and Wildlife Service has interests and activities related to accumulation of toxic substances in fish and waterfowl. Nonetheless, EPA's Clean Lakes Program is the only federal program involved directly in restoring lakes nationwide, and it provides the primary federal support (technical, administrative, and financial) for state lake programs.

State Programs

Obtaining detailed information on state programs is beyond the scope of this study. Information on lake management and restoration activities compiled by the North American Lake Management Society from EPA's 1988 survey of state agencies (U.S. EPA, 1990b) does provide comparative data on state programs, however. Only 42 states reported on lake management activities in the broadest sense, and in most of the states only a few people were involved in the program. More than 30 of the states reported that the federal Clean Lakes Program was the major source of funding for proposed lake restoration projects.

Only a few states are involved in lake restoration activities beyond those associated with the CLP. Notable among these are Florida, Minnesota, South Dakota, Vermont, Washington, and Wisconsin. Data on the numbers of lakes restored through state-sponsored programs are not readily available. Some of the programs are too recent to have established a track record. For example, Florida's Surface Water Improvement and Management (SWIM) Program was initiated in 1987, and to date most of its funds have been devoted to feasibility and planning studies. Minnesota's Clean Water Partnership (also initiated in 1987) provides funding to local governments to improve water resources (lakes, wetlands, streams, ground water aquifers) that have been degraded by activities related to land use (nonpoint source pollution), as well as to protect aquatic systems threatened by such degradation. Operation of the program is somewhat similar to that of the federal CLP. To date the program has awarded funds to 30 projects and has allocated $2.6 million in state funds (which have been matched equally by local units of government).

The NALMS survey also showed that many states have citizen groups actively involved in lake management activities. Programs such as "Water Watch" in Kentucky, "Lake Watch" in Florida, "Volunteer Lake Assessment Program" in New Hampshire, and "Volunteer Lake Monitoring Program in Illinois" all use specially trained volunteers to monitor environmental trends in lakes. These activities include measuring Secchi disk transparency, recording rainfall and lake levels, and collecting and storing water samples for analysis in a central laboratory. These self-help programs are designed primarily to assist lake associations in collecting information and interpreting data. Collection of long-term data sets is essential in determining the need for restoration activities and in evaluating these activities if they are undertaken.

Evaluation of Past Restoration Efforts: Need for Monitoring Data

Without question, state and federal programs affecting the quality of lakes have made major strides in three areas over the past 20 years: (1) eliminating or decreasing pollution sources to lakes (especially point sources of pollutants), (2) cleaning up pollution problems (e.g., removing specific pollutants and contaminated sediments from lakes), and (3) restoring or improving user-oriented qualities of some lakes.

A 1980 study by EPA evaluated economic benefits resulting from the CLP and concluded that the program was highly cost-effective (U.S. EPA, 1980). The analysis was based on 28 projects in 16 states that received a total of $15.35 million in federal grants and an approximately equal sum from state and local sources. Twelve categories of benefits were considered in the assessment, but many benefits could not be quantified in monetary terms. The 1980 value of the benefits that could be quantified was estimated to be $127.5 million (Table 4.6), which represents a return of $4.15 per total project dollar.

However, our ability to assess the effectiveness of restoration projects funded by government programs in quantitative, scientific terms is greatly diminished by a paucity of data on lake quality before and after treatment. For example, in a review of 43 Florida lakes that had been sites of restoration projects, Dierberg et al. (1988a) found that only 7 lakes had sufficient data to permit an evaluation of water quality improvement. Baker and Swain (1989) found similarly dismal statistics in attempting to analyze lake restoration projects in Minnesota. Most restoration projects are required to include some pre- and posttreatment monitoring, but collection of data for an adequate period of time and at sufficient detail before and after restoration has been relatively uncommon. In some cases, sufficient monitoring may have been done, but rigorous analysis and interpretation of the results were not a part of the monitoring effort. All too often,

TABLE 4.6   Comparison of Economic Benefits and Project Costs (federal share only) for 28 Clean Lakes Projects

| | Economic Benefits[a] | | | | | | | | | | | Total Discounted Benefits[b] ($) | Grant Amount ($) |
Lake	Recreation	Aesthetics	Flood Control	Economic Development	Fish/Wildlife	Agriculture	Property Value	Public Health	Water Supply	Education R&D	Miscellaneous		
Annabessacook	+	+		$	+		$	+				23,246,100	497,906
Bomoseen	$	+								+	+	1,830,500	74,640
Buckingham	$		+							+		127,700	23,250
Charles	$	$								+		2,286,600	387,163
Clear	$	$	+	+			+			+		471,500	358,682
Cochrane	$	+		$		+				+		52,500	9,906
Collins Park	$	+								+		51,700	79,355
Ellis	$	$	$	$	+	$		+		+	$	11,123,000	1,625,000
59th St. Pond	$	+	+					+				4,837,000	498,035
Frank Holten	$	$	+					+		+	+	1,862,300	927,000
Henry	$				+	+				+	$	134,200	220,000
Jackson	$	$				+	+					7,309,800	725,663
Lansing	$		+				+			+		1,155,900	800,000
Liberty	$	$					+	+		+		813,000	577,975
Lilly	+	+				+	$			+		2,880,000	350,000
Little Pond		+							$	+		212,200	9,946
Loch Raven	$	$							$	+		11,944,100	150,900
Medical	$	+				+	$					931,700	128,217
Mirror and Shadow	$						+	+		+		312,700	215,000
Moses	$	+								+	+	534,700	3,251,000
Nutting	$	$		$			+	+		+		5,292,100	241,159
Penn	$	$	+	+						+	$	186,000	87,900
Rivanna	$								$	+		923,500	63,835
Steinmetz	$	+	+	+					$	+		126,300	36,680
Temescal	$	$							$	+	$	1,112,500	315,618
Tivoli	$	+	+	+				+		+		240,100	202,645
Vancouver	$	+		+		$		+		+		47,370,000	3,468,328
Washington Park	$	$								+		120,800	23,250
Total												127,488,500	15,349,053

[a]"$" in monetary terms; "+" in qualitative terms.
[b]Total discounted benefits include "$" items only.

SOURCE: U.S. Environmental Protection Agency, 1980.

monitoring data are simply filed away and are not accessible for others to evaluate.

Need for Coordination of Management Efforts

Water quality management and fisheries management have evolved as almost completely separate disciplines. Water quality experts and fisheries experts are trained in separate university departments, belong to different professional societies, attend different scientific meetings, and read different scientific journals. Wildlife and fisheries managers may be trained in the same college or department, but the disciplines remain segregated in many ways. Yet members of these distinct professions find themselves managing nuisance algal blooms, toxic chemicals, fish production, and waterfowl habitat on the same ecosystems. It is not surprising that management programs sometimes work at cross-purposes. For example, in Medical Lake, Washington, fish stocking caused a decline in herbivorous zooplankton, confounding efforts of water quality managers to control nuisance algal blooms (see Box 4.1).

BOX 4.1
MEDICAL LAKE, WASHINGTON

Medical Lake is a large (63 km^2), deep (maximum depth, 18 m; mean depth, 10 m), eutrophic lake in eastern Washington. Prior to 1964, the homes on its small watershed were serviced with septic tanks and cesspools. Dense blooms of blue-green algae continued after wastewater diversion because of high rates of internal nutrient loading from enriched sediments, and this led to frequent curtailments of swimming. More than half of the lake's volume was anoxic, and fish were rare or absent. Attempts by the Washington Department of Game to stock trout were unsuccessful because of high levels of ammonia and hydrogen sulfide (Bauman and Soltero, 1978; Soltero et al., 1981).

In 1977, 936 m^3 of liquid aluminum sulfate were added to remove phosphorus from the water column and retard its recycling. After treatment, phosphorus concentrations and algal density were reduced; blue-green algae were largely replaced with less noxious species of green algae; water clarity increased; and large-bodied zooplankton, *Daphnia pulex*, became the primary regulators of phytoplankton density. The lake once again became attractive for various recreational and aesthetic uses (Mires et al., 1981; Soltero et al., 1981). The significant improvement in water quality encouraged the Washington Department of Game to attempt to establish a rainbow trout (*Oncorhynchus mykiss*) fishery in the lake to provide economic benefits to the area. Between 1978 and 1981, 45,330 trout were stocked. During this period, Knapp and Soltero (1983) determined that all age classes of trout in the lake fed almost exclusively on *D. pulex*, and this led to the near elimination of this algae grazer from the lake in 1981. Water quality again deteriorated; coupled with lack of food, this led to an extensive trout kill in autumn of 1981. *Daphnia pulex* again increased in the years after the trout kill (Scholz et al., 1985).

The Medical Lake restoration project had the rare benefit of long-term monitoring. Not only did the project demonstrate the effectiveness of controlling phosphorus recycling by alum addition, it also showed that the switch in control of algal biomass from "bottom-up" resource limitation (by nutrients) to "top-down" control by grazers can be reversed easily by a poorly planned overstocking with game fish for short-term economic gain (Scholz et al., 1985). A moderate and balanced restocking of several fish species, along with establishment of a refuge for *Daphnia*, might have allowed continued high water quality and a resumption of multiple lake uses. Interestingly, the overstocking with trout may represent a brief (but significant) perturbation, and the lake may yet return to a grazer-controlled, more stable ecosystem.

LAKE RESTORATION TECHNOLOGY

Overview

Lake restoration methods can be divided into two major categories: (1) watershed activities to eliminate stress loadings, and (2) in-lake activities to restore or improve the aquatic ecosystem itself. The former are necessary for the long-term success of restoration programs (except when the stress originates in the lake itself, e.g., lake stage regulation), but they are not restorative actions per se and moreover may not be sufficient. The latter are done to accelerate the return to earlier (more natural) conditions, to remove long-lived contaminants or exotic species, or to reintroduce ecosystem components lost through the impact of stress. Numerous restoration techniques have been developed for lakes over recent decades (Table 4.7). Some methods are applicable to several types of stress: dredging is useful to remove sediments laden with nutrients or contaminated with toxic materials, or to deepen lakes prematurely filled in by excessive erosion. Others apply to only one type of stress: alum treatment is used to remove excess phosphorus from the water and prevent its recycling to the water from the sediments (by forming an aluminum hydroxide barrier at the sediment-water interface). Some methods (e.g., sediment removal) need be used only once to achieve long-term

TABLE 4.7   Restoration Techniques for Major Categories of Lake Degradation

| | Problem Category | | | | |
Technique	Eutrophication	Siltation	Acidification	Exotic Species	Toxic Contaminants
Nutrient source reduction	X	X	X		X
Diversion	X	X			
Land disposal	X				
Product modification	X				X
Wastewater treatment	X				X
Interception of nonpoint sources	X	X			X
Dilution	X		X		X
Flushing	X		X		X
In-lake methods	X		X	X	X
Alum treatment	X				
Sediment skimming	X	X			X
Sediment oxidation	X				
Deep-water discharge	X				X
Biomanipulation	X			X	
Artificial circulation	X				
Biocides (algicides/ herbicides/piscicides)	X			X	
Biocontrol agents				X	
Drawdown/sediment desiccation	X	X			
Bioharvesting	X			X	
Aeration	X				
Dredging	X	X			
Liming			X		

benefits (if the cause of the stress has been eliminated); others must be applied continuously (aeration) or repeatedly (liming of acidic lakes; herbicide treatment of macrophyte problems) to maintain the benefits of treatment.

Lake restoration and management techniques are listed in Table 4.7 according to the type of problems they seek to remedy, and the sections below briefly describe the most important of these techniques. More comprehensive reviews of the techniques are provided by Cooke et al. (1986) and Cooke and Kennedy (1989).

Problems caused by excess nutrients have received the most attention over the past two decades, and more techniques have been developed to address these problems than all other types of stresses combined. Excessive nutrient enrichment manifests itself in several distinct problems; algal blooms, macrophyte proliferation, oxygen depletion, and loss of sport fisheries are the most important.

Eutrophication

CONTROL OF ALGAL BLOOMS

Nutrient Source Reduction

High loading of nutrients to lakes produces algal blooms and other problems. In many cases, oxygen-demanding organic matter, silt, or toxic materials accompany the nutrient loadings. Reduction of nutrient loadings (and related inputs) can be accomplished (Table 4.8) by (1) diverting point sources of nutrients (e.g., municipal sewage effluents) or nutrient-laden streams out of the lake's watershed; (2) modifying products to contain lower amounts of nutrients (mainly phosphorus); (3) removing nutrients from wastewater in engineered treatment systems; (4) intercepting nutrients in pre-lake impoundments (storm water detention and retention ponds, natural or artificial wetlands); (5) decreasing nutrient runoff from agricultural lands by "best management practices"; and (6) instituting land use and management controls.

Diversion   Wastewater effluent is rarely diverted out of watersheds because of the difficulty of finding an alternative disposal site, but a few well-known diversions have occurred. The lower lakes of Madison, Wisconsin (Monona, Waubesa, and Kegonsa) have deteriorated drastically during the twentieth century because of sewage discharges (Sawyer, 1947; Lathrop, 1979). By 1958, most of the effluent had been diverted downstream from the lakes, and all effluent was

TABLE 4.8   Control Methods for External Sources of Nutrients

  I. Stream or wastewater diversion

 II. Municipal wastewater treatment
     (tertiary treatment for N and P removal)

III. Product modification
     (e.g., legislative ban of phosphate in laundry detergents; slow release fertilizers)

IV. Treatment of inflow streams
     A. Diversion
        1. Into wetlands
        2. Over upland vegetation
     B. In-stream methods
        1. Sedimentation basins to remove particulate N and P
        2. Channel aeration
        3. Chemical precipitation
        4. Biotic harvesting

 V. Land use practices
     A. Prospective zoning
        1. On-site storm water retention or detention regulations
        2. Setback and other shoreline restrictions on new construction
        3. Restrictions on shoreline vegetation removal
        4. Restrictive zoning in watershed to minimize development
        5. Minimization of impervious areas in developments
        6. Use of grassy swales instead of curb and gutter drainage
     B. Treatment of urban runoff (best management practices)
        1. Retention/detention basins
        2. Swirl concentrators
        3. First flush diversion (or low flow) to sanitary sewers
        4. Diversion of runoff into wetlands
        5. Street sweeping or vacuuming
        6. Public education (reduce litter accumulation, control lawn fertilizer losses)
     C. Treatment of agricultural runoff (best management practices)
        1. Runoff controls (change volume and peak flow)
           a. No or minimum tillage
           b. Winter cover crop
           c. Contour plowing and strip cropping
           d. Terraces
           e. Grassed outlets; vegetated borders on fields and along waterways
           f. Detention ponds
        2. Nutrient loss controls
           a. Timing and frequency of fertilizer applications
           b. Amount and type of fertilizer used
           c. Control in situ transformation of fertilizer to soluble forms
           d. Crop rotation with legumes
           e. Storage of manure during winter

diverted from the Madison lakes by 1971. The diversion was partially successful. Severe blue-green algal problems were ameliorated, but the lakes remained eutrophic (Sonzogni et al., 1975). Diversion of sewage effluent from Seattle and nearby suburban areas out of Lake Washington in the mid-1960s restored the lake from a state of moderate eutrophy and declining water quality to a mesotrophic or an oligotrophic system with high water quality (see Box 4.2). Effluent from an advanced wastewater treatment plant at South Lake Tahoe, California, was diverted from Lake Tahoe into a man-made impoundment in the Nevada desert to provide added protection to that highly oligotrophic and pristine lake in the Sierra Nevadas (Goldman, 1988).

Land disposal of treated municipal wastewater (a form of nutrient diversion) is becoming common in warm climates; often the water is used for irrigation of agricultural lands (e.g., citrus groves in Florida) or golf courses. This practice is used as much for water conservation purposes (to decrease demands on ground water or surface supplies) as for protection or restoration of lakes or receiving streams.

Product Modification The most important example relative to reduction of nutrient loadings is the reduction or removal of phosphates from laundry detergents; other materials—carbonates, silicates, organic compounds are substituted to achieve the functions that phosphates provide. Laws banning phosphates or requiring lower levels of phosphates in laundry detergents are in effect in at least 10 states along the Great Lakes plus parts of several other states (Maki et al., 1984). Detergent phosphates account for about two-thirds of the phosphate in the municipal sewage of areas without detergent phosphate regulations. Phosphate levels in sewage from areas with laws regulating detergent phosphate levels are typically about 50 percent lower than levels in other areas. Bans are not instituted to reduce phosphorus loadings to a single lake—local ordinances are ineffective because consumers will obtain the products in nearby areas where they are not banned. This approach thus is used to lower phosphorus levels in surface waters on a statewide or regionwide basis. Although limitations on detergent phosphates usually are not sufficient to lower phosphorus loadings to lakes below the levels leading to eutrophication, they do lead to lower costs of removing phosphate from municipal sewage.

Removal of Phosphorus from Wastewater Removal of phosphorus from wastewater in municipal treatment plants, almost unheard of 20 years ago, is now a well-developed and widespread practice (see Box 4.3). It is required for wastewater entering the Great Lakes from all plants

BOX 4.2
LAKE WASHINGTON

Lake Washington, an important recreational lake (area, 87 km^2; maximum depth, 65 m), was first impacted by raw and treated sewage from Seattle early in this century. This first episode of pollution ended in the 1930s with the construction of a sewerage system that diverted the sewage effluent to Puget Sound. A second episode began in the 1940s when suburban growth spread along the lake north and south of Seattle. By 1963, effluent from 11 sewage treatment plants was being discharged to the lake or its tributaries and contributed 63 percent of the phosphorus load entering the lake. By 1955, accelerated nutrient enrichment had progressed to the point that the first bloom of *Oscillatoria rubescens* was observed. This blue-green alga is widely recognized as an indicator of excess nutrient enrichment.

Deteriorating water quality from algal growths was reported in the press, and related issues were addressed by the scientific community (see Lehman, 1986; Edmondson, 1991). By 1957, public concern had resulted in state legislation to form a metropolitan governmental agency to address the problem of water supply and waste management. Establishment of this agency required a public referendum. The first proposal was defeated in March 1958, but a revised proposal passed in September 1958, and provided funds to construct facilities to divert sewage from the lake to Puget Sound. Diversion of effluent began in 1963, and by March 1967, 99 percent of the sewage effluent had been diverted from the lake. The lake responded rapidly to decreased nutrient loading, as limnologists had predicted. Noticeable differences occurred soon after diversion was complete. By the summer of 1971, Secchi disk transparency was greater than it had been in 1950. By 1975, large changes had occurred compared with conditions in 1963: Secchi transparency increased from 1.0 to 4.0 m, total phosphorus decreased from 70 to 16 µg per liter, and epilimnetic chlorophyll decreased from 35 to 4 µg per liter (Edmondson, 1979). Nitrogen was no longer limiting after diversion. It had become a limiting nutrient because of the large biomass of algae produced by increased phosphorus loads. Nuisance blooms of algae were no longer a threat to the lake. The much larger volume of Puget Sound enabled it to assimilate the treated effluent without suffering significant water quality impairment.

The diversion project developed rapidly because of scientific analysis, legislative action, and public support. The project was effective because the scientific community, headed by W. T. Edmondson, convinced the public that water quality would become worse with no action and would be improved only by eliminating sewage inputs to the lake. Scientific projections for rapid improvement in water quality were borne out after diversion had taken place. Scientific information used to support the proposal for sewage diversion included data for nutrients, oxygen, and water transparency. Data from 1955 to 1959 were compared with those describing conditions in 1933 and the early 1950s. Scientific support for action also came from Edmondson's analysis of the problem in relation to earlier eutrophication problems of European lakes, which was simplified by Hasler's general review of cultural eutrophication (Hasler, 1947). In particular, Lake Zürich in Switzerland and other lakes had developed blooms of *Oscillatoria rubescens* as a result of pollution by domestic sewage.

Diversion is more effective than tertiary treatment of sewage effluent for two reasons. Although tertiary treatment removes 90 to 95 percent of the phosphorus (P) from sewage, the effluent may still contain 0.5 to 1.0 mg of phosphorus per liter, which is 10-fold higher than the phosphorus content of Lake Washington when it was most enriched. In addition, diversion reduces inputs of other substances that might be secondary factors in promoting algal growth. Diversion thus completely removes phosphorus, the main contributor to eutrophication, as well as other substances that could promote algal growth but might not be removed completely in sewage treatment.

with discharges exceeding 1 million gallons per day. Most plants remove phosphorus by chemical methods in "tertiary" treatment units, installed after conventional "secondary" (biological) treatment units. Phosphorus removal is achieved by adding alum, calcium, or (more rarely) iron salts to the waste stream. The metal ions form hydrous precipitates, and phosphate is removed by coprecipitation and adsorption onto the surfaces of the precipitates (Hsu, 1975). Removal efficiencies far exceeding 90 percent are common. Effluent standards for phosphorus, where such exist, usually are 1 mg per liter (based on a conventional engineering goal of 90 percent removal efficiency

BOX 4.3
SHAGAWA LAKE, MINNESOTA

In contrast to Lake Washington, Shagawa Lake, Minnesota, is an important recreational lake in which the effects of reduced phosphorus loading were not as great as predicted. Its main external source of phosphorus was the sewage treatment plant at Ely, Minnesota, a community of 5,000 residents. Ely was established around 1900 and attained its largest population (ca. 6,000) in the 1930s. The three-basin lake has a maximum depth of 13 m, mean depth of 5.6 m, mean volume of 5.2×10^7 m^3, and surface area of 925 ha (Larsen et al., 1975). Shagawa Lake has had a long history of water quality deterioration as the result of cultural eutrophication. Wastewater from Ely initially flowed untreated into the lake, then received primary treatment in 1911 and secondary treatment in 1952. A tertiary treatment plant to remove phosphorus was constructed with EPA demonstration grant funds in 1973 and reduced the lake's external phosphorus loading by 80 to 85 percent (Larsen et al., 1979, 1981). The lake's average total phosphorus (P) concentration should have declined from 51 µg of phosphorus per liter (pretreatment) to about 12 µg of phosphorus per liter in 1.5 years, according to a tank reactor model. Instead, only a 40 percent reduction occurred. By 1976, the average total phosphorus concentration had decreased from about 50 to 29 µg per liter, and soluble reactive phosphorus decreased even more, from 21 to 4.5 µg per liter. The only noticeable biological response was a small decline in the annual average chlorophyll concentration, caused by a decrease during May and June to less than half the pretreatment value of 15 µg per liter. However, there was no trend in chlorophyll concentrations during the main recreational season (July to August), when the most severe blue-green algal blooms occurred (chlorophyll levels of 60 µg per liter and sometimes up to 100 µg per liter). The lake's water residence time is 8 to 9 months, and its phosphorus residence time was estimated to be less than 6 months (Larsen et al., 1975). A rapid response to nutrient reduction thus would be expected (based on loading model concepts). Problems with blue-green algal blooms continued in the lake during late summer throughout the 1980s. A large bloom in August of 1987 was thought to be responsible for the death of several dogs (apparently by ingestion of toxins excreted by the algae (B. Wilson, Minnesota Pollution Control Agency, personal communication, 1991).

This demonstration project has been considered a failure by some because a rapid biological response to reduced phosphorus loading was not obtained, as predicted from simple models of phosphorus loading. These models failed to predict lake conditions because they do not account for internal loading of phosphorus, a process whose importance was not widely recognized prior to the Shagawa Lake project. Data collected after external loading was reduced showed that biological productivity is being maintained by release of phosphorus from recent sediments when deep-water oxygen depletion occurs (Larsen et al., 1981). Phosphorus transport from these deep waters to the surface was sufficient to maintain algal blooms. Restoration of the lake may take much longer than anticipated because of feedback from the sediment. Recovery could be accelerated by alum treatment, but the size of the lake makes this approach impractical.

Evaluation of this project possibly should be delayed until it can be established how much internal loading will be reduced with time. Intuitively, one would expect that internal loads over the long term depend on external loads. Thus, with some unknown time lag, the rate of internal loading may decrease and water quality may improve. This appears to have occurred in Lake Sammamish, Washington, after wastewater was diverted from it (Welch et al., 1986). Continued research in Shagawa Lake to determine the long-term interplay between reduced external loading and internal loading would be helpful. Long-term data may be required to establish trends because of interannual variability in biological productivity caused by climatic factors (independent of nutrient loading).

and a typical raw sewage concentration of 10 mg per liter), but well-run tertiary plants routinely produce effluents with 0.1 mg of phosphorus per liter or even less.

Biological techniques also have been developed that remove 80 to 90 percent of the phosphorus from wastewater (Shapiro et al., 1967; Marais et al., 1983). Conventional biological waste treatment plants have low phosphorus removal efficiencies (20 to 40 percent) because phosphorus levels in raw waste far exceed the stoichiometric needs of the microorganisms (activated sludge) that consume the waste's organic matter. Efficient biological removal is obtained by manipulating operating variables such as aeration time, cell residence time, and organic loading rates and preventing the sludge from becoming anoxic, all of which promote "luxury" uptake and retention of phosphorus by the sludge. Biological techniques have the advantages of requiring no chemical doses and no additional treatment units and of costing less than chemical methods. However, chemical methods are easier to control and operate at higher removal efficiencies.

Interception of Nonpoint Sources of Nutrients   Control or elimination of point sources of nutrients usually is insufficient to reverse eutrophication problems, and nonpoint sources must be managed as well. The costs of treating nonpoint sources by engineered systems are high, and diversion of inputs to other ecosystems may have high political and social costs, as well as untoward ecological and hydrological effects. This has led to the development of pre-lake interception systems as "low-technology" supplements to tertiary wastewater treatment and agricultural best management practices to decrease nonpoint sources of nutrient loading. Most interception systems function to remove other contaminants (loading silt, particulate organic matter, toxic metals) as well as nutrients. Three types of interception systems have been used: storm water detention and retention (siltation) basins, artificial or natural wetlands, and in-stream phosphorus precipitation.

Detention basins function by impounding storm water runoff (urban or agricultural) for a long enough time to allow settling of particulate materials. Retention ponds are designed to catch a certain amount of runoff (e.g., the first inch) and retain it until it infiltrates through the soil. Because much of the phosphorus (and other contaminants) in runoff water is associated with suspended particulates, detention and retention basins are effective as low-cost, low-maintenance treatment systems, and such ponds are required for new urban developments in many states. Design criteria and performance standards have been evaluated (Walker, 1987). Detention ponds eventually fill up with silt and organic matter, and must be reconstructed or dredged.

Prereservoir detention basins are a variation on the same idea. They are constructed on natural streams just upstream from their entry to a lake or reservoir. Such basins protect the reservoir from silt, phosphorus, and bacterial loadings, and are effective as long as they remain aerobic to prevent internal loading. Based on loading model calculations, Benndorf and Putz (1987) concluded that basins with a 2-day retention time should achieve about 35 percent phosphorus removal, 5 days should yield about 40 percent removal, and 15 days should yield 50 to 60 percent removal. Actual data from the watershed of Jesenice Reservoir (Czechoslovakia) shows greater removal efficiency. A prereservoir detention basin with a water residence time of 5 days was found to retain 60 to 70 percent of the total phosphorus entering it (Fiala and Vasata, 1982). Removal efficiency information can be used along with a phosphorus budget of the downstream lake to determine the detention time (hence the basin size) required to decrease the phosphorus loading to a lake.

Man-made and engineered natural wetlands (see Box 4.4) have been successful, in some cases, in retaining materials suspended in water flowing through them. Wetlands are effective in retaining suspended solids, given adequate detention time, and most phosphorus removal is associated with this process. They are also highly effective in reducing stream loads of metals such as lead and zinc (Martin, 1988). Temperate wetland systems for this purpose generally em-

---

BOX 4.4
CLEAR LAKE, MINNESOTA

Clear Lake, in Waseca, Minnesota, is an example of the use of an engineered wetland (Barten, 1987). This 257-ha, heavily used recreational lake became eutrophic from sanitary sewage and urban and agricultural runoff. In 1981, about 50 percent of the water load and 55 percent of the phosphorus load were diverted into a modified 21-ha marsh throughout the growing season. All cells of the marsh can be drained to harvest plants. About 40 percent of the lake's annual phosphorus load is retained in the marsh. Nonetheless, the average lake concentration of phosphorus fell only about 30 percent, apparently because excessive phosphorus loading over many years had led to high rates of hypolimnetic oxygen demand and high internal phosphorus recycling in the lake. A whole-lake fish reclamation project (rotenone treatment to remove rough fish) in the fall of 1986 caused only a small decrease in phosphorus concentrations but did reduce chlorophyll levels dramatically for about a year (because fish removal enhanced zooplankton grazing). Finally, hypolimnetic alum treatment in 1988 was successful in lowering the N-lake phosphorus levels and summertime chlorophyl concentrations.

phasize emergent plants such as cattails, whereas floating plants like water hyacinth are most effective in subtropical and tropical wetlands (Reddy and DeBush, 1987).

In contrast, diversion of streams or runoff into unmanaged natural wetlands appears to provide only limited long-term nutrient removal. Although such wetlands may assimilate nutrient inputs during the growing season, a large outflow of nutrients released from dead vegetation the following spring may offset the nutrients stored the previous growing season. In addition, large losses of nutrients from wetlands during high-flow, intensive rain events or through channelization tend to counterbalance the net storage of nutrients during longer periods of low or moderate flow rates (Richardson, 1988).

In a few cases, streams flowing into lakes have been treated by adding phosphorus-precipitating chemicals (iron, aluminum), but because volumes of water that need to be treated generally are large (compared with municipal wastewater), this usually is not a cost-effective approach. Iron is preferred for in-stream treatments because it has fewer toxicity problems than does aluminum, but binding of phosphorus to iron requires continuously aerobic conditions. Success in lowering phosphorus concentrations has been reported when relatively small flows can be treated. An example is the addition of ferric sulfate to water pumped into Foxcote Reservoir (England) to remove dissolved phosphorus. Although internal phosphorus loading in the reservoir has reduced the treatment's effectiveness, the length of time that Foxcote Reservoir cannot be used as a potable water supply during summer months has decreased (Young et al., 1988).

Wahnbach Reservoir, an important municipal water supply for Bonn, Germany, is protected from nutrient, silt, and organic matter loading from its main tributary by a prereservoir detention basin and phosphorus elimination plant (Bernhardt, 1980; Clasen, 1989). Water from the detention basin is treated with iron to remove phosphorus and then filtered through an ion exchanger and a series of activated carbon and sand filters. The plant removes 95 to 99 percent of phosphorus, coliform bacteria, algae, and turbidity; 77 percent of the water's biochemical oxygen demand; and 58 percent of the dissolved organic carbon. Grossly enriched river water is converted into nearly drinkable water before it enters the reservoir. Costs of this project have not been published.

Best Management Practices   Numerous best management practices (BMPs) (Table 4.8) have been developed to decrease losses of soil, nutrients, and other contaminants from agricultural lands and urban areas. (The interception methods described in the section immediately above are essentially one class of BMPs.) Effectiveness in preventing nutrient export, technical feasibility, social acceptability, and cost vary widely among the practices, and quantitative information on these aspects is lacking for many of them, which are still in the developmental stage. Some practices (e.g., restrictive zoning ordinances and setback requirements) are more suitable for new developments than for developed areas. Although BMPs seldom provide the complete solution in restoring degraded lakes, they are key elements in an evolving strategy that recognizes that lakes can be managed and protected effectively only in the context of the watershed in which they exist.

Dilution

Dilution is a procedure that can lower water column phosphorus (P) concentrations by adding water that is low in phosphorus. It will also increase washout of algal cells from a lake. In principle, addition of dilution water to a lake will increase its total phosphorus loading rate but decrease the mean inflow phosphorus concentration. The lake's flushing rate is also increased, and this tends to decrease phosphorus sedimentation. As a result, the water column phosphorus concentration will decrease, although increasing the amount of dilution water will not produce a proportionate reduction in water column phosphorus concentration. The best candidate lakes are those with high flushing rates and moderate problems with high phosphorus concentrations.

Moses Lake, Washington, is the best documented case history of dilution (Welch, 1981; Welch and Weiher, 1987). Columbia River water was diverted through the lake and then to agricultural areas for irrigation. Algal blooms were reduced by 50 percent and water clarity increased by 100 percent during the 9 years that dilution water entered the lake. Nonetheless, from an economic cost-benefit perspective, Moses Lake was one of the few unsuccessful projects noted by EPA (1980) in its analysis of benefits of the Clean Lakes Program (see Table 4.6). The project received about $3.25 million in federal funds (and about the same in local funds), but quantifiable benefits amounted to only $0.53 million.

Dilution does not appear to be associated with negative impacts on the lake, other than those associated with increased flow. Few case histories of its use exist, in part because of the general absence of sufficient supplies of nutrient-poor water to add to a lake's inflow.

*In-Lake Methods to Reduce Phosphorus Concentrations
and Cycling*

Phosphorus Inactivation A significant reduction in nutrient loading to a eutrophic lake is a necessary but sometimes insufficient step
in order to decrease water column phosphorus concentrations enough
to reduce the amount of algae. Phosphorus release from lake sediments at high pH, or when dissolved oxygen in overlying water is
low or zero, can be a major source of phosphorus to the water column. Under certain conditions, phosphorus released from lake sediments will be transported to the upper layers of a lake and stimulate
an algal bloom. This process, in which sediments enriched in organic
and inorganic matter from external loading and in-lake production
cause dissolved oxygen consumption and phosphorus release, is known
as internal loading. It can be great enough to delay or prevent a
lake's recovery from nutrient diversion or interception (see Box 4.3).

Phosphorus inactivation reduces the rate of phosphorus release
from lake sediments by the addition of aluminum salts (sodium alu-
minate, aluminum sulfate) to them (Cooke et al., 1986). Aluminum
hydroxide is formed and appears as a visible floc that settles to the
sediment and binds with phosphate ions to form a solid that is insoluble under low or zero dissolved oxygen. Phosphate ions diffus-
ing from the sediment are trapped by the floc. The process has proved
to be effective and long-lasting. Several Wisconsin lakes treated in
the early 1970s exhibited improved conditions 10 years later (Garrison and Knauer, 1984; see Box 4.5). Treatment of shallow, well-mixed
lakes can also be effective but appears not to have the longevity
found with deep, thermally stratified lakes. A representative case
history is Long Lake, Kitsap County Washington (Welch et al., 1988).

In contrast, Eau Galle Reservoir, a flood control impoundment in
Wisconsin, illustrates the ineffectiveness of phosphorus inactivation
when nutrient loading is not reduced significantly (Kennedy et al.,
1987). The effects of treatment on the quality of this water body were
overwhelmed in a few months by continued nutrient loading. Because reservoirs are difficult to protect from nutrient loading, this
technique is not considered widely applicable to this type of water
body (see Box 4.5).

Aluminum is a potentially toxic metal. At naturally occurring pH
(6 to 8) in waters with carbonate alkalinity, nearly all aluminum is
found as nontoxic aluminum hydroxide. If the pH falls much below
6, toxic forms of soluble aluminum will increase. Several observations of treated lakes with normal pH have failed to demonstrate any

BOX 4.5
WEST AND EAST TWIN LAKES, OHIO

Some lakes have significant internal sources of nutrients from littoral wetland and macrophyte zones, and especially from bottom sediments. The presence of an anoxic hypolimnion greatly increases the rate of release of nutrients from sediments to the overlying water. Under certain circumstances (e.g., high mean depth, large area, exposure to winds), these nutrients are transported from the hypolimnion to the epilimnion and subsidize algal blooms. Diversion of cultural nutrient loading, although essential, may not be sufficient to return these lakes to their undisturbed condition; curtailment of internal loading may also be required. Such lakes may be more common than those in which water column nutrient concentrations are determined by external loading alone. Shagawa Lake, Minnesota, described earlier (see Box 4.3) is another example.

West and East Twin Lakes are small (34 and 27 ha, respectively), thermally stratified lakes of glacial origin in a 335-ha forested, urbanized watershed (including lakes) in northeastern Ohio. Prior to 1973, domestic waste from about 360 homes was discharged to septic tanks and leach fields. Although most of these disposal systems were located in ideal soil, sloping lawns became saturated with effluents, and organic matter, nutrients, and bacteria were washed into the lakes with surface and shallow ground water flows. In 1969, high densities of algae and coliform bacteria caused the lakes to be closed to contact recreation. Between 1971 and 1973, all domestic wastewater was diverted out of the watershed of both lakes. Lake scientists predicted that internal release of phosphorus from anoxic hypolimnetic sediments, followed by vertical entrainment to the epilimnion, would delay recovery of the lakes. It was further predicted that application of aluminum sulfate (alum) to the hypolimnion would accelerate lake recovery by controlling phosphorus release from anoxic sediments (Cooke et al., 1982).

External and internal phosphorus budgets were determined from 1971 to 1976, and changes in lake trophic state were monitored from 1969 to 1976 and at widely spaced intervals through 1989. A basis for the alum dose was determined by field and laboratory toxicity tests and by calculations of expected dissolved aluminum concentrations for various lake alkalinities. A pilot treatment of a small (3-ha) lake was carried out in 1974 as a test of dose and application guidelines. In 1975, West Twin's hypolimnion received 100 tons of liquid alum. East Twin, the downstream lake, served as a reference. Alum treatment sharply reduced phosphorus release from the anoxic hypolimnetic sediments.

Prior to nutrient diversion, the lakes were classified as eutrophic, based on water transparency and on phosphorus and chlorophyll concentrations. In 1989, 14 years after treatment and 16 years after diversion, they were near the mesotrophic-oligotrophic border, a state consistent with expectations in this ecoregion. In laboratory experiments under anaerobic conditions, phosphorus release rates from the treated West Twin sediments were still significantly lower than release rates from untreated East Twin sediments in 1989, showing that the alum treatment retained its effectiveness for 14 years (Cooke et al., 1986; Cooke and Martin, 1989).

The key event to restoration of the two lakes was diversion of nutrient inflows. Alum treatment aided in the recovery of West Twin but would have had little long-term effect if loading had continued. Whereas hypolimnetic phosphorus concentrations in West Twin after treatment remained at less than a third of those in East Twin (and at about 20 percent of pretreatment concentrations through 1986), surface water phosphorus concentrations in the two lakes remained similar and declined over the years 1975 to 1989 in a nearly identical pattern. Several studies demonstrated that phosphorus release into the hypolimnion of West Twin was controlled by alum treatment. However, vertical entrainment appeared not to be as large a source of phosphorus to the epilimnion as was predicted or as has been calculated for deeper lakes with greater exposure to wind mixing (e.g., Lake Mendota, Shagawa Lake). The results demonstrate the importance of controlling the load of nutrients from the watershed and also underscore the importance of long-term monitoring of restoration projects. The initial results appeared to support the hypothesis that control of internal loading was the key to restoration, but longer-term data did not. Use of the reference lake was also an essential component in understanding the mechanism of recovery. Unfortunately, projects with this type of design are rare in lake restoration.

deleterious effects to fish or invertebrate animals living in treated sediments over very long periods (years) of exposure. Phosphorus inactivation, mistakenly classified as an algicide treatment by some agencies, is considered to be safe and cost-effective when the aluminum sulfate dose is below that which will create low-pH conditions. The high water clarity that occurs after treatment can promote the invasion and/or spread of rooted macrophytes in shallow water. An alternative to aluminum salts for such purposes is calcium hydroxide (lime). This material has been used in one successful treatment (Prepas et al., 1990), but there are no data on treatment longevity.

Sediment Skimming   Phosphorus release from lake sediments is greatest from the most recent phosphorus-rich surficial layers. Sediment skimming (see Box 4.6) involves the use of a hydraulic dredge to remove this layer. This procedure, although effective, is more costly than phosphorus inactivation. It does have a restorative effect without the addition of potentially toxic materials, especially when nutrient inflows have been reduced or eliminated. Once the equipment is set up for sediment skimming, it might be reasonable to proceed with a full-scale sediment removal to accomplish both lake deepening and control of internal loading (provided an adequate containment area for the sediment water slurry is available).

BOX 4.6
LAKE TRUMMEN, SWEDEN

Lake Trummen, Sweden, received domestic wastewater and flax mill discharges for many years, and algal blooms and fish winter kill were common. The loading was diverted from the lake, but no improvement occurred (because of internal nutrient loading). A sediment-skimming treatment removed the enriched surficial materials, and the phosphorus content of the remaining sediment was 10 percent of the material that had been removed. This was followed several years later by the removal of carp, which had become abundant, disturbed the sediments, and promoted phosphorus release. The lake improved greatly and remained in this improved state for at least 9 years. Continued removal of the carp was found to be essential to maintaining lake quality (Bjork, 1988).

Sediment Oxidation   Phosphorus release from lake sediments can be controlled by accelerating the oxidation of sediment organic matter and providing a chemical environment that favors the binding of phosphorus by iron in the top 5 to 10 cm of lake sediment. Although the procedure is still in the development stage, it does not involve the addition of a potentially toxic element such as aluminum. Instead, calcium nitrate, $Ca(NO_3)_2$, is injected into the sediments. The nitrate serves as an electron acceptor in the absence of oxygen, and decomposition of organic matter proceeds via denitrification (nitrate is reduced to N_2, which evolves as a gas). At the same time iron sulfide, FeS, is oxidized, and phosphate ions are bound to the resulting ferric hydroxide. In some lakes, calcium hydroxide is added to bring the pH to the optimum for denitrification, and ferric chloride may be added if the lake is iron deficient (Ripl and Lindmark, 1978). Lake Lillesjon, Sweden, received a treatment with ferric chloride, lime, and calcium nitrate. The oxygen demand of the sediment decreased by 30 percent, and release of phosphorus from sediments to water was reduced to 10 to 20 percent of the pretreatment rate.[1] Some have suggested the use of nitrate-rich effluent from wastewater treatment plants to oxidize lake sediments, but field demonstration of this approach has not yet been conducted.

Deep-Water Discharge   The impact of phosphorus release from lake sediments can be controlled by siphoning the nutrient-rich deep (hypolimnetic) water from a lake or discharging the hypolimnetic water of a reservoir through a deep gate in its dam. If release exceeds new external loading, the procedure should gradually deplete the sediments of phosphorus and could reduce the amount of nutrients entrained from deep to surface waters each summer. Summer and early autumn algal blooms should be reduced (Nurnberg, 1987). Continued high nutrient loading to a lake is likely to negate the effects of this technique.

Although deep-water discharge is not widely used, the few recorded attempts are encouraging. It appears that the greater the amount of phosphorus discharged in this way, the greater is the decrease in phosphorus concentration in the upper waters where algae grow. Moreover, the more years deep-water discharge operates, the

---

[1]In contrast, calcium nitrate treatment of sediments in Long Lake (St. Paul, Minnesota) did not produce any noticeable improvement in trophic conditions and water quality. The lack of success in this case can be attributed, at least in part, to failure to control nonpoint source nutrient loadings (Noonan, 1986).

greater is the change in a lake's concentration of nutrients. However, there can be significant negative impacts of discharging nutrient-rich hypolimnetic waters to receiving streams. The dissolved oxygen content of such discharged waters may be near zero, and there will probably be high concentrations of soluble iron, manganese, hydrogen sulfide, ammonium, and phosphate. Treatment of the discharge would probably be required. Also, high discharge rates could induce a midsummer partial mixing of the water column, and this is likely to trigger an algal bloom when nutrient-rich bottom waters are mixed with surface waters.

Management of Symptoms

The techniques and procedures described above for control of algal blooms can be restorative because they produce a lasting decrease in nutrient concentrations. In some situations, however, it is not technically possible or economically practical to control external or internal nutrient loadings enough to prevent degraded water quality conditions. Several in-lake management tools are available to alleviate the symptoms of nutrient overenrichment and improve water quality for lake users. Some of these management tools (e.g., artificial circulation, use of algicides) require continuous or repeated applications (i.e., their benefits are short-lived), but others, such as biomanipulation, potentially can provide long-term benefits.

Biomanipulation Biomanipulation was broadly defined by Shapiro et al. (1975) to include a wide array of biological controls for water quality problems. They distinguished these from the many chemical and engineering approaches that exist for water quality improvement. More recently, a narrower definition, derived from the pioneering studies of Hrbacek et al. (1961), has been adopted by some limnologists: the manipulation of fish community structure to permit large herbivorous zooplankton grazers to flourish and to control nuisance algae (Shapiro, 1990b). This approach to biomanipulation is currently the object of substantial research programs in the United States, Canada, and several European nations (Gulati et al., 1990). Biomanipulation is not regarded as a substitute for reduction of nutrient loads. Important questions revolve around the capacity of biomanipulation to (1) reduce algal biomass where loads cannot be controlled and (2) augment or accelerate the effects of load reductions.

Results are still emerging, and it is unlikely that general principles concerning the efficacy of biomanipulation will be complete for several years (Gulati et al., 1990). However, certain patterns are clear.

Fish removal allows large, generalist grazers to become abundant and commonly reduces algal biomass and production by factors of 10 or more (Henrikson et al., 1980; Reinertsen et al., 1990). However, elimination of fish is neither practical nor desirable in many lakes. The alternative is to establish fish populations dominated by large, piscivorous fish. These predators reduce the biomass of smaller planktivorous fish, allow grazer biomass to increase, and reduce the biomass of algae. In whole-lake experiments, piscivore enhancements have improved water quality (Shapiro and Wright, 1984; Carpenter et al., 1987; Benndorf et al., 1988). Based on these experiments and case histories of fish kills, it appears that (1) the greatest improvements are possible in lakes dominated by planktivores prior to treatment, and (2) piscivore additions must achieve substantial (tenfold or greater) changes in planktivore biomass to influence water quality (Carpenter and Kitchell, 1988; Gulati et al., 1990).

Biomanipulation research is now expanding at two main interfaces. The first is the linkage between water quality and fisheries ecology. Management for large piscivores is a key element of biomanipulation (see Box 4.1). The high variability of fish stocks and the capacity of nutrient loads to destabilize lake food webs are key challenges that demand the best interactive efforts of fisheries ecology and limnology (Carpenter, 1988; Kitchell, 1991). The second interface is that between littoral zone ecology and the pelagic food web (Figure 4.4). In shallow lakes, fish removals that improved water clarity have been followed by expansion of submersed aquatic vegetation (Gulati et al., 1990). Once established, the submersed plants shelter fish that may eliminate grazers, causing declining water clarity and reduction in submersed vegetation. Abrupt transitions between alternate stable states of macrophyte and algal dominance may be triggered by nutrient mitigation, biomanipulation, or aquatic plant management. There is an obvious need for better understanding of the interactive effects of restoration and management of aquatic plants, fisheries, and phytoplankton.

Artificial Circulation   Artificial circulation is a management technique whose goal is to achieve and maintain an isothermal and isochemical water column in a lake or reservoir that otherwise would exhibit stratification during summer. This is accomplished by injecting compressed air into a pipeline tethered at the lake's bottom in the deep zone. The last several meters of the pipe are perforated so that a vigorous bubble curtain is created, with enough energy to mix the water column rapidly. Even on the warmest days, a properly sized system will have a temperature vertical difference of less than 3°C (versus 20°C in typical stratified lakes). Pumps and whirling blades may also accomplish this goal (Cooke et al., 1986).

Expected improvements include (1) habitat expansion; (2) low concentrations of soluble iron and manganese, ammonium, hydrogen sulfide, and other reduced compounds associated with anoxic waters; (3) a reduction of algal biomass (in some cases); and (4) the elimination of surface thermal microstratification, a factor that favors the formation of blue-green algal scums.

Artificial circulation has been successful in improving potable water supplies (by eliminating iron, manganese, and hydrogen sulfide). Blue-green algae have declined in some cases, but not in others, and success in this respect seems to be depend on whether circulation reduces the pH of the surface water (Shapiro, 1973, 1984, 1990b; Shapiro et al., 1975). Even if artificial circulation does decrease the abundance of nuisance blue-green algae, it may not necessarily decrease the total amount of algae in the lake.

Most problems in the use of artificial circulation are associated with an underpowered compressor. A warm, uncirculated layer of water may develop on the lake's surface if the circulator cannot overcome the difference in water density created between surface and subsurface layers in hot weather. This would provide an ideal habitat for blue-green algae, which can regulate their depth with gas vacuoles and create surface scums. Another problem is the creation of turbid water if the bubble curtain disturbs flocculent sediments. This procedure provides little lasting benefit when it is shut off.

Algicides Algicides are chemicals that achieve control of nuisance algae through a toxic effect. The most common algicide is copper sulfate, to which blue-green algae are particularly sensitive. This is a purely symptomatic treatment; no lasting benefits are achieved, and a residue of copper is left in lake sediments. Copper sulfate treatments are effective only as long as the cupric ion (Cu^{2+}) concentration remains sufficiently high in the water, but concentrations usually fall rapidly (within hours or a few days) after treatment because copper adsorbs onto suspended particles and forms organic complexes and insoluble precipitates that settle to the lake's sediments (McKnight et al., 1983). Loss of copper is especially rapid in alkaline waters, and most lakes with algal bloom problems fall in this category.

Significant negative effects may occur as a result of copper sulfate treatment: dissolved oxygen depletion following decay of killed cells, sediment contamination, and toxicity to nontarget species, including fish and algae-grazing zooplankton. Repeated applications are required, making the cost-effectiveness poor. A summary of the chemistry, effectiveness, dose, and negative effects of the use of copper sulfate is found in Cooke and Carlson (1989).

CONTROL OF AQUATIC MACROPHYTES

Overview

Macrophytes are natural and essential components of lake ecosystems. Small-bodied fish species and young fish of many species find food and shelter from predators in the littoral zone. Cultural eutrophication decreases water clarity and leads to the elimination of macrophytes that are important as food for waterfowl (e.g., *Vallisneria americana*). The simplified macrophyte community of eutrophic lakes usually is dominated by one or a few species, often exotics, that thrive in disturbed habitats. These species often concentrate their biomass near the water surface and thus are much more conspicuous to lake users than are native species that grow deeper in the water (Nichols et al., 1991).

To date, macrophyte restoration techniques have been limited to methods for killing nuisance plants. Most lake managers recognize that moderate macrophyte growth is essential for a healthy fishery and thus seek to control macrophytes rather than to eliminate them. Almost nothing is known about replacement of nuisance macrophytes by desirable species. Improved water clarity is probably essential for restoration of desirable macrophytes but may not be sufficient. Further steps such as sediment amendments and planting may be necessary. There is a need for research that moves beyond suppression of nuisance plants to the establishment of diverse macrophyte communities that provide essential habitat for waterfowl and fish (Nichols et al., 1991).

Biological Agents

Biological agents offer the prospect of long-term management of nuisance macrophytes at reasonable cost and minimal environmental impact, but the risks of escape and irruption of the biological control agent itself must be considered carefully (Magnuson, 1976). Many of the plants that cause nuisance problems are exotics, often imported for use in aquaria and inadvertently introduced to lakes. These plants have few pathogens or native animals that graze on them. Biological control research commonly involves the exploration for pathogens (bacteria, fungi, viruses) or predators (herbivorous insects) in the native habitat of the plants. These organisms are imported under con-

trolled conditions, and if found to be safe (i.e., to not attack nontarget organisms) and effective against the target plant, they are released at sites where the plant is a nuisance. Herbivorous fish, especially the white amur (or grass carp, *Ctenopharyngodon idella* Val.),
that consume a variety of plant species also are common biological
control agents. Sometimes biological agents are used with mechanical or chemical treatment for rapid relief while the bioagent develops
to the density required to produce control.

Grass carp are permitted in 26 states. In contrast to phytophagous
insects, which were brought into the United States under strict quar-
antine until their effectiveness and negative impacts could be evaluated, grass carp were introduced to some lakes with little or no prior
testing. Grass carp have voracious appetites for certain plants and in
warm waters may consume 50 to 60 percent of their body weight
each day. Compared with mechanical and chemical procedures, grass
carp are more cost-effective by a factor of 10. If properly stocked,
they are not likely to produce negative environmental impacts
(Cooke and Kennedy, 1989), and unlike the common carp, *Cyprinus
carpio*, they are not likely to become a massive nuisance (Stanley et
al., 1978). Nonetheless, grass carp are controversial for many reasons, and caution in their use is warranted. Not the least among
these reasons is the fact that once grass carp are stocked in a lake,
they are almost impossible to remove, and their effects on vegetation
will remain for many years (Leslie et al., 1987). Their preferred diet
does not include such nuisance plants as Eurasian water milfoil, water hyacinth, or alligator weed (Fowler and Robson, 1978). Preferred
plants include hydrilla and native species such as elodea, and some
pondweeds (*Potamogeton*). Their effectiveness is related to stocking
rate, water temperature, length of growing season, size of fish, and
types of plants to be controlled. If the stocking level is too low,
then only palatable plants will be grazed, which actually may make
problems with macrophytes worse (Leslie et al., 1987). Overstocking
has resulted in eradication of submergent littoral vegetation and
attendant loss in fish habitat, as well as increases in turbidity, algal
blooms, and shoreline erosion (see Box 4.7).

In Florida, where their use is common, many lakes stocked with
grass carp are very turbid (because of algal blooms), and shoreline
erosion is so extensive, due to the absence of a "damping" effect by
submersed plants, that shoreline trees have fallen. Precautions must
be taken to minimize fish movement to habitats where vegetation is
desirable. Infertile hybrid grass carp were used in the 1970s and
early 1980s to avoid potential problems of grass carp reproduction in
open aquatic systems, but the hybrids have lower feeding rates than

---

BOX 4.7
LAKE BALDWIN, FLORIDA

Lake Baldwin is an 80-ha eutrophic lake located on the Orlando Ridge in Orlando, Florida (Canfield et al., 1983). The lake was stocked with 34 grass carp (each >304 mm length, 0.8 kg) per hectare of hydrilla during summer and fall of 1978. An earlier stocking of fingerling grass carp in 1974 failed to control hydrilla, presumably because of high predation pressure on the fingerlings. Shireman and Maceina (1981) reported that hydrilla was nearly eradicated 2 years after the second stocking. According to these authors, hydrilla control was evident when grass carp biomass reached 130 kg of fish per hectare of hydrilla beds. Phytoplankton chlorophyll increased from approximately 5 μg per liter before stocking to levels as high as 30 μg per liter after aquatic plants had been eradicated (Canfield et al., 1983). Secchi disk transparency readings decreased from 6 m during the height of hydrilla infestation to approximately 1.5 m after hydrilla had been eradicated by grass carp. Chlorophyll a and total alkalinity also increased in the lake after hydrilla had been controlled. The long-term effects of eliminating the lake's macrophytes on total fish biomass and species composition could not be determined from a relatively short period of study after treatment but could be significant (Canfield et al., 1983).

---

the parental stock (Osborne, 1982; Shireman et al., 1983). This drawback has been overcome with the development of a sterile triploid form that has about the same feeding and growth characteristics as the fertile diploid form (Wiley and Wike, 1986).

Insect control of alligator weed and water hyacinth has been effective in Florida, Alabama, Mississippi, Texas, Louisiana, and Georgia. Alligator weed is controlled primarily by two insects: *Agasicles hygrophila*, commonly known as the alligator weed flea bettle, and *Vogtia malloi pastrana*, commonly known as the alligator weed stem borer. The species of insects involved include *Sameodes ilbiguttalis* (Warren) (Lepidoptera: Pyralidae) and *Neochetina eichhornia* Warner and *N. bruchi* Hustache (Coleoptera: Curculionidae). Techniques have been developed to concentrate the insects, allowing their reproduction, population growth, and subsequent spread through the lake. A characteristic of the insects used in controlling these plants is that they can complete their life cycle only on the target species. The insects effective against alligator weed and water hyacinth are sensitive to cold weather, limiting their distribution to southern waters (Center et al., 1988), but the plants themselves are also limited to warm climates. Little is known of the effects of bird predation on these insects.

The development of effective plant pathogens has not been as rapid as that of herbivorous insects. Pathogenic fungi have many of the properties of an ideal biological control agent, including target specificity and low or zero pathogenicity to humans. Successful use of plant pathogens has been increased by combining their application with the use of a herbivorous insect and either chemical or mechanical control agents (Charudattan, 1986).

The use of biological controls, including manipulation of food webs to enhance grazing on algae or to reduce nutrient recycling, has been effective. This approach treating the symptoms of eutrophication or the invasion of exotic plant species has promise for providing low-cost improvements with long-term effectiveness, and it avoids the problems associated with chemical and mechanical technologies. More emphasis on research funding for this type of lake management is needed.

Water-Level Drawdown

Some aquatic plants are susceptible to exposure to dry, freezing conditions and can be controlled in temperate latitudes by lowering the lake level in November and refilling in early spring. Three to four weeks of continuous exposure to below-freezing air temperature will kill the roots and reproductive structure of some nuisance plants, including Eurasian water milfoil, coontail, and southern naiad. The procedure is most likely to be effective only in northern and some midwestern areas. Water removal also allows other lake restoration activities to occur, including fish management, sediment removal, and repair of dams and shoreline structures. Other benefits can occur in lakes with flocculent organic sediments through sediment consolidation and compaction. Drawdown in warm weather can also be effective for these purposes, but it may interfere with recreation, irrigation, and water supply.

Some aquatic plants are not affected by drawdown, including water hyacinth, elodea, hydrilla, and bushy pondweed (*Najas flexilis*). Negative effects include a failure to refill (if dry weather persists), the possible stimulation of algal blooms, and the potential for altering or destroying wetlands. Cooke et al. (1986) summarized the use of this technique, including the responses of 74 aquatic plants to drawdown and desiccation.

Harvesting

Harvesting nuisance aquatic plants is a common lake management procedure, particularly in northern climates with short growing seasons and an absence of exotic plants with very high growth rates, such as water hyacinth and alligator weed (Cooke et al., 1986). Harvesting is not a restorative procedure, and its goal is to make a lake more usable for recreation. Harvesters are machines that combine a cutter bar and conveyor system with a large on-board storage area to receive the cut plants. The cutter bar is lowered to the sediment surface, or to a depth of 1.5 to 1.8 m, and the plants are cut, collected, stored on board, and then transported to a disposal site on land. Machines range in storage capacity from about 3 to 23 m^3. At most, several hectares per day can be cut, which precludes use of this machinery to attempt plant eradication. In southern climates, particularly where exotic plant species have successfully invaded, plant densities are high and regrowth rates rapid, making harvesting largely impractical. In northern climates, harvesting once per season is generally adequate, and regrowth is reduced in the following season, especially if harvesting is done late in the growing season (Kimbel and Carpenter, 1981).

Harvesting has several positive features. No toxic materials are used, and the lake can be open for use during harvesting. Cut plants may have agronomic value as mulch and possibly as a supplement to livestock feed. In lakes with a high plant biomass and low external nutrient income, removal of organic matter and nutrients could have some restorative effect.

Harvesting constitutes habitat removal, and with this will come the removal or elimination of organisms living in this habitat. For example, harvesting may remove young-of-the-year fish, as well as larval insect forms associated with plants. Other negative effects include an increased likelihood of algal blooms and short-term increases in water column turbidity and nutrient concentrations (Cooke et al., 1986). Nicholson (1981) suggested that the replacement of native plant species with nuisance exotic species in Lake Chautauqua, New York, was caused by use of herbicides and harvesting. Fragmentation, dispersal, and rerooting of nuisance plants also may occur. Unless the harvester is operated to remove plant root crowns, plant regrowth can occur within weeks. Efforts are under way to develop new machines, including diver-operated dredges and tilling machines to destroy roots.

Herbicides

Herbicides produce plant control through toxic actions. Most modern chemicals are effective and do not leave long-lived toxic residues or accumulate in food webs. In some cases, herbicide use is the only practical way to manage a plant-choked water body. Costs of harvesting and herbicide treatments are comparable in northern and midwestern areas, but herbicides are usually less costly than harvesting in southern areas. The herbicide fluridone has been shown to be highly effective against major nuisance plants such as hydrilla, and at the recommended dose it exhibits very low toxicity to nontarget organisms such as fish, benthic invertebrates, and birds (Hamelink et al., 1986), but high cost may deter use of this compound.

Herbicides have the potential to produce water quality problems. If dead plants are left in the lake to decompose (a common practice), they consume oxygen and release nutrients. This can be avoided if a pelletized form of the herbicide is used before plant emergence. At least one herbicide (diquat) is toxic to some fish-food organisms. There is evidence that 2,4-dichlorophenoxyacetic acid is associated with the development of non-Hodgkin's lymphoma in applicators, and the photodegradation products of fluridone are embryotoxic (Hoar et al., 1986; Kennedy, 1986). (Nevertheless, fluridone is registered for use in potable water supply reservoirs.) Brooker and Edwards (1975) and Newbold (1975) reviewed the use of herbicides in aquatic ecosystems, and Cooke and Kennedy (1989) and Cooke and Carlson (1989) describe costs, effectiveness, and additional negative effects of aquatic herbicides.

LOW DISSOLVED OXYGEN

Hypolimnetic Aeration

The purpose of hypolimnetic aeration is to increase the dissolved oxygen content of the deep, stagnant, cold layer of a lake or reservoir (the hypolimnion) without destratification. Although several methods exist to do this, including direct injection of liquid oxygen (Prepas et al., 1990), the principal technology involves an airlift system called an aerator. This is a large double-sleeved cylinder, open at the bottom and vented from the closed top via a pipe to the atmosphere. The cylinder is placed in the hypolimnion, compressed air is injected at the bottom of the inner cylinder, and the water is aerated as it rises in the cylinder. Gases such as carbon dioxide and methane are vented to the atmosphere via the pipe, and the aerated water returns down the outer cylinder and to the hypolimnion. The number of these units needed per lake depends on the hypolimnion volume and its oxygen demand (Pastorok et al., 1982).

Aerators are effective when properly sized and installed (McQueen and Lean, 1986). If aerators are operated continuously during the stratified period, a cold-water fishery can be restored, and the quality of raw potable water or deep-water discharge can be improved. Hypolimnetic aerators have not been shown to be effective in algal control, but there is evidence from ongoing work on Vadnais Lake, Minnesota, that an addition of ferric iron to the aerator can reduce internal phosphorus loading to the upper water column (D. Shuler, St. Paul Water Utility, personal communication, 1990).

Hypolimnetic aeration is not appropriate for every thermally stratified lake. In shallower systems, the temperature gradient through the metalimnion may not be steep, and the aerator could slowly destratify the lake and introduce low-oxygen, high-nutrient water to the lake's surface. An algal bloom would be likely. Hypolimnetic aeration is a management and not a restoration procedure. Dissolved oxygen consumption in deep-water and bottom sediments will again make this habitat anoxic if the aerator is shut off.

Artificial Circulation

A management technique described above—artificial circulation— will aerate an entire water column through the mixing energy imparted by a curtain of bubbles rising from a perforated pipe at the lake's bottom. Properly sized to maintain isothermal conditions, a circulator will eliminate low dissolved oxygen and problems associated with it, and *may* control nuisance blue-green algae. However, it will also eliminate the cold-water layer and thereby the possibility of a cold-water fishery or the use of cold water for a potable water supply.

Winter Aeration

Shallow, productive lakes in northern climates may experience oxygen depletion during the winter ice-cover period, especially during winters of high snowfall, which eliminates light penetration through the ice and prevents photosynthesis from continuing. Fish kills commonly result. Some otherwise productive northern lakes lack viable sport fisheries because this problem occurs almost annually. Aerators are installed in many lakes in Minnesota and Wisconsin to avoid or reduce winter fish kills. Again, winter aeration is a management technique and does not solve the underlying cause—organic sediments with high oxygen demand. Wirth (1988) evaluated the effectiveness of winter aeration in 29 lakes and found good or satisfactory performance in 26 cases. The three cases judged marginal or failures involved improper operation of aeration systems or inadequate capacity for the size of the lake. Two problems remain to be solved regarding winter aeration. First, the efficiency of oxygen (O_2) transfer needs to be improved to decrease energy consumption (the major cost involved in winter aeration). Second, aeration causes ice to weaken and open water patches to occur near aerators, a dangerous situation for persons using a lake for snowmobiling or ice fishing. Deaths from drowning have been recorded and are a serious concern on lakes with winter aerators, whose sites must be clearly marked as hazardous to lake users. A bubbleless aerator based on hollow fiber membranes has been proposed to solve these problems (Semmens et al., 1990).

Excess Sediment

Volume loss caused by excessive watershed and shoreline erosion and subsequent high sedimentation rates in a lake or reservoir is a common problem. High loading rates of inorganic sediments also reduce water clarity, possibly to the point of inhibiting primary production, which occurred in Lake Chicot, Arkansas, a riverine lake tributary to the Mississippi River (Stefan et al., 1990). Best management practices and land use controls to decrease soil erosion are the long-term solutions for such problems, but where significant volume losses have occurred, sediment removal is the only practical method of restoring the original volume. Buildup of organic sediments from proliferating macrophyte growths also has caused significant volume losses in many shallow lakes. Hydraulic dredges are the usual means of removing excess sediment from lakes and reservoirs (see Box 4.8). These devices remove a mud-water slurry via a floating suction line and deposit the slurry in a containment area. Normally this area is on land; in some cases the recovered lake sediment can serve as a useful amendment to agricultural soils. In large lakes the dredged material may be deposited in a designated lake area. Dewatering occurs from the shore-based containment area, and the elutriate is often returned to the lake (sometimes after treatment). The solids remain in a properly designed containment area.

BOX 4.8
SPRINGFIELD LAKE, ILLINOIS

Springfield Lake, Illinois, is an example of successful dredging to restore an ensilted reservoir (Buckler et al., 1988). Over its 51-year history, this 1,635-ha potable water supply and recreation impoundment lost more than 13 percent of its storage capacity (9.5×10^6 m^3) because of deposition of agricultural soils. This led to increased water treatment costs, loss of shoreline property values, algal blooms, weeds, turbidity, rough fish, and impaired recreation. A material balance study of silt and nutrients identified the major sources of loading, and land management practices were instituted in cooperation with city, county, state, and federal agencies at a cost of $1.6 million. Hydraulic dredging removed 2×10 m^3 of sediment, which was pumped to adjacent farmland and reclaimed for agricultural uses. The water quality of the return flow from the disposal sites was within standards, and no negative effects on lake quality were noted during dredging operations. Dredging costs were $4.1 million.

Several major problems are encountered in dredging projects. One is an inadequately designed containment area that allows turbid, nutrient-rich water to overflow and return to the lake. Normally, in-lake problems such as turbidity or nutrient release are minimal. A more common problem is that land management and shoreline protection steps are not taken to prevent a second episode of erosion and volume loss. Another possible problem is the occurrence of toxic materials in lake sediments, which then require special and expensive sediment disposal procedures. Finding adequate, inexpensive, and environmentally sound disposal sites is a serious problem even when sediments are not contaminated. A large literature exists on dredging, which is widely used to restore or maintain channel depths in rivers and harbors. Cooke et al. (1986) and Cooke and Kennedy (1989) have reviewed this technique in more detail.

Exotic Species

Extirpation of exotic species is far more difficult to accomplish than is their introduction (Magnuson, 1976). Successes are infrequent and few generalizations can be derived from them. Control or management of the exotic species is usually the only practical alternative. Most attempts even to control exotic species in lakes fail. For success, control measures must be specific for the nuisance species and highly effective. The requisite combination of specificity and effectiveness is rarely found. The exotic sea lamprey (*Petromyzon marinus*) in the Great Lakes has been suppressed (but not eliminated) by the chemical 3-trifluoromethyl-4-nitrophenol (see Lake Michigan case study, Appendix A). In Australia and Papua New Guinea, infestations of the exotic kariba weed (*Salvinia molesta*) have been controlled (but not eliminated) by the herbivorous beetle *Cyrtobagous salviniae* (Barrett, 1989; also see discussion of control of aquatic macrophytes, above).

In small lakes and ponds, exotic or nuisance fish are sometimes removed by applying rotenone to kill all fish and then restocking with the desired species (Magnuson, 1976). The risks that accompany this drastic approach make it controversial among lake users, including anglers. The fish community after such treatment has fewer species than the system can support and thus is highly susceptible to invasion. The most likely invaders are undesirable species that lead to long-term degradation of the fishery (Magnuson, 1976). The result is a perpetual cycle of fish removal and restocking, rather than a restored, self-sustaining community.

In some cases, the invading species declines naturally in population after some years, eventually becoming a subdominant member of the community. Natural declines are known for the macrophytes *Elodea canadensis* in Europe (Hutchinson, 1975) and *Myriophyllum spicatum* in North America (Carpenter, 1980; Painter and McCabe, 1988; Nichols et al., 1991). In the case of *M. spicatum*, management by dredging, drawdown, mechanical harvesting, and herbicides may actually prolong infestation (Smith and Barko, 1990). In general, the long-term community consequences of macrophyte control are poorly known.

In other cases, species invasions accompanied by extirpations of native species have permanently altered lake ecosystems. Depending on the outcome of efforts to establish reproducing populations of lake trout, Lake Michigan may be an example of a permanently altered ecosystem (see Lake Michigan case study, Appendix A).

Lake Victoria, East Africa, provides a spectacular and recent example of an ecosystem transformed by species introduction. Introduction of Nile perch (*Lates nilotica*) and Nile tilapia (*Oreochromis niloticus*), combined with heavy fishing pressure, has depleted native cichlid stocks, and the introduced species now are the mainstay of the lake's fishery (Ogutu-Ohwayo, 1990). Although the fishery is economically successful, local processing techniques (frying and smoking) consume large amounts of wood, which is a scarce resource in the region (Bruton, 1990). The initial high productivity of the fishery was due in part to the high biomass of the native cichlid forage, but after depletion of the forage, growth rates and condition factors of the introduced predators declined (Ogutu-Ohwayo, 1990). The productivity of the fishery is likely to decline. Bruton (1990) is pessimistic about prospects for restoring the native fish community. Ogutu-Ohwayo (1990) noted that restoration of ancestral stocks would deprive an impoverished region of an important local industry and source of protein.

Introductions of exotic species sometimes lead to economic benefits (as occurred initially in Lake Victoria). However, economic benefits may be short-lived because of instabilities in the population density of the invader. In general, economic benefits deriving from the invasion must be balanced against the long-term costs of stabilizing the ecosystem. Lake Michigan (see case study, Appendix A) is an excellent example of an ecosystem with a profitable fishery for exotic species, sustained at the cost of perpetual management. Lakes dominated by exotic species tend to be more variable and less predictable than lakes that lack exotics. Unpredictability adds to the cost of management (Walters, 1986).

Acidification

The long-term solution to lake acidification, of course, is to decrease emissions of sulfur and nitrogen oxides to the atmosphere, because these are precursors of the sulfuric and nitric acids that cause acidic deposition. Control actions in the United States over the past 15 years already have had significant effects; emissions of SO_2 and NO_x peaked in the late 1970s and declined nationally by 10 to 20 percent from 1975 to 1985. Much higher percentage decreases were achieved in some states during the 1980s. The recently enacted Clean Air Act of 1990 (P.L. 101-549) mandates a further reduction in SO_2 emissions of 10 million metric tons over the next 10 years (from current rates of nearly 21 million metric tons). These reductions will decrease the amount of acid deposition sufficiently to reverse the acidification process in some impacted lakes but will not be sufficient to restore all acid-sensitive systems (NAPAP, 1990b).

Liming is by far the most common in-lake restoration technique for acidified lakes, and a large amount of experience with this method has accumulated over the past decade. Liming is common in Scandinavia (e.g., Sverdrup and Bjerle, 1983; Wright, 1985) and is used on a regional management scale in Sweden (Lessmark and Thornelof, 1986), but liming projects have also been undertaken in the United States, especially in the Adirondack region of New York (e.g., Porcella, 1989; Young et al., 1989), as well as in Canada (Molot et al., 1986) and Great Britain (Brown et al., 1988; Dalziel et al., 1988).

Most liming projects actually add calcium carbonate (calcite), the major constituent of limestone, rather than powdered lime (calcium oxide) or slurried lime (calcium hydroxide), directly to the lake by helicopter (in remote areas) or by boat (in lakes with road access). Calcite is preferred because it dissolves more slowly than calcium hydroxide and does not cause such extreme increases in pH. However, finely powdered calcite can produce short-term pH values of 8 to 9 (Fordham and Driscoll, 1989), which may cause stress to aquatic organisms acclimated to living in acidic environments. Dissolution of the calcite adds both calcium ions and carbonate alkalinity to the lake water. Calcite that does not dissolve immediately settles to the lake bottom and slowly dissolves at the sediment-water interface. Treatment of acidic lakes with calcium carbonate is not designed to convert them from soft- (low calcium and alkalinity) to hard-water systems (high calcium and alkalinity, pH > 7). Instead, only enough calcite is added to raise the pH to circumneutrality. Conversion of an acidic lake to a well-buffered hard-water lake would not constitute restoration, because acidic lakes inherently are soft-water systems with low buffering capacity.

Calcite treatments are often short-lived because many acidic lakes are in drainage systems with short water residence times (sometimes only several months, often a year or so). Continued input of acidic water from the drainage basin can reinduce acidic conditions in time periods equivalent to a few water residence times. Liming of watersheds (by applying calcium carbonate to the land portion of the watershed) is more costly, but this approach potentially can provide benefits for much longer periods. Watershed liming has been done in a few cases (Brocksen et al., 1988; Brown et al., 1988), but it is too early to judge the longevity of beneficial effects.

Limnological studies on acidic lakes restored by liming have shown that the aquatic ecosystem responds favorably in short periods of time: acid-sensitive species return, fish condition improves, and the symptoms of acidic systems decrease. Nonetheless, several detailed studies have indicated that recovery is incomplete and that not all of the species present before acidification return in a few years of liming.

An innovative approach to restoring acidic lakes currently under investigation in northern Wisconsin involves pumping ground water into the lake (Garrison et al., 1992). Ground water is typically high in hardness and alkalinity. This is especially true of ground water in limestone aquifers, but even surficial (water table) aquifers in unconsolidated soil (e.g., glacial till) have higher hardness and alkalinity levels than do surface waters in acid-sensitive regions. Advantages of this approach involve low cost (no chemicals need be added to the lake) and ease of repeating the additions to maintain lake pH in the desired range. This method would not be useful in areas where ground water supplies are very limited (e.g., where granitic bedrock is near the surface).

Contaminants

Elimination or reduction of the input of contaminants (synthetic organic compounds and heavy metals) is necessary for remediation of contaminant effects. From a technical standpoint, input reduction is straightforward when point sources predominate. Where nonpoint inputs are substantial (see Lake Michigan case study, Appendix A), input reduction is much more difficult and costly.

Decontamination of lake ecosystems is most straightforward when contaminants are locally concentrated in sediments that can be removed by dredging. In other cases, contaminated sediments can be covered to retard recycling of contaminants to the overlying water. Where aquatic macrophytes concentrate metal contaminants, harvesting of the plants provides a means of biological decontamination of the system (Clark et al., 1981).

In many cases, contaminants are widely dispersed in lakes, and sediment or macrophyte removal is impractical. There is considerable interest in developing bioremediation techniques for dispersed organic pollutants. Bioremediation involves development of natural or mutant microbes that metabolize organic contaminants to nontoxic or less toxic compounds. These microbes can then be introduced to contaminated sites to degrade specific pollutants. Considerable research is under way concerning the use of altered microbes in ecosystems (Tiedje et al., 1989).

Remediation of chemical contaminants in lakes relates directly to management of fish and wildlife. Certain contaminants, such as halogenated hydrocarbons and methylmercury, are bioconcentrated, and they accumulate at increasingly higher concentrations in the tissues of organisms higher in the food chain. Consequently, piscivorous fish, birds, and mammals can develop in their tissues concentrations several orders of magnitude greater than those found in water (Thomann, 1989). Fish species differ widely in their tendency to bioaccumulate contaminants because of differences in diet and growth rate. Therefore, contamination may determine which fish are exploited and the composition of the remaining stock. Fisheries management decisions also can affect the amounts of contaminants in fish at the top of the food chain by manipulating the composition of the fish stock (see Lake Michigan case study, Appendix A). Reproduction of piscivorous birds and mammals was severely affected by dichlorodiphenyltrichloroethane (DDT) in some ecosystems (NRC, 1986). Organochlorine contaminants remain a threat to populations of waterfowl raptorial birds, minks, and otters, and wildlife populations depend on the extent to which contaminants can be remediated.

INTEGRATED AQUATIC SYSTEMS

Lake restorations must be viewed in a watershed context. Abatement of eutrophication, siltation, and contaminant problems is far simpler, and generally more effective, when inputs can be controlled or reduced. This chapter has described many in-lake techniques that can ameliorate symptoms of eutrophication. Reduction of inputs enhances the long-term effectiveness of in-lake approaches.

Lake restoration has strong interactions with restoration of other watershed components. Restoration of influent streams affects the input of sediment, solutes (including nutrients and contaminants), and water to the lake. The surrounding wetlands affect water and solute fluxes and habitats for fish spawning. Conversely, lake restoration affects wetlands by influencing macrophyte distribution, water levels, and wave and ice impacts on littoral areas. Lake restorations and stream restorations interact through the life cycles of migratory fish.

From a technical standpoint, the watershed is the most logical scale at which to undertake restoration. However, institutional constraints, and occasional ecological surprises, can make watershed restoration more difficult than it appears. Institutional complexities are best illustrated by the Lake Michigan case study (Appendix A), in which the major participants include international commissions, two U.S. federal agencies, and water quality managers and fisheries managers from five states. The Lake Apopka case study (Appendix A) illustrates unexpected ecological consequences of watershed change. Draining, diking, and canal building left the lake vulnerable to the effects of a 1947 hurricane that uprooted and drastically reduced aquatic vegetation. Subsequent algal blooms left the water so turbid that macrophytes could not be reestablished. Fishery management contributed to water quality problems via deliberate, massive kills of gizzard shad with rotenone. The dead fish were not removed, and this added more phosphorus to the lake water. Another case of miscommunication between water quality managers and fisheries managers occurred in the Medical Lake, Washington, restoration (see Box 4.1). Fish stocking caused a decline in herbivorous zooplankton, confounding the efforts of water quality managers to control nuisance algal blooms.

One consequence of fragmentation in the management of water quality, fisheries, and wildlife is missed opportunities to restore habitat. Aquatic plant management is an important example. Most macrophyte management is aimed at control or suppression of nuisance growths of exotic species. This is an important step, but habitat restoration requires reestablishment of native species important for fish and wildlife habitat. The biomanipulation concept and bioaccumulation of contaminants such as methylmercury and chlorinated hydrocarbons are further examples of links between management of water quality and fisheries.

In sum, restoration of lakes must extend beyond the shoreline to the watershed boundary. The watershed is the natural scale for many restorations. Restorations at this scale are more likely to be self-sustaining than piecemeal restorations. The major barriers to watershed restorations are institutional and educational. Political boundaries seldom correspond to watershed boundaries, and different agencies have responsibility for different ecological components of the watershed. Distinctly different scientific disciplines apply to different watershed components. Effective restoration requires collaboration among this diversity of scientists, economists, managers, and policymakers.

NEEDS IN LAKE RESTORATION

Needs in Federal Lakes Programs

The significance of lakes and reservoirs to the economy of the United States is apparent. Equally apparent are the deterioration of these resources over recent decades and the inadequacy of federal programs to restore lakes. At present, the collective federal water quality program emphasizes streams, rivers, and wetlands. The most recent report of the Council on Environmental Quality (CEQ, 1989) on environmental trends in the nation does not even mention lakes and reservoirs, except for the Great Lakes. Lakes were also neglected in a report by the National Research Council on the nation's water resources (NRC, 1982). The apparent assumption is that lakes will be protected and will restore themselves if the water quality of streams and rivers is improved. This assumption does not recognize that many lakes and reservoirs have lost significant volume from siltation or that their excessive productivity and cycling of toxic materials through food webs are subsidized by contaminated sediments. Some lakes have had components of their biological communities replaced by nuisance species, many of which are exotics, or have lost important species (with no replacement) because of toxic stresses from heavy metals, synthetic organic compounds, or acidity.

Funds for EPA's Clean Lakes Program have not been included in the president's budget since fiscal year 1981, but Congress has recognized the program's importance and annually restored some funds to it. The 1991 budget for the CLP is $8 million. Although this amount will help to maintain or initiate a few restoration programs, it is minuscule relative to the large task of restoration facing the United States. The annual uncertainty in CLP funding has led some states to postpone the development of full-scale lake programs. Moreover, EPA apparently will not recommend the CLP for continuation in the reauthorization of the Clean Water Act, which Congress began to consider in 1991.

The need for an expanded, well-funded Clean Lakes Program to provide the nation with adequate supplies of safe, protected surface waters in the future is apparent. The next century is very likely to witness increased agriculture, urbanization, and release of toxic substances, all of which will add to the current impaired state of the waters we depend upon for potable water supplies, irrigation, recreation, and industrial uses. Nonetheless, this valuable program is scheduled once again for termination, in part due to a (mistaken) philosophical viewpoint that lake restoration is a problem for state and local governments and not a federal responsibility. As noted repeatedly in this chapter, in many cases lakes do not cleanse or restore themselves. They are sinks for incoming contaminants, which recycle and maintain the impaired conditions.

The status of restoration programs within the variety of federal agencies that have responsibilities to protect and manage the nation's lakes and reservoirs needs to be *enhanced, not diminished.* Moreover, a better understanding must be developed within federal agencies of the importance of lakes for the wide range of uses and benefits described in this chapter: potable water supply, recreation, wildlife habitat, irrigation, water storage, and flood control. A failure to protect, manage, and restore these systems is likely to mean that their usefulness for such purposes will be even more diminished in coming decades. Their continued usefulness as economic resources requires an active and continuous federal program. In addition, the knowledge and experience gained from a U.S. program will be vital to developing countries, where a shortage of clean surface water already hinders economic progress.

By far the most widespread problem affecting lakes and reservoirs is agricultural nonpoint runoff of silt and associated nutrients and pesticides. This problem and its manifestations are within the purview of numerous federal agencies, and coordination of nonpoint source control programs would profit from oversight by an interagency task force or committee.

State lake programs are a key to long-term monitoring and assessment of the nation's lakes, as well as to their restoration, protection, and management. Currently, the CLP provides a 50 percent match to state and local funds for lake restoration. Administrators in the relevant state agencies are the best informed and equipped to determine state needs for lakes. In many cases, the existence of a state program is directly dependent on the continued existence of the federal program, in part because states, as well as various federal agencies, often emphasize stream and river quality and protection in their programs. The states need a continuing federal commitment to lake management and restoration to stimulate and support their efforts.

Lake and reservoir water quality standards are needed for nutrients and related parameters, based on ecoregional attainable lake quality. Criteria for toxic substances are also necessary, which must take into account the trapping and recycling capacities of lake systems. The development and enforcement of standards will help to prevent impairment of lake use. A more complete discussion of this issue was given by Duda et al. (1987).

The quality of drinking water withdrawn from surface impoundments is another interagency issue that would benefit from the development of lake standards and from cooperative activities among federal programs or between federal and state programs. The present emphasis on restoring and managing lakes and reservoirs for their recreational value ignores and may conflict with managing lakes to ensure their roles as water supplies. With our growing population and the increasing popularity of aquatic recreation, multiuse conflicts may require that additional interagency efforts be made to achieve resolution. Appropriate standards for raw potable water may be too stringent to allow multiple uses of lakes in some areas.

Project Selection and Design

Because of limited resources, it will be impossible to undertake all lake restoration projects. Criteria thus are needed to set priorities, select projects, and evaluate project design. A "triage" framework is a minimum initial step. In this approach, systems would be divided into three categories: (1) those that will recover without intervention, (2) those that cannot be restored even with extensive intervention, and (3) those that can be restored with appropriate action. Systems in the third group bear further consideration. Selections from that group should be based on criteria such as the likelihood of success, benefits, costs, and technical review of the restoration plan. It is imperative that project selection be based on these criteria, and not on political ones.

Lake restoration is still a developing science. Every project is an opportunity to learn. It is essential that projects be regarded as large-scale experiments (Matson and Carpenter, 1990). We cannot learn from them unless proper baseline and follow-up data are collected, analyzed, and published in a form accessible to others. These experimental aspects of restoration projects are as important as the other technical components, and they should be designed with the same care. A peer-review system is crucial for maintaining the rigor and quality of restoration ecology.

Need for Integration of Management Programs

Effective lake restoration demands an ecosystem perspective. It often depends on land use in the surrounding watershed and interacts with the management of connecting streams and wetlands. Ironically, agency structures frequently dictate a piecemeal approach to management or restoration. Training is similarly fragmented among specialties such as limnology, water chemistry, fisheries science, and wildlife management. Ecosystem scientists are trained at few institutions. Moreover, there is no single governmental agency responsible for ensuring an integrated, ecosystem approach to lake restoration and management.

Better-coordinated efforts to manage water quality, fisheries, and wildlife are needed at both state and federal levels. Mechanisms for coordination will vary among restoration projects. At a minimum, coordinated planning of restoration projects and regular communication among the agencies involved are essential.

CONCLUSIONS AND RECOMMENDATIONS

Use the Ecoregion Concept to Restore Lakes

Morphometric features and hydrologic factors can vary widely from lake to lake even within a small region, but nonetheless the earth can be characterized as containing ecological regions (or "ecoregions") that have broad similarities of soil, relief, and dominant vegetation. Omernik (1987) divided the conterminous United States into 76 ecoregions, or areas of regional similarity in soils, land use, land surface forms, and potential natural vegetation.

• **The committee believes that goals for restoration of lakes need to be realistic and should be based on the concept of expected conditions for individual ecoregions. Further development of project selection and evaluation techniques based on ecoregion concepts and refinement of ecoregion definitions and descriptions should be encouraged and supported by the U.S. Environmental Protection Agency.**

Research Needed

Lake restoration is a relatively new and developing field. This is especially true for holistic approaches that consider lakes as components of a landscape and treat their restoration at the watershed scale. Although numerous techniques are available to restore lakes and manage the consequences of degradation from certain stresses, many of them require further development to improve their efficiency and effectiveness and to identify situations in which they are best applied. For certain kinds of lake problems (e.g., contaminants, macrophytes), suitable restoration techniques are lacking.

In addition, the current base of knowledge about the nation's lakes is grossly inadequate, depending largely on questionnaires characterized by incomplete and qualitative responses.

Therefore the committee recommends the following:

• **The federal government should support research and development for watershed-scale restorations that integrate lake, stream, and wetland components. State agencies and university researchers should participate in planning, implementing, and evaluating restoration projects. In addition, an interagency program under the Federal Coordinating Council for Science, Engineering, and Technology could be formed to coordinate the selection, planning, and evaluation of these demonstration projects. The research and imple-**

mentation of the projects would be managed by the participating agencies.

• Research and development are needed in several areas of applied limnology, and these programs should take an experimental approach (one that emphasizes manipulation of whole-lake systems or large in-lake enclosures in controlled fashion).

• Improved techniques for littoral zone and aquatic macrophyte management need to be developed. Research should go beyond the removal of nuisance macrophytes to address the restoration of native species that are essential for waterfowl and fish habitat. Basic research is necessary to improve understanding of fundamental limnological processes in littoral zones and the interactions between littoral and pelagic zones of lakes.

• Biomanipulation (food web management) has great potential for low-cost and long-term management of lakes, and research in this emerging field must be stimulated.

• Innovative and low-cost approaches to contaminant cleanup in lakes need to be developed, especially for such widespread problems as contamination by mercury and PCBs.

• The relationships between loadings of stress-causing substances and responses of lakes need to be understood more precisely. This is true even for such well-studied phenomena as phosphorus and algal bloom problems. Research should be undertaken to improve predictions of trophic state from nutrient loading relationships. In particular, phosphorus loading should be evaluated in terms of both its biological availability, which can be estimated chemically, and its effects on plant communities in receiving waters.

• Improved assessment programs are needed to determine the severity and extent of damage in lakes and their change in status over time. Innovative basic research is required to improve the science of assessment and monitoring. There is a great need for cost-effective, reliable indicators of ecosystem function, including those that will reflect long-term change and response to stress. Research on indicators should include traditional community and ecosystem measurements, paleoecological trend assessments, and remote sensing.

• Procedures such as food web manipulation, introduction of phytophagous insects and fish, liming, and reintroduction of native species show promise for effective and long-lasting results when used alone or in combination with other restoration measures. Further research and development should be undertaken on these techniques.

• Paleolimnological approaches should be used to infer the past trophic history of lakes and to decide whether lakes should be restored. Paleolimnological approaches also should be used to infer whether a lake has been restored to its predisturbance condition.

Education and Training

The public needs to be better informed about the rationales, goals, and methods of aquatic ecosystem restoration. In addition, scientists with the broad training needed for aquatic ecosystem restoration are in short supply. The committee recommends the following:

• **Public education and outreach should be components of aquatic ecosystem restorations. Lake associations and citizen monitoring groups have proved helpful in educating the general public, and efforts should be made to ensure that such groups have accurate information about the causes of lake degradation and various lake restoration methods.**

• **Funding is needed for both undergraduate and graduate programs in aquatic ecosystem restoration. Training programs must cross traditional disciplinary boundaries such as those between basic and applied ecology; between water quality management and fisheries or wildlife management; and among lake, stream, and wetlands ecology.**

REFERENCES AND RECOMMENDED READING

Association of State and Interstate Water Pollution Control Administrators (ASIWPCA). 1984. America's Clean Water: The States' Evolution of Progress 1972-1982. ASIWPCA, Washington, D.C.

Association of State and Interstate Water Pollution Control Administrators (ASIWPCA). 1985. America's Clean Water: The States' Nonpoint Source Assessment. ASIWPCA, Washington, D.C.

Baker, L. A., and E. B. Swain. 1989. Review of lake management in Minnesota. Lake Reservoir Manage. 5:1-10.

Barrett, S. C. H. 1989. Waterweed invasions. Sci. Am. (October):90-97.

Barten, J. M. 1987. Stormwater runoff treatment in a wetland filter: Effects on the water quality of Clear Lake. Lake Reservoir Manage. 3:297-305.

Bauman, L. R., and R. A. Soltero. 1978. Limnological investigation of eutrophic Medical Lake, Wash. Northwest Sci. 52:127-136.

Benndorf, J., and K. Putz. 1987. Control of eutrophication of lakes and reservoirs by means of pre-dams—I. Mode of operation and calculation of nutrient elimination capacity. Water Res. 21:829-838.

Benndorf, J., H. Schultz, A. Benndorf, R. Unger, E. Penz, H. Kneschke, K. Kossatz, R. Dumke, U. Hornig, R. Kruspe, and S. Reichel. 1988. Food-web manipulation by enhancement of piscivorous fish stocks: Long-term effects in the hypertrophic Bautzen Reservoir. Limnologica 19:97-110.

Bernhardt, H. 1980. Reservoir protection by in-river nutrient reduction. Pp. 272-277 in Restoration of Lakes and Inland Waters. EPA 440/5-81-010. U.S. Environmental Protection Agency, Washington, D.C.

Bjork, S. 1988. Redevelopment of lake ecosystem—A case-study approach. Ambio 17:90-98.

Brezonik, P. L., S. King, and C. E. Mach. 1988. The influence of water chemistry on metal bioaccumulation and toxicity. Chapter 1 in M. C. Newman and A. W. McIntosh, eds., Metal Ecotoxicology: Concepts and Applications. Lewis Publishers, Chelsea, Mich.

Brezonik, P. L., K. E. Webster, and J. A. Perry. 1991a. Effects of acidification on benthic community processes in Little Rock Lake, Wisconsin. Verh. Int. Ver. Limnol. 24:445-448.

Brezonik, P.L., K. E. Webster, W. A. Swenson, B. Shelley, C. J. Sampson, W. A. Rose, J. A. Perry, J. H. McCormick, T. K. Kratz, P. J. Garrison, T. M. Frost, and J. G. Eaton. 1991b. Responses of Little Rock Lake to experimental acidification: Chemical and biological changes over the pH range 6.1 to 4.7. Can. J. Fish. Aquat. Sci. (in review).

Brocksen, R. W., H. W. Zoettl, D. B. Porcella, R. F. Huettl, K-H. Feger, and J. Wisniewski. 1988. Experimental liming of watersheds: An international cooperative effort between the United States and West Germany. Water, Air, Soil Pollut. 41:455-471.

Brooker, M. P., and R. W. Edwards. 1975. Aquatic herbicides and the control of water weeds. Water Res. 9:1-15.

Brown, D. J. A., G. D. Howells, T. R. K. Dalziel, and B. R. Stewart. 1988. Loch Fleet. A research watershed liming project. Water, Air, Soil Pollut. 41:25-42.

Brown, L. R., and E. C. Wolf. 1984. Soil Erosion: Quiet Crisis in the World Economy. Worldwatch Paper 60. Worldwatch Institute, Washington, D.C.

Brown, R. M., N. I. McClelland, R. A. Deininger, and M. F. O'Connor. 1972. A water quality index—Crashing the psychological barrier. Paper No. 29. Proceedings of the Sixth International Conference on Water Pollution Research, Jerusalem. Pergamon, Oxford.

Bruton, M. N. 1990. The conservation of the fishes of Lake Victoria, Africa: An ecological perspective. Environ. Biol. Fishes 27:161-176.

Buckler, J. H., T. M. Skelly, M. J. Luepke, and G. A. Wilken. 1988. Case study: The Lake Springfield sediment removal project. Lake Reservoir Manage. 4:143-152.

Camanzo, J., C. P. Rice, D. J. Jude, and R. Rossmann. 1987. Organic priority pollutants in nearshore fish from 14 Lake Michigan tributaries and embayments, 1983. J. Great Lakes Res. 13:296-309.

Campbell, P. C. G., and P. Stokes. 1985. Acidification and toxicity of metals to aquatic biota. Can. J. Fish. Aquat. Sci. 42:2034-2049.

Canfield, D. E., and R. W. Bachmann. 1981. Prediction of total phosphorus concentrations, chlorophyll *a*, and Secchi depths in natural and artificial lakes. Can. J. Fish. Aquat. Sci. 38:414-423.

Canfield, D. E., Jr., M. J. Maceina, and J. V. Shireman. 1983. Effect of hydrilla and grass carp on water quality in a Florida lake. Water Resour. Bull. 19:773-778.

Carlson, R. E. 1977. A trophic state index for lakes. Limnol. Oceanogr. 22:-361-369.

Carpenter, S. R. 1980. The decline of *Myriophyllum spicatum* in a hardwater eutrophic lake. Can. J. Bot. 58:527-535.

Carpenter, S. R., ed. 1988. Complex Interactions in Lake Communities. Springer Verlag, New York.

Carpenter, S. R., and J. F. Kitchell. 1988. Consumer control of lake productivity. BioScience 38:764-769.

Carpenter, S. R., J. F. Kitchell, J. R. Hodgson, P. A. Cochran, J. J. Elser, M. M. Elser,D. M. Lodge, D. Kretchmer, X. He, and C. N. von Ende. 1987. Regulation of lake primary productivity by food web structure. Ecology 68: 1863-1876.

Center, T. D., A. F. Confrancesco, and J. K. Balciunas. 1988. Biological control of aquatic and wetland weeds in the southeastern United States. Proceedings of the Seventh International Symposium on Biological Control Weeds, Rome, 239-262.

Charudattan, R. 1986. Integrated control of waterhyacinth *(Eichhornia crassipes)* with a pathogen, insects, and herbicides. Weed Sci. 34 (Suppl. 1):26-30.

Clark, J. R., J. H. VanHassel, R. B. Nicholson, D. S. Cherry, and J. Cairns, Jr. 1981. Accumulation and depuration of metals by duckweed *(Lemna perpusilla)*. Ecotoxicol. Environ. Saf. 5:87-96.

Clasen, J. 1989. Wahnbach Reservoir—Long-term experience with phosphorus removal. Pp. 130-132 in W. Lampert and K.-O. Rothhaupt, eds., Limnology in the Federal Republic of Germany. International Association of Theoretical Applications in Limnology, Plon, Germany.

Clean Air Act Amendments. P.L. 101-549, Nov. 15, 1990.

Cooke, G. D., and R. E. Carlson. 1989. Reservoir Management for Water Quality and THM Precursor Control. American Water Works Association Research Foundation, Denver, Colo.

Cooke, G. D., and R. H. Kennedy. 1989. Water Quality Management for Reservoirs and Tailwaters. Rept. 1. In Reservoir Water Quality Management Techniques. Tech. Rept. E-89-1. U.S. Army Corps of Engineers, Vicksburg, Miss.

Cooke, G. D., and A. B. Martin. 1989. Long-term evaluation of the effectiveness of phosphorus inactivation (abstract). Annual Meeting of the North American Lake Management Society, Austin, Tex.

Cooke, G. D., R. T. Heath, R. H. Kennedy, and M. R. McComas. 1982. Change in lake trophic state and internal phosphorus release after aluminum sulfate application. Water Resour. Bull. 18:699-705.

Cooke, G. D., E. B. Welch, S. A. Peterson, and P. R. Newroth. 1986. Lake and Reservoir Restoration. Butterworth, Stoneham, Mass.

Council on Environmental Quality (CEQ). 1989. Environmental trends. Chapter 2 in Water. Washington, D.C.

Czuczwa, J. M., B. D. McVeety, and R. A. Hites. 1984. Polychlorinated dibenzo-p-dioxins and dibenzofurans in sediments from Siskiwit Lake, Isle Royale. Science 226:568-569.

Czuczwa, J. M., B. D. McVeety, and R. A. Hites. 1985. Polychlorinated dibenzodioxins and dibenzofurans in sediments from Siskiwit Lake, Isle Royale. Chemosphere 14:62-626.

Dalziel, T. R. K., M. V. Proctor, and A. Dickson. 1988. Hydrochemical budget calculations for parts of Loch Fleet catchment, before and after watershed liming. Water, Air, Soil Pollut. 41:417-434.

Dierberg, F. E., V. P. Williams, and W. H. Schneider. 1988a. Water Quality Effects of Lake Enhancement Techniques Used in Florida. Water Resources Research Center, University of Florida, Gainesville. 64 pp. plus app.

Dierberg, F. E., V. P. Williams, and W. H. Schneider. 1988b. Evaluating the water quality effects of lake management in Florida. Lake Reservoir Manage. 2:101-111.

Duda, A. M., and F. J. Johnson. 1984. Lakes are losing the battle in clean water programs. J. Water Pollut. Control Fed. 56:815-822.

Duda, A. M., M. L. Iwanski, R. J. Johnson, and F. A. Joksch. 1987. Numerical standards for managing lake and reservoir water quality. Lake Reservoir Manag. 3:1-16.

Edmondson, W. T. 1979. Lake Washington and predictability of limnological events. Arch. Hydrobiol. 13:234-241.

Edmondson, W. T. 1991. The Uses of Ecology: Lake Washington and Beyond. University of Washington Press, Seattle, Wash. 312 pp.

Eisenreich, S. J. 1987. The chemical limnology of nonpolar organic contaminants: Polychlorinated biphenyls in Lake Superior. Pp. 393-470 in S. J. Eisenreich and R. A. Hites, eds., Sources and Fate of Aquatic Pollutants. Adv. Chem. Ser. 216. American Chemical Society, Washington, D.C.

Elton, C. S. 1958. The Ecology of Invasions by Animals and Plants. Methuen, London.

Federal Water Pollution Control Act Amendments of 1972. P.L. 92-500, Section 314, Oct. 18, 1972, 86 Stat. 816.

Fiala, I., and P. Vasata. 1982. Phosphorus reduction in a man-made lake by means of a small reservoir on the inflow. Arch. Hydrobiol. 94:24-37.

Fordham, G. F., and C. T. Driscoll. 1989. Short-term changes in the acid/base chemistry of two acidic lakes following calcium carbonate treatment. Can. J. Fish. Aquat. Sci. 46:306-314.

Forsberg, C. 1987. Evaluation of lake restoration in Sweden. Schweiz. Z. Hydrol. 49:260-274.

Fowler, M. C., and R. O. Robson. 1978. The effects of the food preferences and stocking rates of grass carp (*Ctenopharyngodon idella* Val.) on mixed plant communities. Aquat. Bot. 5:261-276.

Fulmer, D. G., and G. D. Cooke. 1990. Evaluating the restoration potential of Ohio reservoirs. Lake Reservoir Manage. 6:197-206.

Gakstatter, J. H., A. F. Bartsch, and C. A. Callahan. 1978. The impact of broadly applied effluent phosphorus standards on eutrophication control. Water Resour. Res. 14:1155-1158.

Garrison, P. J., and D. R. Knauer. 1984. Long-term evaluation of three alum treated lakes. Pp. 513-517 in Lake and Reservoir Management. EPA 440/5-84-001. U.S. Environmental Protection Agency, Washington, D.C.

Garrison, P. J., W. J. Rose, C. J. Watras, and J. P. Hurley. 1992. Mitigation of acid rain by ground water addition: An alternative to liming. In V. D. Adams and E. Morgan, eds., Acid Rain Mitigation. Lewis Publishers, Chelsea, Mich. (in review).

Goldman, C. R. 1988. Primary productivity, nutrients and transparency during the early onset of eutrophication in ultra-oligotrophic Lake Tahoe, California-Nevada. Limnol. Oceanogr. 33:1321-1333.

Gulati, R. D., E. H. R. R. Lammens, M.-L. Meijer, and E. van Donk. 1990. Biomanipulation: Tool for Water Management. Kluwer, Boston.

Hamelink, J. L., D. R. Buckler, F. L. Mayer, D. V. Palawski, and H. O. Sanders. 1986. Toxicity of fluridone to aquatic invertebrates and fish. Environ. Toxicol. Chem. 5:87-94.

Harkins, R. D. 1974. An objective water quality index. J. Water Pollut. Control. Fed. 46:588-591.

Hartig, J. H., and R. L. Thomas. 1988. Development of plans to restore degraded areas in the Great Lakes. Environ. Manage. 12:327-347.

Hasler, A. D. 1947. Eutrophication of lakes by domestic drainage. Ecology 28: 383-395.

Hawkes, C. L., D. L. Miller, and W. G. Layther. 1986. Fish ecoregions of Kansas: Stream fish assemblage patterns and associated environmental correlates. Environ. Biol. Fishes 17: 267-279.

Heiskary, S. A., C. B. Wilson, and D. P. Larsen. 1987. Analysis of regional patterns in lake water quality: Using ecoregions for lake management in Minnesota. Lake Reservoir Manage. 3:337-344.

Henning, T. 1989. Historical and areal deposition of mercury in NE Minnesota and Northern Wisconsin Lakes. M.S. thesis. University of Minnesota, Minneapolis, Minn.

Henrikson, L., H. G. Nyman, H. G. Oscarson, and J. A. E. Stenson. 1980. Trophic changes without changes in external nutrient loading. Hydrobiologia 68:257-263.

Hoar, S. K., A. Blair, F. F. Holmes, C. D. Boysen, R. J. Robel, R. Hoover, and J. F. Traument. 1986. Agricultural herbicide use and risk of lymphoma and soft-tissue sarcoma. J. Am. Med. Assoc. 256:1141-1147.

Hrbacek, J. M., M. Dvorakova, V. Korinek, and L. Prochazkova. 1961. Demonstration of the effect of the fish stock on the species composition of zooplankton and the intensity of metabolism of the whole plankton assemblage. Verh. Int. Ver. Limnol. 14:192-195.

Hsu, P. H. 1975. Precipitation of phosphate from solution using aluminum salts. Water Res. 9:1155-1161.

Hughes, R. M., and D. P. Larsen. 1988. Ecoregions: An approach to surface water protection. J. Water Pollut. Contr. Fed. 60:486-493.

Hughes, R. M., D. P. Larsen, and J. M. Omernik. 1986. Regional reference sites: A method for assessing stream potentials. Environ. Manage. 10:629-635.

Hutchinson, G. E. 1975. A Treatise on Limnology. Vol. III. Aquatic Botany. John Wiley & Sons, New York.

Janus, L. L., and R. A. Vollenweider. 1981. The OECD Cooperative Programme on Eutrophication. Canadian Contribution Summary Report. Sci. Ser. No. 131. Canada Centre for Inland Waters, Burlington, Ont.

Johnson, B. M., R. S. Stewart, and S. J. Gilbert. 1992. Fisheries Biomanipulation in the Madison Lakes. Wisconsin Department of Natural Resources, Madison, Wis.

Kennedy, G. L., Jr. 1986. Biological effects of acetamide, formamide and the monomethyl and dimethyl derivatives. CRC Crit. Rev. Toxicol. 17:129-182.

Kennedy, R. A., and R. F. Gaugush. 1988. Assessment of water quality in Corps of Engineers reservoirs. Lake Reservoir Manage. 4:253-260.

Kennedy, R. H., J. W. Barko, W. F. James, W. D. Taylor, and G. L. Godshalk. 1987. Aluminum sulfate treatment of a reservoir: Rationale, application methods, and preliminary results. Lake Reservoir Manage. 3:85-90.

Ketelle, M. J., and P. D. Uttormark. 1971. Problem Lakes in the United States. U.S. Environmental Protection Agency. Project EHR 16010. Washington, D.C.

Kimbel, J. C., and S. C. Carpenter. 1981. Effects of mechanical harvesting on *Myriophyllum spicatum* L. regrowth and carbohydrate allocation to roots and shoots. Aquat. Bot. 11:121-127.

Kitchell, J. F., ed. 1991. Food Web Management: A Case Study of Lake Mendota, Wisconsin. Springer-Verlag, New York.

Knapp, S. M., and R. A. Soltero. 1983. Trout-zooplankton relationships in Medical Lake, Washington, following restoration by aluminum sulfate treatment. J. Freshwater Ecol. 2:1-11.

Kratzer, C. R., and P. L. Brezonik. 1981. A Carlson-type trophic state index for nitrogen in Florida lakes. Water Resour. Bull. 17:713-715.

Larsen, D. P., K. W. Malueg, D. W. Schults, and R. M. Brice. 1975. Response of eutrophic Shagawa Lake, Minnesota, U.S.A., to point source phosphorus reduction. Verh. Int. Ver. Limnol. 19:884-892.

Larsen, D. P., J. Van Sickel, K. W. Malueg, and D. P. Smith. 1979. The effect of wastewater phosphorus removal on Shagawa Lake, Minnesota: Phosphorus supplies, lake phosphorus and chlorophyll *a*. Water Res. 13:1259-1272.

Larsen, D. P., D. W. Schults, and K. W. Malueg. 1981. Summer internal phosphorus supplies in Shagawa Lake, Minnesota. Limnol. Oceangr. 26:740-753.

Larsen, D. P., D. R. Dudley, and R. M. Hughes. 1988. A regional approach for assessing attainable surface water quality: An Ohio case study. J. Soil Water Conserv. 43:171-176.

Lathrop, R. C. 1979. Dane County water quality plan. Dane County Regional Planning Commission, Madison, Wis.

Lehman, J. T. 1986. Control of eutrophication in Lake Washington. Pp. 301-316 in Ecological Knowledge and Environmental Problem Solving. National Academy Press, Washington, D.C.

Leslie, A. J., Jr., J. M. Van Dyke, R. S. Hestand III, and B. Z. Thompson. 1987. Management of aquatic plants in multi-use lakes with grass carp (*Ctenopharyngodon idella*). Lake Reservoir Manage. 3:266-276.

Lessmark, O. and E. Thornelof. 1986. Liming in Sweden. Water, Air, Soil Pollut. 31:809-815.

Lodge, D. M., J. J. Magnuson, and A. M. Beckel. 1985. Lake-bottom tyrant. Nat. Hist. 94:32-37.

Mackie, G. L., W. N. Gibbons, B. W. Muncaster, and I. M. Gray. 1989. The Zebra Mussel, *Dreissena polymorpha*: A Synthesis of European Experiences and a Preview for North America. Environment Ontario, Toronto, Canada.

Magnuson, J. J. 1976. Managing with exotics: A game of chance. Trans. Am. Fish. Soc. 105:1-9.

Maki, A. W., D. B. Porcella, and R. H. Wendt. 1984. The impact of detergent phosphorus bans on receiving water quality. Water Res. 18:893-903.

Marais, G. V. R., R. E. Loewenthal, and I. P. Siebritz. 1983. Observations supporting phosphate removal by biological excess uptake—A review. Water Sci. Technol. 15:15-41.

Martin, E. H. 1988. Effectiveness of an urban runoff detention pond-wetlands system. Am. Soc. Civ. Eng, J. Environ. Eng. Div. 114:810-827.

Matson, P. A., and S. R. Carpenter, eds. 1990. Special feature on analysis of response to large-scale perturbations. Ecology 71:2037-2068.

McIntyre, S. C., and J. W. Naney. 1990. Reelfoot lake sedimentation rates and sources. Water Resour. Bull. 26:227-232.

McKnight, D. M., S. W. Chisholm, and D. F. Harlemann. 1983. $CuSO_4$ treatment of nuisance algal blooms in drinking water reservoirs. Environ. Manage. 7:311-320.

McQueen, D. J., and D. R. S. Lean. 1986. Hypolimnetic aeration: An overview. Water Pollut. Res. Can. 21:205-217.

Mires, J. W., R. A. Soltero, and G. R. Keizur. 1981. Changes in the zooplankton community of Medical Lake, Washington, subsequent to its restoration by a whole-lake alum treatment and the establishment of a trout fishery. J. Freshwater Ecol. 1:167-178.

Molot, L. A., J. G. Hamilton, and G. M. Booth. 1986. Neutralization of acidic lakes: Short-term dissolution of dried and slurried calcite. Water Res. 20:757-766.

Mooney, H. A., and J. A. Drake, eds. 1986. Ecology of Biological Invasions of North America and Hawaii. Springer-Verlag, New York.

National Acid Precipitation Assessment Program (NAPAP). 1990a. Current Status of Surface Water Acid-Base Chemistry. State of Science and Technology Rept. 9. NAPAP Interagency Program, Washington, D.C.

National Acid Precipitation Assessment Program (NAPAP). 1990b. Integrated Assessment: Questions 1 and 2. NAPAP Interagency Program, Washington, D.C.

National Research Council (NRC). 1982. Water resources. Pp. 255-286 in Outlook for Science and Technology: The Next Five Years. W. H. Freeman, San Francisco.

National Research Council (NRC). 1986. Ecological Knowledge and Environmental Problem Solving. National Academy Press, Washington, D.C.

National Research Council (NRC). 1987. The Mono Basin Ecosystem: Effects of Changing Lake Level. National Academy Press, Washington, D.C.

Newbold, C. 1975. Herbicides in aquatic ecosystems. Biol. Conserv. 7:97-118.

Nichols, S. A., R. C. Lathrop, and S. R. Carpenter. 1991. Macrophyte community dynamics—A vegetation history. In J. F. Kitchell, ed., Food Web Management: A Case History of Lake Mendota, Wisconsin. Springer-Verlag, New York.

Nicholson, S. A. 1981. Changes in submersed macrophytes in Chautauqua Lake, 1937-1975. Freshwater Biol. 11:523-530.

Noonan, T. 1986. Water quality in Long Lake, Minnesota, following Riplox sediment treatment. Lake and Reservoir Manage. 2:131-137, Proceedings Fifth Annual Conference, Symposium North American Lake Management Society, Nov. 13-16, 1985, Lake Geneva, Wis.

Nurnberg, G. K. 1987. Hypolimnetic withdrawal as a lake restoration technique. Am. Soc. Civ. Eng., J. Environ. Eng. Div. 113:1006-1017.

Ogutu-Ohwayo, R. 1990. The decline of the native fishes of lakes Victoria and Kyoga (East Africa) and the impact of introduced species, especially the Nile perch, *Lates niloticus,* and the Nile tilapia, *Oreochromis niloticus.* Environ. Biol. Fishes 27:81-96.

Omernik, J. M. 1987. Aquatic ecoregions of the conterminous United States. Ann. Assoc. Am. Geogr. 77:118-125.

Omernik, J. M., D. P. Larsen, C. M. Rohm, and S. E. Clarke. 1988. Summer total phosphorus in lakes: A map of Minnesota, Wisconsin, and Michigan, USA. Environ. Manage. 12:815-825.

Osborne, J. A. 1982. The potential of the hybrid grass carp as a weed control agent. J. Freshwater. Ecol. 1:353-360.

Painter, D. S., and K. J. McCabe. 1988. Investigation into the disappearance of Eurasian watermilfoil from the Kawartha lakes. J. Aquat. Plant Manage. 26:3-12.

Pastorok, R. A., M. W. Lorenzen, and T. C. Ginn. 1982. Environmental Aspects of Artificial Aeration and Oxygenation of Reservoirs: A Review of Theory, Techniques, and Experiences. Tech. Rept. E-82-3. U.S. Army Corps of Engineers, Vicksburg, Miss.

Perry, J. A., N. H. Troelstrup, Jr., M. Newsom, and B. Shelley. 1987. Results of a recent whole ecosystem manipulation: The search for generality. Water Sci. Technol. 19:55-72.

Porcella, D. B. 1989. Lake acidification mitigation project (LAMP): An overview of an ecosystem perturbation experiment. Can. J. Fish. Aquat. Sci. 46:246-248.

Postel, S. 1985. Managing freshwater supplies. Pp. 42-72 in State of the World 1985. Worldwatch Institute Report, Washington, D.C.

Prepas, E. E., D. J. Webb, C. L. K. Robinson, and T. P. Murphy. 1990. Impact of liquid oxygen injection on a deep, naturally eutrophic lake: Amisk Lake, Alberta, year one. Ver. Int. Ver. Limnol. 24:320.

Reddy, K. R., and T. A. DeBush. 1987. State-of-the-Art Utilization of Aquatic Plants. In Water Pollut. Control, Water Sci. and Tech. 19:61-79.

Reinertsen, H., A. Jensen, J. I. Koksvik, A. Langeland, and Y. Olsen. 1990. Effects of fish removal on the limnetic ecosystem of a eutrophic lake. Can. J. Fish. Aquat. Sci. 47:166-173.

Richardson, C. J. 1988. Freshwater wetlands: Transformers, filters, or sinks? Forem (Duke University, Durham, N.C.)11:3-9.

Ripl, W., and G. Lindmark. 1978. Ecosystem control by nitrogen metabolism in sediment. Vatten 2:135-144.

Robertson, A., and D. Scavia. 1984. North American Great Lakes. Pp. 135-176 in F. B. Tait, ed., Lakes and Reservoirs. Elsevier, The Netherlands.

Rudd, J. W. M., C. A. Kelly, D. W. Schindler, and M. A. Turner. 1988. Disruption of the nitrogen cycle in acidified lakes. Science 240:1515-1517.

Sawyer, C. N. 1947. Fertilization of lakes by agricultural and urban drainage. J. N. Engl. Water Works. Assoc. 61:109-127.

Schindler, D. W. 1988. Effects of acid rain on freshwater ecosystems. Science 239:149-157.

Scholz, A. T., R. A. Soltero, K. O. McKee, E. Anderson, and J. K. Vehara. 1985. Biomanipulation of a trout fishery and its effect on zooplankton composition, phytoplankton biovolume, and water quality of Medical Lake, Spokane County, Washington, following restoration by treatment with alum. Lake Reservoir Manage. 1:48-56.

Semmens, M. J., T. Ahmed, and M. Voss. 1990. An evaluation of bubbleless membrane aeration for lake aeration. Presented at Second International Conference on Gas Transfer at Water Surfaces, Minneapolis, Minn., Sept. 9-14, 1990. American Society of Civil Engineers and U.S. Corps of Engineers.

Shannon, E. E., and P. L. Brezonik. 1972. Eutrophication analysis: A multivariate approach. Am. Soc. Civ. Eng., J. Sanit. Eng. Div. 98:37-57.

Shapiro, J. 1973. Blue-green algae: Why they become dominant. Science 179:382-384.

Shapiro, J. 1984. Blue-green dominance in lakes: The role and management significance of pH and CO_2. Int. Rev. [Ges.] Hydrobiol. 69:765-780.

Shapiro, J. 1990a. Biomanipulation: The next phase—Making it stable. Hydrobiologia 200/201:13-27.

Shapiro, J. 1990b. Current beliefs regarding dominance by blue-greens: The case for the importance of CO_2 and pH. Verh. Int. Ver. Limnol. 24.

Shapiro, J., and D. I. Wright. 1984. Lake restoration by biomanipulation: Round Lake, Minnesota, the first two years. Freshwater Biol. 14:371-383.

Shapiro, J., G. V. Levin, and H. Zea. 1967. Anoxically induced release of phosphate in wastewater treatment. J. Water Pollut. Control Fed. 39:1811-1818.

Shapiro, J., V. Lamarra, and M. Lynch. 1975. Biomanipulation: An ecosystem approach to lake restoration. Pp. 85-96 in P. L. Brezonik and J. L. Fox, eds., Proceedings of a Symposium on Water Quality Management Through Biological Control, University of Florida, Gainesville.

Shireman, J. V., and M. J. Maceina. 1981. The utilization of grass carp, *Ctenophyaryngodon idella* Val., for hydrilla control in Lake Baldwin, Florida. J. Fish. Biol. 19:629-636.

Shireman, J. V., R. W. Rottmann, and F. J. Aldridge. 1983. Consumption and growth of hybrid grass carp fed four vegetation diets and trout chow in circular tanks. J. Fish. Biol. 22:685-693.

Smith, C. S., and J. W. Barko. 1990. Ecology of Eurasian watermilfoil. J. Aquat. Plant Manage. 28:55-64.

Soltero, R. A., D. G. Nichols, A. F. Gasperino, and M. A. Beckwith. 1981. Lake restoration: Medical Lake, Washington. J. Freshwater Ecol. 2:155-165.

Sonzogni, W. C., G. P. Fitzgerald, and G. F. Lee. 1975. Effects of wastewater diversion on the lower Madison lakes. J. Water Pollut. Control Fed. 47: 535-542.

Stanley, J. G., W. W. Miley, and D. L. Sutton. 1978. Reproduction requirements and likelihood for naturalization of escaped grass carp in the United States. Trans. Am. Fish. Soc. 107:119-128.

Stefan, H. G., S. Dhamothoran, F. R. Schiebe, A. Y. Fu, and J. J. Cardoni. 1990. Dynamic simulation of turbidity and its correction in Lake Chicot, Arkansas. Pp. 193-250 in B. Henderson-Sellers, ed., Water Quality Modeling. Vol. 4. Decision Support Techniques for Lakes and Reservoirs. CRC Press, Boca Raton, Fla.

Sverdrup, H., and I. Bjerle. 1983. The calcite utilization efficiency and the long term effect in several Swedish lakes liming projects. Vatten 39:41-54.

Swain, E. B., and D. D. Helwig. 1989. Mercury in fish from northeastern Minnesota lakes: Historical trends, environmental correlates, and potential sources. J. Minn. Acad. Sci. 55:103-109.

Swain, W. R. 1978. Chlorinated organic residues in fish, water, and precipitation from the vicinity of Isle Royale, Lake Superior. J. Great Lakes Res. 4:398-407.

Thomann, R. V. 1989. Bioaccumulation model of organic chemical distribution in aquatic food chains. Environ. Sci. Technol. 23:699-707.

Tiedje, J. M., R. K. Colwell, Y. L. Grossman, R. E. Hodson, R. E. Lenski, R. N. Mack, and P. J. Regal. 1989. The planned introduction of genetically engineered organisms: Ecological considerations and recommendations. Ecology 70:298-315.

U.S. Environmental Protection Agency (EPA). 1980. Economic Benefits of the Clean Lakes Program. EPA-440-5-80-081. U.S. EPA, Washington, D.C.

U.S. Environmental Protection Agency (EPA). 1985. Clean Lakes Program: A Review of the First Decade. EPA 440/5-85-033. U.S. EPA, Washington, D.C.

U.S. Environmental Protection Agency (EPA). 1989. Report to Congress. Water Quality of the Nation's Lakes. EPA 440/5-89-003. U.S. EPA, Washington, D.C.

U.S. Environmental Protection Agency (EPA). 1990a. The Lake and Reservoir Restoration Guidance Manual. 2nd ed. EPA 440/4-90-006. U.S. EPA, Washington, D.C.

U.S. Environmental Protection Agency (EPA). 1990b. National Water Quality Inventory. 1988 Report to Congress. EPA 440/4-90-003. U.S. EPA, Washington, D.C.

Uttormark, P. D., and J. P. Wall. 1975. Lake classification—A trophic characterization of Wisconsin lakes. EPA-660/3-75-033. U.S. Environmental Protection Agency, Corvallis, Ore, 165 pp.

Vollenweider, R. A., and J. Kerekes. 1981. OECD eutrophication programme. Synthesis Report. Organization for Economic Cooperation and Development, Paris.

Walker, W. W., Jr. 1987. Phosphorus removal by urban runoff detention basins. Lake Reservoir Manage. 3:314-326.

Walski, T. M., and F. L. Parker. 1974. Consumers water quality index. Am. Soc. Civ. Eng., J. Environ. Eng. Div. 100:593-611.

Walters, C. W. 1986. Adaptive Management of Renewable Resources. MacMillan, New York.

Water Quality Act of 1987. P.L. 100-4.

Welch, E. B. 1981. The dilution/flushing technique in lake restoration. Water Resour. Bull. 17:558-564.

Welch, E. B., and E. R. Weiher. 1987. Improvement in Moses Lake quality from dilution and sewage diversion. Lake Reservoir Manage. 3:58-65.

Welch, E. B., D. E. Spyridakis, J. I. Shuster, and R. R. Horner. 1986. Declining lake sediment phosphorus release and oxygen deficit following wastewater diversion. J. Water Pollut. Control Fed. 58:92-96.

Welch, E. B., C. L. DeGasperi, D. E. Spyridakis, and T. J. Belnick. 1988. Internal phosphorus loading and alum effectiveness in shallow lakes. Lake Reservoir Manage. 4:27-33.

Wiley, M. J., and L. D. Wike. 1986. Energy balances of diploid, triploid, and hybrid grass carp. Trans. Am. Fish. Soc. 115:853-863.

Wilson, C. B., and W. W. Walker, Jr. 1989. Development of lake assessment methods based upon the aquatic ecoregion concept. Lake Reservoir Manage. 5:11-22.

Wirth, T. 1988. Lake Aeration in Wisconsin Lakes. Lake Management Program PUBL-WR-196. Wisconsin Department of Natural Resources, Madison. 76 pp.

Wright, R. F. 1985. Liming and reacidification of Hovvatan, a chronically acidified lake in southernmost Norway. Can. J. Fish. Aquat. Sci. 42:1103-1113.

Young, S. N., W. T. Clough, A. J. Thomas, and R. Siddall. 1988. Changes in plant community at Foxcote Reservoir following use of ferric sulphate to control nutrient levels. J. Inst. Water Environ. Manage. 2:5-12.

Young, T. C., J. V. DePinto, J. R. Rhea, and R. D. Scheffe. 1989. Calcite dose selection, treatment efficiency, and residual calcite fate after whole-lake neutralization. Can. J. Fish. Aquat. Sci. 46:315-322.

5

Rivers and Streams

Human activity has profoundly affected rivers and streams in all parts of the world, to such an extent that it is now extremely difficult to find any stream which has not been in some way altered, and probably quite impossible to find any such river. The effects range from pollution to changes in the pattern of flow, and they have become increasingly marked during the past two or three centuries.

H. B. N. Hynes, 1970

There is a phenomenal resiliency in the mechanisms of the earth. A river or lake is almost never dead. If you give it the slightest chance by stopping pollutants from going into it, then nature usually comes back.

René Dubos, 1981

OVERVIEW

Rivers and streams have many of the same economic, recreational, and environmental values and uses as lakes. However, the stresses associated with human use may have begun earlier on rivers because of their importance as transportation routes when roads were few and as sources of power when the Industrial Revolution was in its infancy in the United States. Unfortunately, rivers also served as convenient and inexpensive means of waste disposal because the flow carried away industrial and human waste. During early settlement days in the United States, human communities and factories were widely spaced, and waste discharges relatively minor and nonpersistent, especially when compared to those of today's industrial society. As a consequence of the spacing, volume, and degradability of early wastes, rivers were able to cleanse themselves through natural processes before the water reached the next downstream user. As settlements expanded in size and became more closely spaced, the wastes began to contain a larger percentage of persistent toxicants, the ecological damage became more severe, and the possibility of self-cleansing was more limited. At the same time, agricultural, mining, and timber harvesting activities accelerated, resulting in widespread alteration of watersheds, floodplains, and riparian zones that in turn altered water and sediment regimes in rivers and streams, adversely affecting plant and animal communities. Flow regimes and dilution capacity were reduced or altered by dams, irrigation, and interbasin transfer of water. The cumulative impact of all these changes was frequently missed because of the incremental nature of the changes. Even when their effects became impossible to ignore, the automobile made it easier for a more mobile population to escape to pristine aquatic sites with aesthetic and recreational appeal than to set about repairing those sites damaged by anthropogenic activities.

The changes that have stressed flowing water systems have impaired their value for both human use and environmental services. Stresses arise from (1) water quantity or flow mistiming, (2) morphological modifications of the channel and riparian zone, (3) excessive erosion and sedimentation, (4) deterioration of substrate quality, (5) deterioration of water quality, (6) decline of native species, and (7) introduction of alien species. The locus of the problem can be in the watershed, along the riparian or floodplain zone, or in the channels and pools.

The most extreme form of stress, common in the arid West, is the complete appropriation of water flowing on the surface, either by direct withdrawal or by pumping from the riparian zone (see Box 5.1). Only slightly less extreme is the conversion of reaches of free-flowing rivers to a series of lakelike impoundments (e.g., the Willamette River; see Box 5.2 and Appendix A). In these cases, the free-flowing river no longer exists, and restoration of some semblance of the natural system would require drastic measures such as reduction of water withdrawals or removal of dams. In some cases (the Willamette and Columbia rivers), a few species of migratory sport fish (salmon) are maintained on dammed rivers by using hatcheries and fish ladders, but this is aquaculture, not restoration.

BOX 5.1
THE SANTA CRUZ RIVER, SOUTHERN ARIZONA

The Santa Cruz River is a typical example of many rivers and streams in the valleys of the western United States that have experienced pronounced ecological changes during the past century. It is not an example of a restoration activity, but rather an illustration of how human activities and rapid urbanization of the floodplain can bring about irreversible changes to a stream system.

The Santa Cruz River is a dry, and usually insignificant, stream throughout most of its length. It rises in oak woodlands and grasslands southeast of Tucson. The headwaters of the Santa Cruz are gathered into a shallow, perennial channel that courses southward into Mexico and briefly follows a 56-km westerly course before reentering the United States some 10 km east of the border town of Nogales, Arizona. In Sonora, Mexico, the river's perennial flow is captured by wells and infiltration galleries for agricultural and municipal consumption. Since the late 1960s, effluent discharges from the Nogales wastewater treatment plant have accounted for the permanence of flow for several kilometers north of the border, where all of it infiltrates into the sandy streambed, resulting in a normally dry stream further north. The river is entrenched most dramatically within the San Xavier Indian Reservation, with vertical banks up to 10 m high and 100 m apart, where the river meanders around the base of Martinez Hill. To the north of Martinez Hill, sections of the riverbanks have been soil cemented as a precaution against flood damage in the heavily urbanized floodplain.

Annual flow along the river is extremely variable. During the 68-year period of available records at the Congress Street gauging station, 72 percent of all annual flood peaks occurred during the months of July and August, 19 percent during September and October, and 9 percent November through February. No annual peak flows have been recorded during the months of March, April, May, or June (Betancourt and Turner, 1988). In this century, the greatest geomorphological changes in the Santa Cruz River were caused by floods occurring in 1905, 1915, 1977, and 1983 (the greatest recorded event, which had a peak discharge of approximately 1,500 m^3/s at the Congress Street gauge), and all are associated with El Niño conditions (warmer than average episodes in the tropical Pacific).

Prior to extensive pumpage for agriculture and consumptive use in the Tucson Basin, the amount of water leaving the basin (i.e., stream flow, evaporation, and transpiration) equaled the amount entering, and ground water storage was nearly constant (Betancourt and Turner, 1988).

According to Betancourt and Turner (1991), the radical lowering of the ground water table and channel entrenchment after 1940 helped eliminate native phreatophytes to the advantage of salt cedars (salt cedars commonly survive in habitats where ground water is unavailable). The cottonwood and mesquite bosques south of Martinez Hill, a popular picnic spot for Tucsonans in the 1930s and 1940s, vanished, leaving the floodplain tree-less. Ground water pumpage also eliminated the influence of a near-surface water table by partially controlling downcutting. As a result, channel degradation propagated upstream for kilometers. Downstream of Martinez Hill and within the lim-its of the city of Tucson, the rate of downcutting is most likely influenced by urbanization of the floodplain. Channel bed degradation has been monitored at the site of a bridge (Aldridge and Eychaner, 1984). The elevation of zero flow at this site (Congress Street) dropped 3 to 4.5 m between 1946 and 1980.

Improvement of the Santa Cruz drainage through the city has encouraged urbanization of the floodplain. The proxim-ity of the Santa Cruz River to the inner city has increased the value of the real estate for urban development. Much of this development, however, has occurred piecemeal. Planning seems to have occurred during low-flow years and before local au-thorities could have responded to federal legislation concern-ing floodplain hazards. This problem is not specific to the Santa Cruz floodplain, but to many other communities in the arid and semiarid Southwest as well.

Prior to the beginning of the twentieth century, the 80-km reach of the Santa Cruz River throughout the Tucson Basin was characterized by lengthy segments of unincised alluvium interrupted by short and discontinuous gullies. Marshes and wet meadows are reported to have occupied these short reaches of perennial flow. A near-surface water table prevented lon-gitudinal expansion and coalescence of arroyos.

Today, a continuous channel defines the river's course through the Tucson Basin, and the water table is more than 100 m below the land surface. The disappearance of marshes and wet meadows is the ecological consequence of the lower wa-ter table.

Sloped soil-cemented banks of the Santa Cruz designed to improve flow conveyance through the Tucson Basin will likely result in greater stream power in the downstream reaches and may also result in migration of the headcut in the up-stream reaches. The rate at which this occurs will depend on the frequency and intensity of flood-producing storms in the coming years. Migration of the headcut upstream will increase the amount of sediment transport further downstream.

Another way in which the character of rivers is drastically altered is by cutting off interactions with the riparian zone and floodplains. This may be done directly, by channelization and leveeing (Kissimmee, Illinois, and Mississippi rivers), and indirectly, by regulating the flood regime (navigation dams on the Mississippi). According to the American Rivers Conservation Council (Echeverria et al., 1989), of approximately 3.2 million miles of rivers in the United States, 2.9 million miles remain undammed, while 600,000 miles of river are dammed. The committee could not find a recent national assessment of the number of stream and river miles affected by channelization or leveeing, but the total is probably much greater than the number of miles of river dammed. In the Illinois River, for example, half the floodplain has been leveed (Bellrose et al., 1983), and most of the Lower Mississippi River is leveed (Fremling et al., 1989). Although water resource agencies track their own development projects, the only nationwide inventory of rivers and streams was conducted in the 1970s (U.S. DOI, 1982) in response to passage of the Wild and Scenic Rivers Act of 1968 (P.L. 90-542). The purpose of the inventory was to identify those rivers worthy of the designation wild and scenic, and so narrow were the criteria that less than 2 percent of total river mileage qualified for inclusion on the list. Therefore, there remains a need for a comprehensive up-to-date nationwide assessment of rivers, comparable to the National Wetland Inventory (Tiner, 1984). It would be useful to know how many miles of free-flowing, unchannelized rivers remain in the United States, where these reaches are located, and what the current trends (net gains or losses) are.

Progress has been made in controlling conventional pollution (sewage and other organic wastes) from point sources. In many parts of the United States, water quality has been maintained or restored since the institution of the clean water acts, starting around 1965, although problems remain in some reaches (CEQ, 1989; ORSANCO, 1990). In some cases (e.g., the Willamette and Illinois rivers), water quality in certain critical reaches is maintained only by dilution, and fish and other aquatic organisms are affected by a legacy of toxic substances in sediment deposits. Also, national water quality assessments are based on lake or channel sampling that does not include floodplain pools and backwaters; so the status of these important nursery areas for fish and wildlife is poorly documented.

Since the passage of the Federal Surface Mining Control and Reclamation Act of 1977 (P.L. 95-87), mining companies have been required to restore both land and water affected by mining and acid mine drainage, in most cases to their premining uses. A federal tax on coal provides funds to restore lands abandoned before the act

BOX 5.2
THE WILLAMETTE RIVER

The term *river restoration* is often misunderstood and mis-applied. For example, the Willamette River in northwestern Oregon is a badly perturbed ecosystem—one greatly altered from its original ecological condition—yet it has been described by some as a river restoration success story.

The Willamette River restoration has been directed prima-rily toward water quality restoration, protection of beneficial uses of the river water, and management of certain species of game fish. The restoration also includes reservoir manage-ment and research intended to reduce ecological disturbances in the river occasioned by changes in water temperature caused by the release of water from reservoirs.

Although attention has been given to land use planning in the basin and, in some cases, to stream-bank reclamation, the Willamette River today is in an unnatural condition that re-quires constant management, and no holistic effort has been made to recreate the river's natural antecedent biological or ecological conditions.

Dams on the Willamette and its tributaries have altered the normal temperature and flow regimes of the Willamette and its tributaries, and have led to damaged native wild sal-monid populations. Dams serve not only as barriers to migra-tion of organisms within the river, but also as sediment barri-ers and as obstructions to the flooding of riparian areas and thus to the return of nutrients and sediment to the land.

Much of the Willamette's water quality improvement has been accomplished by augmenting summer water flows with impounded water to dilute pollutants. Point source industrial discharges are also regulated in amount and concentration through a discharge permit system.

As water treatment standards become more rigorous in the future to compensate for increased human population in the Willamette River basin, more treatment of wastewater may be employed, further reducing flow in certain Willamette tribu-taries. This may tend to lower water quality.

Little effort appears to have been made to restore native aquatic life other than anadromous game fish species, and much of the anadromous fish restoration has involved replacement of wild fish by hatchery stock. The river restoration effort has not yet been successful in maintaining natural fish migration routes or in recreating the predisturbance native fish community structure, species by species, to its previous percentage composition.

Without augmentation of river flow when necessary, water quality would be unacceptable. Without hatchery production and release of salmonids, the sport fishery would be severely limited, and without regulation of municipal and industrial waste discharges, the water's high quality could not be guaranteed. The 13 dams on the river, the past riprapping and channelization, and the dredging (in the lower river) are all indications of the inescapable major impacts that human activities have had on the river.

Thus the Willamette River restoration effort does not meet the criteria for restoration used in this report. Rather it is an example of river reclamation in which a severely polluted river was cleaned up so that its beneficial uses could again be enjoyed by the public. Just as clear-cutting a diverse, complex forest ecosystem and replacing it with a stand of Douglas fir produces a tree farm rather than a restored forest, so, too, does taking a highly disrupted and polluted river system and merely abating the pollution fail to suffice to "restore" the river.

Water quality improvement alone, in the absence of a systematic attempt to recreate a fluvial system's diverse and abundant wildlife and plant communities, is not necessarily equivalent to, or sufficient for, restoration.

went into effect and to identify and set aside lands unsuitable for mining in the future. The decision to forgo mining on certain lands will be based on its high value for other uses, including habitat for rare or endangered species.

Although much remains to be done in restoring streams affected by mine drainage and point sources, a variety of federal, state, and local programs are in place to deal with these problems. There is no comparable nexus of programs to deal with restoration of streams, rivers, riparian zones, and floodplains affected by intensification of land use, yet agriculture and urban development are prominent factors in the deterioration of stream habitats, according to a national fisheries habitat survey conducted by the U.S. Fish and Wildlife Service (Judy et al., 1984; Guldin, 1989). In 1985, agriculture was reported by states as the primary nonpoint source of pollution in 64 percent of affected river miles (CEQ, 1989). Existing soil conservation programs are designed to reduce soil erosion on cropland, but they do not necessarily improve or even maintain water quality or habitat in adjacent streams. Greenways along waterways in cities usually serve as parks rather than as a means of restoring the natural functions of rivers, and most urban flood detention basins bear little resemblance in form or function to natural backwaters and floodplain pools.

Increased sediment delivery resulting from deforestation has also increased sedimentation and turbidity in downstream channels, lakes, and reservoirs, with attendant loss of capacity for water storage and conveyance, recreational and aesthetic values, and quantity and quality of habitat for fish and wildlife.

Successful restorations have occurred on smaller rivers and streams where headwaters are either already protected (by being in a national forest, for example) or the riparian zone can be restored so that upstream disturbances do not undo downstream recovery. In the Mattole River (see case study, Appendix A), many sites along the 62-mile length of the stream, from the headwaters to the mouth on the Pacific Ocean, have been the subject of well-focused restoration efforts. An umbrella organization (the Mattole Restoration Council, MRC) coordinates the largely volunteer efforts of 13 member organizations. The MRC has been successful in obtaining grants, expertise, and training for its volunteers, and in monitoring assistance from government agencies. Although the MRC has not delineated specific ecological criteria for success, it is clear that restoration of self-perpetuating native salmonid populations continues to be a major goal. As with most cases of restoration examined for this report, the Mattole story is not yet complete (see case study, Appendix A). Quantitative data are lacking on the extent of watershed and bank treatment and returns of native fish. Salmon must still be maintained by artificial propagation, and after a hopeful start, 5 years of drought brought a resumption of the downward trend in the river's king salmon population.

There may have been many well-meaning but unsuccessful attempts to restore streams, but it is difficult to obtain quantitative data because individuals and agencies are understandably reluctant to publicize failures. In many cases, the original degradation of the stream and the failed restoration were both caused by inadequate analysis of the natural characteristics of the stream: the patterns of water and sediment transport that create and maintain the natural morphometry of the channel and its associated floodplain. Failures in a project reach can trigger degradation that progresses upstream or downstream. The principles and analytical tools of hydrology and fluvial geomorphology need to be applied to a much greater extent than in the past to the planning and execution of projects. Two approaches (see techniques in "Fluvial Restoration," below)—David Rosgen's restoration of the Blanco River in Colorado (Appendix A), and George Palmiter's restoration of several small rivers in Ohio (Box 5.3)—that do make use of these principles should receive wider application elsewhere and should be tested on larger systems.

Restoration in larger river systems is more problematic because of the size and complexity of the systems and the problems. Degradation of a local reach may be caused by intensification of land use over the entire upstream drainage basin, and local citizens and agencies may feel they cannot do much to control problems that are so large scale. Interstate compacts (e.g., ORSANCO on the Ohio River; the joint efforts of Massachusetts and New Hampshire on the Merrimack River, see case study, Appendix A) have worked well in restoring water quality and, in some cases, fisheries. Despite the size of the Merrimack (134 miles of river draining 5,010 square miles), a small group of citizens formed the Merrimack River Watershed Council, which, like the Mattole River Council, mobilized public support and attracted attention and help from a variety of government agencies. Restoration of the Merrimack River has resulted in water quality improvement to the point that benthic organisms have recolonized formerly barren areas, natural resource agencies are working on the reestablishment of anadromous fish, and cities are using the river as a source of drinking water.

These restoration projects (although having much success) are hampered by the lack of baseline and reference data. Baseline data should be collected on a system before restoration, for comparison with data collected during and after restoration. In the case of stream morphology and vegetation, the baseline condition can sometimes be reconstructed from old aerial photographs and maps, or from soil types, which reflect the presettlement vegetation. Reference data come from another reach of the same river or from a similar river. The reference reach may represent the desired goal, a relatively unimpaired, self-maintaining system, or it may represent the unrestored condition. In the first case, judgment of success or failure is based on how closely the restoration approximates the goal; in the second, on how far the system moves from the degraded condition. Thus, baseline data provide comparisons of the same site through time, whereas reference data provide comparisons among sites at the same time. The strongest documentation for success or failure would come from the use of both baseline and reference data in a well-designed, long-term monitoring program. Too often, funding is provided for the restoration, but not for preproject documentation and follow-up, so that the

BOX 5.3
THE PALMITER METHOD

George Palmiter, a railroad switchman and canoeist, devised ways of stabilizing the banks and unclogging the channels of debris- and silt-laden streams in northwestern Ohio (Herbkersman, 1984; Willeke and Baldwin, 1984). The Palmiter method has received nationwide publicity and has been applied to streams in North Carolina, Mississippi, Michigan, and Illinois. Palmiter received the Conservationist of the Year Award from *Outdoor Life* in 1977 and a Rockefeller Public Service Award in 1979.

Palmiter's method provides a way of restoring the hydraulic capacity of streams and reducing low-intensity flooding without resorting to channelization or removal of riparian vegetation. In fact, riparian trees are left in place or planted to shade the stream, to reduce the excessive growth of shrubs and aquatic plants that retard flow, and to increase the frequency of low floods. Shading has the further beneficial effect of lowering the summer water temperature, to the benefit of fish communities (Karr et al., 1986). The living trees anchor the banks and provide a source of food, in the form of leaf litter, for invertebrates and fish to feed on. Downed logs and root wads provide habitat structure for fish and solid substrate for the invertebrates.

The Palmiter method has been applied primarily in low-gradient alluvial streams and small rivers where logjams cause sediment deposition and increased flooding upstream and bank erosion where the stream cuts a new channel around the jam. George Palmiter's guiding principle is "make the river do the work." He makes the midchannel bars upstream of the obstruction vulnerable to erosion by removing any protective layer of woody debris and vegetation, directing flow toward the bar, and creating "starter" channels to initiate scour. The centers of the logjams are cut into smaller pieces and allowed to float downstream, while the buried ends remain as flow deflectors to keep the main current directed away from the bank. These natural deflectors are sometimes supplemented with root wads or fallen trees that are cabled to the bank.

degree of success or failure is poorly quantified, the exact causes of the eventual outcome are difficult to identify, and the science of restoration ecology is not advanced as quickly as it could be.

The deficiencies in documentation are symptomatic of inherent problems in river restoration. The water regime in rivers typically varies seasonally and annually, so that a longer time series of data is required to document pre and postrestoration conditions in rivers than is required for standing waters. Without an adequate time series, the effects of restoration are confounded with the effects of fluctuations in the water regime. The restoration programs themselves must be adaptable and persistent, because high and low flows affect restorative efforts and are not completely predictable or controllable. Vegetative cover is vulnerable to flood scour until roots are well established, so bank restoration may have to be attempted more than once. However, restoration that uses the power of flood flows to reshape channels may not be affected during a drought period.

River restoration and river monitoring must take the structural and functional organization of river systems into account. Rivers and their floodplains (or streams and their riparian zones) are so intimately linked that they should be understood, managed, and restored as integral parts of a single ecosystem. In addition to this lateral linkage, there is an upstream-downstream continuum from headwaters to the sea or basin sink. The entire river-riparian ecosystem is contained within a drainage basin, so restoration must have a watershed perspective. Changes in any segment are communicated dynamically throughout the system. Downstream restoration can be undone by changes in the watershed, riparian zones, or upstream reaches, and the causes of the failure will not be identified if these linkages are not identified and monitored. Restoration of rivers and streams would benefit from greater application of the principles, knowledge, and techniques of the disciplines that treat rivers as integrated systems: hydrology, fluvial geomorphology, and systems ecology.

There is a need for comprehensive, integrated programs that support stream and river restoration at all levels inherent in the drainage hierarchy, from local reaches and tributaries to interstate waterways. Immediate attention should be given to the remnants of large river-floodplain systems that still exist, because there are so few (e.g., there is only one twelfth-order river in the conterminous United States, the Mississippi River). The programs should be designed from a systems perspective, should include habitat restoration as well as water quality, and should focus on the relatively neglected linkage between land use and stream quality. It is especially important in the dynamic river environment that restoration programs be sustained and flexible, that monitoring begin well before restoration is initiated and continue long enough to separate the effects of restoration from the effects of environmental fluctuations, and that results be analyzed and synthesized for the improvement of restoration science.

INTRODUCTION—IMPORTANCE OF RIVERS AND STREAMS

The intensive use of rivers and streams for industrial and municipal water supply, irrigation, transportation, hydropower, cooling of
thermoelectric generating plants, assimilation of human waste, and
commercial fisheries is summarized in a variety of compendia (CEQ,
1989; Guldin, 1989). Many U.S. cities developed along rivers because
of the abundance of fresh water, the ability of rivers to purify human
waste (or at least transport it away from population centers), and
access to river-borne commerce. Thousands of years ago these same
factors, coupled with renewal of the fertility of agricultural lands by
deposition of nutrients and soil during annual floods, allowed humans to concentrate permanently in one place, giving rise to the first
civilizations along the Nile, Tigris, and Euphrates rivers.

The fertilizing effect of floodwaters is utilized today in some developing countries (Welcomme, 1979), and was used at least into the
nineteenth century in England where bottomland fields were diked
for the purpose of directing silt-laden floodwaters into them. This
practice, known as "warping," presumably resulted in increased fertility as well as a rise in the level of the fields. "Warp" referred to the
load of silt and nutrients in river water, and a "fat river" was one
with an especially rich load (Seebohm, 1952; Whitlock, 1965). Today,
dikes in lowland agricultural areas of developed countries typically
are used to keep floodwaters *out* of fields, and chemical fertilizers are
applied to maintain the productivity of the soil. Following their analysis
of the Mississippi River from an energy systems point of view, Odum
et al. (1987) conclude:

> The annual flood is a potential resource that was effectively used by
> the original floodplain and deltaic system. By diking, channelizing,
> and making economic developments that were not adapted to the
> flood cycle, a benefit was often turned into a stress, a drain on part
> of the system, a pathological state.

Some attempts have been made to calculate recreational values of
streams and rivers. The *1985 National Survey of Fishing, Hunting, and
Wildlife-Associated Recreation* (U.S. DOI, Fish and Wildlife Service,
1988) reported that a total of $17.8 billion was spent by 38.4 million
fishermen for non-Great Lakes freshwater fishing in 1985. The survey also reports that 45 percent of these anglers fished in rivers and
streams. If stream fishermen spend amounts comparable to those
spent by pond, lake, and reservoir fishermen, then the economic value
of the recreational fishery along flowing waters amounts to more
than $8 billion per year. This does not include the premium prices
paid for recreational property and residences along rivers and streams,

nor does it include nonconsumptive recreation such as canoeing or wildlife observation.

In contrast to recreational uses, the natural functions of rivers are evaluated in economic terms only when they become so disrupted that they impede human activities. Rivers transport water, sediment, and nutrients from the land to the sea, play an important role in building deltas and beaches, and regulate the salinity and fertility of estuaries and coastal zones. One of several reasons that the Mississippi Delta is experiencing subsidence (and an apparent rise in sea level) and coastal recession is that sediment is no longer allowed to replenish the delta during floods. Instead, the sediment is conveyed between levees to the edge of the continental shelf. As the sediments settle into deep water, they release phosphorus, which stimulates plankton blooms. The blooms may help pull carbon dioxide out of the atmosphere (thereby reducing one of the "greenhouse" gases that contribute to global warming), but they also senesce and sink, using up oxygen in the decay process and perhaps contributing to the spreading zones of oxygen depletion on the bottom, which are adversely affecting Gulf of Mexico fisheries (Turner and Rabalais, 1991).

Another natural function of rivers is the maintenance of biodiversity. Rivers are highways for migratory birds and fish, and home to many unique species of plants and animals (including federally endangered species such as the Colorado squawfish (*Ptychocheilus lucius* Girard) and the Higgin's Eye Pearly Mussel (*Lampsilis higginsi*)). Some freshwater aquatic species, such as representatives of the most ancient orders of fish (sturgeon and paddlefish [O. Acipenseriformes] and gar [O. Semionotiformes]), occur mainly in large rivers, whereas other species are found only in smaller rivers and streams. The north-south orientation of the Mississippi conserved many aquatic species during glacial periods, because it permitted a southward retreat. In contrast, many of the rivers of Europe run east-west, and freshwater fauna was impoverished during the ice ages. The tributaries of the Tennessee River seem to be centers of speciation, where there are more kinds of freshwater mussels and more rapid evolution of darters (a family of small fish [Percidae]) than in any other river system in the United States (Hocutt and Wiley, 1986).

Aquatic fauna are disproportionately imperiled compared to terrestrial fauna, according to the Nature Conservancy (Master, 1991) and the Endangered Species Committee of the American Fisheries Society (Williams et al., 1989). One out of three North American fish, and two out of three of the continent's crayfish are rare or imperiled. One in every 10 species of North American mussel has become extinct in this century, with 73 percent of the remaining species now rare or imperiled. Most of these species will not be protected if restoration and management continue to focus on single species or on a few species of high value for fishing and hunting. In fact, many aquatic species are harmed by management practices that maximize populations of one or a few game species. In the Upper Mississippi River, floodplains are diked and water levels manipulated to maximize seed production on mud flats for the benefit of migratory dabbling ducks. These water control structures and practices can limit access to spawning, feeding, and wintering areas utilized by fish (Nelson, 1991). In the Mississippi Delta, similar impoundments and practices drastically reduce the access of fish and crustaceans to freshwater marshes that are utilized as nurseries (Herke et al., 1987).

Achieving restoration, defined in Chapter 1 as a return of an *ecosystem* to a close approximation of its predisturbance condition, requires having some concept of the predisturbance structural and functional characteristics, to serve as a goal for restoration and as criteria for the design of a restoration project. Important concepts related to the organization and dynamics of river and stream ecosystems include flow and retention, openness, dynamism, patchiness, and resistance and resilience.

CONCEPTS RELATED TO MANAGEMENT AND RESTORATION OF RIVERS AND STREAMS

Flow and Retention

Rivers and streams are characterized by a one-way flow of water, which tends to transport nutrients, sediments, pollutants, and organisms downstream. Various physical and biological mechanisms (retention devices) counteract this natural tendency for energy and materials to wash out of the system. Water and other materials may be constantly added to the system; organic matter and sediments are retained behind natural dams or filters formed by geological features and accumulations of woody debris; and organisms have evolved means of avoiding currents, holding fast, or actively swimming. In a lake, nutrients may cycle between the sediments and the water column, but the same processes in a flowing river tend to be constantly displaced downstream, so that nutrients are said to "spiral" downstream, instead of cycling in one location (Webster, 1975; Elwood et al., 1983). A stream flowing through an old-growth forest has many dams formed by fallen trees, which retain organic matter and nutrients, tightening the spirals and thereby increasing the productivity of the system (Sedell and Froggatt, 1984).

One advantage of flowing water from a management perspective is the constant mixing, which prevents stagnation and increases the capacity for assimilation of organic matter relative to standing water. Mobile organisms such as fish may be able to recolonize disturbed areas rapidly from undisturbed upstream or downstream reaches or tributaries. Disadvantages are that deleterious effects of pollutants tend to propagate downstream, and a single barrier (dam, chronic pollution) may cause the destruction of an entire migratory population (e.g., salmon, which spawn in headwaters and feed as adults in the sea).

Openness

In an open ecosystem, like an open economy, materials and energy are exchanged across the boundaries of the system as well as within the system. In a closed system, the proportion of transboundary exchanges is small in relation to activity within the system. A river or stream is open because a relatively large proportion of the materials and energy come from the surrounding terrestrial system, with the land-water boundary serving as a valve or filter that controls the exchange.

In reality, the terms *open* and *closed* are relative because lakes also are influenced by their drainage basins. Open, sunlit streams and large floodplain rivers both produce a significant amount of the organic matter that is consumed within them (Junk et al., 1989; Wiley et al., 1990). The important concept from a management point of view is that streams are products of their drainage basins and that the terrestrial environment closest to the stream (the riparian zone) has the greatest impact, with the influence diminishing with distance from the stream. Restoration and management of the riparian zone are usually more cost-effective in improving water quality and fish habitat than practices applied farther from the watercourse (Lupi et al., 1988).

Dynamism

With few exceptions (spring-fed streams, drainage from extensive wetlands), flow is highly variable in streams during the course of a year, although the seasonal timing of high flows and low flows may be quite predictable. Because the capacity to scour is a function of flow, most reworking of stream channels occurs during floods of moderate frequency, which may last from a few hours in headwater or desert streams to months in the largest floodplain rivers of the

world (Leopold et al., 1964; Welcomme, 1979). This annual disturbance may be important in maintaining the existing system, just as
fire is important in maintaining prairies. In a gravel-bed stream, the
flood may flush the accumulation of fine particles out of the interstices, thereby restoring the spawning habitat for trout or salmon
(Milhous, 1990). Restoration of the flow regime is one of the most
neglected aspects of stream and river restoration.

True restoration of streams and rivers must take this dynamism
into account by allowing enough spatial and temporal scope for natural
processes, including floods, to occur. Preservation of a river channel
is not sufficient to ensure survival of fish that spawn on floodplains—
both the floodplain and the flood cycle must be maintained. Loss of
a particular side channel due to sedimentation may not be a problem if a river is allowed to create new channels elsewhere; restoration in this case might be scaled to the full width of the meander
zone and a length that would be some multiple of the natural meander length.

The idea that local features of a stream or river are created, undergo change through time, and eventually disappear, while the overall
pattern (e.g., meandering, braiding) remains constant, at least on some
larger spatial scale and longer time scale, is termed *dynamic equilibrium*. Consider the birth and death of oxbows: the river creates, then
abandons a meander loop, which becomes an oxbow lake on the floodplain; eventually, the oxbow fills with sediment and reverts to floodplain. An observer flying over the river valley at 10-year intervals
would see the same pattern: a meandering river channel, flanked by
abandoned meanders in various stages of reversion to floodplain.
However, some of the meanders of 10 years ago would be oxbows,
and some of the old oxbows would be indistinguishable from floodplain. If the observer could view several hundred years of changes
in a few minutes, using time-lapse aerial photography, the river channel
would appear to writhe like a snake, with the meander loops moving
downstream, throwing off oxbows as they go. The dynamic equilibrium in the physical system creates a corresponding dynamic equilibrium in the biological system. Successive plant and animal communities occupy the meander loop as it changes from an active channel
to a contiguous backwater, then perhaps to an isolated oxbow intermittently connected to the main flow during floods, and finally to a
wet depression on the floodplain. As long as the physical system is
creating new cutoffs, there will be habitats suited for each type of
community, and all successional stages will occur within the river-
riparian ecosystem. If the channel is "stabilized" and the floodplain
leveed and developed for agriculture, industry, or housing, the or-

ganisms that utilized sandbars, undercut banks, oxbows, and flood-plain pools will disappear.

The dynamic equilibrium concept contrasts sharply with the concept of "stabilizing" a stream channel to avoid loss or damage to structures or agricultural fields. A farmer may not be comforted by the fact that the soil washed out of his stream bank is building new land, in the form of a point bar, on his neighbor's property downstream. The same highway department that builds a comparatively cheap, narrow span over a stream channel may be preoccupied with the subsequent problem of protecting a bridge abutment from being undermined by scour, rather than considering the more permanent (and more expensive) solution of spanning the entire floodplain width that will be actively reworked by the channel during the life span of the bridge.

The Blanco River case history (Appendix A) is an example of restoration of a predisturbance meander pattern and floodplain terrace. In contrast, the Old River Control Structure on the Lower Mississippi River represents an attempt to forestall the natural 1,000-year cycle of creation and abandonment of deltaic lobes and distributary channels (Penland and Boyd, 1985).

Patchiness

Natural rivers and streams are not uniform environments; rather, they consist of distinct habitats occupied by characteristic biotic communities. Riffles and pools follow one another in sequence in streams (Figure 5.1). Riffle dwellers are adapted to living in swift, shallow water: some species are small and evade the current by hiding in spaces between the rocks; others are adapted to holding on to the substrate. The deeper pools may contain larger-bodied animals that range throughout the water column, as well as organisms adapted to processing the organic matter that settles out.

River-floodplain systems have a lateral structure that begins at the main channel and progresses through unvegetated and vegetated channel borders and floodplain habitats (backwaters and seasonally flooded vegetation types) (Sparks et al., 1990) (Figure 5.2). Backwaters and large-scale eddies provide refuges from the high velocities and colder winter temperatures of the main channel. Within each of the border and floodplain areas, there are distinct patches, usually determined by small differences in land elevation, that in turn determine the period of inundation (or water depth, in permanently flooded areas) and soil saturation (Figure 5.3).

There is a vertical dimension to lotic systems, as well as lateral

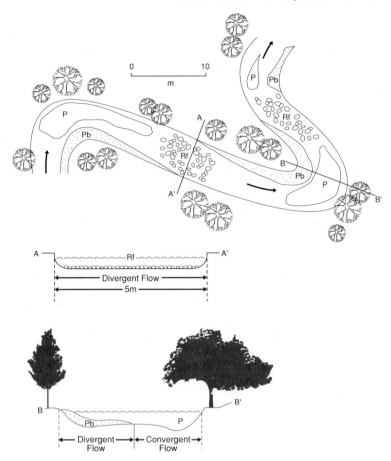

FIGURE 5.1 Idealized natural channel prototype: P, pool; Rf, riffle; Pb, point bar. Source: Reprinted by permission from Brookes, 1988. Copyright © 1988 by John Wiley & Sons, Ltd.

and longitudinal (upstream-downstream) dimensions, and this too can be patchy (Amoros et al., 1987). The area below the bed of the river is known as the hyporheic zone and may have temporary residents (salmon eggs and larvae), as well as permanent residents adapted to life in the interstices between the substrate particles. In many intermittent streams, life retreats to the hyporheic zone when surface flow ceases or when floods threaten to wash organisms out of the water column. The hyporheic zone may extend many stream widths to either side of the channel. Factors such as dissolved oxygen levels, temperature, and interstitial flow rates may vary greatly among patches within the zone.

Modifications such as flow regulation and channelization make the stream environment more uniform, and restoration necessarily involves maintenance or recreation of the original patchiness.

Resistance and Resilience

Stream communities may be more resistant to certain types of disturbance and may recover more quickly from disturbance than lentic communities because they are adapted to a dynamic environment. A record flood may destroy property but have little effect on species that are adapted to flooding; access to the greatly expanded habitat

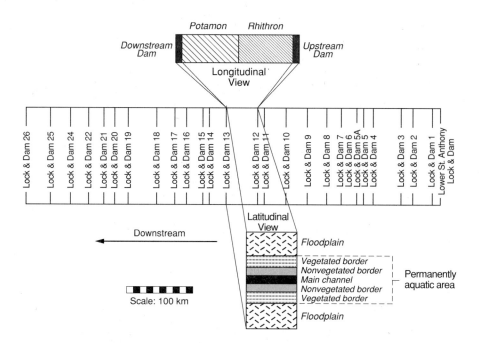

FIGURE 5.2 Longitudinal and latitudinal structure of the Upper Mississippi River. The latitudinal scale is purely schematic and greatly exaggerated relative to the longitudinal scale. The actual width of the river varies from about 500 m (at rocky narrows) to 5 km (including flooded areas). The channel width is also exaggerated: the main channel typically occupies only 3 percent of the floodplain system, or about 6 percent of the total width of the unleveed parts of the river. Source: Sparks et al., 1990.

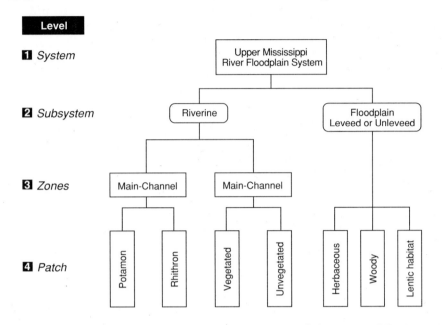

FIGURE 5.3 Hypothesized hierarchical structure of the Upper Mississippi River System floodplain. Source: Reprinted by permission from the Center for Aquatic Ecology. Copyright © by the Illinois History Survey, Champaign, Ill.

created by the flood probably benefits some species, such as floodplain spawners, without doing any permanent damage to other species, such as trees that are capable of surviving temporary inundation.

The hierarchical and patchy structure of streams also contributes to resistance and resilience (ability to recover). Organisms may recolonize a denuded reach from undisturbed upstream and downstream reaches or from tributaries, or they may avoid or survive a disturbance in the main channel by seeking refuge in hyporheic or lateral zones.

THE RIVERINE-RIPARIAN ECOSYSTEM

Integrative Concepts

Various attributes of rivers and streams described above are integral to a discussion of the structure and function of riverine ecosystems (see Table 6.1 for connection with wetland functions). Foremost among these integrative concepts is the idea that rivers and their floodplains are so intimately linked that they should be understood, managed, and restored as integral parts of a single ecosystem. The term "riverine-riparian ecosystem" has been applied by Jensen and Platts (1989) to streams and small rivers (those less than 2 m deep), and the term "river-floodplain system" to large rivers by Sparks et al. (1990) and Junk et al. (1989). To avoid repeating both terms throughout the text, the committee defines the term *riverine-riparian ecosystem* (RRE) as including both small and large systems. To avoid confusion, the committee will use the term *stream-riparian ecosystem* for small systems in which floods are so brief and unpredictable that aquatic organisms have not evolved adaptations for exploiting the riparian zone. The term *river-floodplain ecosystem* is reserved for systems with a predictable, long-lasting flood pulse that is exploited by fish and other aquatic organisms (see Figure 5.3).

The distinction between small and large systems is important because the riparian zone often functions as the donor of nutrients, water, and sediment, and riparian vegetation as a regulator of light and temperature for the recipient stream channel, whereas these functional roles are usually reversed in river-floodplain systems (Swanson and Sparks, 1990). In the larger ecosystem, the channel is usually the donor of water, sediment, and inorganic nutrients to the recipient floodplain, and light penetration and temperature in the inundated floodplain are often influenced by the influx of turbid, cooler channel water. In a stream, almost all the aquatic productivity is concentrated in the channel because the riparian zone is inundated only briefly. In contrast, most of the aquatic productivity in large river-floodplain ecosystems occurs in the floodplain because of (1) the predictable timing and relatively long duration of the annual flood pulse, and (2) the much greater area and volume of the floodplain in comparison to those of the channel (Junk et al., 1989). The channel of a large alluvial river is usually only a fraction of the total area that is seasonally inundated, and the productivity per unit area of the channel may be low because of low light penetration (due to turbidity and depth), high inorganic sediment concentration, and a shifting substrate.

Each riverine-riparian ecosystem contains a riverine subsystem and a riparian subsystem. The riparian subsystem is periodically inundated and is transitional between an aquatic environment and an upland environment (Jensen and Platts, 1989; Junk et al., 1989). The riverine subsystem is composed of the aquatic habitats within the channels. A single RRE occupies a drainage basin.

Jensen and Platts (1989) summarize the arguments for an approach to river restoration that treats the river and its floodplain or the stream and its riparian zone as parts of one ecosystem:

The values of riverine and riparian ecosystems are interdependent. Both riverine and riparian ecosystems are essential elements of fish and wildlife habitat; the riparian ecosystem serves to store and desynchronize peak flow conveyed by the riverine ecosystem; the food chain and nutrient cycling of both ecosystems are intertwined; the cultural and heritage values of riverine and riparian ecosystems are intimately linked.

Riverine and riparian ecosystems also function in an integrated fashion. Impoundment, channelization, and diversion in the riverine system can influence the hydrologic qualities of the riparian ecosystem. Similarly, impacts to the riparian ecosystem such as livestock grazing can cause erosion of streambanks and enlargement of channels, thus influencing the functional qualities of the riverine ecosystem. Since the values and function are interdependent, the approach for restoration of riverine and riparian ecosystems must be integrated.

The above discussion of streams and rivers should not be taken to mean that there is a definable boundary in the RRE, upstream of which is a stream and below which is a river; rather, each RRE is continuous from headwater to oceanic or basin sink. The biological structure and function of the RRE vary in a predictable way along a continuum, in response to variations in physical characteristics (Vannote et al., 1980; Wiley et al., 1990).

Discontinuities (i.e., disruptions in the predictable upstream-downstream patterns), are created when rivers are dammed. A dam may make conditions more like those of the headwaters (an upstream shift), or more like those downstream, or it may have a negligible effect, according to the serial discontinuity concept of Ward and Stanford (1983) (Figure 5.4). In cases where multiple dams create multiple discontinuities in the expected or natural pattern, individual dams could be redesigned and operated to restore some of the conditions (water temperature and dissolved oxygen) that formerly existed at that point in the river (e.g., by releasing epilimnetic water from the upstream reservoir, instead of hypolimnetic water), or a community more suited to the new conditions could be established by stocking. The first option is restoration; the second is creation of a community different from what was there originally, but somewhat representative of another portion of the RRE continuum. The created community may be different from the headwater community because the headwater organisms have other requirements that are not so easily met as those for temperature and oxygen: the quantity and quality of food generated in the reservoir are different from those in the headwaters, and the dam may be a physical barrier to migratory species.

These continuum-discontinuity concepts have important implications for prioritizing and evaluating restoration projects. Restoration goals and the evaluation of success must take into account relative positions along the continuum. This is especially true when using various biotic indices such as species diversity are used. Because the number of species of fish tends to increase downstream and the number of species of aquatic insects increases upstream, it would be inappropriate to expect a restored reach to have the same diversity as reaches located at different points on the continuum. It would be more appropriate to find a reference reach in an adjacent tributary of the same stream order.

The need for reference streams of the same order, against which the success of a restoration can be gauged, raises the complicating issue of regional variation among ecosystems. Although the general idea of an RRE as a continuum is correct, the river continuum concept of Vannote et al. (1980) best describes the type of RRE in which these ideas were first developed: stream-riparian ecosystems that originate in mountainous, forested watersheds of the temperate zone. Other regions differ in physical, chemical, and biological attributes, and the streams draining these different regions would be expected to have different properties (Wiley et al., 1990). Moreover, one RRE can pass through several different regions.

Ecoregions as Applied to Rivers and Streams

Omernik (1987) developed a map that divided the conterminous United States into 76 ecoregions based on regional patterns in land surface form, soil, potential natural vegetation, and general land use (see Figure 4.2). Hughes et al. (1990) evaluated the utility of these ecoregions in accounting for differences in fish communities in relatively undisturbed reference reaches of streams and rivers (1) in statewide case studies in Arkansas, Ohio, and Oregon (Larsen et al., 1986; Rohm et al., 1987; Whittier et al., 1987) and in three separate basin studies in Montana, Ohio, and Oregon (Hughes, 1985; Ohio EPA, 1987); and (2) in unpublished data on the Calapooia River in Oregon from Giattina (U.S. EPA, Chicago). Two analytical techniques were used to evaluate the similarity of the fish communities: detrended correspondence analysis (Gauch, 1982) and the index of biotic integrity (Karr et al., 1986).

Hughes et al. (1990) found that (1) the fish communities did demonstrate ecoregional patterns; (2) ecoregions that differed greatly in landscape attributes supported very different communities; (3) similar ecoregions supported similar communities; and (4) within-region

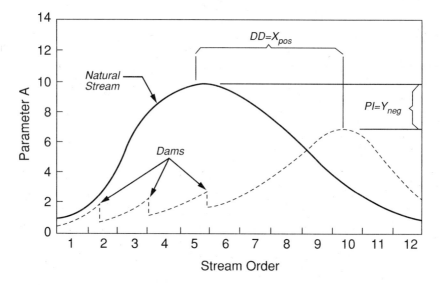

FIGURE 5.4 Theoretical framework for conceptualizing the influence of impoundment on ecological parameters in a river system. Discontinuity distance (DD) is the downstream (positive) or upstream (negative) shift of a parameter for a given distance (X) due to stream regulation. PI is a measure of the difference in the parameter intensity attributed to stream regulation. Source: Reprinted by permission from Ward and Stanford, 1983.

variation was less than among-region variation. The use of reference sites within ecoregions thus appears to be a useful way of establishing criteria for restoration and recovery of RREs. This approach is better than an oversimplistic approach of establishing national standards, which would be unachievable because of natural constraints in some regions and would not recognize the full restoration potential of others. At the same time, using criteria based on reference sites is not as costly as developing site-specific criteria—an impossible task in many cases, where predisturbance conditions are not known.

The natural structural and functional patterns of river-riparian ecosystems are disrupted by a variety of stresses, which are described next.

STRESSES ON RIVERS AND STREAMS

Stresses on the biotic components of river and stream ecosystems arise from (1) changes in the quality, quantity, and seasonal availability of food for organisms; (2) deterioration of water quality, including temperature changes and excessive turbidity and sediment; (3) modifications of the habitat, including the substrate; (4) water quantity or flow mistiming; and (5) biotic interactions (Figure 5.5; Karr et al., 1986). The locus of the problem can be in the watershed, along the riparian or floodplain zone, or in the channels and pools (see Tables 5.1 and 5.2).

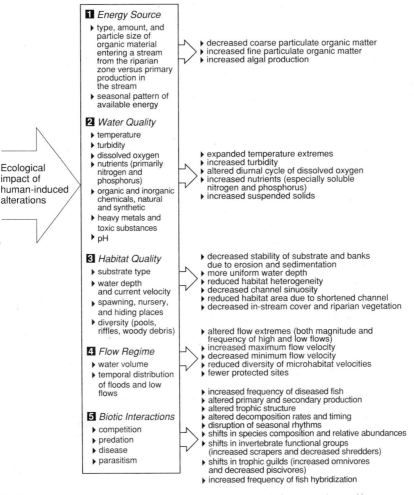

FIGURE 5.5 Five major classes of environmental factors that affect aquatic biota. Arrows indicate the effects that can be expected from human activities, in this case the alteration of headwater streams, excluding small impoundments. Source: Reprinted by permission from Karr et al., 1986. Copyright © by the Illinois Natural History Survey, Champaign, Ill.

TABLE 5.1 Causes of Stream and River Degradation

Dams (hydroelectric, water supply, and navigational aids [locks])
Dredging
Erosion
Filling
Grazing in riparian zone
Industrial point source discharges
Logging
Mining
Municipal point source discharges
Overfishing
Road construction
Urban, suburban, and agricultural nonpoint source runoff

TABLE 5.2 Types of Stream and River Problems

Bank erosion
Blockage of main channel or cool tributaries
Braided channel
Dissolved oxygen deficiency
Excessive flooding
Food scarcity for biota
Genetic deterioration of fish stocks
Gravel unavailable or sediment-covered
Invertebrate deficit
Nutrient loss
Pool deficit
Poor spawning success
Sediment loss
Shelter deficit (for fish resting and refuge)
Siltation
Species (extinct, endangered, threatened)
Stream cover (deficit or overgrown)
Water quality (turbidity and chemical pollution[a])
Water release mistiming
Water supply deficit (from water withdrawals or drought)
Water temperature (too high or too low)
Water velocity (too high or too low)

[a]Includes supersaturation with nitrogen from water passage through hydroelectric facilities (Narver, n.d.).

Point Sources of Pollution

Human activities have had major impacts on streams and rivers. Discharges from population centers and industries are point (end-of-pipe) sources of pollution, whereas human uses of drainage basins (agriculture and silviculture) can cause nonpoint pollution. At first, municipal and industrial wastes simply drained into the nearest watercourse. Later, as populations grew and there were outbreaks of waterborne disease, waste was collected in sewers and diverted away from water intakes. In Chicago, for example (Appendix A), waste was diverted away from Lake Michigan and into the Illinois River, starting on a large scale in 1900. Eventually the assimilative capacity for waste of even the largest rivers was exceeded, and waste treatment plants had to be constructed. Substantial federal assistance for sewage plant construction and upgrading was provided by the Federal Water Pollution Control Act of 1972, and by subsequent legislation, including the landmark Clean Water Act of 1977.

The approach to restoring water quality was to develop criteria for various uses of water and then design waste treatment plants that would achieve effluent standards that in turn would protect or restore the beneficial uses of the stream or river, including fish and wildlife production and use for public water supply. In general, this approach has worked to a substantial degree for conventional pollutants, including oxygen-demanding organic waste, as indicated in the examples of the Illinois and Merrimack rivers, the biennial water quality reports issued by the states under requirements of Section 305(b) of the Clean Water Act (e.g., Illinois EPA, 1990; ORSANCO, 1990) and national water quality summaries (CEQ, 1989; Smith et al., 1987). However, point and nonpoint discharges of toxicants remain a problem; and a legacy of pollutants, including toxicants, remains in sediments and can enter food chains. For example, public health advisories against consumption of certain nonsport fish in the Ohio River were issued by Pennsylvania, Ohio, West Virginia, and Kentucky in 1987 and 1988 because of high levels of chlordane or polychlorinated biphenyls (PCBs; ORSANCO, 1990; see also Merrimack River and Willamette River case studies, Appendix A).

Nonpoint Sources of Pollution

River-riparian systems are products of how their drainages are covered (vegetation type) and how the land is used (grazed, cropped, or urbanized). Over the past 30 years (1960s through 1980s), major land use categories have changed very little (Flather and Hoekstra, 1989). There has been a slight reduction in rangeland and forest (5 percent each) and a 3 percent increase in cropland. There are regional differences: in the Southeast, forest land has increased substantially. Urban land increased 88 percent, from approximately 25 million acres in 1960 to 47 million acres in 1980 (Flather and Hoekstra,

1989). The net result of these changes has been a reduction in land that tends to support natural vegetation (forests and rangeland) and an increase in land heavily modified for human use (cropland and urban land). Although croplands and urban lands probably release more pollutants per acre on average than forests and rangeland, practices such as clear-cutting, fall plowing, and grazing can increase pollution loading of streams and rivers. McElroy et al. (1975) determined that 97 percent of the land in the United States is rural and that *all of it* is a potential source of nonpoint pollution, including sediment, animal waste, nutrients, and pesticides; 64 percent is used for agriculture or silviculture and only 0.6 percent for mining and construction (Table 5.3).

Nutrients and toxicants may be dissolved in water or may ride sediment particles into streams where these materials can wash downstream, accumulate in depositional areas, be ingested by organisms, or be released to the water. Sediments rich in organic matter may release toxic ammonia and hydrogen sulfide, and create low levels of dissolved oxygen in overlying water due to decompositional processes.

According to the Association of State and Interstate Water Pollution Control Administrators (1984), 11 percent of the total river miles in the United States was ranked as having moderately to severely impaired use because of nonpoint sources of pollution. The nonpoint sources and their percentage contribution to total impacted river miles included agriculture (64 percent), mining (9 percent), silviculture (6 percent), urban runoff (5 percent), hydromodification (4 percent), construction (2 percent), and land disposal (1 percent). The

TABLE 5.3   Land Use in the United States

Land Use Category	Percent	Millions of Hectares
Farmland in grass	36	218
Cropland, plus farmsteads and roads	28	167
Construction (annual)	<1	0.59
Commercial forest (includes farm woodlands and forests)	34	202
Annual harvest of forests (growing stock)	<1	4.45
Subsurface mines	<1	2.8
Surface mines	<1	1.2
Active surface mines	<1	0.14
Mineral waste storage	<1	1.2
Total	100	597.38

SOURCE: Modified from McElroy et al., 1975.

major problem was sediment, which accounted for 47 percent of the nonpoint source pollutants in affected river waters.

Practices associated with forestry and farming not only increase the introduction of pollutants into streams, but also alter the physical structure and function of river-riparian ecosystems, as discussed in the sections below on overgrazing and on drainage and channelization.

Overgrazing

The American Fisheries Society recently issued a position statement on the effects of livestock grazing on riparian and stream ecosystems (Armour et al., 1991) from which this summary is largely taken. Overgrazing of livestock in riparian areas is a major problem. Grazing is permitted on 91 percent of the federal land in the 11 contiguous western states, where federal land constitutes 48 percent of the total land area. Thirty-two percent of the land is private rangeland. The best information on the relationship between grazing and stream degradation apparently is available for land administered by the Bureau of Land Management (BLM), but the trends are probably similar for Forest Service and private lands. Fifty-eight percent of the 150 million acres of BLM rangeland is in fair to poor condition, and 19,000 miles of sport fishing streams, 100 million acres of small game and nongame habitat, and 52 million acres of big game habitat have declined in quality as a result of land use practices, including overgrazing.

Armour et al. (1991) point out that because riparian environments are lumped into much broader terrestrial classifications (e.g., "rangeland," as in McElroy et al., 1975, classification), they become unidentifiable for land management purposes, and the problem is probably worse than the above figures indicate. For example, rangeland that is in fair to poor condition probably has river-riparian ecosystems that are in much worse condition because livestock (and wildlife) spend much more time and graze more heavily in the well-watered riparian area.

Overgrazing by livestock can eliminate streamside vegetation directly, or indirectly as a result of caving and trampling of banks, which can lead to channel widening, channel aggradation, lowering of the water table, and decline in water quality downstream because of turbidity, sedimentation, and animal waste. The water may become too turbid, warm, and shallow and the substrate too choked with fine sediment to support native fish and their food base.

Overgrazing on federal land might be reduced if it were not subsidized. The General Accounting Office (U.S. GAO, 1988) reported that the BLM recovered only about 37 percent of the cost of providing grazing on federal land and that the Forest Service recovered only

30 percent. Also, neither agency had current information on range conditions, and only about half the grazing allotments had been evaluated in the last 10 years.

Increasing the fees on federal grazing land would remove the incentive for overgrazing that current low fees provide. Increased fees might help to improve the management and administration of the federal grazing system.

Drainage and Channelization

A U.S. Fish and Wildlife Service manual on stream channelization impacts (Simpson et al., 1982) estimated that as much as 70 percent of the overall riparian habitat associated with streams in the continental United States had been lost or altered and that much of this loss was associated with channelization activities. Unfortunately, there is no reference for the 70 percent figure nor any explanation of how it might have been derived, other than the following discussion, taken from the report. There is little information available on the extent of early (1800s) channelization activities of the U.S. Army Corps of Engineers (COE), but between 1940 and 1971, COE assisted in 889 stream projects totaling 11,077 miles of stream. The Soil Conservation Service (SCS) was involved in the channelization of 21,401 miles as of 1979. The U.S. Department of Agriculture estimated that 189,000 miles of open ditches had been constructed for drainage of agricultural lands by 1959. Bhutani et al. (1975) estimated that channelization and drainage for agriculture would average 6,600 miles per year through 1985. Arthur D. Little (1973) estimated that more than 200,000 miles of stream channel had been modified in the United States by 1972. If streams and rivers in the United States total approximately 3.2 million miles in length (Echeverria et al., 1989), 200,000 miles is approximately 6 percent of the total—quite different from 70 percent! A few states have assessed the extent of channelization. Lopinot (1972) reported that 26.8 percent, or 3,123 miles, of the interior streams in Illinois (excluding the Mississippi, Ohio, and Wabash rivers) had been channelized; this also is a much lower value than the 70 percent average estimated by Simpson et al. (1982), despite the fact that Illinois is a corn belt state where much of the original marshy prairie had to be tiled, ditched, and drained to make it suitable for agriculture. It is clear that an inventory of the nation's streams and rivers is needed, so that their condition can be assessed. This inventory should be updated periodically to track progress in protecting and restoring streams.

Impacts of channelization on habitat for fish and invertebrates include removal or subsequent loss of riparian vegetation, loss of in-stream cover (snags), altered riffle pool sequence, decreased stream sinuosity, altered substrate composition, increased stream velocity, increased bank erosion and bed scour, increased suspended sediment, and increased water temperature (Crandall et al., 1984).

Suspended Sediment

Sediments constitute 47 percent of the materials introduced from nonpoint sources (ASIWPCA, 1984). Particle size ranges from rocks, gravel, and sand to very fine silt. Large particles usually settle to the bottom fairly rapidly, but the fine silt remains suspended for long periods of time, producing turbidity. Because turbidity causes light to be scattered and absorbed rather than transmitted in a straight line, light penetration is reduced, in turn diminishing or even eliminating plant growth (Stern and Stickle, 1978).

When plant beds are eliminated, turbidity problems may worsen. Plant roots anchor the bottom against wave action and disturbance by bottom-feeding fish such as carp. The stems and leaves of floating and emergent plants dampen waves. Jackson and Starrett (1959) showed that wind had little effect on the turbidity of backwater lakes along the Illinois River when plants were present, but that there was a marked effect when vegetation was absent. The loss of aquatic macrophytes leads to the loss of associated "weed fauna" (i.e., the snails and aquatic insects that graze on the plants and in turn provide food for young fish). Smith (1971) indicates that populations of bigeye shiner (*Notropis boops* Gilbert), bigeye chub (*Hybopsis amblops* [Rafinesque]), and pugnose minnow (*Notropis emiliae* [Hay]) have been decimated in Illinois streams because of the disappearance of aquatic vegetation. Predatory fish do not depend directly on plants for their livelihood, but they do depend on good visibility for finding food (and fishermen's lures). Although fish are able to find food using alternate senses, such as the lateral line system, Vinyard and O'Brien (1976) found that turbidity can reduce the feeding of game fish even if there is an abundance of food in the water. However, in many cases food is not abundant because turbid waters also limit the production of zooplankton on which forage fish such as gizzard shad live. Buck (1956) found that the ratio of forage fish to predacious bass and crappie was approximately 1 to 1 in muddy water and 13 to 1 in clear water. He found that when so little food was available, there was only a small population of older, slow-growing bass with very low rates of reproduction. In clear water, he found large bass populations that were reproducing successfully. In addition to sight feeding, many species of game fish exhibit complex reproductive and social behaviors that depend on visual cues. A reduction in visibility interferes with these visual cues and thereby reduces reproduction.

Fish can tolerate short episodes of high levels of suspended sediment, and some species in laboratory bioassays have survived mixtures that can be characterized as slurries of suspended clay particles (Wallen, 1951). Fish exude a protective mucus on their skin and gills that traps and continually flushes particles away. However, this protective mechanism requires metabolic energy and constitutes a stress on the fish at the same time as its ability to find food is reduced.

Other organisms also have similar protective mechanisms. Mussels have a protective mucus on their gills and can close their shells, but these are only temporary measures, and the defenses of mussels against excessive sediment are eventually overwhelmed by long periods of exposure. Because mussels are nonselective filter feeders, the food available to them in silt-laden waters is diluted by the presence of inorganic silt (Widdows et al., 1979), which is rejected as pseudofeces. Laboratory experiments with freshwater mussels kept in water having continuous very high loads of suspended sediment showed that silt interfered with their feeding, because the mussels stayed closed 75 to 95 percent of the time. Mussels dying in these experiments always contained deposits of silt in the mantle cavity and frequently in the gill chambers (Ellis, 1936). The yellow sand-shell (*Lampsilis anondontoides*), a sand-inhabiting species, was most readily killed by silt deposits in Ellis's experiments and has also disappeared from the Illinois River, probably due to increased silt loads (Starrett, 1971). Recent studies have focused on the impact of intermittent exposure to high silt levels, such as might be found in navigable rivers. Payne et al. (1987) found that when freshwater mussels were exposed to intermittent high levels of suspended solids, feeding was disrupted and they shifted to catabolism of endogenous nonproteinaceous energy reserves.

Therefore, although some adult organisms can withstand enormous amounts of sediment in water for several days or weeks, a population may eventually die out due to starvation, reproductive failure, or cumulative stress (Illinois EPA, 1979). Thus, the long-term effect of chronic suspended sediment is to change the species composition of a body of water by changing the habitat and the food supply, and by bringing about differential rates of reproduction in different species.

Sediment Deposition

Diversity in the topography of the bottom of a stream or river is important in maintaining diversity of plant and animal life. In shallow areas with swift waters, gravel beds and riffles provide habitat and spawning areas for many creatures. Where currents are slower, submerged and emergent vegetation becomes established and provides food and shelter for a different group of aquatic animals. In very deep areas, there are holes in the vegetation because rooted plants cannot become established. The edges of these holes are often inhabited by desirable game fish that feed on the forage fish living among the plants.

When sediment deposition exceeds sediment transport, deposits of fine sediment can cover gravel bottoms that many organisms need for feeding and reproduction, and may fill the deep pools and cover the rocks and woody debris where game fish live and feed (Roseboom et al., 1983). Ellis (1936) showed that most of the common freshwater mussels were unable to maintain themselves in either sand or gravel bottoms when a layer of silt from 0.25 to 1 inch deep was allowed to accumulate on the surface of otherwise satisfactory bottom habitats. A study conducted in Idaho showed that when the fine sediment in spawning riffles exceeded 20 to 30 percent by volume, the survival of salmon embryos declined (Bjornn et al., 1977), and in Arizona, salmonid populations were found to be inversely proportional to the fine content of the bottom (Rinne and Medina, 1989). In a Michigan stream, brook trout (*Salvelinus fontinalis* [Mitchill]) populations were reduced by 50 percent when bed load was artificially increased by four to five times (Alexander and Hansen, 1986) and walleye (*Stizostedion vitreum* [Mitchill]) eggs were smothered by fine sediments in Minnesota (Johnson, 1961).

Fine sediments affect invertebrates, as well as fish eggs and larvae, in the hyporheic zone. The hyporheic zone serves as a refuge from predators and swift currents and as a feeding area for early instars. It also functions as a site for nutrient transformation (Stanford and Ward, 1988; Ward, 1989); all these functions are altered when fine sediments fill the interstices between coarser bed particles.

In slow-moving waters, fine sediment deposition may continue unchecked until the bottom becomes so soft and unstable that rooted plants can no longer gain a foothold. Because the deepest holes fill fastest, the end result is a leveling out of the bottom topography and a loss of fish habitat. In extreme cases, sediment can completely fill, and thereby destroy, an aquatic habitat (see Illinois River case study, Appendix A).

In some cases, sediment deposition may create new habitat. Dammed rivers, especially those that carry heavy sediment loads, begin depositing sediment as soon as the dam is completed. As the bottom behind the dam slowly rises, it enters the euphotic zone (i.e., the depth at which sufficient light penetrates to enable plants to grow). The sediments in southern and midwestern rivers that drain agricultural areas serve as sinks for nutrients, particularly phosphorus, which nourishes the new plants once there is sufficient light. In the Mississippi River, for example, extensive new plant beds are located near Montrose, Iowa, just upstream of Lock and Dam 19 (Sparks et al., 1990). In the long-term view, however, these plant beds are only temporary because they will continue to collect sediment until they become higher than mean water level, at which time they will begin providing habitat for terrestrial creatures.

Much of the prime agricultural land in the Midwest and the South is located on alluvial floodplains that developed over thousands of years and supported bottomland hardwood forests. Before the forests were removed, the floodplains served as sedimentation basins and nutrient sinks. Wilkin and Hebel (1982) found that sediment settled in forested floodplains and forested stream borders at the rate of 10 to 20 tons per acre per year. Where the floodplain had been cleared for row crops, sediment was being eroded from the floodplain at a rate of 15 to 60 tons per year. In an agricultural watershed, stream-bank erosion and resuspension of sediment contributed the major portion of annual stream yields of sediment (Sharpley and Syers, 1979). These sediments carry with them the nutrients that make the floodplains desirable for agriculture. By chemically analyzing eroding stream bank soils, Roseboom (1987) determined that bank erosion yielded approximately half of the total phosphate, ammonia, and nitrogen in a channelized floodplain stream in central Illinois.

Once it has returned to the water, sediment can serve as either a source or a sink for nutrients, depending on conditions such as pH, temperature, oxidation-reduction potential, and the amount of nutrients present in the water. For example, phosphorus in the water and phosphorus carried into the water on sediments will come into equilibrium. If plants take up the phosphorus in the water, the sediment can supply more. If there is an excess of dissolved phosphorus, the sediment will take it up (Illinois EPA, 1979; Froelich, 1988). Ammonia supplies nitrogen, another nutrient for aquatic plants; however, it constitutes a greater problem than phosphorus because it is toxic to fish and other animals (Roseboom and Richey, 1977; Thurston et al., 1981).

In extremely turbid waters the presence of these nutrients may not be evident because light is insufficient for plant growth. However, should turbidity be somewhat reduced either by natural processes such as low flow or by reductions in the amount of sediment being introduced, allowing more light to penetrate, algal blooms may occur. Thus, as the Illinois Environmental Protection Agency has pointed out, removing sediments may not be sufficient to ensure high aesthetic enjoyment of water if the nutrients remain in a dissolved state or in sediments on the bottom (Illinois EPA, 1979).

In addition to nutrients, a number of toxic substances are adsorbed on soil particles that move into streams. Among them are metals such as copper, zinc, and lead, which are known to accumulate in sediment. Mathis and Cummings (1973) found that most metals in the Illinois River occurred in sediments at levels several orders of magnitude greater than the levels in water. Organisms that live in the sediment, such as oligochaete worms and clams, contained higher levels of the metals than did organisms such as fish. Because the chemical environment in the gut of a worm or at the gill surface of a clam is different from that in the sediment or water, it is possible that metals and other toxicants can be mobilized from the sediment and taken up by organisms that ingest sediment or live in contact with it.

Pesticides constitute another group of chemicals that can be taken up by organisms. When pesticides are introduced into an aquatic ecosystem they are stored in the bodies of organisms, where biological amplification may take place as the chemicals move through the food web. Although modern pesticides are formulated to degrade, some of the degradation products are not entirely harmless. Also, much agricultural land still contains persistent pesticides or their metabolites from earlier years (Illinois EPA, 1979). Degradation may take days or weeks, and in the meantime pesticides remain deadly to nontarget species as well as those that were targeted. Every year, fish kills caused by agricultural chemicals are reported either to insurance companies or to the EPA. The two most common causes of these fish kills are runoff of insecticides from freshly sprayed fields, usually when spraying is closely followed by heavy rains, and carelessness on the part of applicators who allow leftover chemicals to drip from their tanks (Illinois EPA, 1979).

Eroded silt also often carries with it organic matter that creates an oxygen demand in the water. Ellis (1936) found that the oxygen demand of organic matter mixed with silt lasted 10 to 15 times as long as the oxygen demand created by the same amount of organic matter mixed with sand.

Butts (1974) found that oxygen demand can increase dramatically when sediment containing organic material and bacteria is resuspended by waves or currents. In the Peoria Pool of the Illinois River, for example, he found that under quiescent conditions the sediment oxygen demand was 2.8 g/m^2 per day, but that the demand rose to 20.7

g/m^2 per day when the sediments were disturbed. In some reaches of the river the oxygen demand exerted by sediment was great enough to seriously diminish the oxygen supply in the water, endangering aquatic animals.

Dams

Dams have been placed on every type and size of flowing water, from intermittent headwater streams to the Mississippi River. Croome et al. (1976) suggested that by the year 2000, approximately 66 percent of the world's total stream flow will be controlled by dams. Mermel (1976) used information from the *World Register of Dams* (International Commission on Large Dams, 1973) to conclude that dams are being built on the world's rivers at an average rate of 2 per day. In North America, more than 200 major dams were completed each year between 1962 and 1968 (Beaumont, 1978). The rate of construction of nonfederal dams, which are presumably smaller than the major dams counted by Beaumont, decreased from more than 2,000 per year in the 1960s to about 1,240 per year during the 1970s, according to the 1982 inventory of nonfederal dams conducted by the U.S. Army Corps of Engineers (Johnston Associates, 1989).

Aside from the obvious effects of changing flowing water to standing water, altering downstream flow patterns of both water and sediment, and blocking migrations of aquatic organisms, dams alter water quality and initiate long-term changes in downstream channels, riparian zones, and floodplains. Release of cold, deoxygenated water from the depths of reservoirs adversely affects native stream organisms adapted to warmer, aerated water (NRC, 1987). Flow regulation by dams reduced the area of floodplain wetlands by 67 percent in a 145-km sample reach of the Missouri River (Whitley and Campbell, 1974). Further reductions occurred in other parts of the Missouri when side channels, pools, and wetlands that once supported fish and wildlife were left high and dry after the channel was down-cut. Downcutting was attributable to an increase in the erosive power of the middle Missouri River following storage of sediment in the reservoirs of the upper river (Hesse et al., 1982).

Dams may have an effect on quite distant ecosystems. Closure of the major dams on the upper Missouri River was followed by a step decrease in the sediment load measured on the lower Mississippi at Tarbert Landing, Mississippi, and Simmesport, Louisiana, because the dams trapped sediments (Keown et al., 1981). Hence less sediment is now available for maintaining the Mississippi Delta against coastal erosion and subsidence (Penland and Boyd, 1985). A famous well-documented example (although not located in North America) is the loss of a Mediterranean fishery due to the construction of the Aswan Dam in Egypt. A nursery ground for species of Mediterranean fish had existed behind a sandbar at the mouth of the Nile that paralleled the coast. The loss of particulate material due to currents in the Mediterranean was compensated for by the addition of new sedimentary material from the Nile. When this renewal ceased, the sandbar eroded and the nursery ground was seriously affected (George, 1972).

Alteration of Flow Patterns

Annual flow patterns have been altered not only by dams, but also by diversions, consumptive uses (irrigation, evaporative cooling), and acceleration of runoff in drainage basins. In urban and suburban areas, rain falls on impervious surfaces and is directed into the nearest watercourse via storm sewers. In agricultural areas, drain tiles, ditches, and channelized streams have the same effect. Water drains much more rapidly from logged areas than from the original forest. The end result of these land uses is that flood peaks are higher, and low stages are lower and longer lasting, than in the past because there is less retention of water in the basin itself (Borman et al., 1969; Karr and Dudley, 1981; Herricks and Osborne, 1985). Changes in the flow pattern often trigger unwanted changes in deposition and erosion. Sediments may accumulate in formerly productive channels and backwaters downstream, or a process of headward erosion can begin. During droughts, there may be too little base flow in these modified streams to support aquatic life and other beneficial downstream uses.

Modifications in floodplains, as well as on the upland drainage, have altered flow patterns. Flood protection levees permit the former floodplain to be used for agriculture, industry, or housing, but it is no longer available for fish and wildlife production, production of hardwood timber, recreation, or the storage and conveyance of floods. It is ironic that these levees actually increase flood heights (Belt, 1975). Sedimentation rates increase on the remaining unleveed floodplains to the point that the native vegetation, including valuable hardwoods, may be smothered.

Boat Traffic

Rasmussen (1983) summarized the stresses created by navigation projects and the boat traffic they support. Stresses associated with navigation dams are similar to those described above. Building canals and locks around natural barriers sometimes leads to unwanted introduction of species (the classic example is the invasion of the Great Lakes by the sea lamprey after Niagara Falls was bypassed by the St. Lawrence Seaway). Wing dams are dikes, perpendicular to the shore, that confine the main flow, thus creating a stable channel that tends to maintain adequate depth for navigation by scouring. Sedimentation occurs between the dikes, filling in the productive channel borders. Closing dams across side channels have the same engineering function and the same side effects as wing dams.

Water displacement, propeller wash, and wakes from boats resuspend bottom sediments, increase bank erosion, and can disorient or injure sensitive aquatic species. Aquatic organisms may also be struck by hulls or propellers. Finally, waste discharges and accidental spills from boats or loading facilities can introduce pollutants and exotic species.

Acid

Conducted between 1984 and 1986, the national surface water survey (NSWS) was one of the first activities undertaken by the National Acid Precipitation Assessment Program (NAPAP, 1990).

From a target population of 59,000 stream reaches (211,000 km), overall only 8 percent, or 4,520 reaches (7,900 km), were found to be acidic. Proportions ranged from less than 1 percent in the western United States and southeastern highlands to 39 percent in Florida, but in all other NSWS regions, 12 percent or fewer of the streams were acidic. The major causes of acid streams are acid deposition, acid mine drainage (Box 5.4), and naturally occurring organic acids.

In 47 percent of the chronically acid streams, the dominant acid anions derived from deposition (via acid rain, acid snow, acid fog). In the majority of these streams, sulfate concentrations exceeded base cation concentrations, indicating that the acidic conditions were caused by sulfuric acid. The most likely explanation for the loss of brook trout populations in the Adirondacks is recent acidification caused by high inputs of atmospheric sulfate (NAPAP, 1990).

In 27 percent of the acid streams, organic acids are the main source of acid ions. These streams are located in Florida and the Mid-Atlantic Coastal Plain and are associated with wetlands or organic soils. In Florida, healthy largemouth bass populations are found in waters with pH ranging from 4.0 to 4.5, and there do not appear to be significant population losses. In contrast, the Mid-Atlantic Coastal Plain has experienced a continuing decline of anadromous species since the 1970s. Some streams whose acidity was formerly caused by or-

---

BOX 5.4
ACID MINE DRAINAGE

The United States has a backlog of almost 50 billion tons of old mining and mineral processing wastes (Kleinmann and Hedin, 1990). Therefore it is not surprising that more than 12,000 miles of rivers and streams and 180,000 acres of lakes and reservoirs are adversely affected by mining in the United States (Kleinmann and Hedin, 1990). Acid mine drainage—a fluid generally 20 to 300 times as acidic as acid rain—is responsible for at least a third of this ecological damage (Kleinmann and Hedin, 1990).

In sufficient concentrations, acid mine drainage (AMD) coats stream bottoms with a rust-colored iron precipitate, adds enough sulfuric acid to acidify the water, and kills aquatic life (Kleinmann and Hedin, 1990). Formed from the oxidation of iron pyrite, AMD is associated with coal mining in the eastern United States and with metal mining in the West.

In the past 20 years, the number of rivers and streams adversely affected by AMD has reportedly dropped by about a third, primarily due to perpetual chemical neutralization of mine water before discharge (an expensive process) and by reclamation of abandoned mines (Kleinmann and Hedin, 1990). Some of the improvement, however, has come from natural amelioration by gradual oxidation of the iron pyrite and some by intentional flooding of deep mines to prevent the pyrite from oxidizing.

Still other improvements have been gained by construction of cattail wetlands to purify mine wastewater, usually by bacterial action; more than 400 such wetlands have been constructed in recent years (Kleinmann and Hedin, 1990). Anionic surfactants are also used to inhibit iron-oxidizing bacteria in mine waste piles. Another technique to control AMD caused by fractured streambeds that leak into underground mines is to seal the streambeds by injecting them with polyurethane grout beneath the sediment-water interface to minimize pyrite-water contact.

---

ganic acids have undergone a change in chemical dominance from organic acidity to mixed acid sources, and bioassays indicate that a majority of these streams may be toxic to larval anadromous fish. However, there are so many other contributing factors that it is not possible to link these declines directly to acidification.

Twenty-six percent of the acid streams are the result of acid mine drainage and contain concentrations of sulfate, base cation, iron, and aluminum that are much higher than those found in streams dominated by acid deposition. These streams are found mainly in the Mid-Atlantic Highlands, where 60 percent of the acidic stream length is due to acid mine drainage. Although some relatively low-pH streams contain brook trout, their absence from streams with higher pH may indicate that short-lived acidic episodes can determine the composition of fish communities in some regions.

The effects of low pH on aquatic life are difficult to separate from the effects of other pollutants, physical habitat changes, and changes in stocking patterns that may be occurring simultaneously. Also difficult to sort out are the relative contributions of the various sources of acid anions. It appears, for example, that in some streams, shifts may be occurring in the sources of acidity and the relative proportions of organic and inorganic ions (e.g., streams that formerly derived their acidity from naturally occurring organic acids are becoming more acidic due to deposition of atmospheric sulfate).

Fishing

Sport fish populations appear to be more threatened by habitat loss and pollution than by overharvesting. However, overfishing is a concern in 7 percent of the nation's streams (Flather and Hoekstra, 1989), and Narver (n.d.) includes species reintroduction as one of the nonstructural techniques of river and stream restoration (Tables 5.4 and 5.5). Projections made by Flather and Hoekstra (1989) indicate that as the U.S. population increases, the number of people participating in both cold- and warmwater fishing will increase. An increasing human population implies a further reduction in habitat, resulting in fewer fish per angler. Restricting use of the resource is one way to protect it, but local governments are reluctant to reduce recreational opportunities, and even though state and local governments monitor the population and regulate the catch of important species, there is no way to calculate the illegal harvest. Anecdotal evidence hints at widespread violations of size and creel limits, and a lack of law enforcement (Burgess, 1985). If this is the case, then further regulation may only intensify the illegal fishing pressure.

Releasing hatchery-raised fish is the approach most often used to maintain fishing in many areas that otherwise would not have a sustainable sport fish population. These releases may in themselves constitute a stress when nonlocally adapted strains of fish are released with no understanding of their potential effect on native populations. Highly inbred hatchery stocks may be successfully adapted

TABLE 5.4 Techniques Used in Stream and River Restoration

Structural Methods	Nonstructural Methods
Bank armoring	Flow regulation
Bank overhangs (covers)	Plantings
Brush bundles	Trees
Brush removal	Brush
Channel reconstruction	Herbaceous vegetation
Dams	Grass
Dechannelization	Pollution abatement
Deflectors	Propagation facilities
Fencing	Incubation
Fish passageways	Spawning
Fish screens	Land acquisition
Gabions	Land use regulation
Half-logs[a]	Species reintroduction
Log drop structure	
Meanders	
Riprap	
Rock placement (individual)[b]	
Root wad[c]	
Sediment basins[d]	
Snag placement	
Stream clearance (to remove obstructions)	
Substrate placement	
Trash catchers	
Tree revetement[e]	
Weirs	

[a]Split logs are anchored in the streambed, parallel to the current, with space underneath for salmonids to hide and rest.

[b]Individual rocks are placed in the stream channel to focus the current to protect banks or to provide refuge for fish.

[c]Root wads may be anchored into the stream channel to generate eddy currents for creation of small pools (U.S. FWS, 1984); root wads may also be buried trunk first in reconstructed banks to absorb and dissipate flow energy (see Boxes 5.4 and 5.6 on the Blanco and San Juan rivers, respectively).

[d]Wydoski and Duff (1980).

[e]Tree trunks and branches are angled along banks into current to reduce water velocity (Roseboom and White, 1990).

to a hatchery environment but may not be successful in the wild. Fish originating in different geographic areas may not be able to tolerate conditions, such as low winter temperature, that native stocks tolerate easily. If the introduced fish survive long enough to interbreed with native stocks, their maladaptive genes may not pose a problem until they face an environmental crisis such as an especially

TABLE 5.5 Major Categories of River and Stream Restoration

Bank stabilization
Channel modification
Dechannelization
Fish reintroduction
Flow (volume) augmentation
Organic matter introduction (to increase invertebrate production)[a]
Revegetation
Regulation of land use in watershed
Soil stabilization
Water quality improvement
Water temperature modification

[a]Narver (n.d.).

cold winter; then the entire population (Philipp and Whitt, 1991) and, more importantly, the locally adapted genotype are lost.

Many of the stresses that affect recreational fisheries will also affect commercial fisheries. Salmon populations are closely monitored, and there are already warnings that the salmon harvest is excessive and needs to be restricted to avoid depleting future stocks (Weber, 1986). If restrictions are implemented, it is likely that salmon prices will rise, but the incomes of fishers and of people employed in salmon-dependent businesses will decrease. When a resource such as the salmon fishery has both recreational and commercial value, advocacy groups arise promoting their particular use of the resource. Although sportsmen and commercial fishermen alike recognize that the fishery is a finite resource, allocating the resource appropriately is difficult, with outcomes often based on legalities rather than biological realities (Flather and Hoekstra, 1989).

FLUVIAL RESTORATION

Objectives

Previous sections described the structural and functional characteristics of healthy, undisturbed river-riparian ecosystems and the stresses that have degraded these systems. Here we define and describe the goals and objectives of fluvial restoration.

The goal of restoration is the return of an ecosystem to a close approximation of its condition prior to disturbance (Chapter 1). The essence of a fluvial ecosystem is the dynamic equilibrium of the physical system, which in turn establishes a dynamic equilibrium in the biological components. Therefore, the goal of fluvial restoration should be to restore the river or stream to dynamic equilibrium, not to "stabilize" a channel or bank. The objectives under this broad goal are as follows:

1. Restore the natural sediment and water regime. *Regime* refers to at least two time scales: the daily-to-seasonal variation in water and sediment loads, and the annual-to-decadal patterns of floods and droughts. In arid areas, some organisms depend on rather infrequent (occurring only every few years) and unpredictable flooding. Organisms in large floodplain rivers in tropical and temperate zones depend on highly predictable seasonal flooding.

2. Restore a natural channel geometry, if restoration of the water and sediment regime alone does not.

3. Restore the natural riparian plant community, which becomes a functioning part of the channel geometry and floodplain/riparian hydrology. This step is necessary only if the plant community does not restore itself upon achievement of objectives 1 and 2.

4. Restore native aquatic plants and animals, if they do not recolonize on their own.

Chapter 1 noted that all restorations are exercises in approximation, and fluvial restorations are no exception, given the economic value of water, water-control structures, and structures that are threatened by floods, erosion, and sedimentation. It is unlikely that natural sediment and water regimes, and naturally dynamic channels, can be or will be completely restored throughout the largest river systems of the United States. Benke (1990) found only 42 rivers in the contiguous United States that are more than 120 miles (200 km) long and free flowing. However, there are substantial segments of the Illinois River, Atchafalaya River, and Upper Mississippi River included in public lands (e.g., the Upper Mississippi River Fish and Wildlife Refuge) that do retain floodplains and a flood pulse. The objectives are to add to the existing river-floodplain segments and to restore or rehabilitate degraded segments (see appropriate case histories, Appendix A). In the Illinois and Upper Mississippi rivers, levees are left in place around lands being reclaimed from agriculture or mining for fish and wildlife refuges so that the new refuges are not rapidly degraded by excessive sediment loads carried in by floods. When sediment loads approach predisturbance levels as a result of improved soil conservation in the drainage basin (a process that might take 25-50 years), the levees may be breached. In the meantime, water levels within the refuges approximate a natural cycle in response to seepage through the levees, rainfall, and pumps or gravity drains controlled by refuge managers. The water regime in the restored Kissimmee River will be constrained at the upstream and downstream ends by the need to control water levels in Lake Kissimmee and Lake Okeechobee, respectively. However, this control can be achieved by leaving gates at the lakes and relatively short lengths of the river channelized at the upper and lower ends. In between, the natural flood cycle and dynamic equilibrium will be restored (see case history, Appendix A).

Priorities

Previous sections of this chapter have documented the types and extent of alteration and degradation of the nation's river-riparian ecosystems. Of the nation's total mileage of rivers and streams only 2 percent are high quality, free-flowing segments according to an analysis (Benke, 1990) of the 1982 Nationwide Rivers Inventory (NRI) (but see "Inadequate Information Base," below). According to American Rivers, a conservation organization, approximately 8 percent of the nation's river miles are of sufficient quality to be worthy of special designation and preservation, based on analysis of the NRI and compilation of lists provided by state agencies and conservation groups (Echeverria and Fosburgh, 1988). Only 58 stream segments in 39 states are in the hydrologic benchmark system set up by the U.S. Geological Survey (USGS) to represent streams little changed by man. The point is that 92 to 98 percent of the miles of rivers and streams in the United States are currently so altered that they do not fit legislative criteria for national rivers or wild and scenic rivers, or USGS criteria for a benchmark stream. Estimates of the total river miles in the United States range from 3,120,000 (NRI, as cited in Benke, 1990) to 3,200,000 (Leopold et al., 1964). Given the extent and economic value of water resource development in the United States, it is infeasible to restore 2,870,400 (92 percent of 3,120,000) to 3,136,000 miles (98 percent of 3,200,000) to a "close approximation of [the] condition prior to disturbance" (see Box 1.1).

It does seem reasonable to set a target of restoring as many miles of river-riparian ecosystems as have been affected by point source pollution and urban runoff: 400,000 miles, or 12 percent of the total 3.2 million miles (U.S. EPA, 1990). This target is also commensurate with recommendations of the President's Commission on Americans Outdoors (1986) regarding the need for outdoor recreation and aesthetic environments. The goal should be to move fluvial ecosystems as many steps as possible from the negative side of the habitat quality index toward the positive side (through rehabilitation, creation, or full restoration).

Given a target total of 400,000 miles, what are the priorities for fluvial restoration? Prioritization should be based on both human and ecological values, as suggested in Chapter 3 (see Figure 3.3 and discussion of human influences). Restoration measures that save money or human lives as well as ecosystems should be undertaken as quickly as possible. These include floodplain and riparian zoning, soil conservation in lieu of channel or reservoir dredging, removal of flood-prone structures, razing of unsafe dams, and reduction of government subsidies that promote overgrazing or deforestation of riparian zones. The tax dollars saved by these measures should be applied to other restorations that may not have offsetting economic benefits in the short term, but have high ecological or human values in the long term: chief among these should be preservation of biodiversity through preservation and restoration of critical aquatic habitats. Prioritization on the basis of preserving biodiversity is likely to include a range of stream and river sizes throughout the country. For example, there are springs and small streams in the arid West where populations of several species of endangered desert pupfish occur. The number of species of fishes and mollusks generally increases with stream order, arguing for preservation and restoration of segments of large rivers. One group of small fish, the darters, reach their highest number of species in streams and midsized rivers of the Tennessee drainage, whereas the species richness of aquatic insects is probably greatest in headwater streams.

It is especially important that portions of large rivers be restored, for several reasons. Because many miles of streams coalesce into relatively few miles of mainstem rivers, large rivers are relatively uncommon. Large river-floodplain ecosystems were disproportionately degraded because of their value for a variety of human uses, and the resultant concentration of human populations and development. Of all wetland types, bottomland and hardwood forests along the Lower Mississippi River have suffered the greatest diminution through leveeing, drainage, and clearing (see Chapter 6). Small streams receive some degree of protection by virtue of being located in federal or state forests, parks, and other types of protected land, but there are few programs for the protection of larger rivers, as Benke (1990) points out.

Techniques

Nonstructural techniques can be broadly defined as any restorative method that does not involve either physical alteration (e.g., realignment of the channel, riprapping of the banks) of the river or construction of a dam or some other structure (see Table 5.4).

NONSTRUCTURAL TECHNIQUES

Rivers and streams are resilient and can sometimes recover if the stress is removed and they are simply left alone. The recovery of the Pere Marquette River in Michigan is a good example (Box 5.5). During 30 years of benign neglect following the clear-cutting of the surrounding forest and floating huge volumes of logs down the river to Lake Michigan, the ecosystem began to recover. Although simple neglect has worked in a few instances, it is not likely to achieve much restoration on a national scale, especially on larger streams and rivers where there are multiple stresses, competing uses, and downstream effects from upstream disturbances. Nonstructural techniques include administrative or legislative policies and procedures that stop or regulate some activity, such as withdrawal of water from a river or land use practices that degrade fluvial systems.

Legislative and Administrative Approaches

Reserving flow or reclaiming flow for in-stream uses (fish, wildlife, outdoor recreation) is an example of a legal approach to restoration in regions where water is in short supply and fully committed to withdrawals for crop irrigation, stock watering, or public water supply. Although long regarded as primarily a problem in the arid West, the issue of in-stream flow is being joined elsewhere. Droughts such as the 1988-1989 drought in the Upper Mississippi Basin saw many municipalities asking for permits to withdraw virtually the entire flow of some rivers (e.g., the Mackinaw River in Illinois).

In-Stream Flow

There is a need to amend the appropriative doctrine that is the basis of water law in the West so that flow is reserved for in-stream uses of water for fisheries and other aquatic life, boating and canoeing, aesthetics, and environmental purposes (Lamb and Doerksen, 1990). The existing water laws have two primary principles: (1) first in time is first in right, and (2) beneficial use of water is the basis of the right. Beneficial use in the past meant diversion for agriculture, industry, and municipal water supply. When water is scarce, those who established their appropriative rights last must stop using water until the needs of the more senior users are satisfied. In 1969, Montana became the first western state to provide for the legal acquisition of a water right for in-stream uses; since then, 13 states have followed suit (Lamb and Doerksen, 1990). Although all states except

BOX 5.5
THE PERE MARQUETTE:
A CASE STUDY OF BENIGN NEGLECT

The Pere Marquette (PM) and its tributaries flow through approximately 138 miles of the northern third of Michigan's lower peninsula before emptying into Lake Michigan at Ludington. One of the few remaining free-flowing cold-water rivers in the contiguous United States, the PM has not only never been dammed, but is also extraordinarily clean and free from development despite a spate of ecologically devastating timber practices in the latter part of the nineteenth century. The history of the PM includes multiple use and periodic stress on the ecosystem, yet the watershed has emerged remarkably intact. Indians, timber barons, canoeists, trout fishermen, and others have all used the PM, and the river seems to have evolved both because of and in spite of humanity's changing needs.

The timber industry's exploitation of the Pere Marquette region was so encompassing and voracious that in the early 1900s experts pronounced the river "dead." Among the many repercussions of the widespread clear-cutting were deforestation and its attendant effects on flora and fauna; water warming; siltation and bank erosion due to eradication of cover; and increased damage to banks, fish, and water quality due to the tremendous infusion of logs into the river. Subsequent ramifications included significant changes in runoff due to widespread brush fires and abortive attempts at agriculture.

The land, once cleared, was of little use to the timber industry, so much of it eventually reverted back to state ownership due to tax delinquency. The region's sparse population meant that after the exodus of the loggers, the river suffered little human stress. Remarkably, the ecosystem flouted reports of its demise and began to recover. Local inhabitants and the federal government began taking an active interest in the river's restoration. Mass replanting of cutover lands throughout the area by the Civilian Conservation Corps during the Depression led, in 1938, to the creation of the Manistee National Forest, a federal holding covering a considerable portion of the PM watershed.

With increased use of the river by sportsmen, the federal government assumed a more prominent role in stocking and managing the fishery. The most ecologically significant governmental maneuvers include the planting of salmonids in the PM tributaries and several controversial attempts to control lamprey eel infestation. Salmon were introduced both for their sport-fishing value and to control the overabundance of alewives in Lake Michigan. Many trout fishermen complain, however, that the annual salmon spawning run up the PM destroys trout habitat and leaves the banks of the river strewn with dead and rotting salmon.

The sea lamprey made its way into the river as a result of seagoing shipping traffic on the lake. Eventually the infestation reached such dramatic proportions that the Department of Natural Resources resolved to control proliferation through periodic applications of 3-trifluoromethyl-4-nitrophenol (TFM), an effective lamprey larvicide. Use of TFM did bring the lamprey population under control, but under certain water quality conditions it is also toxic to mammals, fish, and insects. As a result, fish habitat deteriorated and fish abundance decreased. Alternative attempts at lamprey control included the construction of an electric weir to deter lamprey movement upriver, but the weir is currently not in service due to detrimental effects on steelhead migration.

Designation of the PM as both a natural and a scenic river has substantially increased its use by sportsmen and canoeists. Canoe traffic over the last 20 years has risen from perhaps 100 canoes per week to more than 500 per day during tourist season. Expanded human use (including increased fishing) has, in turn, affected the aquatic habitat, and trout and salmon populations have declined further. Creel limits have been drastically reduced over the past 25 years, and canoe traffic is now regulated by the U.S. Forest Service.

The Pere Marquette, though greatly changed, remains free-flowing, clean, and remarkably resilient. Rather than treating it as a resource to be exploited for some human endeavor, most of the PM's management involves maximizing its potential as aquatic habitat and as a scenic river while controlling commercial and residential development.

New Mexico have some sort of in-stream program, acquisition of a right to in-stream use is especially effective because (1) the in-stream use passes all tests of legal legitimacy and the terms of the right are spelled out; (2) the in-stream use has a priority date, so that it is superior to all subsequent rights; and (3) even if the in-stream use is junior in right to other uses, the junior user can legally prohibit a change in stream conditions from those existing when the right was established if the change will damage the junior use (Gould, 1977; Lamb and Doerksen, 1990;).

In the eastern states, water quality rather than water quantity was the problem historically, and the relevant legal principle was "reasonable use" by riparian landowners, if that use did not interfere with the water rights of others along the river or stream (Ausness, 1983). Later, many eastern states moved toward a permit or water-allocation system, to provide water to people who do not own riparian lands (Lamb and Doerksen, 1990). Most eastern states have some statutory provision that can be used to reserve stream flows in time of shortage, but these vary widely in effectiveness and application.

The doctrine of Federal Reserved Water Rights allows the federal government to reserve in-stream flows to fulfill the purposes of certain federal lands (national forests, parks, wildlife refuges, and wild and scenic rivers; Lamb and Doerksen, 1990). The priority date for these uses is the date on which action was initiated to create or change a federal reservation. The doctrine legitimizes in-stream uses of water that might not be recognized under existing state laws, and it gives these uses much earlier priority dates than would most state laws. It applies to future as well as present needs and might cause in-stream uses to supersede other, more senior rights. As of 1987, claims under the doctrine had been for very small amounts of water (Lamb and Doerksen, 1990), and it appears that this relatively new legal tool for maintenance or restoration of in-stream flows could be put to much greater use.

Once the legitimacy of in-stream uses has been established, the next task is to determine what flows those uses require. A relatively simple but crude approach that is appropriate during preliminary planning for a project or to provide a baseline of protection is to determine the minimum flows necessary for fisheries, canoeing, or other in-stream uses. Examples include the lowest flow on record, flows equaled or exceeded 90 percent of the time, or the point at which the wetted perimeter begins to fall sharply with small reductions in flow (Trihey and Stalnaker, 1985). Incremental methods estimate the quality and quantity of fish habitat at each increment of flow and are more suitable where the goal is to restore or upgrade fish populations and where water is in great demand. The Instream Flow Incremental Methodology (IFIM) (Bovee, 1982) is now used by 38 states and is becoming accepted as a "standard" method (Lamb and Doerksen, 1990). It is labor- and data-intensive and requires field measurements and hydraulic modeling, but it provides fairly precise answers to the question: What is gained by a given increment in flow? Its weakness is that it is species specific and inapplicable to multispecies assemblages (Hughes et al., 1990). It is most applicable to western streams where it was first developed. These streams are usually occupied by a small number of highly valued sport species (trout and salmon) whose use of particular habitats under different flow regimes can be visually determined in the relatively shallow, clear waters. Wiley et al. (1987) found that IFIM was a poor predictor of sport fish population density in Illinois, and they recommended collection of habitat preference data for local populations of native species.

Flow Regime

An issue related to in-stream flow is the flow regime, or pattern of high and low flows, particularly below hydroelectric and irrigation supply dams. Daily fluctuations occur below hydroelectric dams, which are often used to supply power during periods of peak demand for electricity. The flow below irrigation storage dams is often the reverse of the normal annual pattern, with minimal flow during the wet months because water is being stored behind the dam, and more flow during dry periods, if there is return flow from the irrigated lands. A nonstructural means of securing more natural flow regimes is to renegotiate release schedules when permits and licenses come up for renewal. Echeverria et al. (1989) have provided a citizens' handbook on how to negotiate more favorable release schedules.

Floodplain Management

The inverse of the water shortage issue is the issue of floods on floodplains. Johnston Associates (1989) describe four eras in the history of floodplain management: (1) the structural era, 1900 to 1960; (2) a turning point in the 1960s; (3) the environmental decade, 1970 to 1980; and (4) maturation in the 1980s. Congressional attitudes have responded to growing urbanization and environmental awareness by shifting emphasis from major flood control and other water resource projects to risk management, environmental improvement, protection of ecosystems, and urban water quality. What started as separate programs for water resource projects, disaster assistance, and environmental quality has become better integrated, and the focus in the 1980s was on implementation of policies and programs rather than new legislation or institutional changes. Landmark events in this evolution were: (1) House Document 465, *A Unified National Pro-*

gram for Managing Flood Losses, August 1966, which has been called the "Magna Carta of contemporary floodplain management planning" (Donovan, 1983); (2) subsequent revisions of H.D. 465 in 1976, 1979, and 1986 to integrate flood insurance and floodplain management objectives and to incorporate executive orders on floodplain management and protection of wetlands; and (3) revision in 1979 of the "Principles and Standards for Planning Water and Related Land Resources" (Johnston Associates, 1989), requiring federal agencies to prepare and consider a nonstructural alternative plan whenever structural water resource projects are proposed and encouraging specific consideration of the ecological values associated with floodplains as part of the evaluation process. The final trend has been decentralization of the federal role and greater sharing of the responsibility for floodplain management with state and local governments, in response to federal deficit reduction policies and growing technical expertise at the state level (Johnston Associates, 1989).

Examples of nonstructural methods of floodplain management that promote preservation or restoration of floodplains are adoption of regulatory floodways, purchase of easements to prevent construction, and purchase of land and removal or relocation of structures. Communities must adopt a regulatory floodway to be eligible for the National Flood Insurance Program (NFIP). Any development within the floodway (including cumulative developments) that would increase the height of the 100-year flood (a flood whose probability of occurrence in a given year equals 1 percent) by more than 1 ft is prohibited. Some states have much more stringent requirements for their floodways: Massachusetts permits no increase in water levels, Wisconsin allows only 0.01 ft, and Illinois and Indiana allow 0.1 ft (Johnston Associates, 1989). Some states and communities have adopted setback standards for structures along designated streams and rivers, but there are no setbacks required by the NFIP.

Another approach is to buy out drainage and levee districts on floodplains and restore the original conditions. The Banner Special Drainage and Levee District on the Illinois River south of Peoria was purchased by the Illinois Department of Conservation, was renamed the Banner Marsh Conservation Area, and is now being restored to lakes and wetlands. Twenty miles downstream of Banner Marsh, the U.S. Fish and Wildlife Service is analyzing the costs and benefits of restoration of Thompson Lake, now the Thompson Drainage and Levee District and part of the largest farm complex (Norris Farms) in Illinois (Roelle et al., 1988).

Establishment of Greenways

Greenways are protected, linear, open-space areas that are either landscaped or left in their natural condition. They may follow a natural feature of the landscape such as a river or stream, or they may lie along a disused railway line or some other right of way.

Little (1990) recognized the existence of five categories of greenways: urban riverside greenways, recreational greenways, ecologically significant natural corridors, scenic and historic routes, and greenway systems or networks. All are intended to provide some degree of protection for nearby natural features; however, only one of these categories, the ecologically significant natural corridor, is of special interest from the perspective of riparian restoration. The importance of the protection that a buffer strip along a stream or river affords to the aquatic ecosystem has been emphasized previously in the section on river-riparian ecosystems. Two examples of ecologically significant natural corridors cited by Little are the Willamette River greenway in Oregon (see Willamette River case study, Appendix A) and the Oconee River greenway along the river's north and middle fork and tributaries, all north of Athens, Georgia.

The purpose of the Willamette greenway as stated by the Oregon legislature is to protect and preserve the natural, scenic, and recreational qualities of the lands along the river, while preserving and restoring features of historic interest. Much land has been protected along the Willamette since the passage of the Greenway Act in 1967. In addition, the uses of public and private lands have been regulated under greenway rules, and five state parks have been established. Despite the achievements, the Willamette greenway should not be considered an ideal plan. Agricultural land is exempt from greenway regulations, and some residential development and destruction of vegetation are occurring along the river within the 150-ft setback zone. Also, addition of new land to the greenway has been slow in recent years after an initial flurry of acquisition activity (JEL, 1989). The Oconee River greenway is essentially a protection plan that controls land use for a mile on either side of the river.

Fencing

In many cases, recovery of riparian vegetation, channel morphology, and fish populations has occurred where livestock were simply excluded from the riparian zone. Sheep Creek in Colorado was fenced to protect it from heavy use by both humans and livestock (Stuber, 1985). Vegetation recovered, the stream became narrower and deeper, and the estimated population of trout in the fenced area was twice that in unfenced areas. Otter Creek in Nebraska was severely degraded by overgrazing until the headwaters were leased by the Nebraska Game and Parks Commission (Van Velson, 1979). Within 3 years, the average width of the stream decreased, pools formed, less sand was deposited on the gravel spawning beds, the water temperature became cooler and more favorable for native fish, and the stream banks stabilized. After 6 years, Van Velson (1979) reported that 20,419 young fish were produced in the 2 miles of stream within the 3.34-acre leased zone.

STRUCTURAL TECHNIQUES FOR FLUVIAL RESTORATION

Simple removal of stresses through legislative or administrative action may not restore stable, degraded systems such as the Blanco River (Box 5.6 and case study, Appendix A) or the Illinois River. Intervention may also be desirable where natural restorative processes can take decades to centuries (Pere Marquette River, see Box 5.5). In the case of a clear-cut old-growth forest, it might take decades for the canopy to close over and shade the streams, and even longer for deadfalls to replace the dams of woody debris that wash out because the basin is relatively barren, and runoff and flooding are consequently greater. In these cases, structural techniques are needed to shift the equilibrium or speed up the restoration process. Amendments to existing man-made structures (dams, spoil banks, levees) can restore some populations and processes. For example, structural modifications of dams range from their complete removal (Box 5.7) to installation of fish ladders, selective water-withdrawal structures (e.g., so that warm, oxygenated water from the surface of a reservoir can be discharged downstream to a warmwater fishery, instead of cold, deoxygenated deep water), and aspirators or other devices in hydroelectric dams to aerate discharge water.

Structural modifications to the river-riparian ecosystems themselves range from the scale of species-specific habitat improvements (in fact, fish biologists use the term *structure* to refer to logs, root wads, or man-made devices that fish use for shelter) to recreation of a preexisting channel morphology (see Box 5.3). Channel or bank modification techniques that use vegetation in a variety of innovative ways are referred to as *biotechnical engineering* (Brookes, 1988). As can be seen in Table 5.6, the costs of traditional bank sloping and riprap greatly exceed the costs of using natural or "soft" engineering approaches. *Soft engineering* (source unknown) refers both to the goal of recreating the natural fluvial system and to the use of locally available

---

BOX 5.6
RESTORATION OF THE BLANCO RIVER

This discussion of the scientific, technological, and administrative aspects of the Blanco River reconstruction project in
southwestern Colorado focuses on the channel stabilization
and fishery problems encountered and the processes used to
solve them.

The case study (see Appendix A) illustrates the use of "soft
engineering" techniques and natural materials to combat stream
and river degradation and bank erosion.   Soft engineering
techniques restabilize river channels and banks without straightening them and without confining water flows in concrete or
riprapped channels.   Instead, this approach requires study of
the river's natural hydrological and hydraulic tendencies and
subsequent use of earth-moving equipment to return the fluvial system to a stable, naturalistic configuration.   The redesigned and repaired stream or river channel is strengthened
with natural materials, such as rocks, logs, root wads, and live
riparian vegetation, to help preserve the new banks and channel.

Before repair work began on the Blanco River in 1987, target sites on both branches of the Blanco River were broad,
shallow, and braided, with no pools. In the course of the 3-
year river reconstruction project directed by hydrologist D. L.
Rosgen (1990, 1991), the river's bank-full width was reduced
from a 400-ft-wide braided channel to a stable, 65-ft-wide
channel with a high pool-to-riffle ratio. Even before conclusion of the project in 1990, major improvements had occurred
in the fishery and in the site's appearance.

The Blanco River project site now has new meanders, deep
pools, new flood terraces, rebuilt floodplains, riparian vegetation, verdant pasture grasses, and banks stabilized with
locally obtained root wads, tree trunks, and boulders.   The
current is focused into the center in the riffle reaches of the
channel by strategic placement of "vortex rocks" in the channel to create cover and spawning habitat. Deep pools were
created on the outside of bends in the channel.   The new
stable channel complex has a natural look compared with cement trapezoidal channels, levees, and riprapped banks.   The
fishing is a delight to landowner and visitors alike.

---

natural materials such as woody debris and alluvium (Box 5.6 and
5.8), in contrast to the "hard," hydraulic engineering approach, which
typically optimizes for one use (flood conveyance, drainage) and utilizes concrete, sheet piling, riprap, or other imported material.

BOX 5.7
DAM REMOVAL

Impoundments have a definite life span because (1) dams deteriorate (concrete material deteriorates in 50 to 100 years), (2) sediments inexorably fill reservoirs, and (3) human technology and human needs change. In establishing a 50-year maximum term for licenses for hydroelectric plants, Congress recognized that public needs and interests change. When a license expires, the Federal Energy Regulatory Commission (FERC) must determine how the public interest is best served (Echeverria et al., 1989). Between 1991 and 1993, more than 200 power projects, representing perhaps more than twice that many dams will be due for license renewal (Echeverria et al., 1989). License renewals and structural deterioration both provide unique opportunities to restore natural functions of rivers by requiring structural or operating changes that allow fish migrations or benefit other in-stream uses. In some cases, restoration of the free-flowing river by removal of the dam may even be feasible, although only a few such examples exist as yet.

Removing a dam may be cheaper than repairing an unsafe dam or one that has failed. Catastrophic dam failures with loss of life and property, such as the Teton Dam failure in 1976, have brought national attention to the safety of large dams. In 1982 the U.S. Army Corps of Engineers identified more than 9,000 high-hazard dams out of 68,000 nonfederal dams inspected. One-third, or 2,925, were evaluated as unsafe, primarily due to inadequate spillway capacity. State estimates of the cost to repair 1,570 unsafe nonfederal dams was $1.22 billion (FEMA, 1985). Extrapolating that figure to all 2,925 unsafe high-hazard dams gives a total estimate for repairs of $2.24 billion (Johnston Associates, 1989).

The Maine legislature passed a resolution in 1990 calling for the removal of the Edwards Dam on the Kennebec River in Maine near Augusta by the year 2000. Despite modification to allow fish passage, state officials say that the dam, which provides power to fewer than 2,000 households, still blocks the migration of Atlantic salmon (Egan, 1990). Eleven other species of fish including shad, smelt, and sturgeon, prospered in the Kennebec before the construction of the dam in 1837 and other developments that impaired fishing. It remains to be seen whether this resolution will be acted on.

Removal of Gilnes Canyon Dam on the Elwha River in Olympic National Park is likely to occur because of the concurrent endorsement of the plan by the two federal agencies involved (the U.S. Fish and Wildlife Service and the National Park Service) following a thorough 7-year study, which included an economic analysis of the costs and benefits. The Elwha River was one of the few in the nation to support all five species of Pacific salmon, including 100-pound king salmon and enough chinook salmon to feed the Lower Elwha Indian Nation all year (Egan, 1990). Constructed in 1924 to produce hydroelectricity—before the land was a national park and before fish ladders were required on many dams—the dam not only blocks the passage of salmon but, along with a lower earthen structure built downstream, has also caused the virtual disappearance from the valley of 22 species of birds and mammals that in some way depended on the salmon (Egan, 1990).

Power from the dam is sold to a paper company. A 7-year study by the federal government found that the dam was costing $500,000 per year in lost revenue from fish runs and tourism. After determining that fishways around the dam would not succeed in restoring the salmon runs, the government concluded that the runs could be restored if the dams were removed. Both the U.S. Fish and Wildlife Service and the National Park Service have endorsed the removal (Egan, 1990).

Restoration of a formerly impounded reach of the Milwaukee River in West Bend, Wisconsin, followed removal of the Woolen Mills Dam, after the Wisconsin Department of Natural Resources (WDNR) ordered the city to rebuild or remove the dam for reasons of public safety (Nelson and Pajak, 1990). The dam was constructed in 1919, impounded 67 acres, and had a head of 14 ft. With intensive community involvement, WDNR developed a10-year plan for dam removal, coupled with restoration of both the riparian zone and a free-flowing river in the 1.5-mile reach that was formerly impounded. Habitat Suitability Index models for smallmouth bass (Edwards et al., 1983), northern pike (Inskip, 1982), and common carp (Edwards and Twomey, 1982) were used to evaluate the impact of dam removal on those key species. The smallmouth bass model was used to plan the type and extent of habitat restoration required to achieve the goal of restoring a riverine sport fishery, subject to the constraints of cost-effectiveness, public safety, and aesthetics in an urban park setting (Nelson and Pajak, 1990).

The Milwaukee River restoration appears to be succeeding, although long-term (10-year) surveillance is needed to quantify changes in fish populations (Nelson and Pajak, 1990). Anglers are catching smallmouth bass, as well as an occasional wall-eye and northern pike. Numerous young-of-the-year smallmouth bass have been observed, indicating that substantial recruitment is occurring. Although most of the original channel had filled with silt and sand, natural scouring removed most of the fine material within 6 months, leaving coarser substrate that now makes up 64 percent of the channel and provides better habitat for smallmouth bass (Nelson and Pajak, 1990).

The removal of the Woolen Mills Dam and restoration of a portion of the Milwaukee River, and the proposed removal of either the Edwards Dam or the Gilnes Canyon Dam, may set precedents that could lead to other dam removals and river restoration efforts.

TABLE 5.6 Costs of Bank Stabilization

Method	Cost per Linear Foot of 12-ft-High Stream Bank[a]
Palmiter tree revetments[b]	$3.73
SWCD[c] tree revetments	$3.00
Willow posts	$3.10
Bank sloping (1:3) and riprap	$12.60

[a]Excluding costs for technical assistance.
[b]See Box 5.2 for details.
[c]Knox County (Illinois) Soil and Water Conservation District.

SOURCE: Reprinted by permission from Roseboom and White, 1990. Copyright © by International Erosion Control Association, Steamboat Springs, Colo.

Relatively modest structural changes may have dramatic beneficial effects if the hydraulic forces of the river are harnessed or carefully directed. George Palmiter describes his techniques as "making the river do the work" (see Box 5.3). Instead of removing mid-channel bars with earth-moving equipment, he directs scouring flows toward the bar and cuts underlying logjams into pieces small enough for the current to carry away. Patience may be required for any project that relies on hydraulic forces to effect restoration. Flows that reshape channels or flush fine particles out of gravel beds may not occur every year, or only a few times a year, so it may take several years before the desired end point is achieved. When the dam was removed on the Milwaukee River in West Bend, Wisconsin (see Box 5.7), it took 6 months for the river to scour much of the accumulated silt and sand, and leave coarser bed material that was better for smallmouth bass. Monitoring and evaluation in these situations should be strongly event dependent, rather than on a fixed schedule. The effectiveness of these types of projects should be evaluated following channel-forming or substrate-flushing flows.

Species-Centered Restoration

Anglers have organized into groups such as Cal Trout, Federation of Fly Fishers, Trout Unlimited, and United Anglers to work for improved fishing and fish habitat improvement. Much of the "restoration" of small streams and rivers has come about as a result of efforts by these groups, often supported by sympathetic government agencies, to manipulate the degraded aquatic habitat in order to maximize production of salmonids or other prized game fish species. Federal agencies, such as the U.S. Forest Service and the U.S. Fish and Wildlife Service, have also been heavily involved in stream habitat rehabilitation. Federal involvement in stream projects dates at least from the mid-1930s, when Civilian Conservation Corps workers installed log and rock dams throughout streams in much of the West. State resource agency involvement and that of private groups date from at least the early 1930s (Wydoski and Duff, 1980).

Much stream work today and in the past has been directed at improving the welfare of salmonids (Table 5.7). An abundance of technical and popular literature attests to the effectiveness of well-planned and well-executed stream improvement projects in increasing the quality and quantity of trout and salmon production (Duff and Banks, 1988). Sometimes, however, this work has been done at the expense of other members of the aquatic community, such as beaver (Flick, n.d.).

Stream improvement projects, as defined by Raleigh and Duff (1980, pp. 66-67) "are attempts to produce, restore, and maintain" stream habitat features essential to trout, such as "clear cold water, a rocky substrate, an approximate pool to riffle ratio of 1 to 1 with areas of slow deep water, a relatively stable flow regime, well vegetated stream banks, and abundant instream cover." Practitioners of species-centered stream management generally introduce artificial structures into stream and river environments to modify banks, channel, bed, or current in hopes of improving salmonid or other game fish productivity.

When this work is done without a profound understanding of the interactions among stream hydrology, fluvial geomorphology, and fish, the least detrimental consequence may be that mechanical structures emplaced in the stream at considerable expense and trouble could be of limited durability and longevity.

Much more serious damage, however, can be done to the stream or river environment by inducing undesirable compensatory adjustments of channel and banks (Raleigh and Duff, 1980; Rosgen and Fittante, 1986; Heede and Rinne, 1990). Stream variables, such as velocity, depth, width, viscosity, parent material, pool-riffle interval, sinuosity, slope, sediment transport, bed-load transport, and bed form are interrelated. Heede and Rinne (1990, p. 257), in a paper that should be required reading for anyone planning stream "improvements," suggest that "the designer should recognize the ongoing physical processes in the river or stream, and, if at all possible, should work with the processes and not against them," using design hints from healthy natural nearby streams. (For an illustration of nature used as a model in river restoration, see Box 5.8; also see Box 5.6.) Changes made in the banks, channel, or gradient by those unable to anticipate either the future natural tendencies of the stream or the probable impact of their intervention on stream hydromorphology may be ill advised.

When stream or river management actions are taken without recognizing whether the aquatic ecosystem is in dynamic equilibrium or disequilibrium, the manager is gambling with the stream or river rather than ensuring improved ecosystem function and dynamic stability (Heede and Rinne, 1990). The well-intentioned but intuitive approach may therefore cause unexpected harm even to species that were meant to be helped. Even when expertly done, trout- or salmon-maximizing stream modifications may result in symptomatic treatment of streams' "defects" from the perspective of salmonid reproduction and survival, rather than a more holistic effort to return the entire stream ecosystem to a biologically healthy condition. Gore (1985) pointed out that even from a fish-centered point of view, restoration of macroinvertebrate communities is essential because they usually are a major portion of the food base for fish. Moreover, benthic community restoration and recovery require the smallest amount of capital investment and least sophisticated structure development. Also, managers should have a better appreciation of the importance of a "keystone species" (Paine, 1966) or "strong interactors" (MacArthur,

TABLE 5.7 Summary Results of 22 Successful Salmonid Habitat Improvement Evaluations

Improvement Technique	Principal Investigator	Application Site	Past Improvement/Biological Changes
Stream-bank fencing	Stuber (1985)	Sheep Creek, Colo.	Biomass of trout (mainly brown trout) was 96% greater in 1983 and 127% greater in 1984 in fenced study zones than in unfenced zones.
	Gunderson (1968)	Knok Creek, Mont.	Average number of brown trout over 6 inches was 27% greater, and average biomass was 44% greater, in ungrazed reach than in adjacent grazed reach.
Boulder groupings	Ward and Slaney (1980)	Keagh River, N.C.	200% increase in coho salmon spoils (to 4,800 per mile).
Stream-bank riprap	Kerr (1985)[a]	Willow Creek, Wis.	Average number of brown trout over 6 inches increased by 35%, and average number over 10 inches increased by 66%.
Half-logs	Hunt (1978)	W. Br. White River, Wis.	Average number of brown trout over 10 inches in April increased by 553%, and average biomass increased by 187%.
	Apelgram and Stewart (1984)[a]	Kinniakinnio River, Wis.	In five study zones the average number of brown trout over 6 inches increased by 41%, the average number over 10 inches increased by 34%, and average biomass increased by 51%.
Stream-bank debrushing	Hunt (1978)	Spring Creek, Wis.	Average number of brook trout over 6 inches in October increased by 53%, and average biomass increased by 34%. Growth ratio of ages 0-2 also improved.
	Hunt (1985)	Lunch Creek, Wis.	Average number of brown trout over 6 inches in September increased by 51%, and average number over 10 inches increased by 82%.
Stream-bank debrushing and half-logs	Cornelius (1984)[a]	Clam River, Wis.	Average midsummer abundance of brook trout and brown trout over 6 inches increased by 65 and 523%, respectively.

Method	Location	Reference	Results
Stream-bank debrushing, brush bundles, and half-logs	Radley Creek, Wis.	Hunt (1986)	Average number of brown trout over 10 inches increased by 41% in one study zone and by 42% in another study zone. Average biomass increased in the two zones by 35 and 50%, respectively.
Bank covers and current deflectors	Lawrence Creek, Wis.	Hunt (1986)	Average number of brook trout over 8 inches increased by 192%, and average biomass increased by 130%. Angler hours increased by 146%, and harvest increased by 191%.
	Big Roche-a-Cri Creek, Wis.	White (1972) and WDNR (1975)	Average biomass of brook trout increased initially by 159% and long-term by 839%. Angler harvest increased initially by 96%. (No long-term measurement was made of angler harvest.)
	MacIntire Creek, Wis.	Thussler (1978)[a]	Average number of brook trout and brown trout over 6 inches in midsummer increased by 84 and 431%, respectively. Average biomass of brook trout increased by 40% and of brown trout by 490%.
"Skyhook" bank cover and current deflectors	Plover River, Wis.	Hauber (1978)[a]	Average number of brook trout and brown trout over 8 inches in midsummer increased by 128 and 200%, respectively. Average number of brown trout over 14 inches increased by 253% (to 67 per mile).
	Prairie River, Wis.	Hauber (1985)[a]	Average number of brook trout and brown trout over 6 inches in midsummer decreased by 40% and increased by 426%, respectively. Average biomass of brook trout decreased by 41%, but average biomass of brown trout increased by 578%.
	Hunting River, Wis.	Hauber (1985)[a]	Number of brook trout and brown trout over 6 inches in June increased by 26 and 91%, respectively. Biomass of brook trout and brown trout increased by 20 and 88%, respectively.

TABLE 5.7 (*Continued*)

Improvement Technique	Principal Investigator	Application Site	Past Improvement/Biological Changes
Sandbag bank cover and current deflectors	Ironside (1984)[a]	Neensh Creek, Wis.	Average number of brown trout over 6 inches in midsummer increased by 181% in Station 1 and by 756% in Station 2. The average number over 10 inches increased by 75% in Station 1 and by 124% in Station 2.
Bank covers, current deflectors, riprap	Glover (1986)	Rapid Creek, S.D.	During the 3rd-5th postdevelopment years, average abundance of brown trout increased by 357% whereas average abundance of NL suckers decreased by 89% and average number of white suckers decreased by 70%.
	Johames (1983)[a]	Ongtown Creek, Wis.	Average number of brook trout over 6 inches in September increased by 105%, and average biomass increased by 65%.
Jauk-dams, tip deflectors	Spotts (1986)	Blockhouse Creek, Pa.	Average biomass of brown trout in late summer during the 3rd and 4th postdevelopment years was 752% greater than predevelopment biomass.
Stream-bank de-brushing, brush bundles, bank cover, and riprap	Johames (1985)[a]	Beaver Brook, Wis.	Average number of brook trout and brown trout over 6 inches in July increased by 65 and 125%, respectively.
Current deflectors, bank cover, and log/rock dams	Hale (1969)	Split Knok Creek, Minn.	Average number of brook trout in September increased by 356%, and average biomass increased by 60%. Average biomass of white suckers decreased by 81%. Angler hours increased by 203%, and harvest increased by 362%.

[a]Personal communication memoranda from principal investigators to R. L. Hunt (in press).

Stream Restoration Recommendations

1. Focus on identifying "limiting factors" at work in each candidate stream for renovation. Try to eliminate or ameliorate those factors that depress salmonid carrying capacity.

2. Maintain or enhance base flow whenever possible (natural flow of a stream when it is not being augmented by surface runoff). Riparian zone and entire watershed management activities should be considered to achieve greater and more stable base flow.

3. Consider species-specific and age-specific requirements of the salmonids present, including both environmental suitability and social interactions with other fish species and/or age groups.

4. Follow a logical sequence of habitat improvement steps. These steps should usually include examination of site, diagnosis of needs, prescription of remedies, planning and organization of work to be done, on-site treatment/development, evaluation of results, and maintenance of development.

5. Disguise artificially man-made structures or modifications of channel shape. Restore aesthetic conditions as quickly as is practical.

6. Tailor management activities to the individual stream.

7. Preserve, restore and accentuate the two most common natural characteristics of streams—the meandered channel profile and the riffle/pool sequence.

8. Work with, not against, the inherent capacity of streams and watersheds to repair their biotic health.

9. Encourage the right kinds of stream-bank vegetation to become dominant, depending on the character of the stream and riparian zone.

10. Make the stream flow work beneficially. Bring the main threads to flow close to hiding/resting/security cover for trout.

11. Integrate habitat management in the stream channel with other terrestrial management activities along the stream's riparian zone and the larger watershed (see also No. 2 above).

SOURCE: Hunt, 1988a.

BOX 5.8
SAN JUAN RIVER RESTORATION

The reconstruction of a mile of the East Fork of the San Juan River in southwestern Colorado illustrates what can be accomplished on very steep, unstable rivers with badly eroding banks through the application of "soft engineering" techniques using natural materials but without resort to channelization or riprap. The work demonstrates that a naturalistic, workable alternative now exists.

Removal of willows by burning and plowing the bottomland along the river in the early 1930s led to the creation of an unstable, braided river channel that migrated back and forth across the valley floor. Without willows to hold riverbank soil, the river eroded its banks, washed away valuable land, and became wide and shallow. Damage was done to roads and irrigation structures. Water quality suffered.

To correct the adverse conditions, hydrologist D. L. Rosgen used nature as a model and—imitating the meander patterns and width-to-depth ratios of stable local stream types of similar gradients, channel bed materials, sediment, and flow regimens—he constructed a new stable river channel adjacent to reconstructed floodplain and river terrace zones. Instead of relying on the creation of a trapezoidal channel built of concrete and steel or armored with uniformly sized imported rock riprap, he used natural materials to reinforce the newly constructed river channel. Where calculations of shear velocity indicated that bank erosion was likely, banks were strengthened with tree trunks, boulders, root wads, and vegetation, all locally obtained.

Since construction 5 years ago, the meander pattern of the new channel has remained stable, and the new channel has proved capable of transporting the sediment supplied by the tributaries, even at full bank discharge, due to a doubling of shear stress values relative to the braided channel. The project suggests that the natural tendencies of rivers are predictable, based on their morphology, substrate, surrounding landforms, and flow rates. So successful was the work on the San Juan River that a new river stabilization project was soon authorized and was undertaken by Rosgen on the nearby Blanco River in southwestern Colorado in 1987 (Rosgen, 1988).

1972). These are important species, but not necessarily top-level carnivores such as game fish that help maintain biological communities by controlling populations of other species, resource (nutrients, substrate) availability, or habitat quality.

Efforts to improve fishing by structural means sometimes also introduce into the ecosystem undesirable, nonbiodegradable materials (e.g., rebar, wire mesh, wire rope, planks, polypropylene, hardware cloth, rubber matting, cyclone fencing, corrugated steel, or fiberglass) (Wesche, 1985) and quarried rock riprap (Hunt, 1988a). Most structural efforts to enhance fish habitat rely on stone or wood dams, current deflectors, and camouflaged wooden bank overhangs (covered with soil and planted with vegetation). One fisheries expert has used selective herbicides along with mechanical brush cutting to make stream habitats more favorable to trout by removing 100 percent of the woody vegetation from both stream banks according to Hunt (1979).

Some fisheries biologists believe that "water and space are going to waste" if they are not used by trout and that ". . . even the best streams could be made better . . ." by producing more trout in them (Hunt, n.d.). To the ecologist interested in stream or river restoration, maximizing the ecosystem for trout, or any single species, is not the same as restoring the biotic structure and function of the stream, which includes optimizing for a number of species.

Ecosystem Restoration

Gore (1985) pointed out that most fluvial restoration projects entail the restoration of habitat, which is soon invaded by pioneering and then colonizing organisms if there are sources of species upstream, downstream, or in tributaries. Restoration of suitable physical conditions is thus of great importance.

The example of the Blanco River (see Box 5.6) shows the importance of taking a systems approach to physical alteration of a stream or river. The U.S. Army Corps of Engineers focused on one function (the capacity of the channel to carry high flows) and on one reach. The trapezoidal channel installed by the COE initiated detrimental changes that propagated downstream. The river became too broad and shallow for fish, and the unstable banks lost riparian vegetation and considerable amounts of sediment. Just as the biological system has critical thresholds for stress (see Illinois River case history, Appendix A), so does the physical system: once the threshold is crossed, dramatic channel modifications may ensue (Hasfurther, 1985).

As the Blanco River example indicates, different disciplines and schools of thought within disciplines (hydraulic engineering, hydrology, fluvial geomorphology) have quite different approaches to understanding fluvial systems and planning structural modifications. The COE approach to the Blanco derives from hydraulic theory that is based on research done in laboratory flumes. According to Hasfurther (1985), even the "regime" equations of Lacey (1930), Blench (1957), and Simons and Albertson (1960) are oriented toward engineering artificial rather than natural channels, although Bhowmik (1981) offered a variation of regime theory that considered geomorphic principles. In contrast, David Rosgen analyzed the system (see Box 5.6 and Appendix A) from the perspective of a fluvial geomorphologist, looking at the degraded reach in the context of what was going on above and below it, and in the context of other similar, but relatively undisturbed streams in the same region.

As explained in the previous section on species-centered restoration, restoring physical characteristics is not a simple undertaking because geologic, hydrologic, hydraulic, and geometric factors interact to develop a given stream system (Hasfurther, 1985). Geologic factors (soil type, topography) influence the nature and amount of sediment production and the water flow pattern (e.g., streams dominated by ground water have much more stable flows than do runoff-dominated streams). Hydrologic factors (climate, land cover, land use) also influence flow and runoff. Hydraulic factors include depth, slope, and velocity and are directly responsible for erosion and sediment transport. Geometric factors include the channel cross-sectional shape, stream pattern (braided, meandering, straight), and the riffle-pool sequence on smaller streams. Changes in sediment load and water flow cause significant adjustments in channel geometry.

Constraints on Fluvial Restoration

CONCEPTUAL LIMITATIONS

In 1989 the U.S. Environmental Protection Agency (EPA) organized a symposium on the application of ecological principles and theory to the recovery of lotic communities and ecosystems following disturbance (Yount and Niemi, 1990). The organizers noted in the preface of their study that environmental decisions are often compromised by a lack of knowledge about the ecosystems and that the ecological theory on which decisions are based may be overly simplistic or outdated (see Thomas, 1989). The conference organizers talked specifically about the outmoded community-as-superorganism analogy and the way it has been used to suggest that water quality criteria can be exceeded once every 3 years on average without unacceptably damaging the exposed biological community (U.S. EPA, 1985). However, these same concerns apply to restoration, which in many cases in fluvial systems amounts to assisting natural restorative processes.

Very few of the concepts described at the beginning of this chapter are utilized in the design of restoration projects. This is unfortunate for both restoration science and the science of ecology, because a good conceptual understanding normally precedes an effective design, and well-designed and well-monitored restorations provide an opportunity to test ecological theory. Chief among conceptual limitations on both management and restoration of fluvial ecosystems is the failure to consider the stream and its riparian zone or the river and its floodplain as components of one ecosystem. Ecologists have lagged behind hydrologists in arriving at this concept. Hydrologists have long considered rivers and their floodplains as one unit because they are inseparable with respect to the water, sediment, and organic budgets. North American hydrologists and flood disaster management agencies define a river's *active floodplain* as the area inundated by a 100-year flood or, stated another way, the flood that has a 1 percent probability of occurring in a given year (Bhowmik and Stall, 1979). An ecological definition of active floodplain was described also in this chapter in the section, "Concepts Related to Management and Restoration of Rivers and Streams." Most of the papers reflecting or based on concepts related to river-riparian ecosystems have been published since the river continuum concept first stimulated debate in 1981 (Vannote et al., 1980), so it is not surprising that more recently published concepts have yet to be applied to the classification and inventory of fluvial systems, let alone to their management and restoration.

INADEQUATE INFORMATION BASE

An example of an inadequate information base is the *Classification of Wetlands and Deep Water Habitats of the United States* (Cowardin et al., 1979), which has very little utility in the assessment of the status of riverine-riparian ecosystems because active floodplains (those still inundated at least annually by their rivers) are not considered part of the riverine system and are not even a category used for classification. Instead, floodplains lose their identity by being broken into smaller units and lumped into the palustrine system with ponds, bogs, fens, prairie marshes, and forested wetlands that can be completely isolated from flowing water throughout the year. There is no way to distinguish an inventoried emergent wetland that retains its function as a spawning and rearing area for migratory fish during the flood from one that is isolated from the river behind a levee. The riverine system as defined for the classification inventory is a channel, and the floodplain is a level plain that may never, or only occasionally, be flooded (Cowardin et al., 1979)—a definition that is not only technically incorrect but does not even agree with the common-sense meaning of the word *floodplain*. The floodplain forests of the Upper Mississippi River at Burlington, Iowa, are flooded by the river for an average of 22 days per year (Swanson and Sparks, 1990), and the average annual flood duration on the Atchafalaya River is 160 days (C. Frederick Bryan, leader, Louisiana Cooperative Fish and Wildlife Research Unit, School of Forestry, Wildlife, and Fisheries, Louisiana State University, Baton Rouge, La., personal communication, May 22, 1990).

In arid regions, arroyos, floodplains, and playa lakes may be flooded less than annually. These are not included in conventional classification systems for wetlands or surface waters, but are extremely important habitats for a variety of plants and animals adapted to unpredictable or sporadic availability of surface water. These areas should be delineated, in either land or wetland classification systems, and their status and trends (including water regime) monitored.

LACK OF APPROPRIATE EXPERTISE

Conceptual deficiencies not only make existing inventories less useful than they should be, but also lead to deficiencies in the planning, execution, and assessment of fluvial restoration projects. A common deficiency includes failure to see the reach of interest as part of a larger river-riparian system and even larger drainage basin. In a survey of stream habitat assessment programs in 10 midwestern states, Osborne (1989) noted that few states incorporate larger-scale habitat characteristics (e.g., sinuosity, gradient) in their field measurements or planning processes. There is a need to develop habitat assessment methodologies appropriate for different regions and different types of fluvial ecosystems (warm- versus cool-water streams, streams versus large rivers).

Another common deficiency is failure to understand that most river-riparian ecosystems are in a dynamic physical equilibrium that can rapidly disequilibrate when a threshold is crossed. These deficiencies have probably contributed to most failed restoration projects, or worse, the need to undo damage wreaked by well-intentioned, but poorly designed "restoration" projects. The type of system-level understanding required is characteristic of those who work in fluvial geomorphology, hydrology, some types of hydraulic engineering, and lotic ecology. However, the state of Missouri has developed a well-integrated program (see Box 5.9).

Too few projects apply available natural-systems-oriented expertise, perhaps because of cost or because the differences in the utility and orientation of the various schools and subdisciplines of hydrology, geology, and engineering are not known to outsiders. For example, hydraulic engineering is usually thought of as part of the problem (e.g., channel alteration) that makes fluvial restoration necessary, rather than as a technical component of the solution.

Without the appropriate conceptual and technical underpinning, restorationists often adopt a trial-and-error approach (Rosgen and Fittante, 1986). Not surprisingly then, the literature on stream habitat enhancement is replete with accounts of the successes and failures of particular types of in-stream structures (Wesche, 1985; Hunt, 1988b; Rivers and Streams Technical Committee, 1990). Commenting on the effects of fish habitat improvement structures, Rosgen and Fittante (1986) report, "Often these structures meet with great success on certain streams and are total disasters on others." It may well be that failures tend to be underreported relative to successes. Many of the accounts and handbooks on stream enhancement structures appear to depend for their authority on the firsthand experience of individual practitioners who may have worked in a particular region on a particular stream type. Many of the recommendations offered in the stream improvement literature appear to be of a "seat-of-the-pants" or rule-of-thumb nature.

In contrast to this descriptive experiential approach, river and stream restorationists should supplement traditional folk knowledge with the systematic application of hydrological principles and hydraulic engineering. The increased use of quantitative descriptions of pre- and posttreatment hydrological conditions is necessary to transform fluvial restoration from an art to a science. Once quantitative measures of "before-and-after" flow regimes are known, these can be more reliably related to the responses of fish and other biota (Heede and Rinne, 1990).

Rosgen and Fittante (1986) propose a planning process and systematic guidelines to minimize use of inappropriate in-stream structures (see Tables 5.8 and 5.9). Use of the procedure in Table 5.8 is advisable in stream restoration projects (and here the term *restoration* is used in explicit contrast to the term *stream enhancement*), provided that the "identification of limiting factors" step is interpreted to mean

BOX 5.9
A SUCCESSFUL STATE PROGRAM IN STREAM RESTORATION

The Missouri Department of Conservation initiated a stream restoration program that is uniquely successful because it (1) is based on management plans developed for each basin, instead of a piecemeal approach; (2) incorporates hydrological and geomorphological principles and information; (3) uses streams and stream corridors on public lands as models of good stream management practices; (4) increases citizen awareness of stream problems and involves local people in stream restoration; and (5) provides technical services and incentives to riparian landowners.

The stream program in the Department of Conservation germinated in 1984 when fishery biologists developed a plan in anticipation of the increased funding that was to come to the states through the Wallop-Breaux amendment to the Sports Fish Restoration Act. The technical services part of the plan was a direct response to a survey of 120 riparian landowners, 80 percent of whom felt they had problems with streams or stream banks. Of those with problems, 95 percent said they would ask for technical assistance if it were available. In 1989, what had been a fisheries program broadened into a department-wide effort, Streams for the Future, dedicated to the management, protection, and improvement of fish, wildlife, and forest resources associated with Missouri streams. The program was broadened because the department recognized that a larger effort was needed to stem the tide of stream degradation. Resource managers sometimes worked at cross-purposes: managers sometimes used practices detrimental to streams to achieve some specific management objective. Streams for the Future ensured that Department of Conservation lands were managed for the benefit of streams. Planners, engineers, and resource biologists began to interact and cross-train one another. Consultants in hydrology and geomorphology were brought in to conduct workshops for the staff and help plan the initial demonstration projects. The department also worked cooperatively with soil and water conservation districts and Soil Conservation Service hydrologists on comprehensive basin plans and local projects.

Although increased public awareness of stream problems and technical assistance to riparian land owners were always important objectives of the program, public participation in stream restoration received a boost in 1988 when a forum of concerned citizens developed a long list of river needs that included litter control, bank stabilization, restoration of fish and wildlife habitat, and water quality monitoring. In response, the Conservation Federation of Missouri, a private, nonprofit umbrella association of most of the state's conservation and environmental clubs and outdoor recreation organizations, and the department worked together to develop Stream Teams that work on segments of local streams and rivers. The Stream Teams include church groups, canoe clubs, 4-H clubs, Boy Scout troops, or single individuals who receive training, assess needs (using an inventory form that is returned to a coordinator at the Missouri Department of Conservation), undertake monitoring of restoration projects, and report results to the coordinator who in turn reports to the federation and the news media. Training has been provided by the Rivers and Streams Committee of the Missouri Chapter of the American Fisheries Society, as well as by the federation and the department.

A unique features of the Missouri program is that it is grounded on a thorough sociological understanding of rural landowners, who control most of the riparian land in the state. Farmers learn new techniques from each other, so one of the goals of the stream program was to establish demonstration projects with cooperative landowners throughout the state. Aside from peer pressure or peer example, other incentives include technical assistance, cost sharing, payment for granting of easements, and loan of equipment and operators (e.g., to drive willow posts or earth anchors in bank stabilization projects). The landowner and the department sign a cooperative agreement, with the stringency of the agreement increasing in direct proportion to the investment made by the department. For example, in return for assistance in revegetation and bank stabilization a farmer might be required to fence livestock off the restored area for at least 10 years. Purchase of permanent riparian easements would require an agreement with strict enforcement and monitoring clauses.

TABLE 5.8 Decision Steps for In-Stream Habitat Structures

1. Inventory streams.
2. Classify stream types.
3. Identify limiting factors.
4. Select candidate structures to correct limitations.
5. Make final selection based on suitability for stream type.
6. Utilize engineering criteria.
7. Determine cost-benefit ratios to make final selection.
8. Implement final design.
9. Monitor and evaluate performance.

SOURCE: Reprinted by permission from Rosgen and Fittante, 1986.

TABLE 5.9 Limitations and Discussions of Various Fish Habitat
Improvement Structures by Stream Types (stream types refer to
Rosgen and Fittante, 1986, Tables 1 and 2)[a]

REARING HABITAT ENHANCEMENT

Low-Stage Check Dam

Rating	Channel Types	Limitations/Discussion
Exc.	B1, B2, C2	No limitations.
Good	C1	Bank erosion due to lateral migration will occur unless bank stabilization is utilized.
Fair	B3, B4, B5, C3, C4, C5, D1, D2	Low dams must be constructed in conjunction with bank stabilization in these channel types. Use in conjunction with confinement measures and bank stabilization to reduce lateral migration.
Poor	B1-1, C1-1	Bedrock streambed limits the development of pools.
N/A	A1, A2, C6	Pools not limiting in these stream types.

Medium-Stage Check Dams

Rating	Channel Types	Limitations/Discussion
Exc.	B1	No limitations.
Good	B2, C2	Stage increase will result in floodplain encroachment. Limit dam height to less than 75% of bankfull stage and select sites with high, stable banks.
Fair	C1	Banks must be adequately protected both up- and downstream of structure.
Poor	B3, B4, B5, C3, C4, C5, D1, D2	Increased stream aggradation accelerated bank erosion, slope rejuvenation and floodplain encroachment can result. Extensive bank stabilization measures must accompany installation. Exceptions are on headwater streams in ephemeral channels to stop gully headcuts.
	B1-1, C1-1	Bedrock streambed limits pool scour depth.
N/A	A1, A2, C6	Pools not limiting factor in these channel types.

TABLE 5.9 (*Continued*)

Boulder Placement

Rating	Channel Types	Limitations/Discussion
Exc.	B2	No limitations.
Good	B1-1, C2	Lower gradient provides more opportunity for bar development up- and downstream of rock—unless placed on meander points (see bank-placed boulder). Use in conjunction with deflectors to increase velocity sufficient to create pools.
Fair	C1-1, C1	Bedrock limits bed scour. Potential bar deposition and lateral migration can be offset by stabilizing the banks and by strategic placement. Due to bed armor and flatter gradients, it is advantageous to create deep pools with a combination of deflectors, boulders, and/or rock clusters.
Poor	B3, B4, B5, C3, C4, C5, D1, D2	The high sediment supply and highly unstable banks limit the effectiveness of boulders placed in the active channel (other than along banks). Bar deposition up- and downstream of boulder and excessive bank erosion often occur. Deflectors can reduce sediment deposition.
N/A	A1, A2, B1, C6	Large boulder and/or pools are not a limiting factor in these channel types.

Bank-Placed Boulder

Rating	Channel Types	Limitations/Discussion
Exc.	B1-1, B2, C1 C1-1, C2	No limitations.
Good	B3, B4, B5, C3, C4, C5	Boulders must be keyed into the bank on "confined" stream types.
Fair	D1, D2	Difficult to locate thalweg channel and where the banks will be inundated from one year to another.
Poor		
N/A	A1, A2, B1	Bank rock and streamside boulders naturally occur and banks are naturally stable.
	C6	Cover and pools not limiting in this channel type.

TABLE 5.9 (*Continued*)

Half-Log Cover

Rating	Channel Types	Limitations/Discussion
Exc.	B2	No limitations.
Good	B1-1, B2, C1, C1-1, C2	Will have to use anchoring techniques compatible with coarse substrate.
Fair	C3	Increased sedimentation may cause bar formation, which results in decreased channel capacity and increased bank erosion. Key is use of deflectors in conjunction with half-log structures.
Poor	B3, B4, B5 C4, C5, D1, D2	Extremely unstable bed conditions—degrading and aggrading reaches that limit the effectiveness of this structure.
N/A	A1, A2, C6	Cover generally not limiting.

Floating Log Cover

Rating	Channel Types	Limitations/Discussion
Exc.	B1, B2, C2	No limitations.
Good	B1-1, C1-1, C1, C3, C4, C5	Overlapping logs reduces bank erosion.
Fair	B3, B4, B5	Undercutting will cause undermining of the anchor and eventual loss of the structure. Take extra precautions to protect banks.
Poor	D1, D2	Shifting active channel makes this structure infeasible.
N/A	A1, A2, C6	In-stream cover generally not limiting. Steep gradient reduces effectiveness. In-stream cover not limiting.

TABLE 5.9 *(Continued)*

Submerged Shelters Located on Meanders

Rating	Channel Types	Limitations/Discussion
Exc.	B1, C2	No limitations.
Good	B1-1, B2, C1, C1-1	Because structures are located on meanders (high-velocity areas of the channel), these channel types may be subject to some bank erosion.
Fair	B3, B4, B5, C3, C4, C5	Need bank stability measures on opposite bank to prevent accelerated bank erosion and lateral migration. Done in conjunction with bank stabilization, this structure can deepen and narrow C3, C4, and C5 channels, in particular.
Poor	D1, D2	Shifting active and thalweg channel makes this structure ineffective.
N/A	A1, A2, C6	Not limited by cover.

Submerged Shelter Located On Straight Reaches

Rating	Channel Types	Limitations/Discussion
Exc.	B1-1, B1, B2, C1-1, C1, C2	No limitations.
Good	C3, C4, C5	Submerged shelters can be placed on straight reaches in these channel types.
Fair	B3, B4, B5	High bedload transport and high stream power of these types limit effectiveness.
Poor	D1, D2	Shifting active and thalweg channel makes this structure ineffective.
N/A	A1, A2, C6	Cover naturally available.

TABLE 5.9 (*Continued*)

Single-Wing Deflector

Rating	Channel Types	Limitations/Discussion
Exc.	B1, B2	No limitation.
Good	C1, C2	May need bank stabilization.
Fair	C3	Must be done with corresponding bank protection.
-	D1, D2	Extensive construction may be needed to gain confinement of the active channel.
Poor	B3, B4, B5,	Channel instability and high sediment supply reduce effectiveness.
	C4, C5	
	B1-1, C1-1	Bedrock bed limits effectiveness.
N/A	A1, A2, C6	Pools not a limiting factor.

Double-Wing Deflector

Rating	Channel Types	Limitations/Discussion
Exc.	B1, B2, C2	No limitations.
Good	C1	May need bank stabilization in conjunction with double deflector.
Fair	C3	Need bank stabilization.
	D1, D2	Extensive construction may be needed to gain confinement.
Poor	B3, B4, B5,	Channel instability and high sediment supply reduce effectiveness.
	C4, C5	
	B1-1, C1-1	Bedrock bed limits effectiveness.
N/A	A1, A2, C6	Pools not a limiting factor.

TABLE 5.9 (*Continued*)

Channel Constrictor

Rating	Channel Types	Limitations/Discussion
Exc.	B2, C2	No limitations.
Good		
Fair	C1	Need bank protection downstream from constrictor.
	C3	Same as C1 except the reduced bed armor may create undercutting that could destroy the foundation of the structure.
	D1, D2	Extensive construction my be needed to gain confinement.
Poor	B3, B4, B5, C4, C5	Bank and bed instability and high sediment supply limit effectiveness.
	B1-1, C1-1	Bedrock bed limits effectiveness.
N/A	A1, A2, B1, C6	Not limiting due to existing low width/depth ratios.

Bank Cover

Rating	Channel Types	Limitations/Discussion
Exc.	B1, B2	No limitations.
Good	B1-1, C1-1, C1, C2, C3	
Fair	C4	Lateral migration may result in undermining the structure.
Poor	B3, B4, B5, C5	Channel instability limits effectiveness.
	D1, D2	Change in annual thalweg position makes these structures impractical.
N/A	A1, A2, C6	Good cover generally available within these channel types.

TABLE 5.9 (*Continued*)

SPAWNING HABITAT ENHANCEMENT

V-Shaped Gravel Trap

Rating	Channel Types	Limitations/Discussion
Exc.	A2, B1	No limitations.
Good	A1, B1-1, B2, C2	
Fair	C1-1, C1	Higher sediment yields make invasion of fines possible. Use with pervious trap so intragravel flow rate is maintained.
Poor	B3	Unstable bank and bed with high sediment supply limit effectiveness.
	B4, B5, C4, C5, C6, D1	No source for suitable spawning gravel.
N/A	C3, D2	Gravel bed stream types.

Note: Downcutting often occurs at the point of the apex, which can undermine the structure. Need bed stabilization in conjunction with this structure.

Log Sill Gravel Traps

Rating	Channel Types	Limitations/Discussion
Exc.	A2, B1, C2	No limitations.
Good	B1-1, B1, B2	
Fair	C1-1	
	C6	Frequent bed scour may inundate gravel with fines.
Poor	B3	High bed-load transport of sand results in unstable channel with both bed and bank instability.
	A1	High velocities and limited gravel source.
	B4, B5, C4, C5, D2	Gravel size bed load unavailable.
N/A	C3, D1	Gravel bed stream types.

TABLE 5.9 (*Continued*)

Gravel Placement

Rating	Channel Types	Limitations/Discussion
Exc.	C2	No limitations.
Good	B2	Must select lower-velocity areas within the reach—transition zones between pool and riffle.
Fair	B1-1, B1	May not be effective given the limited area where critical shear velocities would not be exceeded.
	C1-1, C1	Can cause capacity reduction and increase bank erosion. Treat smaller percentage of the channel area or stabilize banks.
	C6	Potential for fine sediment invasion with minimal disturbance due to frequent bed shifts.
Poor	A1, A2	Ineffective due to steep gradient.
	B3, B4, B5	Will fill in with finer bed material.
	D1, D2	
	C4, C5	Effective for just one year.
N/A	C3	Gravel bed stream type.

Migration Barrier

Rating	Channel Types	Limitations/Discussion
Exc.	A1, A2, B1	No limitations.
Good	B2	Proper site selection must be made within the reach where banks are high and stable.
Fair	B1-1	Erodible banks and moderate confinement limit barrier placement.
Poor	B3, B4, B5	Bank and bed instability can result in structure failure.
	C1-1, C1, C2	Low banks—cannot create adequate height for falls.
	C3, C4, C5	
	C6, D1, D2	
N/A		

[a]Streams are classified according to six factors: gradient; sinuosity; width/depth ratio; dominant particle size of channel bed materials; degree of channel entrenchment; and landform features that indicate stability of banks (e.g., vertical bedrock walls vs. unstable sloping soil banks).

SOURCE: Reprinted by permission from Rosgen and Fittante, 1986.

identification of factors that prevent the reestablishment of predisturbance ecological conditions, rather than merely conditions that limit salmonid production. As suggested by Raleigh and Duff (1980) and Heede and Rinne (1990), successful stream improvement projects require such an integration of hydrologic, hydraulic, and fisheries knowledge. Raleigh and Duff (1980) therefore suggest that, if possible, stream improvement projects should be undertaken by a multidisciplinary team.

CONCLUSIONS AND RECOMMENDATIONS

Establish Reference Reaches for River and Stream Restoration

One of the most effective ways to establish restoration goals and to evaluate the success of stream and river restoration is by comparing the biological communities in a disturbed reach to communities in a set of relatively undisturbed reference streams of the same order in the same ecoregion. The reference streams represent the regional potential for ecosystem restoration and reflect any changes in restoration potential that may occur through time, such as those caused by climate change. The suite of reference streams should include more than one representative of each stream order so that variability among streams of the same order can be quantified. Replication makes it possible to determine whether a restored stream is close enough to the reference standard to be judged a success.

At least 13 states already have formal procedures for designation and management of exceptional waters, and designation of reference streams could be incorporated into, or modeled on, these existing programs. It is particularly important to designate and protect the reference reaches in large rivers and their floodplains, because there are so few left. Some reference streams are already protected because of their location in wilderness areas, national scenic waterways, or parks. The committee recommends that:

• **Reference reaches should be designated and protected in each of the 76 ecoregions of the United States. The reference reaches should include, where possible, representatives of all orders of streams and rivers that occur in the ecoregion. Because remnant large river-floodplain ecosystems are rare, portions of the Atchafalaya River in Louisiana and the Upper Mississippi River Fish and Wildlife Refuge and at least 50 other large rivers (greater than approximately 120 miles or 200 km in length) should be designated as** *reference* *reaches* for use as restoration templates and should be protected as quickly as possible.

In-Stream Flow Requirements and Allocations

The prior appropriations system, which is the basis of water law in the West, should be amended so that flow is reserved for in-stream uses of water for fisheries and other aquatic life, boating and canoeing, aesthetics, and other environmental purposes (NRC, 1992). Most eastern states have some statutory provision that could be used to reserve stream flows in time of shortage, but these vary widely in effectiveness and application. Acquisition of a legal right to in-stream use is especially effective because (1) the in-stream use passes all tests of legal legitimacy and the terms of the right are spelled out; (2) the in-stream use has a priority date, so that it is superior to all subsequent rights; and (3) even if the in-stream use is junior in right to other uses, the junior user can legally prohibit a change in stream conditions from those existing when the right was established if the change will damage the junior use (Lamb and Doerksen, 1990).

Once the legitimacy of in-stream uses has been established, the next task is to determine what flows those uses require. Minimum flows necessary for fisheries, canoeing, or other in-stream uses may be useful for providing a baseline of protection, but may not allow scope for restoration. Incremental methods estimate the quality and quantity of fish habitat at each increment of flow and are more suitable than minimal flows where the goal is to restore aquatic populations and where water is in great demand. The Instream Flow Incremental Methodology (Bovee, 1982) is now used by 38 states and is becoming accepted as a "standard" method (Lamb and Doerksen, 1990). It is labor- and data-intensive, and requires field measurements and hydraulic modeling, but provides fairly precise answers to the question: What is gained by a given increment in flow? Its weakness is that it is species specific and inapplicable to multispecies assemblages.

Opportunities to allocate water to in-stream uses arise (1) when land with water rights is sold or transferred, (2) when municipalities and irrigators decrease water withdrawals through conservation, and (3) when operating permits for dams are considered for renewal.

Thus, the committee recommends that

• **States that have not established a water right for in-stream uses should do so.**
  • **Data on habitat use and methods for incremental flow analysis**

for fish and other aquatic organisms should be developed as quickly as possible. Priority should be given to either sport fish, "keystone" species (those that control populations of other organisms or nutrient cycles), or endangered species. However, in cases where introduced sport fish and endangered species compete, this would pose a problem. Methods of optimizing flows for multiple species should be developed.

• Flow that becomes available as the result of water conservation or lapse of permits should not automatically be reassigned to a consumptive use or to withdrawal. Instead, consideration should be given to assigning the flow to in-stream uses. Operating plans for dams should also consider the annual water regime required by fish and wildlife.

Land Use Management

Rivers are products of their drainage basins, and the biological integrity of stream and river systems is dependent to a large extent on watershed management practices such as grazing, residential and highway construction, flood control, agricultural and irrigation practices, logging, mining, and recreation. In some cases, restoration of the predisturbance flood and sediment regime will reestablish the physical characteristics of the river-riparian system, and the biota will be restored by recolonization, if residual populations occur in other reaches or tributaries.

In the 11 contiguous western states, the federal government owns 48 percent of the total land area, and therefore management practices by federal agencies have a major impact on the streams and rivers that drain those lands. Overgrazing by livestock on the 91 percent of the federal land where grazing is permitted is a major problem, particularly because cattle concentrate in the vulnerable riparian zones. Overgrazing might be reduced if it were not so heavily subsidized: the General Accounting Office (U.S. GAO, 1988) reported that the Bureau of Land Management recovers only 37 percent of the cost of providing grazing on federal land and the Forest Service recovers only 30 percent.

Therefore the committee recommends the following:

• Grazing practices on federal lands should be reexamined and then changed to minimize damages to river-riparian ecosystems and to restore damaged rivers and streams.
• Stream restoration should begin with improved land management practices that will allow natural restoration of the stream to

occur. Structural stream improvement projects should supplement, not supplant, proper land management practices, as recommended by Raleigh and Duff (1980).

• If stream or river erosion control, channel stabilization, streambank protection, or streambed modifications are necessary, "soft engineering" approaches, such as bioengineering techniques for bank stabilization and repairs, should be considered first, where appropriate, in preference to the use of "hard engineering" approaches that rely on dams, levees, channelization, and riprap.

• To effect the restoration of floodplains, bottomlands, and riparian habitats, dikes and levees that are no longer either needed or cost-effective should be razed to reestablish hydrological connections between riparian and floodplain habitats and associated rivers and streams.

• Classification systems for land use and wetlands (i.e., in the *Classification of Wetlands and Deepwater Habitats of the United States* by Cowardin et al., 1979, should explicitly designate riparian environments and floodplains that retain their periodic connections to rivers (and hence their ecological, hydrological, and recreational functions and values as part of river-floodplain ecosystems).

Event-Triggered Sampling and Monitoring

Some types of restoration, characterized as "working with the river" or "letting the river do the work," are effected when a major, channel-altering flood occurs. Other types of restoration are designed to protect against the scouring action of high flows or to provide a refuge for organisms during periods of extreme low flow (droughts). It is important to conduct event-triggered sampling (during the event, in some cases; immediately after, in others) to determine whether the restoration is meeting the design criteria.

• Event-triggered monitoring or surveillance should be planned in advance as part of restoration programs that are designed to convey, resist, or use floods or other extreme events.

Guiding Citizen Participation in Restoration Projects

Some well-intentioned restoration projects have failed because fluvial and biological processes were not adequately taken into account in the design and implementation of the projects. The public has become increasingly aware of the need for aquatic restoration (as can be seen from several case studies in Appendix A), and numerous public and private agencies and citizen organizations are likely to initiate further stream and river restoration projects. These organizations, if properly guided and supported, can be a valuable impetus for effective aquatic ecosystem restoration and, in some cases, a valuable source of volunteer labor to accomplish restoration.

• **A hydrological advisory service should be operated by state or federal agencies to provide technical assistance to groups interested in stream and river restoration. Universities with experts in natural resources or hydrology and/or State Water Resources Centers, based at universities in every state, should also contribute the technical assistance required for the restoration of aquatic ecosystems through free or at-cost expert hydrological and biological advisory services.**

National Rivers and Streams Inventory

The committee could not find a recent national assessment of the number of stream and river miles affected by channelization or leveeing. Although water resource agencies track their own development projects, the only nationwide inventory of rivers and streams was conducted in the 1970s (U.S. DOI, 1982) in response to passage of the Wild and Scenic Rivers Act of 1968 (P.L. 90-542).

• **Therefore, the committee recommends that a comprehensive up-to-date nationwide assessment of rivers and streams be done, comparable to the National Wetland Inventory (Tiner, 1984).**

Training and Education

A new cadre of agricultural specialists, engineers, and biologists is needed, as water resource policies shift away from resource development and exploitation to resource management and restoration.

• **Universities, especially those with federally funded Water Resources Institutes, Agricultural Extension and Research Units, and Cooperative Fish and Wildlife Research Units, should be encouraged to require graduate students in agriculture, environmental engineering, hydraulic engineering, water resource planning and economics, and fisheries management to receive training in hydrology, fluvial geomorphology, and ecology, as well as some practical field experience in natural resource systems.**

REFERENCES AND RECOMMENDED READING

Aldridge, B. N., and J. H. Eychaner. 1984. Floods of October 1977 in southern Arizona and March 1978 in central Arizona. U.S. Geological Survey Water-Supply Paper No. 2223, U.S. Government Printing Office, Washington, D.C.

Alexander, G. R., and E. A. Hansen. 1986. Sand bed load in a brook trout stream. N. Am. J. Fish. Manage. 6:9-23.

Amoros, C., A. L. Roux, and J. L. Reygrobellet. 1987. A method for applied ecological studies of fluvial hydrosystems. Regulated Rivers 1:17-36.

Armour, C. L., D. A. Duff, and W. Elmore. 1991. The effects of livestock grazing on riparian and stream ecosystems. Fisheries 16(1):7-11.

Association of State and Interstate Water Pollution Control Administrators (ASIWPCA), in cooperation with the U.S. Environmental Protection Agency. 1984. America's Clean Water: The States' Evaluation of Progress, 1972-1982. ASIWPCA, Washington, D.C.

Ausness, R. C. 1983. Water rights legislation in the East—A program for reform: Williamsburg, Virginia. William and Mary Law Review 24(4):547-590.

Baker, F. C. 1906. A catalogue of the mollusca of Illinois. Illinois Lab. Nat. Hist. Bull. 7(6):53-136.

Beaumont, P. 1978. Man's impact on river systems: A world-wide view. Area 10:38-41.

Bellrose, F. C., F. L. Paveglio, Jr., and D. W. Steffeck. 1979. Waterfowl populations and the changing environment of the Illinois River valley. Ill. Nat. Hist. Surv. Bull. 32:1-54.

Bellrose, F. C., S. P. Havera, F. L. Paveglio, Jr., and D. W. Steffeck. 1983. The fate of lakes in the Illinois River valley. Ill. Nat. Hist. Surv. Biol. Notes 119:27.

Belt, C. B., Jr. 1975. The 1973 flood and man's constriction of the Mississippi River. Science 189:681-684.

Benke, A. C. 1990. A perspective on America's vanishing streams. J. Am. Benthol. Soc. 9(1):77-78.

Betancourt, J. L., and R. M. Turner. 1988. Historic arroyo cutting and subsequent channel changes at the Congress Street crossing, Santa Cruz River, Tucson, Arizona. In E. E. Whitehead, C. F. Hutchinson, B. N. Timmermann, and R. G. Varady, eds., Arid Lands: Today and Tomorrow, Proceedings of an International Research and Development Conference, Tucson, Arizona, October 20-25, 1985. Westview Press, Boulder, Colo.

Betancourt, J. L., and R. M. Turner. 1991. Tucson's Santa Cruz River and the Arroyo Legacy. The University of Arizona Press, Tucson, Ariz.

Bhowmik, N. G. 1981. Hydraulic considerations in the alteration and design of diversion channels in and around surface mined areas. Pp. 97-104 in D. H. Graves, ed., National Symposium on Surface Mining, Hydrology, Sedimentology, and Reclamation. University of Kentucky, Lexington, Ky.

Bhowmik, N. G., and R. J. Schicht. 1980. Bank Erosion of the Illinois River. Report 92. Illinois State Water Survey, Urbana, Ill.

Bhowmik, N. G., and J. B. Stall. 1979. Hydraulic Geometry and Carrying Capacity of Floodplains. Water Research Center, Research Report No. 145, University of Illinois, Urbana, Ill.

Bhutani, J., R. Holberger, P. Spewak, W. E. Jacobsen, and J. B. Truett. 1975. Impact of Hydrologic Modifications on Water Quality. EPA-600/12-75-007. U.S. Environmental Protection Agency, Office of Research and Development, Washington, D.C. Pp. 46-50; 364-477.

Bjornn, T. C., M. A. Brusven, M. P. Molnau, J. H. Milligan, R. A. Klant, E. Chacho, and C. Shaye. 1977. Transport of granitic sediment in streams and its effects on insects and fish. Bulletin 17, University of Idaho, Forest Wildlife and Range Experiment Station, Moscow, Idaho.

Blench, T. 1957. Regime Behavior of Canals and Rivers. Butterworth Science Publishers, London, England.

Borman, F. H., G. E. Likens, and J. S. Eaton. 1969. Biotic regulation of particulate and solution losses from a forest ecosystem. BioScience 19:600-710.

Bovee, K. D. 1982. A Guide to Stream Habitat Analysis Using the Instream Flow Incremental Methodology. U.S. Fish and Wildlife Service Biological Series Program FWS/OBS-82/26. U.S. Department of the Interior, Fish and Wildlife Service, Washington, D.C.

Briceland, R. H. 1976. Statement of Dr. Richard H. Briceland, Director, Illinois Environmental Protection Agency, to the Task Force on Locks and Dam 26 of the Midwestern Conference, Council of State Governments, Clayton, Mo. November 18.

Brookes, A. 1988. Channelized Rivers Perspectives for Environmental Management. John Wiley & Sons, New York, N.Y. 326 pp.

Brookes, A. 1989. Alternative Channelization Procedures. Pp. 139-162 in J. A. Gore and G. E. Petts, eds., Alternatives in Regulated River Management. CRC Press, Boca Raton, Fla.

Brouha, P. 1987. Wildlife and fisheries management in the USDA Forest Service. Pp. 64-67 in Managing Southern Forests for Wildlife and Fish. USDA Forest Service General Technical Report 50-65.

Buck, D. H. 1956. Effects of Turbidity on Fish and Fishing. Report No. 56. Oklahoma Fisheries Research Laboratory, Norman, Okla.

Burgess, S. A. 1985. Some effects of stream habitat improvement on the aquatic and riparian community of a small mountain stream. Pp. 223-246 in James A. Gore, ed., The Restoration of Rivers and Streams: Theories and Experience. Butterworth, Stoneham, Mass. 280 pp.

Butts, T. A. 1974. Measurements of sediment oxygen demand characteristics of the Upper Illinois Waterway. Report of Investigations, No. 76. Illinois State Water Survey, Urbana, Ill.

Calkins, W. W. 1874. The land & fresh water shells of LaSalle County, Illinois. Ottawa (Illinois) Academy of Natural Science Proceedings. H. McAllaster and Co., Chicago, Ill. 48 pp.

Clean Water Act of 1977. P.L. 95-217, Dec. 27, 1977, 91 Stat. 1566.

Condit, D. 1989. Illinois River Soil Conservation Task Force. Pp. 106-109 in Proceedings of the Second Conference on the Management of the Illinois River System: The 1990's and Beyond. Illinois River Resource Management. A Governor's Conference held October 3-4. Peoria, Ill. 199 pp.

Condit, D., and D. Roseboom. 1989. Stream bank stabilization and the Illinois River Soil and Conservation Task Force. Pp. 110-132 in Proceedings of the Second Conference on the Management of the Illinois River System: The 1990's and Beyond. Illinois River Resource Management. A Governor's Conference held October 3-4. Peoria, Ill. 199 pp.

Conlin, M. 1987. Illinois River fish and wildlife considerations. Pp. 147-154 in Management of the Illinois River System: The 1990's and Beyond. Illinois River Resource Management. A Governor's Conference held April 1-3. Peoria, Ill. 260 pp.

Council on Environmental Quality (CEQ). 1978. Environmental Quality. U.S. Government Printing Office, 76 pp.

Council on Environmental Quality (CEQ), and Interagency Advisory Committee on Environmental Trends. 1989. Environmental Trends. Council on Environmental

Quality, Executive Office of the President, Washington, D.C., and Interagency Advisory Committee on Environmental Trends. 152 pp.

Cowardin, L. M., V. Carter, F. C. Golet, and E. T. LaRoe. 1979. Classification of Wetlands and Deepwater Habitats of the United States. FWS/OBS-79/31. U.S. Department of the Interior, Fish and Wildlife Service. 103 pp.

Craig, N. J., R. E. Turner, and J. W. Day, Jr. 1980. Wetland losses and their consequences in coastal Louisiana. Z. Geomorph. N.F. Suppl. Bd. 34:225-241.

Crandall, D. A., R. C. Mutz, and L. Lautrup. 1984. The Effects of Hydrologic Modifications on Aquatic Biota, Stream Hydrology and Water Quality: A Literature Review. Illinois Environmental Protection Agency, Division of Water Pollution Control, Springfield, Ill.

Croome, R. L., P. A. Tyler, K. F. Walker, and W. D. Williams. 1976. A limnological survey of the River Murray in the Albury-Wodonga area. Search 7(1):14-17.

Cummins, K. W. 1973. Trophic relations of aquatic insects. Annu. Rev. Entomol. 18:183-206.

Demissie, M. 1989. Peoria Lake sedimentation and proposed artificial islands. Pp. 46-57 in Proceedings of the Second Conference on the Management of the Illinois River System: The 1990's and Beyond. Illinois River Resource Management. A Governor's Conference held October 3-4. Peoria, Ill. 199 pp.

Donels, B. 1989. Environmental management program proposals—The Illinois basin. Pp. 77-80 in Proceedings of the Second Conference on the Management of the Illinois River System: The 1990's and Beyond. Illinois River Resource Management. A Governor's Conference held October 3-4. Peoria, Ill. 199 pp.

Donovan, W. J. 1983. The less traveled road: An overview of nonstructural measures in flood plain management planning. In Seminar Proceedings: Implementation of Nonstructural Measures. Policy Study 83-G520. U.S. Army Corps of Engineers, Engineer Institute for Water Resources.

Duff, D. A., and N. Banks. 1988. Indexed Bibliography on Stream Habitat Improvement. USDA-Forest Service Intermountain Region, Wildlife Management Staff, Ogden, Utah.

Echeverria, J. D., and J. Fosburgh. 1988. The American Rivers Outstanding Rivers List. American Rivers, Inc., Washington, D.C.

Echeverria, J. D., P. Barrow, and R. Roos-Collins. 1989. Rivers at Risk. The Concerned Citizen's Guide to Hydropower. Island Press, Washington, D.C. 217 pp.

Edwards, E. A., and K. A. Twomey. 1982. Habitat suitability index models: common carp. U.S. Department of the Interior, Fish and Wildlife Service. FWS/OBS-82/10.12. 27 pp.

Edwards, E. A., G. Gebhart, and O. E. Maughn. 1983. Habitat suitability information: smallmouth bass. U.S. Department of the Interior, Fish and Wildlife Service. FWS/OBS-82/10.36. 47 pp.

Egan, T. 1990. Dams may be razed so the salmon can pass. The New York Times, July 15. Pp. 1 and 14.

Ellis, M. M. 1936. Erosion silt as a factor in aquatic environments. Ecology 17:29-42.

Elwood, J. W., J. D. Newbold, R. V. O'Neill, and W. Van Winkle. 1983. Resource spiraling: An operational paradigm for analyzing lotic ecosystems. Pp. 3-27 in Thomas D. Fontaine III and Steven M. Bartell, eds., Dynamics of Lotic Ecosystems. Ann Arbor Science Publishers, Ann Arbor, Mich. 494 pp.

Federal Water Pollution Control Act Amendments of 1972. P.L. 92-500.

Flather, C. H., and T. W. Hoekstra. 1989. An analysis of the wildlife and fish situation in the United States: 1989-2040. General Technical Report RM-178. U.S. Department of Agriculture, Forest Service, Rocky Mountain Forest and Range Experiment Station, Fort Collins, Colo. 147 pp.

Flick, W. A. n.d. A Stream Improvement Project, Stream Management of Salmonids. Trout Unlimited, Denver, Colo.

Forbes, S. A. 1878. The food of Illinois fishes. Ill. Lab. of Nat. Hist. Bull. 1(2):71-89.

Forbes, S. A., and R. E. Richardson. 1908. The Fishes of Illinois. Illinois Natural History Survey, Urbana, Ill. cxxxvi plus 357 pp.

Forbes, S. A., and R. E. Richardson. 1913. Studies on the biology of the upper Illinois River. Ill. Lab. Nat. Hist. Bull. 9(10):481-574.

Forbes, S. A., and R. E. Richardson. 1919. Some recent changes in Illinois River biology. Ill. Nat. Hist. Surv. Bull. 13(6):139-156.

Fremling, C. R., J. L. Rasmussen, R. E. Sparks, S. P. Cobb, C. F. Bryan, and T. O. Claflin. 1989. Mississippi River fisheries: A case history. Proceedings of the International Large River Symposium (LARS). Can. Spec. Publ. Fish. Aquat. Sci. 106:309-351.

Froelich, P. N. 1988. Kinetic control of dissolved phosphate in natural rivers and estuaries: A primer on the phosphate buffer mechanism. Limnol. Oceanogr. 33(4, part 2):649-668.

Gauch, H. G., Jr. 1982. Multivariate Analysis in Community Ecology. Cambridge University Press, New York. 298 pp.

George, C., Jr. 1972. The role of the Aswan High Dam in changing the fisheries of the southeastern Mediterranean. Pp. 159-178 in M. Taghi Farvar and John P. Milton, eds., The Careless Technology: Ecology International Development. The Natural History Press, Doubleday, New York.

Glover, R. D. 1986. Trout stream rehabilitation in the Black Hills of South Dakota. Pp. 7-15 in The 5th Trout Stream Habitat Improvement Workshop. Pennsylvania Fisheries Commission, Harrisburg, Pa.

Gore, J. A. 1985. Mechanisms of colonization and habitat enhancement for benthic macroinvertebrates in restored river channels. Pp. 81-101 in J. A. Gore, ed., The Restoration of Rivers and Streams. Theories and Experience. Butterworth, Stoneham, Mass. 280 pp.

Gould, G. A. 1977. Preserving instream flows under the appropriation doctrine—Problems and possibilities. Pp. 3-21 in B. L. Lamb, ed., Protecting instream flows under western water law—Selected papers: U.S. Fish and Wildlife Service, Instream Flow Information Paper, No. 2, FWS/OBS-77/47.

Guldin, R. W. 1989. An Analysis of the Water Situation in the United States: 1989-2040. General Technical Report. U.S. Department of Agriculture, Forest Service. Rocky Mountain Forest and Range Experiment Station, Fort Collins, Colo.

Gunderson, D. R. 1968. Floodplain use related to stream morphometry and fish populations. J. Wildl. Manage. 32(3):507-514.

Hale, J. C. 1969. An evaluation of trout stream habitat improvement in a north shore tributary of Lake Superior. Minn. Fish. Invest. (5):37-50.

Hall, J. D., and C. O. Baker. 1982. Influence of forest and rangeland management on anadromous fish habitat in western North America. Rehabilitating and Enhancing Stream Habitat: Review and Evaluation. USDA Forest Service. General Technical Report PNW-133.

Hart, C. A. 1895. On the entomology of the Illinois River and adjacent waters. Ill. Lab. Nat. Hist. Bull. 4(6):149-273.

Hasfurther, V. R. 1985. The use of meander parameters in restoring hydrologic balance to reclaimed stream beds. Pp. 21-40 in J. A. Gore, ed., The Restoration of Rivers and Streams. Theories and Experience. Butterworth, Stoneham, Mass. 279 pp.

Havera, S. P., and F. C. Bellrose. 1985. The Illinois River: A lesson to be learned. Wetlands 4:29-41.

Havera, S. P., F. C. Bellrose, H. K. Archer, F. Paveglio, Jr., D. W. Steffeck, K. S. Lubinski,

R. E. Sparks, W. U. Brigham, L. Coutant, S. Waite, and D. McCormick. 1980. Projected Effects of Increased Diversion of Lake Michigan Water on the Environment of the Illinois River Valley. Report prepared for the U.S. Army Corps of Engineers, Chicago District, Chicago, Ill.

Heede, B. H., and J. N. Rinne. 1990. Hydrodynamic and fluvial morphologic processes: Implications for fisheries management and research. N. Am. J. Fish. Manage. 10(3):249-268.

Herbkersman, C. N. 1984. A guide to the George Palmiter river restoration techniques. Institute of Environmental Sciences, Miami University, Oxford, Ohio. 52 pp.

Herke, W. H., E. E. Knudsen, P. A. Knudsen, and B. D. Rogers. 1987. Effects of semi-impoundment on fish and crustacean nursery use: Evaluation of a "solution." Coastal Zone '87 (May):2562-2576.

Herman, R. J. 1987. National resources inventory and potential stream sediment reductions. Pp. 173-183 in Management of the Illinois River System: The 1990's and Beyond. Illinois River Resource Management. A Governor's Conference held April 1-3, 1987, Peoria, Ill. 260 pp.

Herricks, E. E., and L. L. Osborne. 1985. Water quality restoration and protection in streams and rivers. Pp. 1-20 in J. A. Gore, ed., The Restoration of Rivers and Streams. Theories and Experience. Butterworth, Stoneham, Mass. 280 pp.

Hesse, L. W., G. L. Hergenrader, H. S. Lewis, S. D. Reetz, and A. B. Schlesinger. 1982. The Middle Missouri River. A Collection of Papers on the Biology with Special Reference to Power Station Effects. The Missouri River Study Group, Norfolk, Nebr. 301 pp.

Hocutt, C. H., and E. O. Wiley. 1986. The Zoogeography of North American Freshwater Fishes. John Wiley & Sons, New York. 866 pp.

Hughes, R. M. 1985. Use of watershed characteristics to select control streams for estimating effects of metal mining wastes on extensively disturbed streams. Environ. Manage. 9:253-262.

Hughes, R. M., T. R. Whittier, C. M. Rohm, and D. P. Larsen. 1990. A regional framework for establishing recovery criteria. Environ. Manage. 14(5):673-683.

Hunt, R. L. n.d. In-stream improvement of trout habitat. Stream Management of Salmonids. Trout Unlimited, Denver, Colo.

Hunt, R. L. 1975. A long-term evaluation of trout habitat development and its relation to improving management-oriented research. Trans. Am. Fish. Soc. 105(3):361-364.

Hunt, R. L. 1978. Instream enhancement of trout habitat. Pp. 19-27 in K. Hashagen, ed., Proc. Nat. Symposium on Wild Trout Management. Cal. Trout Inc., San Francisco, Calif.

Hunt, R. L. 1979. Removal of woody streambank vegetation to improve trout habitat. Technical Bulletin No. 115. Department of Natural Resources, Madison, Wisc. 36 pp.

Hunt, R. L. 1985. A follow-up assessment of removing woody streambank vegetation along two Wisconsin trout streams. Wis. Dept. Water Resour. Res. Rep. No. 137.

Hunt, R. L. 1986. An evaluation of brush bundles and half-logs to enhance carrying capacity of two brown trout streams. Pp. 31-62 in The 5th Trout Stream Habitat Improvement Workshop. Pennsylvania Fisheries Commission, Harrisburg, Pa.

Hunt, R. L. 1988a. Management of riparian zones and stream channels to benefit fisheries. In T. W. Hoekstra and J. Capp. eds., Integrating Forest Management for Wildlife and Fish. General Technical Report NC-122. U.S. Department of Agriculture, Forest Service, North Central Forest Experiment Station. St. Paul, Minn. 63 pp.

Hunt, R. L. 1988b. A Compendium of 45 Trout Stream Habitat Development Evaluations in Wisconsin During 1953-1985. Technical Bulletin No. 162. Department of Natural Resources, Madison, Wis. 80 pp.

Illinois Environmental Protection Agency (EPA). 1979. Water Quality Management Plan. Volume III. Nonpoint Sources of Pollution: Soil Erosion and Sedimentation, Livestock Wastes, Fertilizers, Pesticides, Forestry, and Fruit Production. Illinois EPA, Springfield, Ill. 384 pp.

Illinois Environmental Protection Agency (EPA). 1990. Illinois Water Quality Report 1988- 1989. IEPA/WPC/90-160. Illinois EPA, Division of Water Pollution Control, Springfield, Ill. 352 pp.

Injerd, D. 1987. Illinois Lake Michigan water diversion. Pp. 56-64 in Management of the Illinois River System: The 1990's and Beyond. Proceedings of the Illinois River Resource Management. A Governor's Conference held April 1-3. Peoria, Ill. 260 pp.

International Commission on Large Dams. 1973. World Register of Dams.

Jackson, H. O., and W. C. Starrett. 1959. Turbidity and sedimentation at Lake Chautauqua, Illinois. J. Wildl. Manage. 23:157-168.

JEL. 1989. The Willamette River Greenway: A Reawakening is Needed. (Unsigned Oregon government agency chronology.)

Jensen, S. E., and W. S. Platts. 1989. Restoration of degraded riverine/riparian habitat in the Great Basin and Snake River regions. Pp. 377-416 in J. A. Kusler and M. E. Kentula, eds., Wetland Creation and Restoration: The Status of the Science. Vol. I: Regional Reviews. Document No. EPA 600/3-89/038A. U.S. Environmental Protection Agency, Environmental Research Laboratory, Corvallis, Ore. 473 pp.

Johnson, F. H. 1961. Walleye egg survival during incubation on several types of bottoms in Lake Winnibigoshish, Minnesota, and connecting waters. Trans. Am. Fish. Soc. 90:312-322.

Johnston, L. R., Associates. 1989. Interagency task force on floodplain management. A status report on the nation's floodplain management activity (an interim report). Contract No. TV-72105A. Knoxville, Tenn. 465 pp.

Judy, R. D., Jr., P. N. Seeley, T. M. Murray, S. C. Svirsky, M. R. Whitworth, and L. S. Ischinger. 1984. 1982 National fisheries survey. Vol. 1. Technical Report: Initial Findings. FWS/OBS 84/06. U.S. Department of the Interior, Fish and Wildlife Service, Washington, D.C. 141 pp.

Junk, W., P. B. Bayley, and R. E. Sparks. 1989. The flood pulse concept in river-floodplain systems. Proceedings of the International Large River Symposium (LARS). Can. Spec. Publ. Fish. Aquat. Sci. 106:110-127.

Karaki, S., and J. vanHoften. 1974. Resuspension of bed material and wave effects on the Illinois and Upper Mississippi rivers caused by boat traffic. Contract Report No. LMSSD 75-881. Prepared for the U.S. Army Corps of Engineers District, St. Louis, Mo., by Engineering Research Center, Colorado State University, Fort Collins, Colo.

Karr, J. R., and D. R. Dudley. 1981. Ecological perspective on water quality goals. Environ. Manage. 5:55-68.

Karr, J. R., K. D. Fausch, P. L. Angermeier, P. R. Yant, and I. J. Schlosser. 1986. Assessing biological integrity in running waters: A method and its rationale. Special Publication 5. Illinois Natural History Survey, Champaign, Ill. 28 pp.

Kelly, M. H., and R. L. Hite. 1984. An evaluation of empirical correlations between the macroinvertebrate biotic index (MBI) and the STORET water quality index (WQI). Unpublished manuscript. Illinois Environmental Protection Agency, Springfield, Ill.

Keown, M. P., F. A. Dardeau, Jr., and E. M. Causey. 1981. Characterization of the Suspended-Sediment Regime and Bed-Material Gradation of the Mississippi River Basin. Potamology Program (P-I). Report 1, Volume II. U.S. Army Corps of Engineer District, New Orleans, La. 375 pp.

Kleinmann, R. L. P., and R. Hedin. 1990. Biological treatment of mine water: An update. In M. E. Chalkley, B. R. Conrad, V. I. Lakshmanan, and K. G. Wheeland, eds.,

Tailings and Effluent Management. Proceedings of the International Symposium on Tailings and Effluent Management, August 20-24, 1989, Halifax. 28th Annual Conference of Metallurgists of CIM. Pergamon Press, N.Y.

Kofoid, C. A. 1903. Plankton studies. IV. The plankton of the Illinois River, 1094-1099, with introductory notes upon the hydrography of the Illinois River and its basin. Part I. Quantitative investigations and general results. Ill. Lab. Nat. Hist. Bull. 6(2):95-635.

Kusler, J. A., and M. E. Kentula, eds. 1989. Wetland Creation and Restoration: The Status of the Science. Vol. I: Regional Reviews, 473 pp.; Vol. II: Perspectives, 172 pp. EPA 600/3-89/038A. U.S. Environmental Protection Agency, Washington, D.C.

Lacey, G. 1930. Stable channels in alluvium. Proc. Inst. Civil Eng. 229:259-384.

Lamb, B. L., and H. R. Doerksen. 1990. Instream water use in the United States—Water laws and methods for determining flow requirements. Pp. 109-116 in National Water Summary 1987—Hydrologic Events and Water Supply and Use. U.S. Geological Survey, Water-Supply Paper 2350. 553 pp.

Lanyon, R., and C. Lue-Hing. 1987. MSDGC activities in the Upper Illinois basin. Pp. 103-130 in Management of the Illinois River System: The 1990's and Beyond. Illinois River Resource Management. A Governor's Conference held April 1-3. Peoria, Ill. 260 pp.

Larsen, D. P., J. M. Omernik, R. M. Hughes, D. R. Dudley, C. M. Rohm, T. R. Whittiers, A. L. Kinney, and A. L. Gallant. 1986. The correspondence between spatial patterns in fish assemblages in Ohio streams and aquatic ecoregions. Environ. Manage. 10:815-828.

Lee, M. T. 1989. Soil erosion, sediment yield, and deposition in the Illinois River basin. Pp. 718-722 in Proceedings of the International Symposium on Sediment Transport Modeling, American Society of Civil Engineers, August 14-18. New Orleans, La.

Leedy, J. B. 1979. Observations on the Sources of Sediment in Illinois Streams. Report of Investigations, No. 18., Illinois Water Information System Group, University of Illinois, Urbana, Ill.

Leopold, L. B., M. G. Wolman, and J. P. Miller. 1964. Fluvial Processes in Geomorphology. W. H. Freeman, San Francisco, Calif. 522 pp.

Little, Arthur D., Inc. 1973. Statement by John M. Wilkinson, Arthur D. Little, Inc. hearings on Stream Channelization. U.S. House of Representatives, Committee on Government Operations, Conservation and Natural Resources Subcommittee, 92nd Congress, March 20 and 21.

Little, C. E. 1990. Greenways for America. Johns Hopkins Press, Baltimore, Md.

Lopinot, A. C. 1972. Channelized streams and ditches of Illinois. Illinois Department of Conservation, Division of Fisheries. Special Fisheries Report #35. 59 pp.

Lubinski, K. S., M. J. Wallendorf, and M. C. Reese. 1981. Analysis of Upper Mississippi River system correlations between physical, biological and navigation variables. Technical Report in partial fulfillment of Contract No. 895-305. Upper Mississippi River Basin Commission, St. Paul, Minn.

Lupi, F., Jr., R. L. Farnsworth, and J. B. Braden. 1988. Improvement of lake water quality by paying farmers to abate nonpoint source pollution. Project No. G-1420-06. Final Completion Report. U.S. Department of the Interior, U.S. Geological Survey, Washington, D.C. 96 pp.

MacArthur, R. H. 1972. Geographical Ecology. Harper and Row, New York. 269 pp.

Master, L. 1991. Aquatic animals: Endangerment alert. Nature Conservancy 41(2):26-27.

Mathis, B. M., and T. F. Cummings. 1973. Selected metals in sediments, water and biota in the Illinois River. J. Water Pollut. Control Fed. 45(7):1573-1583.

Mathis, B. M., and G. E. Stout. 1987. Summary and Recommendations. Pp. 1-4 in

Management of the Illinois River System: The 1990's and Beyond. Illinois River Resource Management. A Governor's Conference held April 1-3. Peoria, Ill. 260 pp.

McElroy, A. D., S. Y. Chiu, J. W. Nebgen, A. Aleti, and A. E. Vandergrift. 1975. Water pollution from nonpoint sources. Water Res. 9:675-681.

Mermel, T. W. 1976. International activity in dam construction. Water, Power and Dam Construction 28(4):66-69.

Milhous, R. T. 1990. The calculation of flushing flows for gravel and cobble bedrivers. Pp. 598-603 in H.H. Chang and J.C. Hill, eds., Hydraulic Engineering. American Society of Civil Engineering, New York.

Mills, H. B., W. C. Starrett, and F. C. Bellrose. 1966. Man's effect on the fish and wildlife of the Illinois River. Biological Notes, No. 57. Illinois Natural History Survey, Urbana, Ill. 24 pp.

Narver, D. W. n.d. Stream management for West Coast anadromous Salmonids. Stream Management of Salmonids. Trout Unlimited, Denver, Colo.

National Acid Precipitation Assessment Program (NAPAP). 1990. Current Status of Surface Water Acid-Base Chemistry. State of Science and Technology Rept. 9. NAPAP Interagency Program, Washington, D.C.

National Research Council (NRC). 1987. River and Dam Management: A Review of the Bureau of Reclamation's Glen Canyon Environmental Studies. National Academy Press, Washington, D.C.

National Research Council (NRC). 1992. Water Transfers in the West: Efficiency, Equity, and the Environment. National Academy Press, Washington, D.C.

Nelson, R. C. 1991. Draft Fish and Wildlife Coordination Act report for the Swan Lake Rehabilitation and Enhancement Project in Pool 26, Calhoun County, Illinois. Appendix DPR-H in U.S. Army Corps of Engineers, St. Louis District. February 1991. Upper Mississippi River System Environmental Management Program Definite Project Report (SL-5) with Integrated Environmental Assessment.

Nelson, J. E., and P. Pajak. 1990. Fish habitat restoration following dam removal on a warmwater river. Pp. 53-63 in Rivers and Streams Technical Committee, The Restoration of Midwestern Stream Habitat. Proceedings of a symposium held at the 52nd Midwest Fish and Wildlife Conf., December 4-5. Minneapolis, Minn. 117 pp.

Odum, H. T., C. Diamond, and M. T. Brown. 1987. Energy Systems Overview of the Mississippi River Basin. CFW Publication #87-1. Center for Wetlands, University of Florida, Gainesville, Fla. 107 pp.

Ohio Environmental Protection Agency (EPA). 1987. Biological Criteria for the Protection of Aquatic Life. Vol. II. Ohio EPA, Columbus, Ohio. 328 pp.

Ohio River Valley Water Sanitation Commission (ORSANCO). 1990. Assessment of Ohio River Water Quality Conditions, Water Years 1988 and 1989. ORSANCO, Cincinnati, Ohio. 145 pp. plus appendices.

Omernik, J. M. 1987. Ecoregions of the conterminous United States. Ann. Assoc. Am. Geogr. 77:118-125.

Osborne, L. L. 1989. Stream Habitat Assessment in States of the North Central Division, American Fisheries Society. A report to American Fisheries Society, North Central Division, Rivers and Streams Committee. 59 pp.

Paine, R. T. 1966. Food web complexity and species diversity. Am. Nat. 100(910):65-75.

Palmer, A. W. 1903. The pollution and self-purification of the waters of the Illinois River. Water Surv. Bull. 2:62-240.

Parkenson, E. A., and P. A. Slaney. 1975. A review of enhancement techniques applicable to anadromous gamefishes. Fishery Management Report 66. British Columbia Fish and Wildlife Branch.

Payne, B. S., A. C. Miller, and D. W. Aldridge. 1987. Environmental effects of navigation traffic: Laboratory studies of the effects on mussels of intermittent exposure to turbulence and suspended solids. Environmental Impact Research Program Technical Report EL-87-14. Prepared for U.S. Army Corps of Engineers, Washington, D.C., and U.S. Army Corps of Engineers District, Louisville, Ky. 27 pp.

Penland, S., and R. Boyd, eds. 1985. Transgressive Depositional Environments of the Mississippi River Delta Plain: A Guide to the Barrier Islands, Beaches, and Shoals in Louisiana. Guidebook Series No. 3. Louisiana Geological Survey, Baton Rouge, La. 233 pp.

Philipp, D. P., and G. S. Whitt. 1991. Survival and growth of northern Florida, and reciprocal F_1 hybrid largemouth bass in central Illinois. Trans. Am. Fish. Soc., Bethesda, Md.

Platts, W. S., and J. N. Rinne. 1985. Riparian and stream enhancement management and research in the Rocky Mountains. N. Am. J. Fish Manage. 5(2A):115-125.

President's Commission on Americans Outdoors. 1986. Report and Recommendations to the President of the United States. U.S. Government Printing Office, Washington, D.C.

Raleigh, R. F., and D. A. Duff. 1980. Trout stream habitat improvement: Ecology and hydrology. Pp. 67-77 in Proceedings of Wild Trout II. September 24-25, 1979. Trout Unlimited, Vienna, Va.

Rasmussen, J. L. 1983. A Summary of Known Navigation Effects and a Priority List of Data Gaps for the Biological Effects of Navigation on the Upper Mississippi River. U.S. Army Corps of Engineers, Rock Island District, Rock Island, Ill. 96 pp.

Richardson, R. E. 1928. The bottom fauna of the middle Illinois River, 1913-1925. Its distribution, abundance, valuation, and index value in the study of stream pollution. Ill. Nat. Hist. Surv. Bull. 17(12):387-475.

Rinne, J. N., and A. L. Medina. 1989. Factors influencing salmonid populations in six headwater streams, central Arizona, U.S.A. Pol. Arch. Hydrobiol. 35(3-4): 515-532.

Rivers and Streams Technical Committee, North-Central Division, American Fisheries Society. 1990. The Restoration of Midwestern Stream Habitat. Proceedings of a symposium held at the 52nd Midwest Fish and Wildlife Conference, December 4-5. Minneapolis, Minn. 117 pp.

Roelle, J. E., D. B. Hamilton, and R. L. Johnson. 1988. Refuge Management Analyses: Restoration of Thompson Lake as an Alternative to Further Development at Chautauqua National Wildlife Refuge. U.S. Department of the Interior, Fish and Wildlife Service, Research and Development, Washington, D.C. 65 pp.

Rohm, C. M., J. W. Giese, and C. C. Bennett. 1987. Test of an aquatic ecoregion classification of streams in Arkansas. J. Freshwater Ecol. 4:127-140.

Roseboom, D. P. 1987. Case studies of stream and river restoration. Pp. 184-194 in Management of the Illinois River System: The 1990's and Beyond. Illinois River Resource Management. A Governor's Conference held April 1-3. Peoria, Ill. 260 pp.

Roseboom, D. P., and D. L. Richey. 1977. Acute toxicity of residual chlorine and ammonia to some native Illinois fishes. Report of Investigations No. 85, Illinois State Water Survey, Urbana, Ill.

Roseboom, D. P., and B. White. 1990. The Court Creek Restoration Project. Pp. 27-39 in Erosion Control: Technology in Transition. Proceedings of Conference XXI, International Erosion Control Association, Feb. 14-17. Washington, D.C.

Roseboom, D. P., R. L. Evans, J. E. Erickson, and L. G. Brooks. 1983. An Inventory of Court Creek Watershed Characteristics That May Relate to Water Quality in the Watershed. Document No. 83/23-A. Illinois Department of Energy and Natural Resources, Illinois State Water Survey, Peoria, Ill. 95 pp.

Roseboom, D., R. Twait, and D. Sallee. 1989. Habitat restoration for fish and wildlife in backwater lakes of the Illinois River. Pp. 65-68 in Proceedings of the Second

Conference on Management of the Illinois River System: The 1990's and Beyond. Illinois River Resource Management. A Governor's Conference held October 3-4. Peoria, Ill. 199 pp.

Rosgen, D. L. 1988. Conversion of a braided river pattern to meandering—a landmark restoration project. Presented at the California Riparian Systems Conference, September 22-24, Davis, Calif.

Rosgen, D., and B. L. Fittante. 1986. Fish habitat structures—A selection guide using stream classification. Proceedings of Fifth Trout Stream Habitat Improvement Workshop, Lock Haven, Pa.

Ross, P. E., R. E. Sparks, and F. S. Dillon. 1989. Identification of Toxic Substances in the Upper Illinois River. Annual Report. Illinois Department of Energy and Natural Resources, Contract No. WR36. 20 pp.

Ruelle, R., and J. Grettenberger. 1991. A Preliminary Contaminant and Toxicological Survey of Illinois River Sediments. Special Project Report 90-1. U.S. Fish and Wildlife Service, Rock Island, Ill. 17 pp.

Scarpino, P. V. 1985. Great River. An Environmental History of the Upper Mississippi, 1890-1950. University of Missouri Press, Columbia, Mo. 219 pp.

Sedell, J. R., and J. L. Froggatt. 1984. Importance of streamside forests to large rivers: The isolation of the Willamette River, Oregon, U.S.A., from its floodplain by snagging and streamside forest removal. Ver. Int. Limnol. 22:1828-1834.

Seebohm, M. E. 1952. The Evolution of the English Farm. Allen and Unwin, Ltd., London, 246 pp.

Semonin, R. G. 1989. Comments for Illinois River Conference. Pp. 41-45 in Proceedings on the Second Conference of the Management of the Illinois River System: The 1990's and Beyond. Illinois River Resource Management. A Governor's Conference held October 3-4. Peoria, Ill. 199 pp.

Sharpley, A. N., and J. K. Syers. 1979. Phosphorus inputs into a stream draining an agricultural watershed. Water, Air, Soil Pollut. 11:417-428.

Simons, D. B., and M. L. Albertson. 1960. Uniform water conveyance channels in alluvial material. Proc. of the Am. Soc. of Civil Eng. 86(H75):33.

Simons, D. B., R. M. Li, Y. H. Chen, S. S. Ellis, and T. P. Chang. 1981. Investigation of effects of navigation traffic activities on hydrologic, hydraulic, and geomorphic characteristics. Working Paper 2 for Task D submitted to Upper Mississippi River Basin Commission, St. Paul, Minn.

Simpson, P., J. R. Newman, M. A. Keirn, R. M. Matter, and P. A. Guthrie. 1982. Manual of Stream Channelization Impacts on Fish and Wildlife. FWS/OBS-82/24. U.S. Fish and Wildlife Service Contract No. 14-16-0009-80-066. 155 pp.

Smith, P. W. 1971. Illinois streams: A classification based on their fishes and an analysis of factors responsible for disappearance of native species. Biological Notes, No. 76. Illinois Natural History Survey, Urbana, Ill.

Smith, R. A., R. B. Alexander, and M. G. Wolman. 1987. Water-quality trends in the nation's rivers. Science 235:1607-1614.

Sparks, R. E. 1975. Environmental inventory and assessment of navigation pools 24, 25, and 26, Upper Mississippi and Lower Illinois rivers. An electrofishing survey of the Illinois River. Contract Report No. Y-74-4 to U.S. Army Corps of Engineers District, St. Louis, Mo.

Sparks, R. E. 1977. Environmental inventory and assessment of navigation pools 24, 25, and 26, Upper Mississippi and Lower Illinois rivers. An electrofishing survey of the Illinois River. Special Report No. 5. UILU-WRC-77-0005. University of Illinois, Urbana-Champaign Water Resources Center. 122 pp.

Sparks, R. E. 1984. The role of contaminants in the decline of the Illinois River: Implications for the Upper Mississippi. Pp. 25-66 in James G. Wiener, Richard V. Anderson, and David R. McConville, eds., Contaminants in the Upper Mississippi River. Proceedings of the 15th Annual Meeting of the Mississippi River Research Consortium. Butterworth, Stoneham, Mass. 368 pp.

Sparks, R. E., R. C. Thomas, and D. J. Schaeffer. 1980. The effects of barge traffic on suspended sediment and turbidity in the Illinois River. U.S. Fish and Wildlife Service, Rock Island Field Office, Rock Island, Ill.

Sparks, R. E., P. B. Bayley, S. L. Kohler, and L. L. Osborne. 1990. Disturbance and recovery of large floodplain rivers. Environ. Manage. 14(5):699-709.

Spotts, D. E. 1986. Standing stock of fishes before and after a channel relocation in Blockhouse Creek, Lycoming County, Pennsylvania. Pp. 85-91 in The 5th Trout Stream Habitat Improvement Workshop. Pennsylvania Fisheries Commission. Harrisburg, Pa.

Stall, J. B., and S. W. Melsted. 1951. The silting of Lake Chautauqua, Havana, Illinois. Report of Investigations, No. 8. Illinois State Water Survey, in cooperation with Illinois Agriculture Experiment Station. Urbana-Champaign, Ill.

Stanford, J. A., and J. V. Ward. 1988. The hyporheic habitat of river ecosystems. Nature 335:64-66.

Starrett, W. C. 1971. A survey of the mussels (Unionacea) of the Illinois River. A polluted stream. Ill. Nat. Hist. Surv. Bull. 30:267-403.

Starrett, W. C. 1972. Man and the Illinois River. Pp. 131-169 in R. T. Oglesby, C. A. Carlson, and J. A. McCann, eds., River Ecology and the Impact of Man. Academic Press, New York.

Stern, E. M., and W. B. Stickle. 1978. Effects of Turbidity and Suspended Material in Aquatic Environments. Literature review. U.S. Army Corps of Engineers Waterways Experiment Station, Vicksburg, Miss. 117 pp.

Stuber, R. J. 1985. Trout habitat, abundance, and fishing opportunities in fenced vs. unfenced riparian habitat along Sheep Creek, Colorado. Pp. 310-314 in Riparian Ecosystems and Their Management: Reconciling Conflicting Uses. U.S. Forestry Service General Technical Report RM120, Ft. Collins, Colo.

Sullivan, D. J., P. D. Hayes, T. E. Richards, and J. C. Maurer. 1990. Water Resources Data. Illinois Water Year 1989. Vol. 2. Illinois River Basin. U.S. Geological Survey Water Data Report IL-89-2. 467 pp.

Surface Mining Control and Reclamation Act of 1977. P.L. 95-87.

Swanson, F. J., and R. E. Sparks. 1990. Long-term ecological research and the invisible place. BioScience 40(7):502-508.

Thomas, L. M. 1989. Strategies for research in the U.S. Environmental Protection Agency. Environ. Toxicol. and Chem. 8:273-275.

Thompson, J. 1989. Case studies in drainage and levee district formation and development on the floodplain of the lower Illinois River, 1890s-1930s. Special Report 016. University of Illinois at Urbana-Champaign, Water Resources Center. 152 pp.

Thurston, R. V., G. R. Phillips, R. C. Russo, and S. M. Hinkins. 1981. Increased toxicity of ammonia to rainbow trout *(Salmo gairdneri)* resulting from reduced concentrations of dissolved oxygen. Can. J. Fish. Aquat. Sci. 38:983-988.

Tiner, R. W., Jr. 1984. Wetlands of the United States: Current Status and Recent Trends. U.S. Government Printing Office, Washington, D.C. 59 pp.

Trihey, E. W., and C. B. Stalnaker. 1985. Evolution and application of instream flow methodologies to small hydropower developments—An overview of the issue. Pp. 176-183 in F. W. Olson, F. W., R. G. White, and R. H. Hamare, eds., Symposium on Small Hydropower and Fisheries Proceedings. American Fisheries Society, Bethesda, Md.

Turner, R. E., and N. N. Rabalais. 1991. Changes in Mississippi River water quality this century. Implications for coastal food webs. Bioscience 41(3):140-147.

U.S. Army Corps of Engineers. 1990. Upper Mississippi River System Environmental Management Program. Fifth Annual Addendum. U.S. Army Corps of Engineers, North Central Division. Chicago, Ill. 221 pp.

U.S. Department of Agriculture, Forest Service. 1988. T. W. Hoekstar and J. Capp, comps. Integrating Forest Management for Wildlife and Fish: 1987 Society of American Foresters national convention Oct. 18-21, 1987. Minneapolis, Minn. General Technical Report NC-122. USDA Forest Service North Central Forest Experiment Station. 63 pp.

U.S. Department of the Interior (DOI), Fish and Wildlife Service. 1988. 1985 National Survey of Fishing, Hunting, and Wildlife-Associated Recreation. Washington, D.C. 167 pp.

U.S. Department of the Interior, National Park Service. 1982. The Nationwide Rivers Inventory. U.S. Government Printing Office, Washington, D.C.

U.S. Environmental Protection Agency (EPA). 1985. Technical Support Document for Water Quality-Based Toxics Control. Appendix D. Duration and Frequency. Office of Water Enforcement and Permits, and Office of Water Regulations and Standards, U.S. Environmental Protection Agency, Washington, D.C.

U.S. Environmental Protection Agency (EPA). 1990. The Quality of Our Nation's Water. EPA 440/4-90-005. Washington, D.C.

U.S. Fish and Wildlife Service (FWS). 1982. Nonconsumptive use of wildlife in the United States. Pp. 10-13 in Resource Publication 154. U.S. Department of the Interior, Fish and Wildlife Service, Washington, D.C.

U.S. Fish and Wildlife Service/Canadian Wildlife Service. 1986. North American Wildlife Management Plan. U.S. Department of the Interior, Fish and Wildlife Service, Washington, D.C. 19 pp.

U.S. General Accounting Office (U.S. GAO). 1988. Rangeland Management: More Emphasis Needed on Declining and Overstocked Grazing Allotments. GAO/RCED-88-80, U.S. General Accounting Office, Washington, D.C.

van Heerden, I. L., and H. H. Roberts. 1980. The Atchafalaya delta—Louisiana's new prograding coast. Trans. Gulf Coast Assoc. Geol. Soc. 30:497-506.

Vannote, R. L., G. W. Minshall, K. W. Cummins, J. R. Sedell, and C. E. C. Cushing. 1980. The river continuum concept. Can. J. Fish. Aquat. Sci. 37:130-137.

Van Velson, R. 1979. Effects of livestock grazing upon rainbow trout in Otter Creek. Pp. 53-55 in O. B. Cope, ed., Forum—Grazing and Riparian/Stream Ecosystems. Trout Unlimited, Vienna, Va.

Vinyard, G. L., and W. J. O'Brien. 1976. Effects of light and turbidity on the reactive distance of bluegill (Lepomis macrochirus). J. Fish. Res. Bd. Can. 33:2845-2849.

Vonnahme, D. R. 1989. Progress in the Illinois River watershed since the First Illinois River Conference. Pp. 8-14 in Proceedings of the Second Conference on the Management of the Illinois River System: The 1990's and Beyond. Illinois River Resource Management. A Governor's Conference held October 3-4. Peoria, Ill. 199 pp.

Wallen, E. I. 1951. The direct effect of turbidity on fishes. Bulletin No. 48, Series 2. Oklahoma Agricultural and Mechanical College, Stillwater, Okla.

Ward, J. V. 1989. The four-dimensional nature of lotic ecosystems. J. Am. Benthol. Soc. 8(1):2-8.

Ward, B. R., and P. A. Slaney. 1980. Evaluation of in-stream enhancement structures for the production of juvenile steelhead trout and coho salmon in the Keagh River. Pp. 8-15 in Proceedings of the Trout Stream Habitat Improvement Workshop, November 3-7, 1980, Asheville, N.C.

Ward, J. V., and J. A. Stanford. 1983. The serial discontinuity concept of lotic ecosystems. Pp. 29-42 in T. D. Fontaine III and S. M. Bartell, eds., Dynamics of Lotic Ecosystems. Ann Arbor Science Publishers, Stoneham, Mass. 494 pp.

Weber, M. 1986. Federal marine fisheries management. Pp. 267-344 in R. L. DiSilvestro, ed., Audubon Wildlife Report 1986. The National Audubon Society, New York.

Webster, J. R. 1975. Analysis of Potassium and Calcium Dynamics in Stream Vegetation. Ph.D thesis. University of Georgia, Athens.

Welcomme, R. L. 1979. Fisheries Ecology of Floodplain Rivers. Longman, Inc., New York. 317 pp.

Wesche, T. A. 1985. Stream channel modifications and reclamation structures to enhance fish habitat. Pp. 103-163 in J. A. Gore, ed., The Restoration of Rivers and Streams. Theories and Experience. Butterworth, Stoneham, Mass. 280 pp.

Wetmore, F. 1987. Flood damage protection programs. Pp. 89-102 in Management of the Illinois River System: The 1990's and Beyond. Illinois River Resource Management. A Governor's Conference held April 1-3, 1987, Peoria, Ill. 260 pp.

White, R. J. 1972. Response of trout populations to habitat change in Big Roche-a-Cri Creek, Wisconsin. Ph.D. thesis. University of Wisconsin, Madison. 278 pp.

White, R. J. 1978. Principles of trout stream habitat management. Mimeo Report presented at Workshop on Trout Stream Habitat Management, University of Wisconsin, Stevens Point, Wisconsin. August 1, 1978. 22 pp.

White, R. J., and O. M. Brynildson. 1987. Guidelines for management of trout stream habitat in Wisconsin. Wis. Dept. Nat. Resour. Tech. Bull. 30.

Whitley, J. R., and R. S. Campbell. 1974. Some aspects of water quality and biology of the Missouri River. Trans. Mo. Acad. Sci. 8:60-72.

Whitlock, R. 1965. A Short History of Farming in Britain. John Baker Ltd., London. 246 pp.

Whittier, T. R., D. P. Larsen, R. M. Hughes, D. M. Rohm, A. L. Gallant, and J. M. Omernik. 1987. The Ohio Stream Regionalization Project: A Compendium of Results. EPA/600/3-87/025. U.S. Environmental Protection Agency, Corvallis, Ore. 66 pp.

Widdows, J., P. Fieth, and C. M. Worral. 1979. Relationship between seston, available food and feeding activity in the common mussel *Mytilus edulis*. Mar. Biol. 50:195-207.

Wild and Scenic Rivers Act. P.L. 90-542, October 2, 1968.

Wiley, M. J., L. L. Osborne, R. W. Larimore, and T. J. Kwak. 1987. Augmenting Concepts and Techniques for Examining Critical Flow Requirements of Illinois Stream Fisheries. Aquatic Biology Section Technical Report 87/5. Final Report F-43-R. Illinois Natural History Survey. 138 pp.

Wiley, M. J., L. L. Osborne, and R. W. Larimore. 1990. Longitudinal structure of an agricultural prairie river system and its relationship to current stream ecosystem theory. Can. J. Fish. Aquat. Sci. 47(2):373-384.

Wilkin, D. C., and S. J. Hebel. 1982. Erosion, redeposition, and delivery of sediment to midwestern streams. Water Resour. Res. 18(4):1278-1282.

Willeke, G. E., and A. D. Baldwin. 1984. An evaluation of river restoration techniques in Northwestern Ohio. U.S. Army Corps of Engineers, Water Resources Support Center, Institute for Water Resources, Ft. Belvoir, Va. Contract Number DACW 72-79-C-0043. 80 pp.

Williams, J. E., J. E. Johnson, D. A. Hendrickson, S. Contreras-Dalderas, J. D. Williams, M. Navarro-Mendoza, D. E. McAllister, and J. E. Deacon. 1989. Fishes of North American endangered, threatened, or of special concern: 1989. Fisheries 14(6):2-20.

Wisconsin Dept. of Natural Resources (WDNR). 1975. Follow-up inventory of trout in a developed portion of Big Roche-a-Cri Creek. Mimeo report in Cold Water Group Waters Inventory File.

Wydoski, R. S., and D. A. Duff. 1980. Stream Management Improvement as a Potential Management Tool in the Intermountain West. Proceedings of the Bonneville Chapter, American Fisheries Society, Bethesda, Md.

Yount, J. D., and G. J. Niemi, eds. 1990. Recovery of lotic communities and ecosystems from disturbance: Theory and application. Environ. Manage. 14(5):515-762.

6

Wetlands

OVERVIEW

Wetlands occupy a special position in restoration ecology, because they have been affected by so many disturbances and because they fall under regulations that require mitigation of future damages. Unlike lakes and streams, wetlands have not always had recognized value. In recent years, public attitudes have changed from a general disregard of wetlands to a widespread desire to protect and restore them. A major policy forum has recommended "no-net-overall loss" and "net gain" in the quality and quantity of the nation's wetland resources (The Conservation Foundation, 1988). Thus, there have been numerous attempts to restore degraded wetlands, and there are many opinions about the status of wetland restoration.

The Bush administration has espoused the concept of no-net loss of wetland acreage and functioning. However, attempts to implement such a policy have proved difficult, because wetlands often stand in the way of development. Alaska wetlands were given special status (exemption) in the agreement to mitigate damages to wetlands (memo of agreement between EPA and COE, 1990). At present the area of protected wetlands may be reduced by modifying the delineation manual that is used to identify wetlands that are under the Clean Water Act of 1977 (P.L. 95-217), Section 404 jurisdiction. For example, seasonal wetlands would need to be wetter longer; peripheral areas would need to have vegetation classified as wetland "obligates" (species confined to wetlands) not just "facultative" species (those that occur in both wetland and upland habitats). Obviously, if less of the wetland is under regulatory domain, development can continue without a net loss (in the legal sense).

Although the delineation of wetlands is outside the scope of this chapter, one question is central to the committee's charge: Can damaged wetlands be restored? If so, then restoring one wetland might compensate for damaging another. The answer often depends on how good the wetland science is. Determining whether a damaged wetland has been restored requires good information on wildlife, vegetation, soil, and hydrology.

This chapter discusses the functional values of wetlands and describes historic losses and damages. Current wetland restoration technology is summarized, along with constraints on achieving restoration goals, problems encountered during restoration, opportunities for major restoration projects, programs for wetland restoration, and reasons for varying opinions on the success of wetland restoration. Conclusions, recommendations, and research needs complete the chapter; however, recommendations on wetlands policy and institutional changes pertaining to wetlands are included in Chapter 8.

Definition of Wetlands

In the scientific view, wetlands are transitional areas between terrestrial and open-water systems. In the legal view, wetlands are discrete units subject to regulatory jurisdiction. The diversity of wetland types makes it difficult to have a single definition for a wetland.

According to the U.S. Fish and Wildlife Service (FWS), "wetlands are lands transitional between terrestrial and aquatic systems where the water table is usually at or near the surface or the land is covered by shallow water" (Cowardin et al., 1979). The FWS lists three attributes that help identify wetlands: the presence of hydrophytes, hydric soils, and saturated or inundated substrate. The temporal nature of some wetlands is acknowledged—hydrophytes and hydrologic indicators need only be present periodically. This definition is more inclusive than that used by the U.S. Environmental Protection Agency (EPA) and the U.S. Army Corps of Engineers (COE) (Clean Water Act, Section 404 (b)(1) guidelines) for regulatory purposes. The major federal agencies involved in wetland regulation have adopted a uniform manual for delineating wetland boundaries (Federal Interagency Committee for Wetland Delineation, 1989).

The diversity of wetland habitat types and the diversity of species they support are impressive. The classification system of Cowardin et al. (1979) for U.S. wetlands includes 5 system types, 8 subsystems, 11 classes, 28 subclasses, and a large number of dominance types. Included within the category of wetlands are vegetation types that range from early colonizing (i.e., pioneer communities dominated by species such as cattails [*Typha* spp.]), to ancient, self-maintaining (i.e., old-growth, forested wetlands dominated by species such as bald cypress *(Taxodium distichum)* in the South and black spruce *(Picea mariana)* in the northern United States). The disturbances these systems have experienced likewise vary, as does the degree of restoration success.

The dynamic nature of wetlands also makes them ecologically complex. Along the edges of rivers, newly deposited sediments will be readily invaded by opportunistic plants and animals. Initial colonists are unlikely to be the same species as those of the floodplain forest that eventually develops. Along the edges of continents, mud flats are formed by alluvial outwash and are gradually colonized by salt marsh grasses and succulents, which in turn trap sediments that raise the topography and attract additional plant and animal species. Along the edge of an acidic lake, sphagnum moss and herbaceous plants develop a mat that eventually supports bog shrubs and bog forest trees. In all these habitats, the nutrient content of the soil and the biomass of plants and animals increase through time, along with increases in species diversity and ecosystem complexity. The development of open substrates into persistent ecosystems is often called primary succession, a process that may occur over centuries or millennia. However, the process is not unidirectional, and Niering (1989) suggests that the term *succession* be replaced by *vegetation development* or *biotic change* to reflect the complex changes that ecosystems undergo in response to gradual and catastrophic events.

Historical Perspectives on Wetlands

Until the last two decades, wetlands were considered to be wastelands, having little productive use to society and no direct economic value to private landowners. They needed to be "reclaimed" through draining, ditching, diking, or filling to enhance their benefit to the public. Some federal, state, and local governmental policies actually provided incentives for destruction of wetlands. The purpose of the first "official" federal acts dealing with wetlands—the Swamp Lands Acts of 1849, 1850, 1860—was to convey to 15 states along the Mississippi River and to Oregon all swamp and overflow lands unfit for cultivation so that the states could reclaim the land for agriculture (adapted from Reitze, 1974). The drainage and destruction of wetlands continued to be the accepted and often encouraged practice in the United States until the mid-1970s.

By the early to mid-1900s, negative impacts, such as declining waterfowl populations, were becoming apparent. The Fish and Wildlife Coordination Act of 1934, coupled with the U.S. Fish and Wildlife Wetlands Inventories of 1954 and 1973 (Reitze, 1974), prompted the realization that the loss of wetland habitat was causing a decline in fish and waterfowl populations. The public has begun to realize that wetlands are valuable systems providing many benefits to society. However, the conflict between private ownership of wetlands (and limited private benefits) and the desire to preserve social and economic values continues to contribute to the loss and degradation of wetlands.

Functional Value of Wetlands

Wetlands have properties of both aquatic and terrestrial ecosystems. Their most widely valued function is providing habitat for fish, birds, and other wildlife (Table 6.1), that is contributing to the maintenance of biodiversity (Table 6.2). In addition to this "food chain support" function, wetlands carry out hydrologic functions (e.g., flood-peak reduction, shoreline stabilization, ground water recharge) and water quality improvements (sediment accretion, nutrient uptake), all of which are recognized as valuable to society as a whole (Adamus and Stockwell, 1983). For individuals, wetlands provide recreational, educational, research, and aesthetic functions (see Table 6.1).

FOOD CHAIN SUPPORT

Although wetlands within the conterminous United States constitute only about 5 percent of the land surface (more than 40 million hectares, or about 104 million acres; Tiner, 1984; Dahl, 1990), many wetlands are among the most productive of natural ecosystems, exceeding the best agricultural lands and rivaling the production of tropical rain forests (Mitsch and Gosselink, 1986; Niering, 1986; The Conservation Foundation, 1988; CEQ, 1989). They provide habitat for a rich variety of native species. Riverine wetlands also serve as corridors for large, far-ranging species such as the Florida panther and black bear, as well as wetland-dependent species such as amphibians (Harris, 1988). More than one-third of the federally endangered and threatened plants and animals require wetland habitats during some portion of their life cycle (T. Muir, U.S. Fish and Wildlife Service, personal communication, June 1990).

TABLE 6.1 Wetland Functions

Flood conveyance—Riverine wetlands and adjacent floodplain lands often form natural floodways that convey floodwaters from upstream to downstream areas.

Protection from storm waves and erosion—Coastal wetlands and inland wetlands adjoining larger lakes and rivers reduce the impact of storm tides and waves before they reach upland areas.

Flood storage—Inland wetlands may store water during floods and slowly release it to downstream areas, lowering flood peaks.

Sediment control—Wetlands reduce flood flows and the velocity of floodwaters, reducing erosion and causing floodwaters to release sediment.

Habitat for fish and shellfish—Wetlands are important spawning and nursery areas and provide sources of nutrients for commercial and recreational fin and shellfish industries, particularly in coastal areas.

Habitat for waterfowl and other wildlife—Both coastal and inland wetlands provide essential breeding, nesting, feeding, and refuge habitats for many forms of waterfowl, other birds, mammals, and reptiles.

Habitat for rare and endangered species—Almost 35 percent of all rare and endangered animal species either are located in wetland areas or are dependent on them, although wetlands constitute only about 5 percent of the nation's lands.

Recreation—Wetlands serve as recreation sites for fishing, hunting, and observing wildlife.

Source of water supply—Wetlands are becoming increasingly important as sources of ground and surface water with the growth of urban centers and dwindling ground and surface water supplies.

Food production—Because of their high natural productivity, both tidal and inland wetlands have unrealized food production potential for harvesting of marsh vegetation and aquaculture.

Timber production—Under proper management, forested wetlands are an important source of timber, despite the physical problems of timber removal.

Preservation of historic, archaeological values—Some wetlands are of archaeological interest. Indian settlements were located in coastal and inland wetlands, which served as sources of fish and shellfish.

Education and research—Tidal, coastal, and inland wetlands provide educational opportunities for nature observation and scientific study.

Source of open space and contribution to aesthetic values—Both tidal and inland wetlands are areas of great diversity and beauty, and provide open space for recreational and visual enjoyment.

Water quality improvement—Wetlands contribute to improving water quality by removing excess nutrients and many chemical contaminants. They are sometimes used in tertiary treatment of wastewater.

SOURCE: Adapted from Kusler, 1983.

TABLE 6.2   Wetland Attributes That Assist in the Maintenance of Biodiversity

• Persistence of habitat for mating, nesting, and protection from predators during extreme environmental conditions.

• Resilience, the ability to recover from natural or human disturbances (e.g., environmental extremes, such as tidal closure and drought), often conferred through marsh soils.

• Ability to maintain plant populations. Regions with high environmental variability need refuges for long-term maintenance of populations and to ensure resilience (ability to recover rapidly) following extreme events.

• Resistance to invasive species (exotic to the region or alien to the habitat). The continual threats of disturbance to topography and hydrology lead to the need for constructed wetlands to resist invasive species.

• Ability to support nutrient transformations (microbial and chemical processes controlling the concentrations of nutrients and other compounds and facilitating the biogeochemical cycling of nutrients and the flow of energy). Nutrient transformations are not well known for all wetland types. Plant productivity of freshwater marshes is often phosphorus limited, whereas that of coastal marshes is often nitrogen limited; thus, these elements have been the focus of most assessments of nutrient dynamics. In coastal wetlands, the nitrogen dynamics are very important; both fixation and denitrification rates are linked to availability of organic matter in the soil.

Wetland production is important to both aquatic and terrestrial food webs, as summarized by the Council on Environmental Quality (CEQ, 1989):

> Wetlands provide cover, freedom from disturbance, food, and other vital habitat factors. It is estimated that over one-half of all the saltwater fish and shellfish harvested annually in the United States, and most of the freshwater game fish, use wetlands for feeding areas, spawning grounds, and nurseries for young. About one-third of the North American bird species are wetland associates. In addition to supporting resident birds year-round, wetlands are important breeding grounds, overwintering areas, and feeding areas for migratory birds, particularly waterfowl. Of the 10 to 20 million waterfowl that nest in the conterminous 48 United States, 50 percent or more reproduce in the Prairie Pothole wetlands of the Midwest. Bald eagles, ospreys, hawks, egrets, herons, kingfishers, and a variety of shore, marsh, and passerine birds are other components of the wetland avifauna.

Wetland-dependent mammals include muskrats, beaver, marsh rice rats, and swamp rabbits, and otter, mink, raccoon, bobcat, meadow mouse, moose, and white-tailed deer use wetlands as feeding areas.

Our knowledge of how food webs are modified as wetland habitat diminishes is not extensive, nor is our understanding of how trophic structure responds to declines in predatory species, such as the bald eagle and peregrine falcon. The native food web is no doubt essential to the maintenance of community structure. Power (1990) studied one stream system and demonstrated that communities with and without fish have contrasting structures: Where fish are absent, smaller predators increase in abundance and reduce the numbers of chironomids (midge larvae); thus algae are released from chironomid grazing and they develop a tall, thick turf. In the presence of fish, each trophic level reverses in abundance; the fish reduce numbers of smaller predators, so chironomids increase and in turn consume the algal turf, reducing it to a prostrate form. Elsewhere, introductions of exotic animals are known to have caused major changes to the wetland ecosystem (e.g., nutria alter plant successional processes and ecosystem structure in Louisiana coastal marshes; M. Rejmanek, University of California-Davis, personal communication, September 1990). The introduction of foreign plants can lead to vegetation growth that "swamps" native food chains (e.g., water hyacinths clog southern waterways). Until food chain functions are well understood, restoration projects will be jeopardized by the inability to ensure the reestablishment of critical links.

HYDROLOGIC FUNCTIONS

Their position in the landscape, whether as isolated wetlands or floodplains contiguous with rivers and streams, gives wetlands a major role in storage of floodwater and abatement of flooding. Wetlands intercept storm runoff and release floodwaters gradually to downstream systems. Because it is usually the peak flows that contribute to flood damage, wetlands reduce the impact of flooding (Novitzki, 1979). When wetlands are converted to systems that are intolerant of flooding (drained agricultural lands, filled developed lands), their storage capacity decreases and downstream flooding occurs.

The cost of lost flood storage and abatement functions is substantial, and it is borne almost exclusively by taxpayers. Riverine wetlands along the Charles River in Massachusetts were deemed effective in protecting Boston from flooding, and purchasing them was less expensive than building flood control structures (U.S. Army Corps of Engineers, 1972). In what is now a classic study of wetland hydrologic values, the COE determined that losing 3,400 ha of wetlands in the Charles River basin would increase flood damage by $17 million per year (equivalent to $5,000/ha per year, or about $2,000 per acre per year). That such flood protection values are real is supported by experiences where flood protective functions have been lost. Along the Mississippi River, constructing levees and draining the flood-plain have reduced floodwater storage from an estimated 60 days to 12 days (Gosselink et al., 1981) because waters can no longer spread out and be absorbed by the broad floodplain. The result has been annually recurring floods along the lower Mississippi River; the costs include flood damages and construction of extensive structures to abate flooding.

WATER QUALITY IMPROVEMENT FUNCTIONS

The value of wetlands for improving water quality is often over-looked, yet wetlands can remove and transform both organic and inorganic materials—including human waste, toxic compounds, and metals—from inflowing waters (Tuschall, 1981; Best et al., 1982; Best, 1987). Wetland attributes that make them effective in improving water quality include the following (adapted from Mitsch and Gosselink, 1986):

• As water floods into wetlands from rivers and streams, its veloc-ity decreases, causing an increase in sedimentation. Thus, chemicals sorbed to sediments are removed from the water and deposited in the wetlands.

• A variety of anaerobic and aerobic processes function to precipi-tate or volatilize certain chemicals from the water column.

• The accumulation of organic peat that is characteristic of many wetlands can ultimately lead to a permanent sink for many chemi-cals.

• The high rate of productivity of many wetlands can lead to high rates of mineral uptake by, and accumulation in, plant material with subsequent burial in sediments.

• Shallow water coupled with the presence of emergent vegetation leads to significant sediment-plant-water exchange.

HUMAN VALUES

As discussed above, wetlands play an active part in hydrologic functions, water quality improvement, and food chain support func-tions that serve human needs. Because of their importance in flood-peak reduction, shoreline stabilization, ground water recharge, sedi-ment accretion, nutrient removal, toxic material removal, and support of commercially important fish, shellfish, ducks, and geese, wetlands have received special protection under federal and state laws and many local ordinances. Wetlands serve a number of purposes that translate into economic values through reduction in flood and storm damage, conservation of water supplies, treatment of wastewater, and production of food.

Moreover, wetland recreation often leads to private profits. Consider, for example, anglers who buy special gear and clothing; the growing numbers of bird watchers who purchase books, cameras, and binoculars; the publication of sport and wildlife magazines; and the tourism that is generated by aquatic reserves and a new generation of visitor centers in places as unlikely as wetlands constructed to treat urban wastewater (e.g., Arcata, California). Habitats that provide opportunities for research and education contribute additional human values, with increasing numbers of programs for field experiences at the elementary and high school levels, as well as college and graduate course work.

Because wetlands are extremely valuable natural resources, their degradation or loss results in real costs to society. As Dahl (1990) concluded, "Environmental and even socioeconomic benefits (i.e., ground water supply and water quality, shoreline erosion, floodwater storage and trapping of sediments, and climatic changes) are now seriously threatened." However, these values are principally societal values, whereas private wetland owners receive few direct economic benefits from wetlands—and the ownership of wetlands is largely private. Of the acres of wetlands that remain in the United States, almost three-fourths (74 percent) are privately owned (CEQ, 1989).

Restoring damaged wetlands should be a high priority, now that wetlands are recognized as valuable environmental and socioeconomic systems. However, restoration is often very expensive—with estimates as high as $10 million to $50 million for a small (260-acre), urban wetland in Los Angeles, depending on the degree of restoration selected. Restoring farmlands to wetlands may be inexpensive and easier to accomplish. Whereas the costs of wetland draining and filling were borne largely by private owners seeking to achieve a direct personal increase in economic benefits, the restoration of wetlands will be borne almost entirely by the public. Exceptions are wetlands restored within the regulatory process: landowners who disturb or destroy existing wetlands often propose to mitigate the damages by restoring or creating degraded wetlands.

King (1990) has begun an analysis of the cost effectiveness relationship for wetland restoration projects. His approach is to model combinations of tasks that will speed wetland restoration (e.g., site contouring, vegetation planting, soil augmentation, control of exotic species) and the degree of functional equivalency achieved with each additional task and cost. His intent is to determine the point at which it is no longer economical to spend more on a project (i.e., when the additional ecological benefits would be minimal). A major shortcoming of his analysis is, of course, lack of data on functional equivalency. For the San Diego Bay project (Box 6.1), the use of 11 measured values suggested that, at the age of 5 years, the constructed wetland had less than 60 percent of the functional equivalency of reference wetlands. A second data point (obtained at perhaps 10 years) is needed to determine if site development has leveled off or if the site can eventually achieve great similarity with natural wetland functioning.

A second major shortcoming of cost-benefit analyses in general is the inability to put dollar values on ecosystem attributes. Although one can estimate the cost of many of the human values described above, one can never predict all that might be derived from wetland restoration. An endangered plant that might be rescued by a marsh restoration project may some day be found to produce an important pharmacological chemical; restoration of a coastal wetland may prevent real estate damage should sea level rise at unexpected rates; wetland plants may become horticulturally or agriculturally important (Glenn et al., 1991); a habitat-dependent bird may be shown to be effective in controlling mosquito and malarial outbreaks; the presence of open space may be shown to be essential to mental health.

LOSS OF WETLANDS

Trends in historical losses of wetlands in the United States were recently summarized in a report to Congress (Dahl, 1990):

> At the time of Colonial America, the area that now constitutes the 50 United States contained an estimated 392 million acres (about 160 million hectares) of wetlands. . . . Over a period of 200 years, the lower 48 states lost an estimated 53 percent of their original wetlands. . . . On average, this means that the lower 48 states *have lost over 60 acres* (about 25 hectares) of wetlands *for every hour* between the 1780's and 1980's [emphasis added].

By the 1980s, wetlands constituted only 5 percent of the landscape, down from an original 11 percent. The distribution and abundance of wetlands have also changed significantly since the 1780s (Figure 6.1). The midwestern farm belt states of Illinois, Indiana, Iowa, Michigan, Minnesota, Ohio, and Wisconsin lost more than 36 million acres (about 15 million hectares) of wetlands—roughly one-third of all wetlands lost in the history of our nation. All states, except for Alaska, Hawaii,

BOX 6.1
SWEETWATER MARSH NATIONAL WILDLIFE REFUGE, SAN DIEGO BAY, CALIFORNIA

Southern California's best-studied wetland restoration site is in the Sweetwater Marsh National Wildlife Refuge, which includes 128 ha of wetlands (mostly intertidal salt marsh) and some uplands along the eastern side of San Diego Bay, California (32°38'N, 117°6'W). The site and the restoration project are both significant—the wetland provides habitat for endangered species and thus is critical for maintaining regional biodiversity; the project has exceptionally high criteria for judging success and thus serves as a model for future restorations.

Protection of the site and strict standards for restoration came about only after a lengthy court battle. The new refuge was designated after a federal district court (Thompson, 1988) settled a lawsuit filed by the Sierra Club and the League for Coastal Protection against three federal agencies. Wetland habitat had been damaged by construction of a wider freeway, a new freeway interchange, and a flood control channel. Endangered species had been jeopardized, and mitigation measures had not been implemented. The lawsuit also led to reinitiation of consultations and a new biological opinion by the U.S. Fish and Wildlife Service (1988), which included strict criteria for successful mitigation. The requirements were expanded to include *functional* wetlands that would support *persistent* populations of three endangered species, the light-footed clapper rail *(Rallus longirostris levipes)*, the California least tern *(Sterna antillarum browni)*, and the salt marsh bird's beak *(Cordylanthus maritimus* ssp. *maritimus).*

The current shoreline of San Diego Bay bears little resemblance to what was once the natural landscape. The bay entrance to Paradise Creek marsh has been filled, and tidal flushing has been rerouted through a channel dredged straight south to the Sweetwater River. A railroad and Interstate 5 cross the landward edge of the refuge and wetlands.

The alterations preceding the restoration/mitigation project included widening of Interstate 5, construction of a new freeway interchange, and excavation of a new flood control channel through existing wetland. Restoration began in fall 1984 with the excavation of about 4.9 ha of disturbed upper intertidal marsh, including areas previously used as an urban dump. Eight lower intertidal islands and adjacent channels were constructed in fall 1984 and planted with cordgrass *(Spartina foliosa)* in winter 1985. The goal (DeWald and Rieger, 1982) was to create nesting habitat for the light-footed clapper rail, and foraging habitat for the California least tern, which commonly nests on the nearby dredge spoil.

Cordgrass plants that would have been destroyed by construction were salvaged from Paradise Creek and placed in a small (0.1 ha or 0.23 acre) intertidal nursery that was constructed to hold and propagate cordgrass. Additional cordgrass plants and other species were moved to pots for propagation off-site. The California Department of Transportation (Caltrans) used the advice of salt marsh ecologists to design habitats with the appropriate intertidal elevations. The island configuration was Caltrans's attempt to accommodate a large area of channels and cordgrass marsh, as well as higher marsh refuges, all within a small, linear site (of approximately 5 ha). Cordgrass was transplanted at 6- and 3-ft intervals in early 1985, with survival somewhat reduced by rough handling of potted plants (Swift, 1988). Plants were fertilized with urea four times during 1985-1986 (H. Hunt, Caltrans, personal communication, June 1990).

After settlement of the lawsuit, Caltrans began a monitoring program to assess plant cover and faunal use. They also funded a propagation research program to improve transplantation methods and to develop techniques for establishing the endangered salt marsh bird's beak, which is a hemiparasitic annual plant.

Comparisons of constructed and natural wetland functioning were initiated in 1987 (after three growing seasons for the marsh transplants), with research funding from the California Sea Grant College and monitoring funds from Caltrans (Cantilli, 1989; Rutherford, 1989; Zalejko, 1989; Langis et al., 1991; Zedler, 1991; Zedler and Langis, 1991). Paradise Creek was selected as a reference system because part of that wetland had been filled, and it represented the lost habitat. In addition, it was hydrologically connected to the northern four islands of the restoration site.

Three wetland functions were evaluated in the lower-marsh cordgrass habitat:

1. *Food base for top carnivores.* Rutherford (1989) found differences in epibenthic invertebrates based on quarterly sam-

pling, mostly in year 4 of the marsh development. The constructed marsh averaged only one-third as many individuals as were present in Paradise Creek marsh. The seven most abundant invertebrate taxa were a fly larva of the genus *Pericoma*, capitellid polychaetes, the isopod *Ligia*, the amphipod *Orchestia*, a small snail *Assiminea*, a midge, and a Grapsid crab. A native anemone *(Diadumene franciscana)* was common in the natural marsh but rare in the constructed marsh. Its functional role is unknown. An exotic mussel *(Musculista senhousia)* was more common in the constructed marsh than in the natural marsh. (Sacco, 1989, also found similar species but lower densities of infaunal invertebrates (mostly oligochaete and polychaete worms) in cordgrass marshes of North Carolina. The presence of less soil organic matter was suggested as a reason for the low density.)

2. *Plant growth.* Clapper rails prefer dense cover of tall cordgrass (Jorgensen, 1975), which protects them and their nests from aerial predators. Although transplants expanded their cover between 1987 (after three growing seasons; Swift, 1988) and 1989 (year 5; Pacific Estuarine Research Laboratory [PERL], unpublished data), biomass and plant height were not equivalent in the constructed and natural marshes. Aboveground biomass was about one-half as great (Langis et al., 1991), and plants were about 20 cm shorter in the constructed marsh in July 1988 (PERL, unpublished data). Shorter cordgrass provides poorer cover and lacks the vertical refuge that many marsh insects require at high tide.

Debris deposition, erosion and accretion, and poor handling of transplants explain the delayed vegetative expansion of cordgrass, but lower biomass and height are more likely due to differences in nitrogen (Covin and Zedler, 1988). Foliar nitrogen concentrations were 16 percent lower in the constructed marsh than in the natural marsh, whereas phosphorus concentrations were similar (Langis et al., 1991).

3. *Nitrogen supply function.* Nitrogen-fixation rates were lower on the soil surface and were limited by low concentrations of soil organic matter (Zalejko, 1989). Soil nitrogen concentrations were lower in the constructed marsh, with less

than one-third as much in the sediment and about one-tenth as much in the pore water (Langis et al., 1991). (Others have also found lower concentrations of nutrients in constructed salt marsh soils in North Carolina [Craft et al., 1988] and New Jersey [Shisler and Charette, 1984].)

At San Diego Bay, soil nitrogen concentrations did not increase during the 2-year study (Langis et al., 1991); thus it is hard to predict when or if the site will be functionally equivalent to the reference wetland. Reasons for low concentrations of organic matter in the soil and for low concentrations of nutrients are related to the site's history; the sandy substrate was part of the alluvial outwash of Sweetwater River.

To suggest an overall comparison, Zedler and Langis (PERL, 1990) presented a "functional equivalency index" based on 11 marsh attributes that indicated less than 60 percent equivalency when the marsh was 4 to 5 years of age. For each attribute, the mean value for the constructed marsh was expressed as a percentage of the mean value for the reference wetland (organic matter content, 51 percent; sediment inorganic nitrogen, 45 percent; sediment nitrogen total Kjeldahl nitrogen (TKN), 52 percent; pore-water inorganic nitrogen, 17 percent; nitrogen fixation (surface centimeters), 51 percent; nitrogen fixation (rhizosphere), 110 percent; biomass of vascular plants, 42 percent; foliar nitrogen concentration, 84 percent; height of vascular plants, 65 percent; epibenthic invertebrate numbers, 36 percent; epibenthic invertebrate species lists, 78 percent).

Although cordgrass cover in the reconstructed marsh is expanding to fill in bare areas, there is no evidence that nutrient conditions are improving. It is not yet possible to predict if or when plant heights and biomass will be equivalent to those of the reference wetland.

Because a disturbed high-marsh wetland was excavated to construct the cordgrass marsh, there was a net loss of acreage. Because the disturbed site was not studied, and because there are few data on the role of higher marsh habitats, those functional losses will go unrecorded.

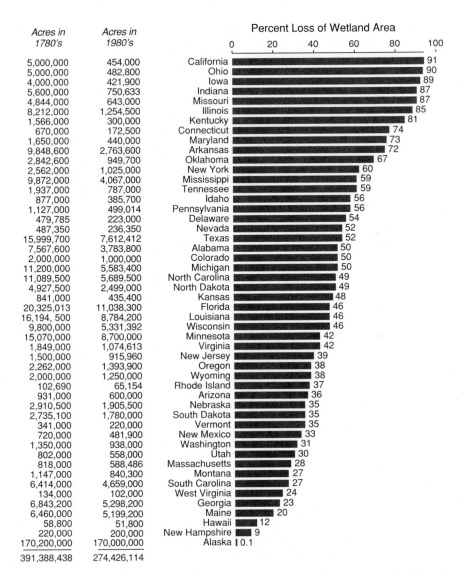

Acres in 1780's	Acres in 1980's	State	Percent Loss
5,000,000	454,000	California	91
5,000,000	482,800	Ohio	90
4,000,000	421,900	Iowa	89
5,600,000	750,633	Indiana	87
4,844,000	643,000	Missouri	87
8,212,000	1,254,500	Illinois	85
1,566,000	300,000	Kentucky	81
670,000	172,500	Connecticut	74
1,650,000	440,000	Maryland	73
9,848,600	2,763,600	Arkansas	72
2,842,600	949,700	Oklahoma	67
2,562,000	1,025,000	New York	60
9,872,000	4,067,000	Mississippi	59
1,937,000	787,000	Tennessee	59
877,000	385,700	Idaho	56
1,127,000	499,014	Pennsylvania	56
479,785	223,000	Delaware	54
487,350	236,350	Nevada	52
15,999,700	7,612,412	Texas	52
7,567,600	3,783,800	Alabama	50
2,000,000	1,000,000	Colorado	50
11,200,000	5,583,400	Michigan	50
11,089,500	5,689,500	North Carolina	49
4,927,500	2,499,000	North Dakota	49
841,000	435,400	Kansas	48
20,325,013	11,038,300	Florida	46
16,194,000	8,784,200	Louisiana	46
9,800,000	5,331,392	Wisconsin	46
15,070,000	8,700,000	Minnesota	42
1,849,000	1,074,613	Virginia	42
1,500,000	915,960	New Jersey	39
2,262,000	1,393,900	Oregon	38
2,000,000	1,250,000	Wyoming	38
102,690	65,154	Rhode Island	37
931,000	600,000	Arizona	36
2,910,500	1,905,500	Nebraska	35
2,735,100	1,780,000	South Dakota	35
341,000	220,000	Vermont	35
720,000	481,900	New Mexico	33
1,350,000	938,000	Washington	31
802,000	558,000	Utah	30
818,000	588,486	Massachusetts	28
1,147,000	840,300	Montana	27
6,414,000	4,659,000	South Carolina	27
134,000	102,000	West Virginia	24
6,843,200	5,298,200	Georgia	23
6,460,000	5,199,200	Maine	20
58,800	51,800	Hawaii	12
220,000	200,000	New Hampshire	9
170,200,000	170,000,000	Alaska	0.1
391,388,438	274,426,114		

FIGURE 6.1 A comparison of wetland area in the 1780s and 1980s (left), and the percentage of wetland area lost by the 1980s (bar graph; numbers = percent loss in 200 years). All data courtesy of the U.S. Fish and Wildlife Service.

and New Hampshire, have lost more than 20 percent of their original wetland acreage. The highest percentage of loss, 91 percent, has taken place in California, where only 454,000 acres (ca. 183,700 ha) remain, while the highest acreage loss, approximately 9,286,713 acres (ca. 3,759,800 ha) has occurred in Florida (a 46 percent loss; Figure 6.1).

Damages to Wetlands

Alterations to wetlands fall in three sometimes overlapping categories—biological, chemical, and physical (Table 6.3). The biological alterations result from management to maximize one or a few specific wetland values, from harvesting or removal of natural vegetation or animals, and from introduction of nonnative plants and animals. Like lakes and streams, wetlands are also subject to chemical alterations through point and nonpoint nutrient runoff from surrounding uplands; discharge of toxic, hazardous, or other chemical waste; and increased nutrient loadings.

For wetlands, the most destructive alteration has been physical, often eliminating the topography and hydrology that supports the wetland ecosystem. The most significant historical loss of wetlands has resulted from agricultural practices (87 percent; Tiner, 1984), with most due to wetland drainage (Figure 6.2). The construction of dams and the dredging of river, stream, and coastal waterways eliminate wetlands at the project site and also affect downstream systems. The stabilization of water levels for rivers or lakes eliminates the vital pulsing function that flooding provides, thus interrupting nutrient and sediment delivery (Loucks, 1989). The natural maintenance and expansion of wetlands often depend on sedimentation events, which are interrupted by dams (long-term reduction in sediment load) and dredging projects (short-term increase in sediment release). In urban areas, filling has accounted for significant wetland loss. This continues to be the case where restoration of one wetland is proposed as mitigation for additional development of other wetlands (Fonseca et al., 1988; Zedler, 1988a). In most physical alterations of the habitat, the wetland ecosystem is obliterated, and there is little opportunity for restoration once the habitat has been flattened or replaced by urban development, reservoirs, ports, or marinas.

Biological, chemical, and physical alterations often occur together, and the result is a cumulative impact that may well exceed the "sum" of the individual disturbances. A wetland that has been bulldozed to remove vegetation would recover more rapidly and more completely if it were not also contaminated with toxic materials or affected by hydrologic alterations. In addition, the alteration of one wetland

TABLE 6.3 Types of Alterations to Wetlands

Biological

• Grazing—consumption and compaction of vegetation by either domestic or wild animals
• Disrupting natural populations

Chemical

• Changing nutrient levels—increasing or decreasing levels of nutrients within the local water or soil system; forcing changes in wetland plant community
• Introducing toxics—adding toxic compounds to a wetland either intentionally (e.g., herbicide treatment to reduce vegetation) or unintentionally, adversely affecting wetland plants and animals

Physical

• Filling—adding any material to change the bottom level of a wetland or to replace the wetland with dry land
• Draining—removing the water from a wetland by ditching, tiling, pumping, and so on
• Excavating—dredging and removing soil and vegetation from a wetland
• Diverting water away—preventing the flow of water into a wetland by removing water upstream, lowering lake levels, or lowering ground water tables
• Clearing—removing vegetation by burning, digging, application of herbicides, scraping, mowing, or otherwise cutting
• Flooding—raising water levels, either behind dams or by pumping or otherwise channeling water into a wetland
• Diverting or withholding sediment—trapping sediment through construction of dams, channelization, or other types of projects, thereby inhibiting the regeneration of wetlands in natural areas of deposition such as deltas
• Shading—placing pile-supported platforms or bridges over wetlands, causing vegetation to die
• Conducting activities in adjacent areas—disrupting the interactions between wetlands and adjacent land areas, or incidentally impacting wetlands through activities at adjoining sites

SOURCE: Reprinted, by permission, from World Wildlife Fund and The Conservation Foundation, 1988. Copyright © by The Conservation Foundation, Washington, D.C.

affects the condition of neighboring wetlands. If an exotic plant gains a foothold in one disturbed system, its seeds are more likely to disperse to nearby wetlands, increasing chances of spread into less disturbed systems. Changes in hydrologic regimes, such as those brought on by ground water depletion from wells or by dams, levees, or drainage, can effectively eliminate wetland characteristics and the values associated with those characteristics. The cumulative effect of both local and regional disturbances is reduced potential for wetland restoration.

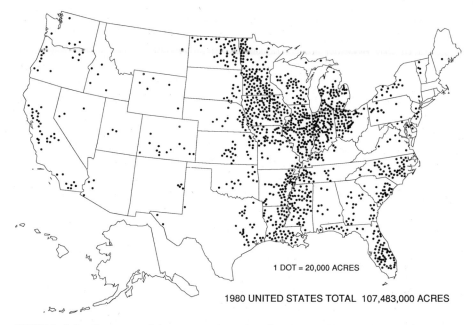

1 DOT = 20,000 ACRES

1980 UNITED STATES TOTAL 107,483,000 ACRES

FIGURE 6.2   Extent and location of artificially drained agricultural land in the United States in 1985.   Data courtesy of U.S. Department of Agriculture Economic Research Service.   Source: Dahl, 1990.

Effects of Cumulative Impacts

Recently, it has become apparent that the cumulative impact of many individual actions, no single one of which is particularly alarming, threatens the integrity of entire wetland landscapes. Cumulative impact is defined in the Council on Environmental Quality regulations as "the impact on the environment which results from the incremental impact of the action when added to other past, present, and reasonably foreseeable future actions regardless of what agency (federal or nonfederal) or person undertakes such other actions. Cumulative impacts can result from individually minor but collectively significant actions taking place over a period of time" (40 C.F.R. 1508.7 and 1508.8; Gosselink et al., 1990a).

Although the Clean Water Act and regulations for implementation of its Section 404 by both EPA (44 C.F.R. 230) and COE (33 C.F.R. 320-30) require consideration of cumulative impacts, they are seldom evaluated in permit review processes (Gosselink et al., 1990a). The Section 404 permit process focuses on the impact of a proposed activity at an individual wetland permit site.   In contrast, cumulative impacts are landscape-level phenomena that result from decisions at many individual permit sites, as well as activities that are not regulated under Section 404 (Gosselink and Lee, 1989).

Conversion of a wetland forest to agricultural use results in a typical cumulative impact (Gosselink et al., 1990b). Historically, the incremental clearing of 10 ha to as much as 2,000 ha in an individual permit has been perceived to have no significant ecological impact on a total forest system of several million hectares, and the cumulative effect of many such permitted activities has been ignored (Gosselink et al., 1990b). Currently no methodology for cumulative impact assessment is generally accepted by scientists and managers (Gosselink et al., 1990a). Furthermore, wetland restoration has generally been undertaken on an ad hoc basis, and the potential role of restoration in a cumulative context or on a landscape scale is not usually considered.

Gosselink and Lee (1989) described a methodology for cumulative impact assessment and management that incorporates a process of ecological characterization, goal setting, and planning. The method also focuses attention on the landscape level and bases planning on landscape ecology principles. Researchers used the Tensas River basin, an area of approximately 1 million hectares in northeastern Louisiana, as a case study to test this general approach to cumulative impact assessment and management (Gosselink et al., 1990a).

Potential Impacts of Global Climate Change and Sea Level Rise

Most ecosystems will be affected by global warming over the next century, and planning for restoration of self-sustaining ecosystems must consider the potential impact of climate change. The problem is predicting how the climate will change at specific locations (will it get wetter, drier, warmer, or cooler?) and what measures might offset any negative impacts of changing weather and hydrology. For biodiversity reserves, Peters (1988) has suggested that a number of management actions may be needed to prevent species extinctions as climate changes. For example, much larger reserves may be necessary to perpetuate populations, so that heterogeneous topography and soils are included, thus protecting a wider range of habitat types (and presumably also ecotypes). We may need to plan to control environmental conditions in reserves (e.g., through irrigation or drainage). Control of predators and exotic species may become more critical. Peters (1988) also suggests that creating reserves outside the normal climatic regions of an ecosystem type may be needed. Finally, he mentions the possibility of moving reserves as the climate changes, depending on the ability of restoration ecology to reconstruct communities in new locations.

For coastal wetlands, the future is more easily predicted, and management recommendations are clearer. Global warming will result in a rise in sea level, with the effect that existing coastal wetlands will be flooded, salt marshes will be lost and species will be unable to "migrate" inland because of urbanized or otherwise developed landscapes. Because the rate of rise will accelerate, not all species will be able to move inland fast enough to keep up with the rate of loss at lower elevations (Park et al., 1989) Thus, species diversity may decline along with the area of each wetland plant community. Park et al. (1989) have calculated that a 1-m rise in sea level would result in the loss of 65 percent (6,441 square miles) of the coastal marshes in the contiguous United States. Such an increase would result from an average temperature increase of 3°C and is probable by the year 2100 (Park et al., 1989). A global warming of 6°C could result in a 2.3-m rise and a 77 percent (7,423 square miles) loss of the coastal wetlands of the contiguous United States (Park et al., 1988). The specific predictions stemming from a 1-m rise in sea level are summarized by Park et al. (1989, p. 78):

> Some sites, such as Charleston, South Carolina, are well buffered for sea level changes up to one meter, due to high tidal ranges and high sedimentation and accretion rates. Other sites, especially in the Gulf of Mexico, are quite vulnerable to small changes in sea level. Although the model is perhaps too simple for representing complex deltaic dynamics it projects a continuation of current trends for the Louisiana coast south of New Orleans, with the entire undiked area shown to be at risk without any acceleration in sea level rise.
>
> With protection of all existing residential and commercial developments, by the year 2100 salt marshes and freshwater swamps would gradually decline in area, mangrove swamps would gradually increase, and freshwater marshes would gradually decline until a sea level increase of 0.8 m is reached. Freshwater marshes would then disappear rapidly (reflecting a pattern seen in both the Gulf and Mid-Atlantic Coasts); 35 percent of vegetated wetlands would be lost with a 0.5-m rise, 49 percent with a 1-m rise, 56 percent with a 2-m rise, and 68 percent (almost 9,000 mi^2) with a 3-m rise.

Several recommendations for restoration of coastal wetlands follow from the above predictions for a more rapidly rising sea level. The first is that a broad margin or buffer needs to be preserved adjacent to the restoration site, so that high marsh communities will have space to migrate inland. An additional consideration is that changes in water levels may occur at a rate too rapid for marsh flora and fauna to colonize buffer areas, unless these areas are aided by restoration-creation activities. The second recommendation is that long-term management will need to include censuses of species composition across elevation gradients that take note of changing abundances and disappearance of species. Because restoration sites may lack some of the resilience of natural marsh communities (i.e., persistent seed banks, highly organic soils to ameliorate changes in soil moisture, ability to resist exotic plant invasions), constructed wetlands may be more susceptible to rising sea level than their model ecosystems. Comparative long-term censuses of species abundances are needed to ensure that biodiversity is conserved.

RESTORATION OPPORTUNITIES

Wetland Settings and Their Effects

Opportunities and requirements for restoring wetlands differ depending on whether the wetland occurs in a coastal-estuarine, riverine, depressional, or large-scale freshwater setting.

COASTAL-ESTUARINE WETLANDS

Coastal-estuarine wetlands are characterized by the rise and fall of the tides and resulting salinity gradients, which determine, in large measure, the distribution of plant and animal species. Coastal and estuarine wetlands have been diked, filled (in some areas), and ditched for mosquito control or other purposes. Most coastal wetland restoration efforts have involved either the breaching of dikes to allow the reentry of tides or the restoration of vegetation (mangroves, *Spartina alterniflora*) on denuded areas after development projects. In some instances, fills have been removed, primarily as the result of regulatory violations and subsequent enforcement actions. Canals have also been filled to restore coastal or estuarine wetlands, and sediment diversions are proposed for the lower Mississippi.

Most of the wetland restoration and creation literature concerns coastal and estuarine wetlands. Perhaps the greatest success in at least restoring look-alike vegetation is with coastal and estuarine marshes. This success is due not only to the breadth of experience to date but also to the relative ease of determining appropriate elevations by using tide records and elevations and adjacent reference sites. In general, fewer wetland plant species occur in these systems (due to the need for salt tolerance) than in comparable freshwater systems. However, the habitat value of restored coastal wetlands is not fully documented.

STREAM AND RIVERINE WETLANDS

Stream and riverine wetlands are often severely altered. Not only have rivers been the depository for most liquid pollutants, but their hydrologic regimes have also been altered by dams, pumping, dikes, channelization, dredging, bank stabilization, and watershed development. Wetlands in headwater areas, in oxbows, and in low-velocity channels have not been extensively restored, although rather large-scale floodplain forest restoration programs are proposed for areas in the lower Mississippi. Many small-scale restoration projects have been undertaken as part of local greenway and stream restoration programs such as the Urban Streams program in California.

Efforts to restore riverine wetlands are complicated by the hydrologic and sediment regime changes typical of most rivers, which make it impossible to return wetlands to their natural condition without massive removal of dams, channelization, and so on. Nevertheless, these wetlands are increasingly recognized as having great value for water quality protection, fish and wildlife habitat, flood control, and bank stabilization.

DEPRESSIONAL WETLANDS

Many small, depressional wetlands were formed by glaciers 9,000 to 12,000 years ago. Hundreds of thousands of these wetlands exist in the northern tier of states (prairie potholes). Other depressional wetlands were created by solution (karst topography), the wind (sand hills of Nebraska), or other processes (Carolina bays of the Southeast). These wetlands have not, in most states, been modified as substantially as have riverine wetlands. However, many have been drained or are used during dry years for agriculture. Others have been used as landfills or filled for urban development.

Except for efforts to remove drainage tiles and restore natural drainages in the prairie pothole region, few efforts have been made to restore such depressional wetlands. Restoration efforts may be quite inexpensive if cessation of agriculture or blockage of drainage ditches is the primary activity. However, restoration is potentially expensive where fill must be removed or where extensive removal of drainage tiles is involved.

LARGE FRESHWATER WETLANDS WITH POOR DRAINAGE

Many of the larger, altered wetland complexes consist of relatively flat, poorly drained lands. Examples include former lake bottoms formed during glacial periods (e.g., lands adjacent to the Great Lakes), wide floodplains or deltas (lower Mississippi), coastal areas of low relief (e.g., pocosins and pine flatwoods of the Southeast), the Florida Everglades, and alluvium in front mountain ranges. Many of these lands have been at least partially drained for agricultural purposes. However, the wetland soils and hydrology are otherwise largely intact.

These lands constitute the largest area of potential wetland restoration sites, not only because of their large acreage but also because wetland restoration may be less expensive and may be achieved through the filling of drainage ditches or installation of minor water-control structures. Land values may also be relatively low, particularly where such lands are no longer being used for agriculture and there are no alternative uses.

RESTORATION OF FORMER WETLANDS
IN AGRICULTURAL USE

More than 80 percent of historic wetland loss in the 48 conterminous states has been due to agricultural conversions. Such conversions have involved clearing of the natural forested or herbaceous wetland vegetation, introduction of drainage, and flood control. Federally funded agricultural flood control and drainage projects have facilitated private conversions. In arid parts of the country, massive federally funded irrigation projects have promoted conversions of wetlands to agriculture.

Any national program designed to restore millions of acres of former wetlands will have to focus primarily on wetlands converted to agricultural use. In many cases, the technical problems of reconverting such agricultural lands to functioning wetland systems are not as challenging as those encountered in the urban context, where the physical components of natural wetlands have been severely altered by chemical pollution, fill, barriers to water movement, and vast changes in the watershed. Former wetlands now in agricultural use can typically be reestablished by cessation of planting crops or domestic animal foraging, breaking drainage tiles, filling in drainage ditches, and if feasible, removing flood control structures. In the prairie pothole region of the country, particularly in Minnesota, the U.S. Fish and Wildlife Service, the Soil Conservation Service, and the state of Minnesota have achieved partial restoration of thousands of acres of prairie potholes in this manner. Wetland vegetation has quickly reappeared, and ducks have returned in profusion (see Prairie Pothole case study, Appendix A). However, marsh vegetation may not replicate the historic community, and some animals have not returned. In small-scale restorations of the lower Mississippi valley, forested wetlands are being reestablished themselves on the alluvial soils that retain moisture for long periods following precipitation, even where federal flood control levees and channels block riverine overbank flooding. However, these small-scale wetland restorations are not often subject to rigorous scientific evaluation.

The Swampbuster program of the 1985 Food Security Act, as amended in the 1990 Farm Act, changes in the tax treatment of agricultural drainage in the Tax Reform Act of 1986 (P.L. 99-514), and the Clean Water Act Section 404 program, as well as lower grain prices, have reduced significantly the rate of agricultural conversions of wetlands since the early 1980s. Yet the rate of restoration has been very slow. The major impediments are more often legal, institutional, and financial rather than technical. Former wetlands now in agricultural use are almost all in private ownership, and restoration can occur only when financial arrangements with landowners are available to promote reconversion. In some cases, federal flood control projects will have to be modified. The small-scale restorations described above have occurred in the last 3 to 4 years because such programs have become available.

Small-scale wetland restorations have occurred in the last few years as a result of foreclosures on former wetlands by the Farmers Home Administration (FmHA) and the Resolution Trust Corporation (RTC) in response to the bankruptcies of farms and savings and loan institutions, respectively. The U.S. Department of Agriculture (USDA) and the Resolution Trust Corporation have an arrangement with the U.S. Fish and Wildlife Service to assess these lands for their value as wetland or other wildlife habitat before they are auctioned, resold, or in the case of some FmHA foreclosures, returned to their original owners, typically at bargain prices. In some cases, the U.S. government through the FWS has retained title to these former wetlands that are now being restored.

In the 1990 Farm Act, Congress established, for the first time, a significant wetland restoration program. Known as the Agricultural Wetland Reserve Program (AWRP), it provides that, of the remaining 6 million acres of cropland eligible for the Conservation Reserve Program (CRP), up to 1 million acres may be wetlands for inclusion in the AWRP over the next 5 years. The AWRP has a limit of 200,000 acres per year. However, in contrast to the CRP, easements are to have terms of 30 years or longer. Although this program may cost the taxpayers as much as $100 million annually, once 1 million acres are enrolled, this acreage will no longer be eligible for USDA grant and subsidy programs; its net cost may therefore be significantly less than the appropriation for the program suggests. The AWRP is the largest wetland restoration program in the nation's history, and it has the potential to play a significant role in the restoration of some aquatic ecosystems. For this to happen, environmental quality criteria must have at least as much weight in the selection of AWRP candidate parcels as production control and cost criteria do. Environmental quality criteria should include consideration of opportunities to restore natural vegetation corridors along streams and rivers, as well as corridors between isolated but large wetland tracts. Within large ecosystems, such as the prairie pothole region or the floodplains of the Mississippi River and its tributaries, for example, federal resource agencies with support from state agency and academic experts and the public should be able to identify high-priority wetland restoration sites. If the AWRP can be used to foster restoration of such sites, it will become a credible, powerful tool for aquatic ecosystem restoration that should be expanded when the Food Security Act next comes up for reauthorization in 1995.

Federal Water Projects

Federal agricultural and urban flood control, navigation, and irrigation projects have done extensive damage to wetland systems. One careful, economic study has shown that at least 30 percent of the loss of millions of acres of bottomland hardwood wetlands in the lower Mississippi valley from 1934 to 1984 was caused directly by federally funded COE and Soil Conservation Service flood control projects. The pace of these projects has slowed considerably in the last 10 years as a result of budget constraints, completion of many projects, and passage of the Water Resources Development Act (WRDA) of 1986 (P.L. 99-662) that, for the first time, imposed significant cost-sharing requirements on states or local interests for major water projects. However, the role of federal agencies in designing and implementing wetland restoration projects has been exceedingly limited to date. In coastal Louisiana, COE has used some dredged materials from maintenance of navigation channels to construct deltaic wetlands and has constructed the Carnaevon water diversion project, which is designed primarily to divert Mississippi River water and sediments to offset saltwater intrusion and build new wetlands.

At least four actions of Congress in 1990 suggest that federal water development agencies may begin to become actively involved in actual wetland restoration projects.

1. Section 306 of the 1990 WRDA provides that "environmental protection" is now a central mission of COE, coequal with its traditional missions of navigation and flood control. Although environmental protection is not synonymous with aquatic system restoration, it is a step in the right direction. Section 307 of the 1990 WRDA also calls upon COE to develop a wetland plan "within one year." The COE water resource development program, has as an interim goal no overall net loss of the nation's remaining wetland base, as defined by acreage and function, and as a long-term goal increasing the quality and quantity of the nation's wetlands, as defined by acreage and function. The COE is also directed to carry out wetland restoration and creation demonstration projects.

2. The Congress appropriated $6 million for COE to help finance part of the restoration of the Kissimmee River and floodplain wetlands in Florida, to begin the process of undoing the channelization of the river that COE completed 20 years ago. In addition, Congress appropriated $500,000 to the National Park Service to conduct a hydrologic study of the Everglades system that could serve as a basis for a larger water redistribution program for South Florida, with the objective of restoring the Everglades ecosystem.

3. Through the Coastal Wetlands Planning, Protection and Restoration Act of 1990 (P.L. 101-646), Congress has directed COE, EPA, and other federal agencies to work with the state of Louisiana to identify and construct wetland restoration projects. These agencies will also assist Louisiana in devising and implementing a management plan to abate wetland loss based on the no-net-loss principle and to achieve wetland restoration in coastal Louisiana, an area that has 40 percent of the nation's coastal wetlands and 80 percent of its coastal wetland erosion. Congress has appropriated approximately $50 million per year toward this major initiative.

4. Finally, Congress enacted the Fallon Paiote Shoshone Truckee Carson Pyramid Lake Settlement Act, which includes an appropriation to buy 50,000 acre-feet of water rights from willing farmers to help to restore water flows to and in the Stillwater wetlands in Nevada.

PROGRAMS FOR WETLAND RESTORATION

Although the nation has programs under the Clean Water Act to restore the chemical quality of water in aquatic ecosystems by reducing point source pollution, it has not had any systematic programs to restore the physical characteristics of such ecosystems. Programs such as that provided by Section 404 of the Clean Water Act and the Swampbuster program are intended to retard loss of wetlands, not restore them. Various federal programs encourage small-scale, nonsystematic restoration programs. These include the Section 404 program, the Conservation Reserve Program, FWS restoration projects, and a limited number of COE and Bureau of Reclamation projects.

Section 404, Clean Water Act

Clean Water Act regulations establish procedures for permitting many activities in wetlands. Consistent with Council on Environmental Quality (CEQ) regulations implementing the National Environmental Policy Act, Section 404 permit writers typically seek to avoid or minimize impacts on wetlands resulting from those activities. Where impacts are unavoidable, however, other forms of mitigation such as enhancement, restoration, or creation of replacement habitat are often employed to offset losses. In many cases, restoration should be the preferred method. Although many small-scale wetland restoration projects are being conducted as required by conditions of Section 404 permits, COE maintains no systematic record of these projects.

Conservation Reserve and Agricultural Wetland Reserve Programs

Congress established the Conservation Reserve Program in the 1985 Food Security Act to provide incentives to farmers to revegetate highly erodible lands. Under the program, farmers enter into 10-year contracts and receive annual payments as long as they remain in the program. Some drained wetlands have been restored through this program. Most of these are prairie pothole wetland restoration projects in north-central states; CRP has funded a smaller number of wetland restoration projects in the South. The USDA estimates that millions of acres of cropland have been taken out of production through this program.

As much as the CRP has achieved, it has severe limitations as a wetland restoration program. The USDA views the CRP primarily as a crop production control program. Wetlands were included almost as an afterthought. Although restoration of scattered wetlands in the prairie pothole region successfully reestablishes duck breeding, feeding, and nesting habitat, such opportunistic, small-scale restoration projects may have fewer benefits in other kinds of wetlands systems, such as broad floodplains of southern rivers. Furthermore, CRP contracts are for only 10 years, enough time for the establishment of herbaceous wetland species but not for the reestablishment of wetland forests. Also, farmers may withdraw the contracts after 4 years.

The Agricultural Wetland Reserve Program of the 1990 Farm Bill is directed at wetland systems and provides for conservation easements of 30 years or longer. With these features, the AWRP has the potential to be used to promote restoration of large, forested wetland systems. The 1990 bill authorizes funding that could support the restoration of one million or more acres of former wetlands.

Farmers Home Administration and Resolution Trust Corporation Lands

The Farmers Home Administration and the Resolution Trust Corporation have been taking over and assuming title to hundreds of thousands of acres of land due to foreclosures on farmland and to the bankruptcies of savings and loan banks. Some of this acreage is former wetlands, cleared and drained for agricultural or other use.

In general, these small wetland restoration projects provide very useful opportunities to restore individual wetland parcels. By themselves, however, they are unlikely to restore larger wetland systems that have been seriously degraded or to restore wetlands throughout a large landscape. The federal agencies responsible for water resource development projects, in particular COE, the Tennessee Valley Authority, and the Bureau of Reclamation, have, until very recently, paid little heed to opportunities for aquatic ecosystem restoration. Even EPA's Office of Wetland Protection does not have a program for supporting active wetland restoration. Its policy emphasis is more on wetland creation than on wetland restoration (W. Sipple, U.S. Environmental Protection Agency, personal communication, October 1990).

In the absence of federal programs designed to support large-scale restoration of wetlands, such projects are thus left to a mixture of grassroots, local, and state initiatives.

STATUS OF WETLAND RESTORATION RESEARCH AND TECHNOLOGY

Research on wetland restoration has focused chiefly on techniques of species establishment and on development of species composition and wetland community structure. The functional values of wetlands, although widely recognized, are seldom evaluated. Mitigation efforts cannot yet claim to have duplicated lost wetland functional values. It has not been shown that restored wetlands maintain regional biodiversity and recreate functional ecosystems (Zedler and Weller, 1989). There is some evidence that created wetlands can look like natural ones; there are few data to show that they behave like natural ones. In many cases, scientific knowledge of how to restore degraded wetland systems is limited, and wetland creation has been largely a matter of trial and error.

The largest single wetland restoration and creation research program has been the Dredged Materials Research Program of the U.S. Army Corps of Engineers (e.g., Woodhouse et al., 1974; Johnson and McGuinness, 1975; Parnell et al., 1978; Saucier et al., 1978; Newling and Landin, 1985). However, the construction of wetland habitat on dredge spoils is not considered restoration. Establishment of wetland biota and functions on spoils may involve unique problems, such as high erosion rates, extreme hypersalinity, or toxic contaminants that often go beyond those of on-site restoration. However, the research from dredge spoil sites has helped develop restoration ecology, as evidenced by the literature on individual species requirements, contaminant impacts, successional patterns, and wildlife establishment (e.g., Parnell et al., 1978).

Importance of Setting Goals for Functional Value

To set goals for the functional values of wetlands, we must understand how each type of natural wetland performs (i.e., in support of food chains, hydrologic functions, and water quality improvement). Success could then be measured by the degree to which those functions are achieved. Functional performance is much more difficult to assess than are structure or composition. It is also much more difficult to characterize in the permit process. Goals of wetland restoration have seldom included functional performance, and functional values have not been followed over time. An exception is a restoration site along the San Diego Bay (see Box 6.1), where functions such as providing persistent vegetative cover, self-maintaining populations, and nitrogen fixation were required as part of a legal settlement involving endangered species habitat.

Techniques for Restoring Wetlands

Techniques for restoring wetlands fall into three broad categories: (1) reestablishing or managing wetland hydrology, (2) eliminating or controlling chemical or other contaminants affecting wetlands, and (3) reestablishing and managing native biota (may include control of nuisance species). The restoration technique required depends on the type of disturbance (see Table 6.3). At the simplest level, restoration may involve elimination or control of overgrazing, allowing for ultimate reestablishment of native wetland biota. At a more complex level, restoration may require a combination of techniques ranging from removing fill, reestablishing proper hydrology, reintroducing native flora and fauna, and controlling exotic species. Some of the more commonly used wetland restoration techniques follow:

REESTABLISHING RIVER FLOW

Numerous wetlands have been modified through diversion of water away from wetlands. The first step in restoration is reestablishing flow into the wetland. The Kissimmee River and Marsh Restoration project in Florida is an excellent example of reestablishing the marsh hydrology (Loftin et al., 1990; see the Kissimmee Riverine-Floodplain case study, Appendix A). Initial restoration techniques consisted of installing structures to redivert water flow back into old river channels and adjacent marsh. Because some of the area still supports the original marsh communities, restoring the hydrology is allowing residual plant communities to reinvade the reflooded marsh.

RESTORING FLOOD REGIMES

Alteration of surface wetland hydrology occurs in several forms, ranging from excessive flooding to excessive draining. Excessive flooding of wetlands often occurs in dammed and diked rivers, streams, lakes, and wetlands. Providing control structures, or removing the structures that cause flooding, is the first step in restoration. If sedimentation in the flooded area has altered the original or "designed" elevation of the wetland, sediment removal may be required.

HALTING DRAINAGE

Loss of surface waters also commonly results from subsurface drainage, or lowering of ground water. General dewatering of the landscape through ditching even if not directly in streams or wetlands, ground water pumping, subsurface tile drainage systems, or other mechanisms can lead to surface dewatering of wetlands. As an example, installation of tile drainage systems to improve agricultural production in the prairie pothole region of the north-central United States (see case study, Appendix A) has generally resulted in the conversion of wetlands (in this case, seasonally wet systems) to agricultural lands. Restoration of the hydric condition and, ultimately, of the prairie pothole wetlands can be accomplished by destroying the drainage tile system.

The solution is more complicated when sediment loading to the depressions has raised the wetland's elevation relative to the landscape. Where sediment loading is significant, it may be necessary to remove sediment and reestablish original wetland elevation, in addition to plugging tiles, or it may be necessary to create a low level berm to impound water in the wetland.

REESTABLISHING TOPOGRAPHY

When wetland landscapes have been altered through filling or substrate removal (dredging), reestablishing the proper landscape contour concomitant with surface and subsurface hydrology is an essential first step in wetland restoration. Needed are removal of materials from filled wetlands and replacement of materials in dredged wetlands. Finding a source of proper-quality fill material may be difficult.

CONTROLLING CONTAMINANT LOADINGS

The simplest technique for restoring chemically contaminated systems is removal of contaminant inflow. Depending on the type and concentration of contaminants, removal of already deposited material may be necessary to achieve restoration. If removal of material is necessary, it may be difficult to locate a proper disposal site for contaminated sediment.

REESTABLISHING BIOTA

The final step in restoration generally involves establishing the "proper" species or at least enhancing the process of ecological succession, especially of the native plant community or communities. The level of effort depends on the ability of surrounding wetlands to provide a source for natural invasion of flora and fauna, the size of the area being restored, and the potential for invasion of exotic species.

Although the ultimate goal of restoration should be a self-sustaining ecosystem, some management (or control) may be necessary in the initial phase of restoration (e.g., stabilizing hydrology to assist in plant community establishment). Introduction of fauna and replanting or additional planting may be necessary to achieve the desired ecosystem. The length of time dedicated to "initial management" depends in part on the degree of success achieved in reestablishing ecological complexity and wetland functions.

Most of the research related to wetland restoration has focused on planting vegetation, on the assumption that establishment of the fauna will then follow naturally. Research on introducing animal species to restored wetlands is extremely limited.

CONSTRAINTS ON ACHIEVING RESTORATION GOALS

The restoration of wetlands has the same general goal as that for all aquatic ecosystems: *returning the system to a close approximation of the predisturbance ecosystem that is persistent and self-sustaining* (although dynamic in its composition and functioning). The more degraded a site, the harder restoration becomes. A degraded wetland, surrounded by an industrial metropolis, may achieve its highest restoration state as an impaired system that requires active management to support native species. Even the smallest urban remnants, however, can serve an educational role, informing residents of their landscape heritage.

The sections below discuss several constraints on achieving 100 percent success in returning wetland ecosystems to their ideal predisturbance conditions. The most important of these appears to be the degree of disturbance to the site and its landscape, as illustrated by restoration projects visited by the committee or reviewed by others. Additional limiting factors include ecological, biological, and institutional constraints.

Degree of Disturbance to the Site and Its Landscape

In large urban settings (see Box 6.2), the entire hydrologic regime has been altered due to fills, drainage, ground water extractions, dams, dikes, levees, and other alterations. Water quality is often degraded by both point and nonpoint source pollutants. Sediment yields from the watershed are often great. Ecotones and buffers are often absent or degraded. Exotic plant and animal species are common.

Wetlands in rural contexts may be similarly affected. However, the number of impacts is often much lower for rural wetlands, where (because there is less development) there is often greater opportunity for restoring natural hydrology and functions. It may not be necessary to start with a comprehensive watershed and/or land use plan.

A simple model (Figure 6.3) shows the contrast between the restoration potential of a somewhat disturbed community (little damage, of short duration) that exists in a landscape with many other wetlands in good ecological condition and the restoration potential of a much-disturbed wetland in a highly disturbed (e.g., urban) landscape.

RELATIVELY UNDISTURBED SITE

The Kissimmee River site (see case study, Appendix A) is a *relatively* little-disturbed site in a landscape that retains many wetlands of a similar type. The flow of the river was diverted 20 years ago,

BOX 6.2
SAN FRANCISCO BAY, CALIFORNIA

The ecological condition of San Francisco Estuary has national relevance because much of what has happened there has also occurred in the world's other large urbanized estuaries. The Sacramento and San Joaquin rivers drain California's Central Valley—40 percent of the state's land area. They empty into a 3,000-km² marsh region, known as the "delta." The estuary has a surface area of 1,240 km² (San Francisco Estuary Project, 1990).

Since the mid-nineteenth century, one-third of the bay has been converted to dry land by filling, 90 percent of the bay's wetlands have been destroyed, and more than 60 percent of its freshwater inflow has been diverted, mainly for irrigation. The reduction in fresh water has been accompanied by a massive, although gradual, infusion of toxic chemicals into the bay from refineries and other industries, municipal sewage discharges, return irrigation water, accidental oil spills, and oil and grease washed by rain through storm drains into the bay.

Records of commercial fish catches are the earliest systematic data on the wildlife of the bay region (Skinner, 1962). So productive were the bay's fisheries that in 1875, the bay region supplied 93 percent of the state's commercial fishery products. In 1892, the bay shrimp catch exceeded 5 million pounds; the salmon catch peaked at 10 million pounds for the year, and the oysters catch at 15 million. All fishing production in the bay peaked between 1870 and 1915 except for sealing and whaling, which had declined earlier. By 1950, however, the entire fishery was almost depleted. A similar story could be told of the waterfowl, shorebirds, and game that once abounded in the San Francisco Bay region.

Prior to the Gold Rush of 1849, the 2,200-km² tidal marshes of the bay covered twice the area of the bay's open water. About 1,400 km² of these marshes were in the delta; the remaining 800 km² fringed the bay proper. Today only about 5 percent of the aboriginal marsh remains (Atwater, 1979) due to diking, filling, and anthropogenic sedimentation.

Before gold mining and flood control came to the Sacramento River Valley, nearly 60 percent of the delta was flooded by ordinary tides, and high tides covered almost the entire Delta (Kahrl, 1979). The delta's 2,023 km² were also inundated by the Sacramento River's flood waters. The bay was contiguous with the open water of the delta, which was itself contiguous with an enormous expanse of tule marshes many kilometers wide, consisting of tule (Scirpus validus), common tule *(Scirpus acutus)*, and other dominant and subdominant marsh species. The tule marshes followed the courses of the Sacramento River bottomland and extended along the San Joaquin River through the Central Valley. Extensive levees on the delta eventually excluded tides and floodwaters from 90 percent of the delta, vastly changing its physical character, vegetation, and fauna.

Changes in the bay-delta are well documented in Nichols et al. (1986). Nineteenth-century hydraulic mining in the Sacramento-San Joaquin drainages deposited tens of millions of cubic meters of earth and rock in the bay, reducing its depth, and changing its shape and circulatory patterns. Combined with the loss of marshes and the reduction in freshwater inflow to less than 40 percent of historic totals, deleterious changes detected in planktonic abundances had repercussions throughout the aquatic food web.

Dams above the delta have cut off anadromous fish from their spawning grounds, and alterations in freshwater flow regimes and salinity have contributed to the demise of bay fisheries. These problems have been compounded by massive discharges of agricultural wastewater, much of it containing toxic elements, such as selenium, leached from the soil, along with sulfate and nitrate from fertilizers and soil amendments. Untreated urban runoff, containing substantial quantities of oil and grease, and spills of industrial chemicals add further to the stresses being placed on the estuary.

Damage to aquatic life from untreated discharges was first documented in the early 1950s, about the time municipalities around the bay began giving their sewage primary treatment (San Francisco Estuary Project, 1990). Secondary treatment of sewage began in the mid-1960s, along with the consolidation and relocation of discharges to deeper water (San Francisco Estuary Project, 1990). Effluent discharge standards were tightened as a result of the state's Porter-Cologne Water Quality Act of 1969 and the federal Clean Water Act of 1977. Today 37 percent of the publicly owned treatment works perform tertiary treatment on waste streams, and the remainder deliver secondary treatment (San Francisco Estuary Project, 1990). Over the past 30 years, more than $3 billion has been spent on improvements in wastewater treatment or discharge (Condit, cited in San Francisco Estuary Project, 1990).

Numerous wetland restoration projects have been conducted on San Francisco Bay (Berger, 1990) and on the West Coast (Josselyn and Buchholz, 1984). For a comparison of natural and restored eastern coastal marshes with respect to fish and wildlife habitat value, see Roberts (1989).

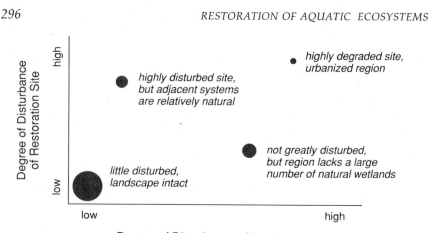

FIGURE 6.3 Hypothesized model of restoration potential for wetlands differing in degree of disturbance and landscape condition. Large dot indicates high potential for successful restoration; smaller dots indicate comparatively lower potentials.

but much of the meandering riverbed was left intact. Flows were restored to a portion of the natural river channel 5 years ago. The 15-year disturbance was sufficient to allow upland species to invade the old river shoreline, to eliminate the food base for invertebrates, and to modify the substrate. However, after 5 years of restored flows, many of those damages are being repaired—predisturbance wetland vegetation is returning, the substrate and food base are recovering, and native invertebrates are reestablishing themselves. The abundant wetlands in and around the Kissimmee River retained seed banks and provided a reservoir of native invertebrate, fish, and bird species able to reinvade the restored river channel. The river flow is a hydrologic connection that serves as a wildlife corridor, allowing dispersal of wetland plant propagules and migration of aquatic animals.

HIGHLY DISTURBED SITES

Examples of highly degraded systems (Figure 6.3) are the urban salt marshes of Southern California (see Box 6.1) and New Jersey (see Box 6.3). In both these cases, the damage to individual sites has been severe, and large portions of the landscape have been modified, leaving few refuges for native species and thus reducing opportunities for reestablishment of the total plant and animal community.

In San Diego Bay, the historic wetlands have been subjected to both dredging and filling. The natural shoreline is no longer visible. Excavation for cordgrass restoration unearthed sandy, rather than

BOX 6.3
RESTORATION IN THE HACKENSACK RIVER
MEADOWLANDS: SUMMARY

The Hackensack Meadowlands is a 21,000-acre estuarine area of fresh- and saltwater marshes and meadows situated in the lower Hackensack River basin amidst the New York-northeastern New Jersey metropolitan area. Almost 18,000 acres of the Hackensack Meadowlands was originally wetland (M. Thiesing, Marine and Wetlands Protection Branch, U.S. Environmental Protection Agency, personal communication, 1991), but extensive development, drainage, diking, filling, garbage dumping, and sewage pumping have disturbed many of the Meadowlands' natural ecological processes.

The Hackensack Meadowlands Development Commission (HMDC) was set up by an act of the New Jersey legislature in 1969 to improve the management of the Meadowlands by providing for the reclamation, planned development, and redevelopment of the Hackensack Meadowlands within Bergen and Hudson Counties (HMRDA, 1968). Since 1969, HMDC has overseen economic development worth $1.5 billion in the Meadowlands while helping to bring about major environmental improvements. Some of the new development, however, has impinged on natural areas. From 1969 to 1984, more than 863 acres of wetlands were filled in accordance with HMDC's masterplan; little filling has occurred since then (Smith, 1991), and habitat enhancement work on 190 acres of wetlands has been performed in mitigation.

Detailed records have been reviewed on a recent significant mitigation conducted by Hartz Mountain Industries (see Hackensack Meadowlands case study, Appendix A). Ecological studies dating back to the late nineteenth century indicate that in the last phase of its natural succession, the Hartz Mountain project site may have been high brackish marsh dominated by salt hay (*Spartina patens*) and salt grass (*Distichlis spicata*), with an Atlantic white cedar (*Chaemecyparis thyoides*) bog at its southern upland edge before the whole area was ditched and then diked between 1914 and 1950 (HMDC, 1984; Kraus and Smith, n.d.; Smith, 1991).

The resulting altered hydrology quickly led to major changes in vegetation. With tidal flow excluded and water salinity reduced, the common reed (*Phragmites australis*) invaded the area and became the dominant vegetation. Construction of the Oradell Dam across the Hackensack River upstream from the project site in 1922 reduced the freshwater flow, further altering ecological conditions.

The Hartz Mountain project, a mall and office complex, was allowed to proceed and fill 127 acres of wetlands, with the stipulation that the company would have to mitigate its impacts by construction of a 63-acre brackish marsh in Secaucus, New Jersey on Mill Creek. The mitigation goals were to enhance wildlife diversity and abundance by converting the site from a common reed-dominated high marsh community to a cordgrass (*Spartina alterniflora*) intertidal marsh.

Between 1985 and 1987 the high marsh was sculpted into channels and open water; lower-elevation intertidal zones, and raised areas (berms) from +5.73 to +10.33 National Geodetic Vertical Datum (NGVD), built up of excavated materials. More than 80 percent of the site is now inundated during part of the mean tide cycle, and a vigorous growth of cordgrass has become established. The intertidal cordgrass marsh created out of high marsh at the mitigation site appears to have met its goals of enhancing habitat heterogeneity, vegetational diversity, and wildlife utilization, principally by birds. However, the project should be viewed as a habitat enhancement and conversion rather than an ecosystem restoration for the following reasons:

1. The mitigation did not endeavor to recreate the particular estuarine ecosystem that existed on the site prior to the damming of the Hackensack River and prior to other significant environmental modifications that have occurred in the Meadowlands.

2. Because of the limited areal scope of the mitigation work and limited goals, the mitigation project had virtually no impact on the regionwide ecological degradation of the Meadowlands. The resulting habitat cannot be considered "restored," because of the influence of these intractable conditions on the mitigation project site.

3. Where once there was probably a high marsh of *Spartina patens, Distichlis spicata*, and other species, the contractors produced an intertidal marsh with mud flats and raised islands of woody vegetation. There is no evidence that the ecosystem created on the mitigation site has existed there within human memory.

The regulated development of HMDC is far better than the indiscriminate dumping and haphazard development that preceded HMDC. Water quality in the Hackensack River appears to be far better than the sewerlike conditions reported 20 years ago. Certain aquatic organisms, such as grass shrimp and mummichog, are now thriving in vast numbers, and some species of waterfowl and fish have returned. However, as this committee has pointed out elsewhere, ecosystem restoration involves more than water quality improvement and increased wildlife use.

native, marsh soils, as well as broken glass and other buried trash. Within San Diego Bay and in San Diego County as a whole, only about 10 percent of the natural salt marsh acreage has escaped urban development. The Hackensack Meadowlands site is a second example of restoration efforts undertaken in an extremely disturbed urban setting (see Box 6.3) characterized by widespread changes in tidal flow patterns, extensive deposits of toxic materials, human disturbances, and invasions of undesirable species (Phragmites). In both these cases, the sites have experienced great damage both locally and to their biological support systems.

SITES WITH INTERMEDIATE DISTURBANCE

Systems with an intermediate degree of disturbance (Figure 6.3) also exist, where either the site or the landscape (but not both) is still intact. Several examples illustrate the variety of challenges facing restoration projects in such sites. Carolina Bay wetlands of the southeastern United States are abundant throughout the southeastern Atlantic Coastal Plain (Prouty, 1952). These isolated elliptical wetlands range in size from less than 50 m to more than 8 km in diameter and may be either only temporarily inundated or permanently flooded. Although they occupy a small areal portion of the landscape, their ecological importance to wetland and semiaquatic organisms is great (Sharitz and Gibbons, 1982). Most of these wetlands have been disturbed, chiefly through ditching and draining to support agricultural usage. Many have been repeatedly plowed and planted or continually grazed by livestock; permanent ponds have been dug in others. Most of these bays are surrounded by agricultural land or managed forests; very few are physically connected with other wetlands (Sharitz and Gibbons, 1982). Reestablishing the hydroperiod by closing ditches or filling artificial ponds and cessation of agricultural use may allow these bays to resume their wetland function (Schalles et al., 1989).

On the Savannah River site in South Carolina, characteristic biotic communities have recovered in the 40 years since agricultural usage ceased and natural hydroperiod was reestablished. The persistent seed bank allows rapid recovery of the vegetation of these wetlands if disturbance has not been excessively severe (e.g., soil removal) or long. Dispersal of wetland plant species to recovering bays may be facilitated by waterfowl that use these wetlands when they are inundated. Thus, functional recovery of these systems may occur relatively rapidly after restoration of a more natural hydrologic state.

The Salmon River salt marsh in Oregon (Frenkel and Morlan, 1990) is also an example of a site with intermediate restoration potential—the effects of disturbance are not easily reversible, yet the site exists within a region that has large reserves of natural habitat. In this case, a 21-ha salt marsh was diked in 1961 and used for grazing, as fresh water gradually diluted the salts and allowed pasture vegetation to become established. During the 17-year diked period, the topography subsided 30 to 40 cm due to compaction and loss of soil buoyancy. The U.S. Forest Service breached the dike in 1978 to restore the salt marsh to a functional ecosystem. However, the topographic subsidence was not readily reversible. Ten years after dike breaching, accretion had raised the topography only 2 to 5 cm throughout most of the site. Native salt marsh plants have now reestablished themselves at their appropriate intertidal elevations, but the area of low salt marsh is much larger than that present before diking. From a functional standpoint, aboveground biomass indicates a high level of similarity with productivity in reference wetlands. Thus, the restoration process is under way, but at the current rate of accretion, reestablishment of the predisturbance high salt marsh could take several decades. The case is an example of a site that would fall in the upper left portion of the model shown in Figure 6.3.

In the northern United States and Canada, prairie pothole wetlands were altered by European settlers to facilitate farming (see case study, Appendix A). The poorly drained depression soils were drained by elaborate and extensive tile fields, seasonal inundation was eliminated, and the potholes were converted to tillable land. Approximately one-half of the 20 million acres was drained, with dramatic effects on fish and wildlife habitat. Restoration efforts supported by federal, state, local, and private programs are now under way. Physical measures to restore the natural hydrology, such as breaking the tile fields and filling ditches, followed by natural recovery of wetland plant communities, have restored waterfowl habitat. The chief obstacle is development of a program to persuade individual landowners to take pothole areas out of agricultural production and allow these restoration measures to take place. Many of these pothole sites could be described as falling in the lower right portion of Figure 6.3.

Bottomland hardwood forests of the southeastern United States have undergone rapid reductions in area and changes in composition (Box 6.4). The lower Mississippi River floodplain is an example of a large-scale disturbance in which the physical condition of the wetland area has been altered and cumulative impacts have occurred. Large areas of mature forested wetlands have been removed or totally disrupted through diking, draining, and clearing for conversion to agriculture, and through urban encroachment. Extensive federal water management and flood control programs in the Mississippi watershed have altered the basin hydrology on a large scale and have allowed major changes in land use to occur. Thus, within this landscape, only 23 percent of the area of floodplain forest remains (Tiner, 1984). The converted agricultural lands typically retain the natural alluvial soils even though the hydrology of these floodplain and backwater areas has been substantially altered. If crop cultivation ceased, wetland vegetation (including forests) would be reestablished eventually in much of the area because the poorly drained alluvial soils hold sufficient moisture. Blockage of small drainage ditches and canals and breaching of levees or dikes would enhance recovery of riverine overflow hydrologic conditions. These former forests could be characterized as falling somewhere in the middle of the model shown in Figure 6.3 and would require a longer time for regrowth than would marshes.

Reforestation with bottomland species is being undertaken in several large-scale efforts on federal lands, with wildlife habitat and increased timber values as the goal. Much of the converted wetland area in the lower Mississippi valley is privately owned, however. Furthermore, removal of large water-control structures is not politically or economically realistic. Thus, the major constraints on wetland restoration in the lower Mississippi valley are not so much physical or technical as institutional and economic.

If the model shown in Figure 6.3 is correct, then the two disturbance variables (for the site and the surrounding landscape) can be extremely useful in predicting the restoration potential of various wetland systems. Understanding the factors that limit restoration potential can lead to setting realistic goals for systems that will be most difficult to restore and to making better decisions when restoration is proposed as mitigation for further destruction of wetland habitat.

Because some landscapes have lost the majority of their wetland area to irreversible uses, restoration opportunities may be few in these areas. The remaining degraded systems may be highly stressed, yet

BOX 6.4
BOTTOMLAND HARDWOOD WETLAND RESTORATION
IN THE MISSISSIPPI DRAINAGE

At the time of European settlement, approximately 80 million hectares of forested wetlands existed in the contermi- nous United States. By the mid-1970s, this area had been reduced to about 24.4 million hectares (Gosselink and Lee, 1989). Drainage, logging, and conversion to agriculture ac- counted for most of the losses, especially in the Mississippi floodplain, where 78 percent of the forested wetlands have been lost (MacDonald et al., 1979). Clearing for cotton fields was extensive in the early 1800s. In the 1920s, following major floods, congressional actions began to increase federal flood control project construction in the lower Mississippi River valley. River channelization and flood control structures al- tered the natural hydrologic regimes of vast floodplain areas. Further conversion for soybeans has had substantial impacts on the bottomland hardwood resource in recent decades, as have urbanization and industrial development.

The drained and converted agricultural lands of the Missis- sippi floodplain retain the natural alluvial soils; it is the hy- drology that has been greatly altered. In these cases, restora- tion of large areas could be initiated by halting row crop planting and other farmland activities and by closing drain- age ditches and small canals. In many areas, the clay soils retard drainage and hold sufficient moisture, so that wetland vegetation would become reestablished. Planting of wet- land forest species can accelerate the return to a bottomland forest. It is not realistic, however, to anticipate that true res- toration to the original geologic, hydrologic, and biological conditions is possible except in very limited areas.

Most bottomland forest restoration projects focus on tech- niques of planting and establishing forest species (i.e., refor- estation). Restoration success is commonly judged, at least in the early phases, by the success of tree seedling establish- ment, and increased timber values and wildlife habitat are often the major goals. Typical costs of direct seeding in 1989 were about $40 to $60 per acre (Allen and Kennedy, 1989); planting seedlings costs two or three times as much. The most successful technique is to plant mixtures of species in blocks or rows. This approach enhances the establishment of species that grow more slowly and compete poorly.

A small number of restoration projects have come under the Conservation Reserve Program. Most of these projects began during the late 1980s. Although some may appear prom- ising in terms of species composition and structure, it is too soon to assess the recovery of other wetland functions.

they can still perform critical functions associated with wetlands. Restoration measures must be carefully designed to retain any remaining values, including functional attributes, wetland soils, biota, and gene pools. In such landscapes, the creation of new wetlands should be explored as an interim measure, to provide habitats for mobile organisms, to determine how well restoration efforts might work in degraded areas, and to retain genetic diversity.

Ecological Constraints

Ecologists predict that early succession communities should be easiest to restore. Hartman (1988) calls the smooth cordgrass (*Spartina alterniflora*) a good colonizer species, with a ready ability to expand vegetatively. Seneca and Broome (in press) have evaluated a large number of restoration projects that were studied over a period of several years; for some, the vegetation, invertebrates, and birds are very similar to those found in natural cordgrass marshes. Broome (1989) reviewed wetland restoration projects in the southeastern United States and stated that "a smooth cordgrass stand established on sand in an area where natural marshes are relatively young will likely be comparable to the natural marsh for most measurements in a few years." In contrast, Broome predicts that it will take several years to restore more mature marshes that have accumulated peat and have highly organic and nutrient-rich soils.

Similarly, ecosystems dominated by short-lived plants might be more quickly restored than those dominated by long-lived perennials. Not many wetlands are dominated by annuals, although the vernal pools of California are, and these are persistent communities. Because one of these annual species is endangered (the mesa mint, *Pogogyne abramsii;* cf. Zedler, 1987), there are now attempts to restore its habitat and recover the population to levels that would allow it to persist. Restoration involves the scraping of shallow impoundments where soils already have a clay layer to retain rainwater. Four years after construction of several dozen pools at Del Mar Mesa (San Diego) in 1986, successful establishment of mesa mint and its other plant associates is still not at hand (P. Zedler, San Diego State University, personal communication, June 1990). A series of dry years (with less than the average 25-cm annual precipitation) has reduced mesa mint densities in natural pools, as well as in artificial ones, so it is not clear if declines would continue in wetter years while a seed bank is still present. The project might not fail; however, its success seems to depend on the weather—a most unfortunate circumstance in a region with a history of prolonged droughts.

Biological Constraints

The native biota may also set limits on the degree to which predisturbance conditions can be restored. On the one hand, cattail marshes are a pioneer community with a few plant species that spread rapidly into open areas, including disturbed areas. Such species are preadapted to colonize newly graded sites. A mature spruce bog, on the other hand, requires decades, if not centuries, to develop in nature. Reconstruction of a forest would take at least as many years as the age of the oldest trees, and replacement of the peat that had accumulated over centuries would take considerably longer.

CONSTRAINTS ON THREATENED SPECIES

Large numbers of wetland species have received special attention because their populations have dwindled to levels that mark them as endangered. Careful study of endangered species indicates that they have complex requirements and narrow ecological limits. One particular plant, the salt marsh bird's beak (*Cordylanthus maritimus* ssp. *maritimus*), is a good example of how difficult it can be to restore such populations (cf. Fink and Zedler, 1990). This hemiparasitic annual lives only along the upper margin of salt marshes. Although it can live independently in the greenhouse (with water and nutrients added), it grows best when its roots can attach to those of grass species hosts. The plant is not very salt tolerant during germination or growth, nor are the seedlings very tolerant of inundation. It needs partial shade because plants grow poorly under either open conditions or a dense canopy. Attempts to reestablish populations of this species to restored salt marsh habitats have not yet succeeded (B. Fink, San Diego State University, personal communication, June 1990). In addition, field trials suggest that pollinators are limiting and that the small-scale disturbances that create openings in the marsh canopy may be lacking. Small burrowing rodents may have been the natural factor that maintained open patches in the salt marsh.

We cannot yet rely on our ability to restore wetlands that support remnant populations of endangered species, desirable as that may be. Therefore it is not an acceptable mitigation policy to allow further damage to ecosystems that cannot be fully restored to compensate for further losses. Until restoration and creation activities can guarantee full replacement of wetland functions, no further modifications of endangered species habitat should be allowed, because the risk to biodiversity is too great. Where restoration is intended to compensate for further damages to wetlands, the activity should be restricted to systems that clearly can be replaced.

CONSTRAINTS IMPOSED BY EXOTIC SPECIES

The susceptibility of a site to invasion by exotic or undesirable species is of primary concern because aggressive plant species such as *Phragmites australis, Lythrum salicaria,* or *Typha* species may dominate sites intended for other vegetation (Larson, 1988; Odum, 1988). Of recent and urgent concern is the spread of three *Spartina* species *(S. alterniflora, S. patens,* and *S. townsendii/anglica)* along the West Coast of North America. All three species occur in Washington, Oregon, and California's San Francisco Bay. It is ironic that at least some of these introductions were made deliberately for the purpose of marsh "restoration" (Spicher and Josselyn, 1985). The exotic plant problem is now considered urgent in the Pacific Northwest. The state of Washington formed a multiagency working group in 1989 to address the problems of habitat alteration and impacts on fisheries and wildlife. They funded a literature review (Aberle, 1990) and sponsored a scientific workshop in November 1990 (T. Mumford, Department of Ecology, Seattle, personal communication, 1990) to develop a research, management, and control program.

Exotic animal species are of equal concern, although their ecology and control mechanisms are little understood. The zebra mussel invasions of the Great Lakes (see Chapter 4) show how quickly a lake system can become dominated by an exotic brought in with ballast water. The recent invasion of San Francisco Bay by an Asian clam, *Potamocorbula amurensis,* and its current dominance of subtidal habitats (F. Nichols, U.S. Geological Survey, personal communication, July 1990), verify the need for concern. Estuaries are particularly susceptible because of their "seascape linkage" to other coastal areas that are used as ports and where ships disperse foreign organisms and larvae as they discharge ballast water. Carlton (1989) indicates the magnitude of the problem by calling attention to 32 known exotic species that have become established at South Slough, a National Estuarine Research Reserve (NERR) in Coos Bay, Oregon. The exotics include two plant species *(Zostera japonica* and *Sargassum muticum),* two fish species, four bryozoans, nine crustaceans, three molluscs, several polychaete species, two coelenterates, and one sponge. Further south, Nordby and Zedler (1991) reported the occurrence of two exotic fish species in Tijuana Estuary, another NERR, and Rutherford (1989) documented the occurrence of a Japanese mussel, *Musculista senhousia,* in a San Diego Bay salt marsh restoration site. In the lower

Mississippi drainage area, the introduced nutria *(Myocastor coypu)* has spread throughout swamps and marshes and has retarded reforestation of bald cypress wetlands by feeding on planted cypress seedlings (Conner and Toliver, 1990).

Although some efforts are under way to control invasive plant species (e.g., *Spartina* species; T. Mumford, Washington State Department of Ecology, personal communication, July 1990), there is little possibility of controlling invasive animals, other than by preventing their introduction. As Carlton (1989, p. 271) said of their impact,

> These have been so fundamental and pervasive that we may never fully know what the biota of the continental margins looked like before ships and before the movements of commercial fishery products. But with increased awareness of the scale and rate at which introduced species are being transported and released today, we may be able to develop more specific and enforceable controls on the movement and release of species for mariculture, for scientific research, or by ballast water.

Such invaders may pose much greater problems for restoration sites than for natural systems, because disturbed substrates have few defenses against germinating seeds or settling larvae.

Institutional Constraints

Legal, political, and economic constraints often govern where restoration can be done and how effective it will be. Legal issues concern land ownership and regulatory processes, discussed below. Political decisions may determine whether funds are available. Federal land management, water resource development, agricultural, and environmental agencies lack a clear mandate for wetland restoration. Although the lengthy list of restoration strategies posed by the U.S. Fish and Wildlife Service (1990) is encouraging, funding for restoration is not high on the nation's environmental priority list.

Wetlands are sensitive to small changes in water supply, hydroperiod, flood cycles, and sediment regimes because they are shallow-water or wet-soil systems. A persistent change in water depth of 0.3 m can greatly affect the functioning and species composition (e.g., marsh versus shrub versus forest) of a wetland. Wetlands are thus very susceptible to hydrologic manipulations caused by dams and dikes that reduce inflows, and by irrigation runoff, urban drainage, and wastewater discharges that increase inflows. Thus, restoration of wetland hydrology must be integrated institutionally, as well as technically, with manipulation of water level and management of hydroperiod. Manipulation of water levels is often the task of one agency (e.g., U.S. Army Corps of Engineers), whereas restoration efforts are carried out by another (e.g., U.S. Fish and Wildlife Service or a state's natural resources department). Integrated water and land use planning is needed to restore the essential surface water elevations, flood cycles, water velocities, and (for coastal systems) salinity regimes so that the desired wetland ecosystem can be achieved.

Implementing the president's goal of no net loss of wetland acreage and function will not be cheap. Economic constraints will always limit the location, number, and types of projects that can be implemented. The techniques used to restore wetlands vary widely in cost. Wetlands that are near urban centers are the most expensive to restore because of the cost of land. For example, the city of San Diego recently paid $3.5 million for a 20-acre semitidal wetland that will still need hydrological restoration and topographic modifications. Restoring the 250-acre Ballona wetland (near Los Angeles airport) may cost $50 million, in part because a major roadway through the marsh would have to be elevated so that it would not be inundated by the restored tidal flows. At state and local levels, other demands for land use and funds usually take precedence. Thus, in general, the potential for major restoration projects is currently lower in urban than in rural areas.

REGULATORY VERSUS NONREGULATORY CONTEXT

Wetlands have been restored in several nonregulatory contexts such as the creation of waterfowl impoundments using water-control structures, the removal of dikes from coastal and estuarine marshes, the blockage of drainage on partially drained agricultural lands no longer used for agricultural purposes, and the grading of gravel pits and other strip-mined lands.

The majority of wetland restoration efforts, however, have occurred as a result of federal, state, or local regulatory actions. In these contexts, private or public landowners seeking permits for various types of development are required either to create, to enhance, or to restore wetlands on-site after damage or to restore wetlands at other sites to compensate for wetland damage at a development site. Few of the Clean Water Act, Section 404 mitigation projects have constituted wetland restoration as defined in this report, and COE and EPA—the two agencies responsible for implementation of the Section 404 program—do not have systematic information about the number of acres of wetlands restored or the effectiveness of particular restoration projects. In the few states for which Section 404 permit records have been surveyed by EPA, the general finding is that mitigation was not accomplished. The site had not been examined to see if restoration had been done, was not monitored, or was shown to differ substantially from what was required in the permit (U.S. EPA Environmental Research Laboratory, Corvallis, Oregon, manuscripts in preparation).

For several reasons, the requirements for successful restoration are often unfulfilled in regulatory (mitigation) contexts:

1. Wetland restoration projects undertaken as mitigation are, quite often, poorly designed by individuals lacking multidisciplinary expertise. Hydrology, sediment regimes, control of exotics, and protection of buffers or ecotones are often inadequately addressed.

2. Landowners often prepare the least expensive and least time-consuming restoration plan acceptable to the regulatory agency. The owners and managers of mitigation wetlands are rarely motivated to complete the restoration or make corrections except those mandated by regulatory requirements. This often means half-hearted attempts to restore or, in some instances, failure to complete a restoration. In contrast, the owners and managers of wetlands in nonregulatory contexts usually have long-term wetland protection and management goals.

3. Wetlands restored in regulatory contexts are often small in size, widely separated from other wetlands, and threatened by adjacent land uses. Wetlands in nonregulatory contexts are often larger, more closely tied in with existing wetland and aquatic systems, and at least partially buffered from adjacent activities.

4. Wetlands restored in regulatory contexts often receive little management after initial restoration because private and public landowners, who are not motivated to provide such management, may move on or have no legal obligation for such management. Similarly, the responsible federal agencies do not have staff to assess the adequacy of restoration projects and do not monitor or require permittee monitoring of permit mitigation conditions for sufficient time periods (10 years or longer). As a result, such wetlands may be overrun by exotic species, quickly filled by sediment, polluted, or otherwise misused.

For these reasons, wetlands restored in regulatory contexts are much less likely to achieve the restoration goals, and the risk of failure is much greater. Many things can be done to reduce the risk of failure in a wetland restoration project (see Box 6.5). This is particularly true where landowners are allowed to destroy or damage an existing wetland based on a "promise" of future restoration. However, in many instances the goals and success criteria for such restoration are not clearly articulated; mechanisms are not incorporated in the regulatory permit to ensure compliance with restoration plans

BOX 6.5
WAYS TO REDUCE RISKS OF FAILURE IN
WETLAND RESTORATION PROJECTS

• Adherence to goal of no net loss in wetland acreage and function
• More detailed assessment of *function* prior to wetland damage or destruction
• More detailed plans
• Higher standards for success
• More expertise
• Larger buffers
• More detailed and longer-term surveillance and monitoring
• Greater midcourse correction capability
• Longer-term and greater maintenance responsibilities
• More detailed reports with broader distribution
• Larger bonds
• Complete restoration or creation before allowing damages (in mitigation projects)
• Require 3:1, 5:1, or 10:1 habitat replacement ratios (depending on functional value of habitat loss) when projects are part of compensatory mitigation)

and to provide for mid-course corrections if plans fail to achieve their intended results; and no attempt is made to relate individual, piecemeal restoration efforts to broader hydrologic and ecosystem management goals.

The standards for "worthwhile efforts" will differ in nonregulatory and regulatory contexts. In rural, agricultural areas, the cessation of agriculture in floodplains and potholes will be worthwhile, even if preagricultural plant and animal communities are not restored. An attempt to achieve 100 percent success in a restoration is a desirable, but not essential, criterion for undertaking projects. Such projects would be considered more successful if measures were taken to recreate native ecosystems. The important decision is to begin the highly worthwhile process of restoring wetlands.

In the regulatory context, and particularly in highly disturbed urban settings, what is worthwhile depends entirely on the functions retained by degraded sites and on the likelihood that a more desirable system can be provided. It must be ensured that the restored system will provide more functions than were carried out before restoration and that any functions retained by the degraded wetland will not be lost. Otherwise, the requirements of no net loss will be violated. Careful and detailed assessment of functions before and after restoration will be needed to justify restoration as mitigation for additional wetland loss.

Problems Encountered In Restoration Projects

A recent EPA publication (Kusler and Kentula, 1989) on the status of wetland creation and restoration efforts includes a wide range of experiences, opinions, case descriptions, and data that are useful to the committee's evaluation of restoration science efforts. Sixteen chapters, contributed by authors from around the nation, review the state of the art of restoring wetlands along the eastern, southern, and western coasts of the United States; in marsh, pothole, and wooded wetlands of the interior; and in riparian systems along the streams of the Midwest, West, and Southwest.

In reviewing wetland restoration projects in the coastal plain of Florida, Lewis (1989) listed several problems encountered during the early years following restoration of salt marshes, mangrove forests, salt barrens, and brackish marshes. These problems included incorrect elevations for target plant species, improper drainage (ponding and transplant mortality), wave damage, inappropriate transplanting material, human disturbance, and insufficient monitoring and reporting. He also indicated the need for buffers, control of exotic species, and long-term management of restored sites. Because mangrove ecosystems are dominated by trees, it is clear that it will take many years to restore their forest character. Attempts to plant older (>1 year) seedlings were considered risky due to lower rates of survival and higher cost.

Broome's (1989) review of tidal wetland restoration in the southeastern United States echoed many of Lewis's concerns, covering in addition experiences with sites that were too saline (hypersalinity restricts plant growth); or had the wrong soil physical properties (wrong texture) or soil chemistry (low nutrients); that had improperly timed construction (too late for planting); that lacked maintenance (e.g., replanting, fertilization, wrack removal); and that had potential problems (loss of genetic integrity) if transplanted materials were not obtained from areas close by.

For restoration projects carried out in the northeastern United States, Shisler (1989) identified additional problems associated with the colder climate. He noted that ice rafting can damage marsh vegetation, that thick peaty soils cannot be restored easily (sand is a poor replacement), and that migrating Canada geese and snow geese can damage wetlands. He called attention to invasive species (*Phragmites australis, Lithrum salicaria*) and native *Typha* species that may need to be controlled to facilitate restoration of postpioneer plant communities. He also noted that in restoration projects, "endangered species" needs might not be met, because so few of their requirements are known.

Reviews of seagrass ecosystems (Fonseca et al., 1988; Fonseca, 1989) added three important problems to the growing list of challenges encountered in restoration projects: the difficulty of replacing climax species (e.g., turtlegrass, *Thalassia testudinum*; Williams, 1990), the inability of some restored sites to attract animal communities equivalent in abundance and composition to the communities of predisturbance sites, and the possibility of major transplant failure (200 acres in Biscayne Bay). On the Pacific Coast, eelgrass restoration is in a similar state, with little assurance that this wetland type can be replaced, due to poor quality of the overlying water and the underlying substrates, especially in urban areas (S. Williams, San Diego State University, personal communication, June 1990). Fonseca (1989) concludes that every eelgrass mitigation project results in a reduction in habitat. Problems that may be unique to this intertidal and subtidal habitat include damage from smothering by macroalgae, grazing by fish, clouding of water by decapods, and disturbance by motorboat propellers.

Attempts to restore forested wetlands of the Southeast (e.g., bottomland hardwoods and cypress swamps) have encountered difficulties related to the time required to replace mature trees, the lack of material to transplant, the lack of knowledge of how and when to carry out seeding or transplantation (Clewell and Lea, 1989), and altered hydrology (drainage for conversion to agriculture) of the wetland area. Natural forested wetlands may support hundreds of plant species, many of which thrive in the understory (91 percent of 409 species in one riverine forest were understory species). Old-growth forests are dominated by trees that gradually achieve a dominant role in the canopy and that are self-sustaining through their ability to reproduce in their own shade. It is not clear that such climax species can be successfully established in open sites, or whether their introduction must await development of seral (intermediate successional stage) plant communities. Clewell and Lea (1989) noted the need for intensive site preparation to reduce competition between weeds and transplanted tree seedlings. Their review was the first to mention insect herbivory and fire as potential problems. In many cases, restoration of suitable hydrologic conditions will be necessary. The short time period within which forest restoration attempts have been monitored precludes an evaluation of their functional equivalency with natural reference systems.

In the arid and semiarid Southwest, restoration of riparian habitat has the unique problem that transplants need to be irrigated (Carothers et al., 1989). Because riparian forests became established during times of flooding, the hydrology necessary to recreate them cannot be produced at will. In many cases, dams prevent the restoration of the hydrology necessary to sustain riparian woodlands. Thus, although the tree canopy can be recreated through transplantation and irrigation, the system cannot maintain itself unless flooding regimes are also restored. In addition, invasion of exotics such as salt cedar (Tamarisk), damage to plants from grazing practices and from beaver, and permanent lowering of water tables due to ground water depletion by wells used for irrigation are other obstacles to the restoration of western riparian sites.

Hollands's (1989) discussion of pothole wetlands included problems that may be most frequent in wetlands that lack an outflow. These systems have highly variable water levels, and plant distributions can shift up or down the slopes within a year as establishment and mortality occur. In transplanting such sites, it may be difficult to predict the elevation that will have the appropriate hydrology in the following growing season. A swamp restoration project in Massachusetts experienced 80 percent mortality of planted shrubs due to a year of unusual flooding.

Wetlands along the fringes of lakes and rivers experience similar water-level fluctuations, as well as erosion due to wave action. Levine and Willard (1989) documented an Ohio case of dike undermining during a year of high water and transplant failure due to drought. Either water-level extreme can restrict restoration success.

Additional problems were noted for a highly urbanized wetland in San Diego Bay, where a combined freeway widening, new freeway interchange, and flood control project has been under way for six years. Such large, complicated projects are likely to experience problems with contaminated substrates and construction errors (see Box 6.6). The most recent error (July 1990) was the accidental drainage of an important lagoon used by birds.

Summary

The problems encountered during restoration include every aspect of construction—site selection, topographic contouring, transplantation, inadequate nutrient supplies for plant growth, and pest inva-

BOX 6.6
UNFORESEEN PROBLEMS

The problems encountered while restoring salt marsh habitat at San Diego Bay (see Box 6.1) were numerous. They indicate the need for constant surveillance during project construction as well as a mechanism to repair damage and errors as they develop. Problems will occur, and project managers will need to be on call to solve and correct them.

Toxic Materials

During the excavation of Connector Marsh, bulldozers unearthed an old landfill that contained toxic materials (discarded lead paint). The sediment was sufficiently contaminated with lead that large quantities had to be trucked to a toxic waste dump at a large and unanticipated cost.

Preproject Vegetation Salvage

An attempt was made to preserve the gene pool of cordgrass plants to be covered by the highway detour road. A 10,000 square feet intertidal nursery was excavated to receive the salvaged cordgrass. During excavation of the nursery site, just south of the Connector Marsh, an old city dump was uncovered, and the substrate was full of broken glass—an unforeseen hazard for personnel who did the transplantation work. Then, the initial transplant failed, apparently due to insect herbivory (transplants carried the larvae and pupae of the Dipteran, *Incertella* sp., to the nursery). A second transplantation effort worked well, and the nursery plants thrived for about 5 years (1985-1990). Then, during a second restoration phase, bulldozers excavating channels adjacent to the nursery also bulldozed the nursery. The nursery had not been mapped as an ecologically sensitive area, and biologists were not on-site the day it happened.

Removal of Temporary Fill

In constructing the temporary detour road, contractors laid down a fabric layer so that the integrity of marsh soils could be preserved. Five years later, when the material was to be removed, some sections had buckled and sunk below the planned excavation grade. While the construction crew was on hold,

BOX 6.6 (*continued*)

biologists had little time to decide whether the buried fabric could be left or the excavation plans altered. Plans were immediately revised, so that all fabric could be excavated, and habitat intended for mid-upper marsh was replaced by lower marsh and tidal pond habitats.

Tidal Modifications

Excavation of two marshes caused major and sudden changes in the tidal flushing regime of the reference wetland at Paradise Creek. Early during construction, a tide gate was erroneously installed midway along the Connector Marsh channel. The gate was in place for more than a year, and it muted tidal flows to the northern four marsh islands and the Paradise Creek Marsh. Its removal, following the advice of researchers, led to increased growth of cordgrass in the northern four islands during 1988.

Early in 1990, three culverts were placed in the same channel, and a temporary access road was built with fill, in order to excavate an adjacent 17-acre marsh, known as Marisma de Naci<n. Tidal flows to Paradise Creek and the north islands were again muted. A hydrologist (Haltiner, 1990) advised that six culverts would be needed to provide adequate tidal flushing to the marshes upstream. Contractors suggested a cheaper alternative, the partial excavation of a planned flood control channel to San Diego Bay. This alternative was accepted, and the Connector Marsh restoration site was then subdivided hydrologically, with the northern half connected directly to San Diego Bay (providing strong tidal circulation) and the southern half connected through Sweetwater River. The access road was in place for several months until Marisma de Nación was completed. Removal of the access road led to a third major change in tidal flushing, with connection of the southern islands to the bay as well as to Sweetwater Marsh. Most recently, in November 1990, the direct tidal flows to San Diego Bay were terminated because the flood control channel was again under construction.

None of the physical impacts of these tidal modifications has been assessed. Hydrologists predicted that the new tidal regime would alter the sedimentation dynamics and, as a result, gradually modify the topography of the channels and marsh habitats (Haltiner, 1990). Both the constructed marsh and the natural reference wetland were affected by muted

---

BOX 6.6 (*continued*)

tidal flows and sudden shifts in tidal regimes. In late 1990, after the cordgrass marsh had experienced greater tidal flushing for about one growing season, the above-ground biomass of plants had increased significantly in both the natural and the constructed marsh sites (J. Zedler and R. Langis, San Diego State University, unpublished data). This finding supports the argument that impaired tidal flushing has been detrimental throughout the multiyear construction process. The lesson is that large-scale, long-term restoration projects will not provide "stable" hydrology until well after construction has been completed.

---

sion. The result is that many projects fall short of the goal of returning ecosystems to the predisturbance condition, and there is indeed considerable controversy over whether or not wetlands can actually be restored. The arguments are particularly important when wetland restoration is undertaken within the mitigation context, and the *promise* of full restoration of a degraded site *allows* a natural wetland to be destroyed.

Experience with wetland restoration varies with region and wetland type. Many coastal and estuarine mitigation projects have been constructed along the Atlantic Coast and have been monitored and evaluated in the scientific literature. Fewer projects have been followed along the Gulf of Mexico and Pacific coasts, and there is less information in the literature about the success of restoration in these regions. Less is known about how to restore inland as opposed to coastal wetlands. The most common types of freshwater wetland restoration projects are impoundments to create waterfowl and wildlife habitat and the establishment of marshes on dredged spoils along major rivers. Although there is much literature describing waterfowl abundances, there are few critical studies of the success of these restoration efforts as persistent, naturally functioning ecosystems. The problems encountered encompass every aspect of wetland restoration. If all were to coincide, the resulting worst-case scenario (see Box 6.7) would clearly be a total failure.

Worst-case restoration projects have happened, but the lack of reports makes documentation difficult, and describing the failures of projects that powerful entities consider successful is not without risk. Even with evaluation and reports of projects, there are often large differences in opinion about how well efforts succeeded.

BOX 6.7
CHARACTERISTICS OF A WORST-CASE
WETLAND RESTORATION PROJECT

Most restoration projects include at least one major short-coming, and the worst cases have all the following attributes:

• No specific goal—Vague generalities, such as "optimize ecosystem integrity," may take the place of testable objectives. If objectives are listed, they describe the proposed structure, not the functioning, of the site.

• The proposed restoration of one area justifies destruction of another (project is part of a mitigation agreement). In many cases, the destruction proceeds, but the restoration project never occurs (M. Kentula, U.S. Environmental Protection Agency, personal communication, June 1990). Mitigation may be out-of-kind and off-site, making it difficult to provide the lost functions. Net losses in both acreage and function result.

• Innumerable problems are encountered during the restoration process, making it impossible to follow plans; decisions are made on-site and without consultation. Biologists are not required to be present at critical times. Construction crews make errors that are uncorrectable.

• There is no follow-up to determine if the project was carried to completion or whether it achieved the restoration goals (or complied with mitigation requirements).

• Access is denied when an agency seeks permission to evaluate a restoration project on private land.

• There are no reports that can be reviewed by either agencies or scientists.

• The regulatory agency lacks the manpower to enforce requirements.

• Photographs of vegetation growing at the site are presented at local or national meetings proclaiming the project a success, without measurable criteria or data to support these claims.

CONTROVERSIES ABOUT THE SUCCESS OF
RESTORATION PROJECTS

Whether we can succeed in restoring wetlands is a controversial subject. The authors of the EPA review papers (Kusler and Kentula, 1989) were asked to evaluate the projects they described. Owing to a lack of standard criteria for measuring success and inadequate statements as to the goals of restoration, the degree of success achieved by many projects could not be characterized. The range of opinions on those that were judged covered the spectrum from success to total failure. At conferences dealing with restoration, creation, and mitigation, a similar range of opinions can be heard, at times with controversy over a single project that has been evaluated by different individuals (J. Zedler, personal observation). Many ecologists believe that it is a major challenge to recreate in a short period of time what nature has produced over centuries (Bradshaw, 1988).

Vested Interests

Some individuals insist that wetlands can be restored to fully functional ecosystems. They point to greenbelts, parks, and mitigation sites and to lovely pictures of wetland vegetation. Because so much wetland restoration is done within the regulatory context (i.e., as a mitigation requirement for damage done elsewhere), the stakes are incredibly high, and the controversy is a heated one, as indicated in the following example.

A San Diego developer hoped to convince resource agencies that he could build 416 multifamily units on 10 acres of a 20-acre urban wetland (Famosa Slough) without having a negative impact, even though he would need to fill intertidal flats and marsh habitats. He promised to restore an equal area of city-owned wetland and to increase the value of the remaining lagoon. Wetland biologists rebutted his claims, and the city decided to buy the property as a wildlife reserve. The developer then asked $15 million for the site, claiming that the project could obtain a Section 404 permit from COE. Although there were many legal and political factors, the issue of whether or not lost wetland functions could be replaced played a role in determining the property value. In September 1990, after years of controversy and negotiation, the city of San Diego purchased Famosa Slough for $3.5 million. The difference in sale price more than $0.5 million per acre—is some measure of the value of being able to replace lost wetlands through compensatory restoration.

Within the mitigation context, proponents of new developments are extremely eager to believe that habitat functions can be moved about at will. It is not surprising that entrepreneurs promise the desired success and that considerable effort is made to promote completed projects as successful. At the same time, there is little incentive, in the form of funding, to evaluate projects independently, thoroughly, or scientifically. In particular, the funding needed to examine how mitigation sites *function* (not just what they look like) is nearly always lacking. Project proponents do not want to know, and regulatory agencies cannot afford to find out.

Different Standards of Evaluation

For projects that EPA review authors (Kusler and Kentula, 1989) labeled as successful, the overwhelming majority were judged on the basis of vegetation establishment. Neither the wildlife present nor the functional capacity (i.e., for hydrologic functions, water quality improvement, food chain support) of the ecosystems played an important role in evaluation for the simple reason that data were not available.

For restoration projects in general, examination of a few characteristics may indicate success, whereas detailed studies would reveal shortcomings. Although restoration sites that develop dense plant cover or support native plant species may be considered a success by some evaluators, others would require data on fish, invertebrate, bird, and mammal uses and would want to examine the site's long-term persistence before providing a stamp of approval.

Evaluations of the food chain support function are rare, although several EPA review authors (Kusler and Kentula, 1989) indicated that animal components would be established on their own. Determining whether or not the presence of several animal species constitutes equivalence of food chain support functions requires detailed studies of the kind not usually funded by project proponents or required by regulatory agencies. For an indication of standards to be applied in assessing the functional equivalency of animal communities, some well-studied Atlantic coastal marsh restoration sites provide some insights.

Cammen (1976a,b) reported significant differences in the invertebrate infauna of constructed wetlands along coastal North Carolina. At a 1-year-old marsh, the dominant taxa were insect larvae, whereas polychaetes had dominated the natural wetland. Total densities and calculated secondary production also were markedly lower than in the natural system. Similar observations were made at a 2-year-old site. However, Sacco et al. (1987) revisited the latter site when it was approximately 15 years old and reported a 10-fold increase in densities and high similarity with the natural indigenous fauna. Moy (1989) and Sacco (1989) found that the fauna of a site graded to support *Spartina alterniflora* had fewer nematodes, ostracods, copepods, and oligochaetes than did natural marshes.

Species composition in natural and man-made wetlands does not always become more similar through time. Moy (1989) demonstrated a convergence of species lists for one pair of adjacent marshes but a divergence in composition over a 2-year period in another pair receiving the same source water. Sacco (1989) evaluated the composition, density, and trophic structure of infauna in coastal North Carolina marshes of age 1 to 17 years. Although variable, the data showed that the six constructed marshes had similar faunal components and trophic groupings (deposit feeders, suspension feeders, and carnivores) but that densities were always lower than in natural marshes.

Duration of Evaluation Period

A judgment of success shortly after site preparation and transplantation may not tell the whole story. Long-term success may increase or decrease, although so few examples of long-term evaluation are available that it is difficult to know why some restoration projects improve and others disintegrate. One factor might be the presence of pest species. Clewell and Lea (1989) call attention to weed species, which are abundant in the early stages of a restored forested wetland. Problems with weeds should diminish in time, with development of an overstory canopy and with the use of control measures, but other problem species may become more prevalent as restoration sites age. In San Diego Bay, a transplanted cordgrass marsh that grew well for 3 years developed an outbreak of native scale insects in its fourth year (Zedler, 1988c). The once-lush cordgrass has declined to a level that would now be considered a failure. The reason for the scale population explosion appears to be the lack of native predatory beetles, which owe their rarity at the restoration site to its isolated location (away from natural marshes) and the lack of tall cordgrass stems to provide a refuge from inundation by high tides. The reason for the short plants is not entirely clear, but low concentrations of soil nutrients are a probable cause.

The principle in both these examples is that the total ecosystem was not transplanted—projects always start with something less (e.g., fewer canopy strata, fewer trophic levels), expecting that the rest will follow in time. There is no guarantee that it will, and there is not enough experience to predict when or why it might not. Some species are keystones of the communities, and we do not yet know which ones they are. One scientifically useful aspect of restoration failures is the information they provide about the role of species that are absent.

NEEDS

Need for Wetland Restoration

Despite the constraints on achieving 100 percent success, despite our inadequate knowledge of how to restore wetlands, and despite the prob-lems that will occur during implementation, there is an urgent need to restore large areas of wetland throughout the nation. This need derives from the severe losses of acreage and function that have occurred over the past two centuries (Dahl, 1990), as well as from the potential for substantial human benefits from restored wetlands.

Where wetlands have been drained for agriculture or silvicultural purposes, wetland restoration must be balanced against both the need for agricultural commodities and the landowners' property rights. Because landscape drainage affects both surface and subsurface waters, the halting of drainage may increase crop yields in surrounding lands, while only moderately reducing crop loss in the restored wetland. If restoration of wetland hydrology is shown to cause unreasonable losses in production, then programs to offset personal losses (e.g., CRP, AWRP crop subsidy programs) may be needed.

Perhaps the greatest short-term return can be realized by initially restoring isolated wetlands, wetlands within minimally developed watersheds, and wetlands in the upper reaches of watersheds. Elsewhere, the higher priority may be acquisition of property and/or flood easements in critical parts of the landscape to control hydrology in downstream portions of the landscape before restoration is undertaken. For example, the Russell Sage National Wildlife Refuge System near Monroe, Louisiana, is located in the midreaches of its watershed. Although reforestation of some original bottomland hardwood forest is possible, off-site liabilities related to watershed drainage and flood "management" limit the potential for restoration of a functional bottomland hardwood forest.

Setting a Wetland Restoration Goal

Setting an acreage goal for wetland restoration in the United States requires several assumptions about the extent and condition of damaged wetlands. If only the wetlands along rivers and streams are considered, and if a 100-ft-wide wetland along each side of the waterway is restored, one estimates a potential wetland area of 77,578,758 acres (3.2 million miles, 0.038 mile wide, 640 acres per square mile).

If 30 percent of these waterways are within agricultural areas (and thus are damaged), there should be approximately 23,272,727 acres of potential buffer. Also, if as much as 50 percent of the area is undamaged, there is still the potential for more than 10 million acres of wetland restoration along streams and rivers alone.

Because more than 100 million acres of wetland habitat have been lost in the past 200 years (Dahl, 1990), the restoration of 10 million acres would constitute less than one-tenth of the lost habitat. (See Chapter 8 for recommended goals.)

Need for Adaptive Management

Wetland ecosystems are complex and incompletely understood. The shortcomings of restored and constructed wetlands are only beginning to be known. Whether or not restored wetlands will be successful depends in large part on their hydrology, which controls their structure and functioning (Larson, 1988). Nutrient supply functions are especially important. Microbial and chemical processes, which are determined by hydrologic conditions, control the concentrations of nutrients and other compounds, and facilitate the biogeochemical cycling of nutrients and the flow of energy. Restored coastal wetlands need to be able to support nitrogen dynamics, including the processes of nitrogen fixation and denitrification, both of which are influenced by soil organic matter. Freshwater wetlands need to have long-term capability for phosphorus retention by sediments, which is related to the amount of extractable aluminum in the wetland soil (Richardson, 1985).

Because our limited knowledge of how to restore ecosystems makes it impossible to guarantee that objectives can be fulfilled, projects should include an *initial experimental phase* for the more risky aspects. The restoration plan should be *adaptive*, supporting the acquisition of new information and its use in project implementation and long-term management. A plan that has *flexibility* and relies on *informed decision making* should best be able to accommodate difficulties during construction and follow-up monitoring. As restoration proceeds, managers can *reevaluate* whether the degree of restoration achieved is acceptable or if new techniques should be introduced. For those situations in which unexpected benefits develop (e.g., nesting of the endangered California least tern on part of a San Diego Bay site intended for salt marsh; see Box 6.1), an adaptive management program is essential in order to capitalize on those events and to expand and revise restoration objectives appropriately.

Need for Improved Assessment

The recognized importance of natural wetland, together with the development of a policy for no net loss in wetland acreage and function, places a large burden of proof on wetland restoration projects. The emphasis in evaluation techniques should thus be on the *functioning* of restored systems. Ewel (1987) listed five criteria for successful ecosystem restoration: sustainability, low susceptibility to invasion by exotics, productivity similar to that of a natural counterpart, ability to retain nutrients, and biotic interactions similar to those of reference systems. It is also important to take into account the ability of restored wetlands to resume their unique role in global hydrologic and nutrient cycles.

Wetland assessment protocols have been developed for extant wetlands (e.g., the U.S. Fish and Wildlife Service's Habitat Evaluation Procedure (HEP) and Wetland Evaluation Technique (WET); Adamus and Stockwell, 1983), but these procedures can have a very narrow focus (HEP) or assume existing knowledge of the system (WET). Neither was developed to compare the functioning of restored and natural wetlands. Based on what is known of restored salt marsh systems, suggestions have been made for southern California coastal wetlands (Zedler et al., 1988; PERL, 1990). Here the committee provides general recommendations for functional assessment.

Determining whether a restoration site has developed the natural functions requires a long-term assessment. Judging its success or failure after the first year or two is inappropriate. As Odum (1988, p. 67) points out, "Unfortunately, dramatic unanticipated changes may occur over the ensuing years . . . it is not uncommon for the plant community to become invaded and dominated by aggressive 'disturbance species'. . . . The long-term result may be a wetland environment which has limited functional value for wildlife habitat support or nutrient processing and which lacks aesthetic attractiveness to the degree originally planned." Likewise, Broome et al. (1987, p. 197) conclude that monitoring a constructed wetland for four growing seasons was insufficient to determine "if the created marsh reaches equivalent levels of production for all plant species and remains self-sustaining."

The committee recommends periodic evaluations, to determine the rate and extent of restoration, with the frequency of comparison declining through time. Perhaps after several long-term assessments have been done, that information can be used to select critical assessment time periods. Until then, evaluations at 1, 2, 3, 5, 10, 15, and 20 years are recommended. These will provide valuable information on the differential rates of development for various functions. For example, the establishment of full canopy coverage by vegetation may take a year or two, but development of the appropriate canopy architecture (necessary height profiles) may take much longer. Nitrogen fixation may begin immediately, but nutrient concentrations (e.g., pore water [NH_4] comparable to that of natural marshes) may take decades to build up. A constructed marsh that is 60 percent "equivalent" at 5 years of age (cf. San Diego Bay; see Box 6.1) may stagnate, or it may continue to gain functional values.

The committee believes that the standards for assessing restoration projects should be high and that field sampling protocols should be able to withstand scientific peer review. Sound evaluations should have two beneficial outcomes—results will indicate (1) the rate of the development of ecosystem structure and function, and (2) causal factors responsible for functions that are lacking or impaired. If, for example, plants are growing poorly, data on nutrient concentrations and soil redox potential can help identify corrective measures for the site (e.g., fertilization, drainage improvements) as well as help prevent similar problems in the future (Broome et al., 1987).

The absence of sufficient funding of long-term monitoring efforts is a major obstacle to evaluating the success of restoration sites. Providing financial incentives, such as bonding of restored sites in a fashion similar to that specified in current regulations for restoration of mined lands, is one possible approach. Bond monies are returned to the project developer after standards for restoration have been achieved over a period of time. Milestones for restoration success can be established, and a percentage of the bond could be returned when each milestone is reached.

Need for High Standards in Assessing Functional Equivalency (Restored Versus Natural Wetlands)

Structural and functional attributes should form the basis for evaluating the degree of restoration achieved. The committee suggests that restoration projects be evaluated against a range of specific objectives (e.g., Table 6.4), resulting in a judgment that could be as negative as *total failure*, where no progress is made in replacing lost values, or as positive as *relatively successful* in returning lost structural and functional attributes. No project should be considered a failure just because it did not reset the clock. Instead of resulting in a yes or no judgment, evaluation procedures should provide a graded scale of achievement.

Two factors will thus influence the rating of a restoration project: (1) the specific criteria used to evaluate achievements and (2) the reference data or sites used for comparison. From the ecological perspective,

• Assessment criteria should include both structural and functional attributes of the ecosystem.
• Criteria should be established well before the assessment takes place and should be linked to the specific objectives of the restoration project.
• Several criteria should be included in the evaluation; the more information, the less chance there will be for an error in judgment.
• Criteria may need to be regionalized to reflect special concerns (e.g., in regions where wetland species are threatened with extinction, more strict standards may be required if the restoration goal is to retain biodiversity).
• A reasonable range of reference systems and long-term data sets should be available with which to compare the restored system's attributes.
• Assessment measurements should take into account both temporal variation and spatial heterogeneity. Attributes that are known to be patchy in either time or space would need widespread and long-term characterization.
• There should be an a priori indication of how similar the restored system could ever be to the reference system(s). We know that 100 percent similarity to presettlement conditions is impossible to achieve. If multiple samples are taken within a community type, they may average around 85 percent similarity with one another; such an approach indicates a reasonable target for the similarity of restored and natural sites. Two parts of the same pristine aquatic ecosystem would never be 100 percent similar in either structure or function. Furthermore, two samples of the same ecosystem would not produce identical data sets.
• There should be an a priori time frame within which the system is expected to achieve the required similarity to reference systems. For regions in which environmental conditions are highly variable from year to year, the time frame will probably be long. At least one project in San Francisco Bay has a 20-year monitoring requirement; this is appropriate for a wetland subject to highly variable inflows of fresh water.
• Criteria and methods used to assess the restoration site should be able to stand up to peer review.

Detailed standards for assessing successful restoration have not been developed. The need is most urgent where restoration is carried out as mitigation—that is where functional equivalency must be

TABLE 6.4 Ecosystem Attributes to Consider in Assessing the Functional Equivalency of Constructed and Natural Wetlands Based on Experience in Coastal Salt Marshes

Function	Suggested Measures
Hydrologic function	• Ground water recharge: Monitor water level in nearby wells. • Shoreline stabilization: Map shorelines from aerial photographs or install and monitor markers. • Flood-peak reduction: Monitor water levels in relation to flow velocity. • Restoration tidal flows: Monitor water levels over tide cycles; determine amplitude; lags; monitor salinity of water and soil. • Development of hydrologic equilibria: Measure erosion and accretion of channels and marsh. • Sample inflowing waters for nutrient concentrations (N, P) and flow rates.
Nutrient supply functions and their limiting factors	• Analyze soil texture and organic matter content. • Determine nutrient concentrations (N, P) in soil and pore water. • Survey for toxic substances (heavy metals, selenium, and others.)
Persistence of the plant community	• Determine cover of dominant species and map using aerial photographs and ground truthing. • Survey populations of sensitive species quantitatively. • Determine the life history characteristics of sensitive plant populations to predict their ability to persist in the restored wetland (e.g., numbers, flowering, seed production, seed germination potential, seedling establishment, and successful recruitment).
Plant growth[a] and its limiting factors	• Measure end-of-season live standing crop (EOSL); estimate biomass by measuring total stem length (meters per square meter) of species such as cordgrass. • Measure redox potential in soil profiles, and measure pH. • Assess/monitor organic matter decomposition. • Assess cover of floating or epibenthic algae by dominant type. • Determine nutrient content of inflowing waters.

TABLE 6.4   (*Continued*)

Function	Suggested Measures
Persistence of consumer populations	• Arthropods: Document outbreaks; document presence of carnivores that could control potential pest species. • Fish and aquatic invertebrates: Sample community composition (seasonal sampling probably needed). • Birds: Survey for abundance seasonally. • Record activities (habitat use and movements between habitats) in relation to changes in water levels (e.g., tidal inundation); identify areas used for feeding, nesting, and refuge during adverse conditions.
Resilience	• Follow the recovery of populations that die back during periods of environmental extremes.
Resistance to invasive exotics	• Map the occurrence of weedy plants, and rank their abundance by species. • Census exotic animals, and determine if populations are increasing, stable, or declining.
Other items	• Monitor trash so that the area can be cleaned up at appropriate intervals. • Document any visual disturbances or noise problems that are correctable.

[a]Productivity rates of algae and vascular plants are highly variable—the former on a weekly basis, the latter yearly. Measures of peak biomass are, however, useful.

SOURCE: PERL, 1990.

met to comply with the emerging policy of no net loss in wetland acreage and function. Recommendations have been put forth for southern California (PERL, 1990), where mitigation is driving many restoration projects and where several species are endangered due to past losses of habitat.

Need to Accelerate the Restoration Process

Wherever wetlands are restored to mitigate damage to mature wetlands, there will likely be a net loss in functioning while the younger system develops biomass and complexity. There should be ways to speed the development of ecosystem functions (often called ecological succession) in restoration sites, thus shortening the time required to attain maturity. Mature vegetated wetlands have fine-textured soil that is rich in organic matter and nutrients, a well-developed rhizosphere and dense canopy of native vegetation, and a wide range of native arthropods and larger animals. In contrast, a newly developing marsh, such as would be found on a sandbar or an alluvial outwash, would have mineral soil, low plant cover, and few species of animals. The quality of the substrate in graded sites (i.e., soils that are coarse, low in nutrients, and lacking in organic matter) may slow succession.

Most transplantation attempts have concentrated on dominant species of plants. Interestingly, data from the Corvallis EPA Environmental Research Laboratory (M. Brown, University of Florida, and S. Gwin, U.S. Environmental Protection Agency Corvallis Laboratory, personal communications, June 1990) show that almost all of the vegetation present on created freshwater marshes in Oregon and Florida did not include species on the transplant lists. Although it is not clear whether the listed species were ever planted, it is clear that planners did not know what species would succeed on the sites. Note that plants acquired from the wild to accelerate restoration may degrade the site they were taken from, resulting in a net loss of wetland quality.

Various techniques for accelerating the rate of ecosystem development are currently emerging:

• sculpturing the site to add topographic heterogeneity and provide microsites for enhanced microbial, arthropod, fish, and bird activities, at least in tidal marshes;

• mulching the site with litter, detritus, seed, and root materials from native wetlands to provide organic matter and propagules (cf. Clewell and Lea, 1989); and

• augmenting the soil (tilling in organic matter or mineral nutrients) to stimulate plant growth and shorten the time needed to provide dense cover. This should also enhance microbial processes (e.g., nitrogen fixation) and nutrient recycling rates. Potential pitfalls are high foliar nitrogen concentrations that may stimulate excessive herbivory. Research is proceeding through the California Sea Grant Program and the National Oceanic and Atmospheric Administration's Coastal Ocean Program to develop this ecotechnology for coastal wetlands.

Need for Research

Several regional reviews and conferences have considered the status of restoration knowledge (Josselyn, 1982; Lewis, 1982; Strickland, 1986; Larson and Niell, 1987; Kusler et al., 1988; Zelazny and Feierabend,

1988; Kusler and Kentula, 1989; Conners et al., 1990). This committee's review of research needs highlights the broad areas requiring further study and indicates the opportunities afforded by restoration projects. A more detailed statement would require substantial regionalization and consideration of the diversity of wetland types. For example, the review carried out by Conners et al. (1990) is a 60-page book that deals with only one wetland type in one part of the nation for one purpose—research needed to mitigate salt marsh that will be lost as highways are improved. The authors provide lists of research topics in each of the following areas: hydrodynamic processes, soil development, vegetation establishment, monitoring, spatial requirements, construction and equipment, and storm water treatment.

APPLIED RESEARCH

The practice of restoration must move from a trial-and-error process to a science—it needs to become predictive (Zedler and Weller, 1989). We need to know what ecosystem functions can be restored under various conditions (i.e., constraints) and how rapidly restoration can proceed. The curve for functional development over time cannot be plotted, because there are not enough data to indicate either its slope or its asymptote. A variety of approaches can work together to bring restoration into the area of science, given access to restoration (and reference) sites as well as funding opportunities:

• The requirement for careful long-term monitoring of selected restoration projects will fill gaps in information on how ecosystem functions develop. An understanding of how early and late restoration conditions compare will lead to predictive power that will help in future projects.

• A major need is to develop scientifically defensible standards for assessing the degree of restoration achieved. Not only is it important to have appropriate comparison data sets (reference wetlands), it is also essential that suitable sampling protocols be developed. Restoration success is usually interpreted as the establishment of vegetation that covers a specified percentage of the site for a defined period of time. Such measures do not evaluate whether a project is functioning properly or if it will persist. A mandate to develop standard assessment methods will stimulate comparisons of different sampling and measurement techniques, with the result that science will gain deeper understanding of how methods affect data, as well as how data are influenced by the type of ecosystem under study.

• The gradual accumulation of baseline studies of wetland ecosystem functioning will provide opportunities for comparison of different system types among various regions, and at different ages of development. Regional and national data bases would facilitate such comparisons. Syntheses of existing knowledge, especially papers in the "gray" literature (permit files, records of managers), are also needed.

• The availability of restoration sites for experimental work would have great potential for answering research questions (Gross, 1987). Determining the relationship between various plant and animal species is very approachable on sites where single-species vegetative cover can be planted with or without animal associates. Testing the development of ecosystems with different hydrologic, substrate, and biologic features is feasible where these attributes can be part of the restoration design. The potential is limited only by one's scientific creativity.

• The insistence on an adaptive management approach would facilitate studies of how management alternatives reduce or compensate for the impacts of sedimentation, erosion, eutrophication, contamination, and human disturbance.

CONCLUSIONS AND RECOMMENDATIONS

Planning

The piecemeal approach to restoration has not been efficient (economically or ecologically) or effective in achieving broader goals for restoring aquatic systems. The specific goals and objectives for a restoration site are often established in a vacuum. The following recommendations are directed to any person, state, or federal agency involved in a wetland restoration effort.

• **Use a landscape-level approach (at the biogeographic region or watershed scale), with advance identification of sites that will provide the largest gains after restoration.**
• **Inventory biogeographic regions and establish sites with restoration potential. Develop priorities for restoration within each region.**
• **Integrate the restoration project with the rest of the landscape; use regional and watershed approaches in setting restoration objectives.**
• **Incorporate long-term studies, new findings, and local expertise into an adaptive management program for the restoration site within its regional context. Expect problems and be prepared to**

provide rapid responses and solutions (e.g., for trash removal, exotic species invasions, sedimentation and erosion events).

• Funding priority should be given to programs for restoration of damaged wetlands over wetlands creation because of the superior chances of success. An exception would be where restoration is part of a mitigation agreement that would result in a net loss of acreage.

• Give whole-ecosystem restoration and restoration for nongame species priority over restoration to support game species. The goal should not espouse single-species management.

• Retain habitat remnants (e.g., small, fragmented wetlands) in urban areas, for use in education, research, recreation, and aesthetic enjoyment.

• An appropriate federal agency should prepare periodic "state of the wetlands" reports.

Research Needs and Techniques

Traditional research on wetlands and ecosystem development should also be continued, using both natural and restored wetlands. Examples of this traditional research include the following topics mentioned by several EPA authors (modified from Kusler and Kentula, 1989):

• the hydrologic needs and requirements of wetland plants and animals, including minimum water depths, hydroperiod, velocity, dissolved nutrients, the role of large-scale but infrequent events such as floods, and the effects of long-term fluctuations in water levels;

• the importance and functional significance of substrate to wetland plants and animals, and to chemical and biological functions;

• characteristics of development rates for natural successional vegetation;

• recolonization of restored sites by invertebrate and vertebrate fauna;

• functions of wetlands, with special emphasis on habitat values for a broad range of species, food chain support, and water quality enhancement;

• evaluation of the stability and persistence of wetland ecosystems; and

• evaluation of the impact of sediment deposition or erosion, nutrient loading or removal, toxic runoff, pedestrian and off-road vehicle use, grazing, and other impacts on wetland structure and function.

Additional research needs concern characterizing the role of microbes. Bacteria and fungi are important to nutrient cycles, acting as both facilitators and competitors for plant nutrient uptake. Cooke and Lefor (1990) sampled planted and natural salt marsh soils at Indian River marsh in Connecticut and observed that vesicular-arbuscular mycorrhizae (fungi that help plants take up nutrients) were absent in the transplanted site but present in nearly every natural marsh sample. The role of mycorrhizae in wetland soils is not at all clear, and restoration sites offer opportunities to carry out field experiments in this area. The potential for improving transplant growth by manipulating soil microbial communities needs to be explored. Other bacteria perform important "bioremediation" functions. By decomposing contaminants, they may reduce or eliminate the effects of toxic waste spills, such as gasoline seepage from aging underground tanks (R. Gersberg, San Diego State University, personal communication, 1990). Finally, the microbes of wetlands are important to global carbon, sulfur, and nitrogen cycles, and their ability to perform these roles in restored wetlands has barely been explored (Cantilli, 1989).

We also need to know what ecosystem functions can be restored under various constraints and how rapidly restoration can proceed. A variety of specific techniques may be used to restore physical features and ecological functions of wetlands, depending on wetland type, disturbance condition, and other constraints.

• **Develop innovative methods of accelerating the restoration process (e.g., better propagation techniques for native plant species), surveys and protocols for obtaining adequate genetic diversity in the transplant material, soil augmentation procedures to shorten time to obtain the desired vegetative cover, ways to use microbes to detoxify contaminants (bioremediation) and enhance nutrient availability, and methods for controlling exotic species.**

• **Design and conduct experimental research programs to examine restoration techniques and functional development over time in different system types.**

• **Use restoration sites for scientific experiments that are designed to accelerate the restoration process.**

• **Support baseline studies of wetland ecosystem functioning to provide comparisons of different system types among various regions and at different stages of development. Establish regional and national data bases.**

REFERENCES AND RECOMMENDED READING

Aberle, B. 1990. The Biology, Control, and Eradication of Introduced *Spartina* (Cordgrass) Worldwide and Recommendations for Its Control in Washington. Draft report. Washington State Department of Natural Resources, Seattle. 88 pp.

Adamus, P. R., and L. T. Stockwell. 1983. A Method for Wetland Functional Assessment. Vol. 1. Federal Highway Administration Report No. FHWA-IP-82-23. U.S. Department of Transportation, Washington, D.C.

Allen, J. A., and H. E. Kennedy, Jr. 1989. Bottomland Hardwood Reforestation in the LowerMississippi Valley. U.S. Fish and Wildlife Service, Slidell, La. and U.S. Forest Service, Southern Forest Experiment Station, Stoneville, Miss. 28 pp.

Atwater, B. F. 1979. History, landforms, and vegetation of the estuary's tidal marshes. In T. J. Conomos, ed., San Francisco Bay, The Urbanized Estuary. American Association for the Advancement of Science, Pacific Division, San Francisco, Calif.

Belt, C. B., Jr. 1975. The 1973 flood and man's constriction of the Mississippi River. Science 189:681-684.

Berger, J., ed. 1990. Ecological Restoration in the San Francisco Bay Area: A Descriptive Directory and Source Book. Restoring the Earth, Inc., Berkeley, Calif.

Best, G. R. 1987. Natural wetlands-southern environment: Wastewater to wetlands, Where do we go from here? Pp. 99-120 in K. R. Reddy and W. H. Smith, eds., Aquatic Plants for Water Treatment and Resource Recovery. Magnolia Publishing, Orlando, Fla.

Best, G. R., J. R. Tuschall, P. L. Brezonik, J. R. Butner, W. F. DeBusk, K. C. Ewel, A. Hernandez, and H. T. Odum. 1982. The Fate of Selected Heavy Metals in a Forested Wetland Ecosystem. Report to U.S. Environmental Protection Agency. Center for Wetlands, University of Florida, Gainesville, Fla.

Bradshaw, A. D. 1988. Alternative endpoints for reclamation. Pp. 69-85 in J. Cairns, Jr., ed., Rehabilitating Damaged Ecosystems. Vol. II. CRC Press, Boca Raton, Fla.

Broome, S. W. 1989. Creation and restoration of tidal wetlands of the southeastern United States. Pp. 37-72 in J. A. Kusler and M. E. Kentula, eds., Wetland Creation and Restoration: The Status of the Science. Vol. I. U.S. EPA/7600/3-89/038. U.S. Environmental Protection Agency, Environmental Research Laboratory, Corvallis, Ore.

Broome, S. W., C. B. Craft, and E. D. Seneca. 1987. Creation and development of brackish-water marsh habitat. Pp. 197-205 in J. Zelazny and J. S. Feierabend, eds., Increasing Our Wetland Resources. Conference Proceedings, National Wildlife Federation Corporate Conservation Council, October 4-7, 1987. National Wildlife Federation, Washington, D.C.

Cairns, J., ed. 1988. Rehabilitating Damaged Ecosystems. Vol. 1, 192 pp.; Vol. 2, 222 pp. CRC Press, Boca Raton, Fla.

Cammen, L. M. 1976a. Abundance and production of macroinvertebrates from natural and artificially established salt marshes in North Carolina. Am. Midl. Nat. 96:244-253.

Cammen, L. M. 1976b. Macroinvertebrate colonization of *Spartina* marshes artificially established on dredge spoil. Coastal Mar. Sci. 4:357-372.

Cantilli, J. F. 1989. Sulfide Phytotoxicity in Tidal Salt Marshes. M.S. thesis, San Diego State University.

Carlton, J. T. 1989. Man's role in changing the face of the ocean: Biological invasions and implications for conservation of near-shore environments. Conserv. Biol. 3:265-273.

Carothers, S. W., G. S. Mills, and R. R. Johnson. 1989. The creation and restoration of riparian habitat in southwestern arid and semi-arid regions. Pp. 359-376 in J. A. Kusler and M. E. Kentula, eds., Wetland Creation and Restoration: The Status of the Science. Vol. I. U.S. EPA/7600/3089/038. U.S. Environmental Protection Agency, Environmental Research Laboratory, Corvallis, Ore.

Chabreck, R. H. 1989. Creation, restoration, and enhancement of marshes of the northcentral Gulf Coast. Pp. 127-144 in J. A. Kusler and M. E. Kentula, eds., Wetland Creation and Restoration: The Status of the Science. Vol. I. U.S. EPA/7600/3089/038. U.S. Environmental Protection Agency, Environmental Research Laboratory, Corvallis, Ore.

Clean Water Act of 1977. P.L. 95-217, Dec. 27, 1977, 91 Stat. 1566.

Clewell, A. F., and R. Lea. 1989. Creation and restoration of forested wetland vegetation in the southeastern United States. Pp. 199-238 in J. A. Kusler and M. E. Kentula, eds., Wetland Creation and Restoration: The Status of the Science. Vol. I. U.S. EPA/7600/3089/038. U.S. Environmental Protection Agency, Environmental Research Laboratory, Corvallis, Ore.

Coastal Wetlands Planning, Protection and Restoration Act of 1990. P.L. 101-646, Nov. 29, 1990.

Conner, W. H., and J. R. Toliver. 1990. Observations on the regeneration of baldcypress (*Taxodium distichum* (L.) Rich.) in Louisiana swamps. South. J. Appl. For. 14:115-118.

Conners, D. H., F. Riesenberg, R. D. Charney, M. A. McEwen, R. B. Krone, and G. Tchobanoglous. 1990. Research Needs: Salt Marsh Restoration, Rehabilitation, and Creation Technique for Caltrans Construction Projects. Department of Civil Engineering, University of California, Davis. 60 pp.

Cooke, J. C., and M. W. Lefor. 1990. Comparison of vesicular-arbuscular mycorrhizae in plants from disturbed and adjacent undisturbed regions of a coastal salt marsh in Clinton, Connecticut, USA. Environ. Manage. 14:131-137.

Council on Environmental Quality (CEQ). 1989. Environmental Trends—Chapter 5: Wetlands and wildlife. Office of the President, Council on Environmental Quality, Washington, D.C. 152 pp.

Covin, J. D., and J. B. Zedler. 1988. Nitrogen effects on *Spartina foliosa* and *Salicornia virginica* in the salt marsh at Tijuana Estuary, Calif. Wetlands 8:51-65.

Cowardin, L. M., V. Carter, F. C. Golet, and E. T. LaRoe. 1979. Classification of Wetlands and Deepwater Habitats of the United States. Office of Biological Services, Fish and Wildlife Service, U.S. Department of the Interior, Washington, D.C.

Craft, C. B., S. W. Broome, and E. D. Seneca. 1988. Nitrogen, phosphorus and organic carbon pools in natural and transplanted marsh soils. Estuaries 11:272-280.

Dahl, T. E. 1990. Wetland Losses in the United States: 1780's to 1980's. U.S. Department of the Interior, Fish and Wildlife Service, Washington, D.C. 21 pp.

DeWald, J. M., and J. P. Rieger. 1982. Restoration of a Degraded Salt Marsh: Objectives and Techniques. Unpublished report. California Department of Transportation, Caltrans District 11, San Diego. 10 pp.

Erwin, K. L. 1989. Freshwater marsh creation and restoration in the southeast. Pp. 239- 272 in J. A. Kusler and M. E. Kentula, eds., Wetland Creation and Restoration: The Status of the Science. Vol. I. U.S. EPA/7600/3-89/038. U.S. Environmental Protection Agency, Environmental Research Laboratory, Corvallis, Ore.

Ewel, J. 1987. Restoration is the ultimate test of ecological theory. Pp. 31-33 in W. Jordan, M. Gilpin, and J. Aber, eds., Restoration Ecology: A Synthetic Approach to Ecological Research. Cambridge University Press, New York.

Federal Interagency Committee for Wetland Delineation. 1989. Federal Manual for Identifying and Delineating Jurisdictional Wetlands. Cooperative technical publication. U.S. Army Corps of Engineers, U.S. Environmental Protection Agency, U.S.

Fish and Wildlife Service, and U.S. Department of Agriculture. Soil Conservation Service, Washington, D.C. 76 pp. plus appendices.

Fink, B., and J. Zedler. 1990. Endangered plant recovery: Experimental approaches with *Cordylanthus maritimus* ssp. *maritimus*. Pp. 460-468 in H. G. Hughes and T. M. Bonnicksenm, eds., Restoration '89: The New Management Challenge. Society for Ecological Restoration, Madison, Wis.

Fonseca, M. S. 1989. Regional analysis of the creation and restoration of seagrass systems. Pp. 175-198 in J. A. Kusler and M. E. Kentula, eds., Wetland Creation and Restoration: The Status of the Science. Vol. I. U.S. EPA/7600/3-89/038. U.S. Environmental Protection Agency, Environmental Research Laboratory, Corvallis, Ore.

Fonseca, M. S., W. J. Kenworthy, and G. W. Thayer. 1988. Restoration and management of seagrass systems: A Review. Pp. 353-368 in D. D. Hook, W. H. McKee, Jr., H. K. Smith, J. Gregory, V. G. Burrell, Jr., M. R. DeVoe, R. E. Sojka, S. Gilbert, R. Banks, L. H. Stolzy, C. Brooks, T. D. Matthews, and T. H. Shear, eds., The Ecology and Management of Wetlands. Vol. 2. Timber Press, Portland, Ore.

Frenkel, R. E., and J. C. Morlan. 1990. Restoration of the Salmon River Salt Marshes: Retrospect and Prospect. Final Report to the U.S. Environmental Protection Agency. Seattle, Wash.

Glenn, E. P., J. W. O'Leary, M. C. Watson, T. L. Thompson, and R. O. Kuehl. 1991. *Salicornia bigelovii* Torr.: An oilseed halophyte for seawater irrigation. Science 251(4997):1065-1067.

Gosselink, J. G., and L. C. Lee. 1989. Cumulative impact assessment in bottomland hardwood forests. Wetlands 9:83-174.

Gosselink, J. G., S. E. Bayley, W. H. Conner, R. E. Turner. 1981. Ecological factors in the determination of riparian wetland boundaries. Pp. 197-219 in J. B. Clark and J. Benforado, eds., Wetlands of Bottomland Hardwood Forest. Elsevier Science Publishing Co., New York.

Gosselink, J. G., W. H. Conner, J. W. Day, Jr., and R. E. Turner. 1989. Classification of wetland resources: Land, Timber, and Ecology. Pp. 28-48 in D. Jackson and J. L. Chambers, eds., Timber Harvesting in Wetlands. Division of Continuing Education, Louisiana State University, Baton Rouge, La.

Gosselink, J. G., L. C. Lee, and T. A. Muir. 1990a. Ecological Processes and Cumulative Impacts. Lewis Publishers, Chelsea, Mich. 708 pp.

Gosselink, J. G., G. P. Shaffer, L. C. Lee, D. M. Burdick, D. L. Childers, N. C. Leibowitz, S. C. Hamilton, R. Boumans, D. Cushman, S. Fields, M. Koch, and J. M. Visser. 1990b. Landscape conservation in a forested wetland watershed. BioScience 40(8): 588-600.

Greeson, P. B., J. R. Clark, and J. E. Clark, eds. 1979. Wetland Functions and Values: The State of Our Understanding. Proceedings of the National Symposium on Wetlands. American Water Resources Association, Minneapolis, Minn. 674 pp.

Gross, K. L. 1987. Mechanisms of colonization and species persistence in plant communities. Pp. 173-188 in W. Jordan, M. Gilpin, and J. Aber, eds., Restoration Ecology: A Synthetic Approach to Ecological Research. Cambridge University Press, New York.

Hackensack Meadowland Reclamation and Development Act (HMRDA). 1968. State of New Jersey Statutes. Chapter 17, Section 13:17-1 to 13-17-86.

Haltiner, J. 1990. Sweetwater Marsh: Morphology and tidal circulation. Philip Williams & Assoc., Ltd. Report Prepared for Caltrans District 11, San Diego, and Pacific Estuarine Research Laboratory, San Diego State University, San Diego, Calif.

Harris, L. D. 1988. The nature of cumulative impacts on biotic diversity of wetland vertebrates. Environ. Manage. 12:675-693.

Hartman, J. M. 1988. Recolonization of small disturbance patches in a New England salt marsh. Am. J. Bot. 75:1625-1631.

Hollands, G. G. 1989. Regional analysis of the creation and restoration of kettle and pothole wetlands. Pp. 287-304 in J. A. Kusler and M. E. Kentula, eds., Wetland Creation and Restoration: The Status of the Science. Vol. I. U.S. EPA/7600/3-89/ 038. U.S. Environmental Protection Agency, Environmental Research Laboratory, Corvallis, Ore.

Hook, D., D., W. H. McKee, Jr., H. K. Smith, J. Gregory, V. G. Burrell, Jr., M. R. DeVoe, R. E. Sojka, S. Gilbert, R. Banks, L. H. Stolzy, C. Brooks, T. D. Matthews, and T. H. Shear, eds. 1988. The Ecology and Management of Wetlands. Vol. 1, 592 pp.; Vol. 2, 394 pp. Timber Press, Portland, Ore.

Hunt, C. E. 1988. Down by the River: The Impacts of Federal Water Projects and Policies on Biological Diversity. Island Press, Washington, D.C.

Johnson, L. E., and W. V. McGuinness, Jr. 1975. Guidelines for Material Placement in Marsh Creation. Contract Report D-75-2 for Environmental Effects Laboratory, U.S. Army Corps of Engineers Waterways Experiment Station, Vicksburg, Miss. 189 pp.

Jorgensen, P. D. 1975. Habitat preference of the light-footed clapper rail in Tijuana Estuary Marsh, Calif. M.S. thesis, San Diego State University.

Josselyn, M., ed. 1982. Wetland Restoration and Enhancement in California. Sea Grant Technical Report No. T-CSGCP-007. Tiburon Center for Environmental Studies, Tiburon, Calif.

Josselyn, M. N., and J. W. Buchholz. 1984. Marsh Restoration in San Francisco Bay—A Guide to Design and Planning, Tech. Rept. #3, Tiburon Center for Environmental Studies, San Francisco State University, Tiburon, Calif.

Kahrl, W. L. 1979. The California Water Atlas. Governor's Office of Planning and Research and California Department of Water Resources, Sacramento, Calif.

King, D. M. 1990. A Framework for Evaluating Wetland Restoration Projects. April 1990 Draft report to Office of Policy, Planning and Evaluation, U.S. Environmental Protection Agency, Washington, D.C.

Kraus, M. L., and D. J. Smith. n.d. Competition and Succession in a Perturbed Urban Estuary: The Effects of Hydrology. Monograph. Hackensack Meadowlands Development Commission, Lyndhurst, N.J.

Kusler, J. A. 1983. Our National Wetland Heritage: A Protection Guidebook. Environmental Law Institute, Washington, D.C.

Kusler, J. A., and M. E. Kentula, eds. 1989. Wetland Creation and Restoration: The Status of the Science. Vol. I-II. U.S. EPA/7600/3-89/038. U.S. Environmental Protection Agency, Environmental Research Laboratory, Corvallis, Ore.

Kusler, J. A., M. L. Quammen, and G. Brooks, eds. 1988. Proceedings of the National Wetland Symposium: Mitigation of Impacts and Losses. Association of State Wetland Managers, Berne, N.Y.

Langis, R., M. Zalejko, and J. B. Zedler. 1991. Nitrogen assessments in a constructed and a natural salt marsh of San Diego Bay, California. Ecol. Appl. 1:40-51.

Larson, J. S. 1988. Wetland creation and restoration: An outline of the scientific perspective. Pp. 73-79 in J. Zelazny and J. S. Feierabend, eds., Increasing Our Wetland Resources. Conference Proceedings, National Wildlife Federation Corporate Conservation Council, October 4-7, 1987. National Wildlife Federation, Washington, D.C.

Larson, J. S., and C. Niell, eds. 1987. Mitigating Freshwater Wetland Alterations in the Glaciated Northeastern United States: An Assessment of the Science Base. Publ. 87-1. Environmental Institute, University of Massachusetts, Amherst.

Levine, D. A., and D. E. Willard. 1989. Regional analysis of fringe wetlands in the Midwest: creation and restoration. Pp. 305-332 in J. A. Kusler and M. E. Kentula, eds., Wetland Creation and Restoration: The Status of the Science. Vol. I. U.S. EPA/ 7600/3-89/038. U.S. Environmental Protection Agency, Environmental Research Laboratory, Corvallis, Ore.

Lewis, R. R. III. 1982. Creation and Restoration of Coastal Plant Communities. CRC Press, Boca Raton, Fla.

Lewis, R. R. III. 1989. Creation and restoration of coastal plain wetlands in Florida. Pp. 73-102 in J. A. Kusler and M. E. Kentula, eds., Wetland Creation and Restoration: The Status of the Science. Vol. I. U.S. EPA/7600/3-89/038. U.S. Environmental Protection Agency, Environmental Research Laboratory, Corvallis, Ore.

Loftin, K. A., L. A. Toth, and J. T. B. Obeysekera. 1990. Kissimmee River Restoration: Alternative Plan Evaluation and Preliminary Design Report. South Florida Water Management District, West Palm Beach, Fla. June.

Loucks, O. L. 1989. Restoration of the pulse control function of wetlands and its relationship to water quality objectives. Pp. 55-65 in J. A. Kusler and M. E. Kentula, eds., Wetland Creation and Restoration: The Status of the Science, Vol. II. U.S. EPA/7600/3-89/038. U.S. Environmental Protection Agency, Environmental Research Laboratory, Corvallis, Ore.

Lowry, D. J. 1989. Restoration and creation of palustrine wetlands associated with riverine systems of the glaciated Northeast. Pp. 273-286 in J. A. Kusler and M. E. Kentula, eds., Wetland Creation and Restoration: The Status of the Science, Vol. I. U.S. EPA/7600/3-89/038. U.S. Environmental Protection Agency, Environmental Research Laboratory, Corvallis, Ore.

MacDonald, P. O., W. E. Frayer, and J. K. Clauser. 1979. Documentation, Chronology, and Future Projections of Bottomland Hardwood Habitat Losses in the Lower Mississippi Alluvial Plain. U.S. Department of the Interior, Fish and Wildlife Service, Washington, D.C., 2 vols.

Mitsch, W. J., and J. G. Gosselink. 1986. Wetlands. Van Nostrand Reinhold, New York. 539 pp.

Moy, L. D. 1989. Are *Spartina* Marshes Renewable Resources? A Faunal Comparison of a Man-Made Marsh and Two Adjacent Natural Marshes. M.S. thesis, North Carolina State University, Raleigh.

Newling, C. J., and M. C. Landin. 1985. Long-Term Monitoring of Habitat Development at Upland and Wetland Dredge Material Disposal Sites, 1974-1982. Tech. Rept. D-85-5. U.S. Army Corps of Engineers Waterway Experiment Station, Vicksburg, Miss.

Nichols, F. H., J. E. Cloern, S. N. Luoma, and D. H. Peterson. 1986. The Modification of an Estuary. Science 231:567-573.

Niering, W. A. 1986. Wetlands. The Audubon Society. Alfred A. Knopf, New York. 538 pp.

Niering, W. A. 1989. Vegetation dynamics in relation to wetland creation. Pp. 67-74 in J. A. Kusler and M. E. Kentula, eds., Wetland Creation and Restoration: The Status of the Science. Vol. II. U.S. EPA/7600/3-89/038. U.S. Environmental Protection Agency, Environmental Research Laboratory, Corvallis, Ore.

Nordby, C. S., and J. B. Zedler. 1991. Responses of fish and macrobenthic assemblages to hydrologic disturbances in Tijuana Estuary and Los Pe:asquitos Lagoon, California. Estuaries 14:80-93.

Novitzki, R. P. 1979. Hydrologic characteristics of Wisconsin's wetlands and their influence on floods, stream flow, and sediment. In P. E. Greeson, J. R. Clark, and J. E. Clark, eds., Wetland Functions and Values: The State of Our Understanding. American Water Resources Association, Minneapolis, Minn. 377 pp.

Odum, W. E. 1988. Predicting ecosystem development following creation and restoration of wetlands. Pp. 67-70 in J. Zelazny and J. S. Feierabend, eds., Increasing

Our Wetland Resources. Conference Proceedings, National Wildlife Federation Corporate Conservation Council, October 4-7, 1987. National Wildlife Federation, Washington, D.C.

Onuf, C. P., and M. L. Quammen. 1985. Coastal and riparian wetlands of the Pacific region: The state of knowledge about the food chain support. Pp. 98-173 in Proceedings of National Wetland Technical Council Workshop, April 14-16, Mill Valley, Calif.

Onuf, C. P., M. L. Quammen, G. P. Shaffer, C. H. Peterson, J. W. Chapman, J. Cermak, and R.W. Holmes. 1978. An analysis of the values of central and southern California coastal wetlands. Pp. 186-199 in P. E. Greeson, J. R. Clark, and J. E. Clark, eds., Wetland Functions and Values: The State of Our Understanding. American Water Resources Association, Minneapolis, Minn.

Pacific Estuarine Research Laboratory (PERL). 1990. A Manual for Assessing Restored and Natural Coastal Wetlands with Examples from Southern California. California Sea Grant Report No. T-CSGCP-021. California Sea Grant Program, La Jolla, Calif. 105 pp.

Park, R. A., M. S. Trehan, P. W. Mausel, and R. C. Howe. 1988. The Effects of Sea Level Rise on U.S. Coastal Wetlands. HRI Report No. 142. Holcomb Research Institute, Butler University Indianapolis. 60 pp.

Park, R. A., M. S. Trehan, P. W. Mausel, and R. C. Howe. 1989. Coastal wetlands in the twenty-first century: Profound alterations due to rising sea level. Pp. 71-80 in Wetlands: Concerns and Successes. Proceedings of the American Water Resources Association, September, Tampa, Fla.

Parnell, J. F., D. M. DuMond, and R. N. Needham. 1978. A Comparison of Plant Succession and Bird Utilization on Diked and Undiked Dredged Material Islands in North Carolina Estuaries. U. S. Army Corps of Engineers Waterways Experiment Station, Vicksburg, Miss.

Peters, R. L., II. 1988. The effect of global climatic change on natural communities. Pp. 450-461 in E. O. Wilson, ed., Biodiversity. National Academy Press, Washington, D.C.

Pontius, F. W. 1990. Federal laws protecting wetlands. J. AWWA 82:12-16.

Power, M. E. 1990. Effects of fish in river food webs. Science 250:811-814.

Prouty, W. F. 1952. Carolina bays and their origin. Bull. Geol. Soc. Am. 63:167-224.

Quammen, M. L. 1986. Measuring the success of wetlands mitigation. Natl. Wetl. Newsl. 8:6-8.

Reitze, A. W., Jr. 1974. Wetlands, special problems in land use control. Chapter 2 in Environmental Planning: Law of Land and Resources. North American International, Washington, D.C. 941 pp.

Richardson, C. J. 1985. Mechanisms controlling phosphorus retention capacity in freshwater wetlands. Science 228:1424-1427.

Roberts, T. H. 1989. Habitat value of man-made coastal marshes in Florida. Pp. 154-179 in F. J. Webb, ed., Proceedings of the 16th Annual Conference on Wetlands Restoration and Creation. Hillsborough Community College, Tampa. Fla.

Rutherford, S. E. 1989. Detritus Production and Epibenthic Communities of Natural Versus Constructed Salt Marshes. M.S. thesis, San Diego State University.

Sacco, J. N. 1989. Infaunal Community Development of Artificially Established Salt marshes in North Carolina. M.S. thesis, North Carolina State University, Raleigh, N.C.

Sacco, J. N., F. L. Booker, and E. D. Seneca. 1987. Comparison of the macrofaunal communities of a human-initiated salt marsh at two and fifteen years of age. Pp. 282-285 in J. Zelazny and S. Feierabend, eds., Increasing Our Wetland Resources. Conference Proceedings, National Wildlife Federation Corporate Conservation Council, October 4-7, 1987. National Wildlife Federation, Washington, D.C.

San Francisco Estuary Project. 1990. Status and Trends Report on Pollutants in the San Francisco Estuary. Prepared under EPA Cooperative Agreement CE-009496-01 by the San Francisco Bay-Delta Aquatic Habitat Institute, 180 Richmond Field Station, 1301 South 46th Street, Richmond, CA 94804.

Saucier, R. T., C. C. Calhoun, Jr., R. M. Engier, T. R. Patin, and H. K. Smith. 1978. Executive Overview and Detailed Summary: Dredged Material Research Program. Tech. Rept. DS-78-212. U.S. Army Corps of Engineers Waterway Experiment Station, Vicksburg, Miss.

Schalles, J. F., R. R. Sharitz, J. W. Gibbons, G. J. Leversee, and J. N. Knox. 1989. Carolina Bays of the Savannah River Plant. SRO-NERP-18.U.S. U.S. Department of Energy, Washington, D.C. 70 pp.

Seneca, E. D., and S. W. Broome. In press. Proceedings of a NOAA Restoration Symposium. Invited Talk. September 1990.

Sharitz, R. R., and J. W. Gibbons. 1982. The Ecology of Southeastern Shrub Bogs (Pocosins) and Carolina Bays: A Community Profile. FWS/OBS-82/04. Division of Biological Services, U.S. Fish and Wildlife Service, Washington, D.C. 93 pp.

Shisler, J. K. 1989. Creation and restoration of the coastal wetlands of the northeastern United States. Pp. 145-174 in J. A. Kusler and M. E. Kentula, eds., Wetland Creation and Restoration: The Status of the Science. Vol. I. U.S. EPA/7600/3-89/038. U.S. Environmental Protection Agency, Environmental Research Laboratory, Corvallis, Ore.

Shisler, J. K., and D. J. Charette. 1984. Evaluation of artificial salt marshes in New Jersey. New Jersey Agricultural Experiment Station Publ. No. P-40502-01-84. New Brunswick, N.J. 160 pp.

Skinner, J. 1962. A Historical Review of the Fish and Wildlife Resources of the San Francisco Bay Area. Report 1. Water Projects Bureau, California Department of Fish and Game, Sacramento, Calif.

Smith, R. L. 1980. Ecology and Field Biology. 3rd ed. Harper and Row, New York. 835 pp.

Spicher, D., and M. Josselyn. 1985. Spartina (Gramineae) in northern California: Distribution and taxonomic notes. Madroño 32:158-167.

Stanley, J. T. 1989. Riparian wetland creation and restoration in the far west: A compilation of information. Pp. 417-464 in J. A. Kusler and M. E. Kentula, eds., Wetland Creation and Restoration: The Status of the Science, Vol. I. U.S. EPA/7600/3-89/038. U.S. Environmental Protection Agency, Environmental Research Laboratory, Corvallis, Ore.

Strickland, R., ed. 1986. Wetland Functions, Rehabilitation, and Creation in the Pacific Northwest. Washington State Department of Ecology, Olympia, Wash.

Swift, K. L. 1988. Salt marsh restoration: Assessing a southern California example. M.S. thesis, San Diego State University. 84 pp.

Tax Reform Act of 1986. P.L. 99-514, 100 Stat. 2085 (1986), 49 U.S.C. § 16 et seq.

The Conservation Foundation. 1988. Protecting America's Wetlands: An Action Agenda. Final Report of the National Wetlands Policy Forum. The Conservation Foundation, Washington, D.C. 69 pp.

Thompson, Justice Gordon, Jr. 1988. Memorandum Decision and Order. Sierra Club and League for Coastal Protection v. John O. Marsh, Secretary of the Army, et al. U.S. District Court, Southern District of California. Civil No. 86-I942-GT(IEG), May 13, San Diego, Calif.

Tiner, R. W., Jr. 1984. Wetlands of the United States: Current Status and Recent Trends. National Wetlands Inventory. U.S. Department of the Interior, Fish and Wildlife Service, Washington, D.C. 59 pp.

Tripp, J. T. B., and M. Herz. 1988. Wetland preservation and restoration: Changing federal priorities. Va. J. Nat. Resour. Law 7:221-275.

Tuschall, J. R., Jr. 1981. Heavy Metal Complexation with Naturally Occurring Organic Ligands in Wetland Ecosystems. Ph.D. dissertation, University of Florida, Gainesville, Fla. 212 pp.

U.S. Army Corps of Engineers. 1972. Charles River Watershed, Massachusetts. New England Division, Waltham, Mass. 65 pp.

U.S. Army Corps of Engineers and U.S. Environmental Protection Agency. 1990. Memorandum of Agreement Regarding Mitigation for Dredged or Fill Material Disposal in Wetlands. February 6.

U.S. Fish and Wildlife Service (FWS). 1988. Biological Opinion 1-1-78-F-14-R2. The combined Sweetwater River food control and highway project, San Diego County, California. Letter from Walley Steuke, Acting Regional Director. U.S. FWS, Portland, Oregon, to Colonel Tadahiko Ono, District Engineer, Los Angeles District. U.S. Army Corps of Engineers, March 30. 28 pp. plus appendices.

U.S. Fish and Wildlife Service (FWS). 1990. Wetlands: Meeting the President's Challenge. 1990 Wetland Action Plan. U.S. FWS, Washington, D.C. 64 pp.

Walters, C. 1986. Adaptive Management of Renewable Resources. Macmillan, New York. 374 p.

Water Resources Development Act of 1986. P.L. 99-662, Nov. 17, 1985, 100 Stat. 4082.

Willard, D. E., V. M. Finn, D. A. Levine, and J. E. Klarquist. 1989. Creation and restoration of riparian wetlands in the agricultural midwest. Pp. 333-358 in J. A. Kusler and M. E. Kentula, eds., Wetland Creation and Restoration: The Status of the Science. Vol. I. U.S. EPA/7600/3-89/038. U.S. Environmental Protection Agency, Environmental Research Laboratory, Corvallis, Ore.

Williams, S. L. 1990. Experimental studies of Caribbean seagrass bed development. Ecol. Monogr. 60:449-469.

Woodhouse, W. W., Jr., E. D. Seneca, and S. W. Broome. 1974. Propagation of *Spartina alterniflora* for Substrate Stabilization and Salt Marsh Development. Coastal Engineering Research Center, U.S. Army Corps of Engineers, Fort Belvoir, Va.

Zalejko, M. K. 1989. Nitrogen Fixation in a Natural and a Constructed Southern California Salt Marsh. M.S. thesis, San Diego State University.

Zedler, J. B. 1988a. Restoring diversity in salt marshes: Can we do it? Pp. 317-325 in E. O. Wilson, ed., Biodiversity. National Academy Press, Washington, D.C.

Zedler, J. B. 1988b. Salt marsh restoration: Lessons from California. Pp. 123-138 in J. Cairns, ed., Rehabilitating Damaged Ecosystems. CRC Press, Boca Raton, Fla.

Zedler, J. B. 1988c. Why it's so difficult to replace lost wetland functions. Pp. 121-123 in J. Zelazny and J. S. Feierabend, eds., Increasing Our Wetland Resources. Conference Proceedings, National Wildlife Federation Corporate Conservation Council, October 4-7, 1987. National Wildlife Federation, Washington, D.C.

Zedler, J. B. 1991. The challenge of protecting endangered species habitat along the southern California coast. Coast. Manage. 19:35-53.

Zedler, J. B., and R. Langis. 1991. Urban wetland restoration: A San Diego Bay example. Proceedings: Third Annual "Country in the City" Symposium, Portland, Ore., April 1990. Audubon Society of Portland.

Zedler, J. B., and M. W. Weller. 1989. Overview and future directions. Pp. 465-473 in J. A. Kusler and M. E. Kentula, eds., Wetland Creation and Restoration: The Status of the Science. Vol. I. U.S. EPA/7600/3-89/038. U.S. Environmental Protection Agency, Environmental Research Laboratory, Corvallis, Ore.

Zedler, J. B., R. Langis, J. Cantilli, M. Zalejko, K. Swift, and S. Rutherford. 1988. Assessing the functions of mitigation marshes in southern California. Pp. 323-330 in J.

A. Kusler, S. Daly, and G. Brooks, eds., Proceedings of the National Wetland Symposium, Urban Wetlands, Oakland, Calif. Association of State Wetland Managers, Berne, N.Y.

Zedler, P. H. 1987. The Ecology of Southern California Vernal Pools: A Community Profile. Biological Report 85 (7.11). U.S. Fish and Wildlife Service, Washington, D.C. 136 pp.

Zelazny, J., and J. S. Feierabend, eds. 1988. Increasing Our Wetland Resources. Conference Proceedings, National Wildlife Federation Corporate Conservation Council, October 4-7, 1987. National Wildlife Federation, Washington, D.C.

7

Integrated Aquatic
Ecosystem Restoration

INTRODUCTION

The goal of this chapter is to explain the need to consider proposed restoration projects in their landscape and watershed contexts on a scale appropriate to the needs of affected plant and animal species. Concepts of landscape ecology are useful in this endeavor. In landscape ecology, one studies the influences of the landscape on biotic and abiotic processes. The focus of landscape ecology is on the effects of the landscape's spatial heterogeneity, geometry, and areal extent on ecological processes. Although still a new field, landscape ecology has demonstrated persuasively that both the temporal and the spatial scales of many ecological studies are too small (Karr, 1991). Principles of landscape ecology help to provide theoretical and empirical underpinnings for resource management and other applied sciences (Risser et al., 1984).

Integrated resource management is the term this committee uses to indicate resource management that seeks to restore the structure and function of whole ecosystems by striving to understand and respond holistically to cumulative ecological impacts. The integrated approach to aquatic restoration tries to consider the major ecological interactions in a watershed and seeks to nurture the watershed's restoration to a functioning system, rather than to manage for a single species or for a resource commodity such as game fish.

Lakes, streams, rivers, ponds, ground water, estuaries, and wetlands are interconnected parts of larger landscapes. Stabilization of

341

lake levels or stream flows may hamper wetland restoration, which depends on variable water levels. Conversely, wetland restoration may increase bird or fish populations in ways that affect stream or lake restoration efforts. In addition to taking such ecosystem interconnections into consideration, aquatic ecosystem restoration also requires that cumulative impacts to ecosystems be considered. Regulating the input of each chemical pollutant to the Great Lakes independently, for example, without considering the chemicals' synergistic and cumulative impacts, is an example of fragmentary management. By contrast, the Great Lakes Water Quality Agreement of 1978 requires signatories to consider the cumulative influence of each chemical, a more integrated approach (NRC/RSC, 1985). However, aquatic ecosystem restoration requires more than water quality management.

Restoration of an aquatic ecosystem requires that the management of all significant ecological elements be coordinated in a comprehensive approach, often on a watershed or other landscape scale. This is a practical approach to resource management. Many state agencies have model watershed programs, and many states have excellent regional planning programs organized by watershed basin, such as the Interstate Commission on the Potomac River Basin and the Tahoe Regional Planning Agency. The U.S. Geological Survey and many state or regional water resource agencies have organized water data by watersheds for years. The Soil Conservation Service's Watershed Program is also concerned with landscape-level processes.

Renewed attention to ecological questions posed on large spatial scales is evident in the science of landscape ecology (Turner, 1987; Dale et al., 1989) and in approaches to population dynamics on continental scales (Brown and Maurer, 1989). More attention to ecological research on large spatial scales is arising from new technological developments in remote sensing and geographic information systems that have expanded research opportunities. Most of this research has dealt with terrestrial systems. However, applications to large, complex freshwater systems include studies of archipelagoes of lakes connected by streams and ground water (Tonn and Magnuson, 1982; Magnuson et al., 1990) and studies of the effects of beaver on extensive lake, stream, and wetland complexes (Naiman et al., 1988). In Europe, where landscape ecology was developed, geographic areas on the scale of 10 to 10,000 km^2 were used in studies of water movement patterns and changes in water quality (e.g., Naveh and Lieberman, 1984; Forman and Godron, 1986).

INSTITUTIONAL BARRIERS TO INTEGRATED
AQUATIC RESTORATION

Fragmentation of ecosystem management is common in U.S. governmental organizations and in industry. Watershed and political boundaries often overlap. Furthermore, different components of a watershed are usually administered by different agencies. As an example, the International Joint Commission established joint U.S. and Canadian goals for the Great Lakes. However, at the national level in the United States, responsibilities are divided among the U.S. Environmental Protection Agency (EPA; water quality and contaminants) and the Fish and Wildlife Service (animal communities and habitats). Independent water quality managers and fishery managers exist in each of six states (Minnesota, Ohio, New York, Pennsylvania, Michigan, and Wisconsin) bordering the Great Lakes. Important constituencies, such as anglers, environmentalists, and industry, are frequently at odds over basinwide issues such as the effects of organochlorine contaminants. The politics and consensus building required for integrated management of the resource are often as complex as the ecosystem itself.

IMPORTANCE OF INTEGRATED AQUATIC
ECOSYSTEM RESTORATION TO WILDLIFE

The amount and timing of water fluctuations and changes in water quality constitute the most important variables limiting the viability of many species of plants and animal populations. For such species, the rate and distribution of environmental change are essential determinants of their survival. Natural succession must be allowed to operate to continue providing diverse landscapes with heterogeneous niches for wildlife, but the tendency of humans is to build static structures (e.g., geographically fixed wildlife refuges and wetlands with immovable borders) that inhibit species' survival.

Some animals exist as subpopulations on patches of aquatic habitat scattered across a watershed. The animals move or migrate to the most favorable sites as the habitat becomes less suitable for them. Subpopulations may become reduced on other patches as the resources there become less usable. This is analogous to ducks moving south to open water as the last pond in their vicinity freezes over. These patches of exploitable habitat, whatever their character, are often the result of natural changes. Changes in landscape should be anticipated to maximize the effectiveness of restoration programs, by recognizing that restoration or management of one part of a watershed will affect other parts of the aquatic ecosystem.

Natural ecosystems are spatially and temporally dynamic, expanding and shrinking with the prevailing rainfall, stream flow, or tidal regimes. For example, wetlands around the Great Salt Lake are widespread during years of heavy rainfall and runoff; marshes along the Colorado River expand after months of river flooding; prairie potholes expand in size and number during wet years; adjacent lakes that are separate in dry years may merge in wet ones; stagnant waters become flowing with increased runoff. A site designed as a marsh might eventually become a pond or a lake, or vice versa.

Watersheds contain arrays of habitats and sites particularly suitable for certain species. In theory, each site may at some time provide a refuge for a species during a period of stress. The importance of a single site to regional biodiversity is variable—ranging from highly critical during years of restricted habitat to redundant during years of expanded habitat. Marginal (redundant) sites during one set of conditions may be the best or the only sites when conditions change. These sites have often been converted to another habitat and are not available when needed. For many species, opportunistic site use allows them to exist in regions not generally adequate for them. Restoration plans thus must be developed with a landscape perspective—an understanding of how specific sites are related to the remaining resources in the watershed or region.

Restoration of aquatic ecosystems to sustain mobile, migratory, or opportunistic species can be immensely complex. Many of these species select what is the best available habitat at a particular time. Individual plants or animals invade or retreat from sites as a result of changes in their range over time. A site may contain few individuals for long periods of time but may provide an essential refuge for the population during periods of stress.

Migration is a seasonal effort to find optimal habitat. Many desert aquatic species, such as spadefoot toads (*Scaphiopus couchi, S. hammondi*), appear after rain in normally dry streambeds, where they feed and reproduce. These streams carry water irregularly, and each of the streams in the toads' population range may vary its flood regime independently. The toads have adapted to this scattered multiwatershed system. An effort to provide habitat for this species might require coordination of restoration activities over more than one watershed.

Like the toad, many opportunistic plants exploit ephemeral niches that occur periodically. These plants have multiple reproductive and dispersal adaptations that allow them to spread through the landscape and find appropriate places to exist. Many herbivorous insects survive by playing a kind of hide-and-seek game within a landscape; they must feed on ephemeral plants before some other herbivores eat the plants and must also try to stay one jump ahead of their predators.

Many declining species—such as the California least tern, the mangrove cuckoo, many frogs and salamanders, and some puddle ducks—are broadly distributed but are rare or threatened at individual sites. These species have evolved to find and use specialized habitat that appears at regionally scattered sites. Regional long-term climate trends and human conversion of these sites have significantly reduced the animals' opportunities to find suitable habitats. A minimum number of sites is needed to maintain a viable population. Most resource management agencies, however, focus on individual sites, rather than on the regional distribution of aquatic types and their summed functions. Future restoration projects would benefit from a large-scale integrated management approach that seeks to help managers understand what attracts and supports mobile species within a target area.

APPROPRIATE SCALE FOR AQUATIC ECOSYSTEM RESTORATION

Gosselink and Lee (1989) discuss the question of the appropriate scale for managing aquatic ecosystems for specific species, focusing on the problem of understanding and evaluating cumulative losses to the ecosystems. They suggest that the area of concern for any given species should be its range. For example, they propose the term *duckshed* as analogous to *watershed* for the unit of management for a local population of ducks. A duckshed would include any area where individuals of the population might have to go to survive under the worst conditions, as well as the ecosystem that supports the population. The range may have within its boundaries considerable space that is used regularly by the particular species. Other areas may serve as occasional habitat and still others as refuges in which the species can survive during periods of natural stress. These "stress shelters" may be of marginal use to the species most of the time, but because these refuge areas appear marginal, they are often not properly protected and are lost. Much of the nation's waterfowl habitat has been lost in this way.

A method that may be useful for planning aquatic ecosystem restoration programs is the Adaptive Environmental Assessment (AEA) approach of Holling (1978) (Walters, 1986). The AEA is a process for involving scientists, resource managers, policy analysts, and decision makers interactively in designing resource management programs.

The focus of AEA workshops is shared responsibility for the development of a simulation model of the system to be managed (e.g., restored). However, the benefits of AEA usually derive from the interactions among participants during the process of model building and not from the model itself (Walters, 1986). The process attempts to take into account, at the earliest stages of the assessment, all relevant social, economic, and environmental considerations, addressing conflicts directly and developing a framework for evaluating trade-offs. The variability and uncertainty common to all environmental systems are explicitly recognized. The key to AEA's success may be its flexibility and lack of prescriptions for problem solving (Holling, 1978).

USE OF HISTORICAL RECORDS IN
RECONSTRUCTING WATERSHEDS

To obtain a broad historical perspective on a watershed prior to planning its restoration, one must gather its ecological history, often through the use of old maps, old newspaper articles, and interviews with area residents. Studying available ecological information and correlating it with available historical information suggest how people have changed a watershed over time and what management tools might best accomplish the restorative changes desired.

Geography provides a unifying focus for such studies. By examining original land survey data, early U.S. Geological Survey maps, and early soil maps, experts can develop maps of a watershed's stream flow and land use patterns at various times. Aerial photographs of most watersheds have been available since about 1940, and in some places, aerial photos are available at relatively frequent intervals from 1940 to the present. In the 1970s, remotely sensed satellite photography also became available for many watersheds, to add to the geographic record. Careful study of existing conditions and of the photographic record over time demonstrates changes in land use in watersheds. In general, not enough research support has been available for comprehensive assessments of ecological change in watersheds combined with evaluation of resource policy options there. One exception was the study by Gosselink et al. (1990) concerning the assessment and management of cumulative effects on wetland resources in the Tensas Basin of Arkansas and Louisiana.

The authors examined a wide variety of existing environmental data about the Tensas watershed and then mapped those that had geographic elements, such as land use, drainage, habitat value, and the distribution of plant communities. Bears are the largest animal species in the area and require a large home range to survive. The study chose this home range as the minimum useful unit of the landscape to acquire. Gosselink et al. (1990) scanned the watershed for areas of restorable and available "bearshed"-sized property to acquire. They also looked for bearsheds that connected larger already-protected habitat units. Their recommendations, if followed, will greatly increase the habitat value of the entire watershed. This EPA-sponsored study did bridge the gap between assessing ecological conditions and making policy recommendations, by setting priorities for land acquisition.

Interpretation of Historical Data for Restoration Purposes

Resource maps for various time periods show the geographical array of resources and suggest patterns of change. Existing hydrological and ecological models then can be used to describe the processes that link resources dynamically in the landscape. By understanding these processes well enough, scientists can show the general outcomes of various potential management policies so that decision makers are able to compare these outcomes critically.

Many states and water basins have resource maps and historical records currently available. These records can serve as inputs to a geographical information system (GIS) data base so that a wide variety of parameters can be examined simultaneously. Sets of related data when superimposed on a GIS map often reveal opportunities for restoring aquatic ecosystems so as to maximize a number of functions, such as reduction of flood damage, erosion control, habitat for species of concern, and water quality improvement. In some areas, sufficient data have been accumulated to approach landscape-scale watershed restoration planning. Other areas may lack even organized inventories of resources or planning capabilities. The EPA has attempted to develop advanced identification systems in some regions to establish wetland restoration priorities. However, the committee knows of no region, watershed, or state that has used its full resource planning capabilities to designate restoration priorities.

CONCLUSION

Wherever possible, decisions about the management and restoration of aquatic resources throughout the United States should not be made on a small-scale, short-term, site-by-site basis, but should instead be made to promote the long-term sustainability of all aquatic

resources in the landscape. Whereas restoration on the large landscape scale is therefore definitely preferable to piecemeal restoration,
small restoration efforts are not necessarily worthless or ineffective.
Success in recreating a self-sustaining ecosystem is more likely, however, when the restoration is planned within the context of the target
ecosystem's larger landscape. Nonetheless, any shift of a damaged
ecosystem to a superior ecological condition is preferable to allowing
the system to remain damaged or to suffer further degradation. Restoration efforts should not be postponed by those using the complexity of a landscape-scale restoration as a rationale for inaction.

REFERENCES AND RECOMMENDED READING

Brown, J. H., and B. A. Maurer. 1989. Macroecology: The division of food and space
    among species on continents. Science 243:1145-1150.
Cairns, J., Jr. 1979. Academic blocks to assessing environmental impact of water sup-
    ply alternatives. Pp. 77-79 in A. M. Blackburn, ed., The Thames Potomac Seminars.
    Interstate Commission on the Potomac River Basin, Bethesda, Md.
Cairns, J., Jr. 1990. Gauging the cumulative effects of developmental activities on com-
    plex ecosystems. Pp. 239-256 in J. G. Gosselink, L. C. Lee, and T. A. Muir, eds.,
    Ecological Processes and Cumulative Impacts: Illustrated by Bottomland Hardwood
    Wetland Ecosystems. Lewis Publishers, Inc., Chelsea, Mich.
Cairns, J., Jr., and T. V. Crawford, eds. 1991. Integrated Environmental Management.
    Lewis Publishers, Chelsea, Mich. 214 pp.
Dale, V. H., R. H. Gardner, and M. G. Turner, guest eds. 1989. Predicting across scales:
    Theory development and testing. Landscape Ecol. 3:147-252.
Forman, R. T. T., and M. Godron. 1986. Landscape Ecology. John Wiley & Sons, New
    York.
Gosselink, J. G., and L. E. Lee. 1989. Cumulative impact assessment in bottomland
    hardwood forests. Wetlands 9:83-174.
Gosselink, J. G., G. P. Shaffer, L. C. Lee, D. M. Burdick, D. L. Childers, N. C. Leibowitz,
    S. C. Hamilton, R. Boumann, D. Cushman, S. Fields, M. Koch, J. M. Visser. 1990.
    Landscape Conservation in a Forested Wetland Watershed. Bioscience 40:588-600.
Hill, R. D., and E. C. Grim. 1977. Environmental factors in surface mine recovery. Pp.
    290-302 in J. Cairns, Jr., K. L. Dickson, and E. E. Herricks, eds., Recovery and
    Restoration of Damaged Ecosystems. University Press of Virginia, Charlottesville,
    Va.
Holling, C. S. 1978. Adaptive Environmental Assessment and Management. John
    Wiley & Sons, New York.
Johnston, C. A., N. E. Detenbeck, and G. J. Niemi. 1990. The cumulative effect of
    wetlands on stream water quality and quantity. A landscape approach. Biogeo-
    chemistry 10:105-141.
Karr, J. R. 1991. Landscapes and ecosystem management. In K. C. Kim and G. L.
    Storm, eds., Biodiversity and Landscapes: Human Challenges for Conservation in
    the Changing World. Center for Biodiversity Research, Pennsylvania State Univer-
    sity, University Park, Pa.
Knight, D. H., and L. L. Wallace. 1989. The Yellowstone fires: Issues in landscape
    ecology. BioScience 39(10)700-706.

Magnuson, J. J., B. J. Benson, and T. K. Kratz. 1990. Temporal coherence in the limnology of a suite of lakes in Wisconsin. U.S.A. Freshwater Biol. 23:145-159.

Maguire, L. 1988. Decision analysis: An integrated approach to ecosystem exploitation and rehabilitation decisions. Pp. 105-122 in J. Cairns, Jr., ed., Rehabilitation of Damaged Ecosystems. Vol. II. CRC Press, Boca Raton, Fla.

Matson, P. A., and S. R. Carpenter, eds. 1990. Special feature on analysis of ecological response to large-scale perturbations. Ecology 71:2037-2068.

Naiman, R. J., C. A. Johnston, and J. C. Kelley. 1988. Alteration of North American streams by beaver. BioScience 38:753-763.

National Research Council (NRC). 1991. Opportunities in the Hydrologic Sciences. National Academy Press, Washington, D.C.

National Research Council (NRC) and Royal Society of Canada (RSC). 1985. The Great Lakes Water Quality Agreement: An Evolving Instrument for Ecosystem Management. National Academy Press, Washington, D.C.

Naveh, Z., and A. S. Lieberman. 1984. Landscape ecology: Theory and ecological processes in the time and space of farmland mosaics. Pp. 121-133 in I. S. Jonneveld and R. T. T. Forman, eds., Changing Landscapes: An Ecological Perspective. Springer-Verlag, New York.

Risser, P. G., J. R. Karr, and R. T. T. Forman. 1984. Landscape ecology: Directions and Approaches. Special Publication No. 2. Illinois Natural History Survey, Champaign, Ill.

Tonn, W. M., and J. J. Magnuson. 1982. Patterns on the species composition and richness of fish assemblages in northern Wisconsin lakes. Ecology 63:1149-1166.

Turner, M. G., ed. 1987. Landscape Heterogeneity and Disturbance. Springer-Verlag, New York.

Walters, C. 1986. Adaptive Management of Renewable Resources. Macmillan, New York. 374 pp.

8

A National Restoration Strategy: Basic Elements and Related Recommendations

INTRODUCTION

Although human management of aquatic ecosystems has yielded many benefits, the ability of the nation's rivers, lakes, and wetlands to serve contemporary social objectives has been of increasing public concern. Massive investments in wastewater treatment have made some improvements in water quality. However, there have been declines in fish populations, waterfowl numbers, species diversity, and other indicators of an aquatic ecosystem's sustainability. Indeed, some question the ability to continue to realize the flood protection, water supply, and other benefits aggressively pursued in past decades. Recognition of these trends has made environmental restoration a central concern for the next decade.

Restoration ecology represents an acknowledged shift in what society demands from its aquatic ecosystems: more environmental services are expected. Restoration also is viewed as a way to gain some of the traditional economic services of reduced flood hazard and reliable water supply. Achieving restoration requires planning on an aquatic ecosystem basis and emphasizing the interdependence of hydrologic, biologic, and chemical processes. Restoration management, which seeks to make the structure of an existing aquatic ecosystem better conform with some prior condition, requires taking one or more of the following actions: (1) reestablishing flooding and flow regimes, and restoring wetlands and riparian areas; (2) reducing the delivery of sediments and chemical contaminants, and initiating their removal from the waters and sediments; and (3) revegetating areas and reintroducing native species. Management may mean manipulation of the structure of the existing aquatic ecosystem, an engineering problem illustrated in the case of the Kissimmee River in Florida (see case study, Appendix A).

It is essential to place physical restoration of lakes, wetlands, and rivers on a par with the current commitments to chemical and biological restoration activities. Physical restoration of aquatic ecosystems is in many cases the most cost-effective strategy for meeting the increasing public demand for certain kinds of services, including clean water without excessive loadings of sediments and nutrients; populations of fish, ducks, wading birds, shorebirds, and other wildlife; contact and noncontact aquatic recreation; and flood control and natural landscape amenities.

However, the current structure of programs and policies is fragmented and may not adequately approach restoration as a challenge requiring management of a whole system. Management responsibilities and disciplinary expertise are separated according to components of the aquatic ecosystem. Agencies and scientists focus separately on wetlands, river hydrology, or wastewater management, or best management practices for nonpoint source control. Rarely is the focus on aquatic ecosystems as a unit.

The focus during the early years of this century on the river basin as a unit of planning needs to be reinvented with new goals and new approaches to make it work more effectively than it did in the past. In making this effort the decline in federal domination of water management must be acknowledged. Indeed, there is no need to reestablish this domination: leadership can now be shared with the states. However, federal programs have been significant determinants of past modifications to aquatic ecosystems, and this past requires a continuing federal presence and further reforms at the federal level as the new era of restoration is approached.

In his budget message for 1991, President Bush stated,

> [T]oday, a consensus is emerging in our society: investments in maintaining and restoring the health of the environment can now be seen as responsible investments for the future.

This presidential sentiment demands a national, but not solely federal, aquatic ecosystem restoration strategy to meet the changing social demands on aquatic resources. Of course, restoration is not warranted simply because it recreates a historic condition. At the most general level, restoration is justified whenever restored ecosystem processes serve socially desired goals for human or ecologic functions and services. In some cases these goals may be achieved only by restoration of an aquatic ecosystem's natural processes. In other cases, using the natural processes of aquatic ecosystems may be the least costly way to meet these goals. In specific instances, the justifiable intent of restoration must be discovered through planning processes that define and evaluate restoration alternatives and then execute restoration projects stressing an integrated landscape perspective.

For the next decade, new policies and program approaches are needed to set restoration priorities and to implement that integrated perspective. Such a strategy should be designed to reflect consideration of the definition of restoration explained in Chapter 1, a systems perspective, and the state of the scientific knowledge as reported in Chapters 4 through 6. Of equal importance, organizational approaches to implementation must be consistent with the substantial policy changes already under way and must take into account emerging federal-state relationships and current financial conditions. The challenge is to direct the multiplicity of programs of government and nongovernment entities toward aquatic ecosystem restoration.

The call for a national restoration strategy may imply a call for federal emphasis with large increases in federal spending on a new program. This is not the intent. Of course, restoration of aquatic ecosystems may, in some instances, be very expensive, particularly where fills or dams are to be removed or other major physical alterations are needed as part of the restoration process. In such cases it may be necessary to have federal leadership and a combination of financial resources from all levels of government. However, it also must be recognized that watershed water use and land use activities are under the primary control of private landowners and of local and state governments, not the federal government.

For this reason, many restoration programs will build on the interests, opportunities, and cost-sharing possibilities provided by private landowners, local governments, and states. For example, most lake restoration efforts to date have been undertaken at the initiative of particular citizen groups (e.g., lake improvement districts) or by local or state agencies interested in solving a problem on a particular lake. Most stream restoration efforts to date have been undertaken and financed at least in part by citizen or local stream protection and restoration coalitions. Local groups have also often provided much of the labor. Many wetland restoration efforts have been initiated by duck clubs, land trusts, waterfront renewal organizations, or local or state agencies (see case studies, Appendix A).

In some instances, restoration of a river segment or wetland can be undertaken at relatively modest cost if undertaken "opportunistically" as part of a broader activity. For example, it is often prohibitively expensive to clear a riparian zone of houses and to reestablish natural vegetation and contours if the single goal is restoration of the system. However, restoration may be possible after a severe flood event in which houses have been destroyed or damaged, landowners wish to move from the area, and disaster and flood insurance benefits are available to pay for much if not all of the costs of relocation and restoration. Similarly, it may be prohibitively expensive to restore urban riverine wetlands along a creek if this were undertaken as a separate program, but it might be possible at very low additional cost if undertaken as part of an effort to upgrade sanitary sewers, which are often buried beneath small creeks. Regrading of the entire creek becomes necessary as the sewers are dug out and replaced.

Large restoration projects require the sort of subwatershed- or watershed-level planning often not possible for individual, piecemeal projects, as well as a selection process for projects to optimize use of limited funds. The committee recognizes the important and innovative efforts of states across the nation to restore aquatic ecosystems —ranging from the California Coastal Conservancy, to the Reinvest in Minnesota Program, to the coordinated state efforts to restore large systems such as the Great Lakes and the Chesapeake Bay. However, because much of the modification to aquatic ecosystems has been directed by federal programs or in response to federal incentives, a federal emphasis has been given to the recommendations that follow.

The central purpose of this chapter is to present options to the federal government to take the lead in stressing this larger context. These options are proposed as building blocks for a national aquatic ecosystem restoration strategy but are developed only to the extent that they provide direction for innovation; details of program design would be developed during implementation. These options have been organized into four elements considered essential to a national restoration strategy:

1. national restoration goals;
2. principles for priority setting and decision making;
3. policy and program redesign for federal agencies; and
4. innovation in financing and in use of land and water markets.

NATIONAL RESTORATION GOALS

1. A National Restoration Strategy Should Be Directed Toward Broad-Based and Measurable Goals

Although restoration goals should be subject to revision as new knowledge becomes available, the definition of goals does provide a necessary reference point for evaluating policies and programs. A current example of how the statement of a goal can direct policy and programs is the no-net-loss of wetlands goal stated by President Bush and now included in recent legislation—Water Resources Development Act of 1990 (P.L. 99-662). Ideally, goals should be established in relation to aquatic ecosystem processes and to the social and ecological values that are achieved as those processes are restored. However, as a practical matter, goals will be more useful in policy design if they are stated in terms of the aquatic ecosystem components used in this study—lakes, rivers and streams, and wetlands—as is the case with the no-net-loss of wetlands goal.

As stated earlier in this chapter, a call for a national restoration strategy and the ability to reach the recommended goals may, in some instances, be very expensive and include economic impacts, particularly where major physical alterations are needed as part of the restoration process. In such cases it may be necessary to have federal leadership and a combination of financial resources from all levels of government. For each of these aquatic ecosystem components, both near-term and longer-term goals are suggested. The cost of achieving these goals needs to be constantly examined and justified in light of the willingness of society to bear such costs.

• Inland and coastal wetlands should be restored at a rate that offsets any further loss of wetlands and contributes to an overall gain of 10 million wetland acres by the year 2010, largely through reconversion of crop- and pastureland and modification of existing water-control structures. This represents a tenfold increase in the wetlands restoration target included in the Wetlands Reserve Program of the 1990 Farm Bill. This number represents less than 10 percent of the total number of acres of wetlands lost in the last 200 years. The committee further recommends that this acreage be expanded in the long term to restore more of the approximately 117 million acres of the wetlands that have been lost in the United States over the past 200 years.

The 10 million wetland acres specified in the restoration goal might be allocated to ecological regions in proportions equal to those regions' shares of the nation's original wetland acres. Within regions, restoration efforts should concentrate on areas identified as environmentally sensitive (e.g., floodplain and coastal zones), ecologically critical (e.g., habitat for endangered and threatened species), and socially important (e.g., recreational corridors).

• The committee further recommends that a total of 400,000 miles of streams and rivers be restored within the next 20 years. Initial efforts should concentrate on those reaches that can best be protected from future disturbance and therefore need some isolation from point and nonpoint source discharges and from flood control and water supply reservoirs. The recommended magnitude of restoration represents approximately 12 percent of the 3.2 million miles of streams and rivers in the United States, and it is recommended because it is comparable to the miles of streams and rivers affected by point source and urban runoff (EPA, 1990). Although control of point source discharge and nonpoint runoff must be part of this strategy, physical restoration of riparian zones and restoration of river hydrology (fluctuating flows) need to be given priority if successful restoration is to occur. It is highly probable that this will require relaxing the engineering controls on some rivers and reducing the consumptive use of water, primarily in irrigated agriculture, in some cases.

• Excluding the Great Lakes, as well as flood control and water supply reservoirs, 1 million acres of lakes should be restored in association with wetland and river restoration by the year 2000, increasing to 2 million acres in the long term. The long-term lake restoration goal includes only one-half of the lake acreage that is now identified in federal reports as being distressed.

Because of the high degree of interdependence of these aquatic ecosystem components, the best results can be achieved if restoration efforts are coordinated and linked. The restoration initiative proposed here should be undertaken in accordance with the definition, criteria, and procedures described in Chapters 1 through 7. Thus, the restoration should be undertaken on the appropriate landscape scale and should extend to the management of hydrology, water quality, and the plants and animals in the restored systems.

2. A National Aquatic Ecosystem Assessment Process Should Monitor the Achievement of the Nation' Goals for Wetlands, Rivers, and Lakes

To devise effective physical restoration programs for different kinds of aquatic ecosystems, it would be very useful for Congress, federal and state agencies, and the public to have up-to-date assessments regarding the physical, chemical, and biological status of the nations' wetlands, rivers, and lakes in different ecoregions. Federal agencies such as the U.S. Environmental Protection Agency (EPA), the U.S. Geological Survey (USGS), and the U.S. Fish and Wildlife Service should devise statistical profiles that can provide an up-to-date picture of what is happening chemically, physically, and biologically to aquatic ecosystems in the country. Such environmental monitoring programs have already begun and include the USGS National Water Quality Assessment Program (NAWQA) and EPA's Environmental Monitoring and Assessment Program (EMAP). In addition, new remote sensing techniques such as those being developed by NASA's proposed Earth Observing System (EOS) have potential value for precisely the kind of watershed and regional scale restoration programs advocated in this report. Such remote sensing may facilitate the assessment and comparison of environmental quality indicators in wide areas of the country. These assessments may eventually provide a picture of the usefulness and effectiveness of ongoing programs designed to achieve physical restoration and reductions in chemical pollution in aquatic ecosystems.

PRINCIPLES FOR PRIORITY SETTING AND DECISION MAKING

3. Policies and Programs for Aquatic Ecosystem Restoration Should Emphasize a Landscape Perspective

It is possible to isolate components of the aquatic ecosystem—lakes, rivers, wetlands, and uplands—as this report does, and to describe restoration techniques for these separate components. In addition, it is likely that government agency missions will continue to be organized around components of an ecosystem and that the disciplinary expertise needed for restoration will remain fragmented along similar lines. Finally, many restoration actions will be opportunistic and limited to specific components of the aquatic ecosystem. The technical success of aquatic ecosystem restoration requires integrated consideration of the various structural components of the watershed in order to reestablish the matrix of chemical, hydrologic, and biological processes that have been compromised by human actions. Successful restoration will be achieved only if these individual actions and individual perspectives recognize the system (ecoregion and more immediate landscape) within which the action is taking place. The landscape approach or perspective (see Glossary) to aquatic ecosystem management is not a new idea. The arguments for watershed, or river basin, planning for water resources have long been made. What is different is the new goals to be met by these planning efforts and the attention to the land, wetlands, and riparian areas as an integral element for achieving these new goals.

4. Restoration Policies and Individual Restoration Projects Should Be Designed and Executed According to the Principles of Adaptive Planning and Management

Ten years ago Gilbert White of the University of Colorado at Boulder noted (White, 1980),

> [T]he sobering prospect is that most of the major public decisions about resource use and environmental management will be made in the face of large uncertainty deriving from ignorance of physical and biological systems and from evolving techniques and social values.

This statement, which is supported by the historical breakdown of expert-based watershed planning processes for water development projects, is truer today. As more is learned about aquatic ecosystems and the social context within which they exist, we realize how much more there is to know.

Adaptive planning and management involve a decision-making process based on trial, monitoring, and feedback. Rather than developing a fixed goal and an inflexible plan to achieve the goal, adaptive management recognizes the imperfect knowledge of interdependencies existing within and among natural and social systems, which requires plans to be modified as technical knowledge improves and social preferences change. In effect, adaptive planning and management constitute a learn-by-doing approach to decision making.

Central to the success of adaptive planning and management are surveillance and monitoring of restoration policy, programs, and individual projects. Therefore, more than simply observing a restoration project (surveillance) and measuring the effects of restoration in relation to specific goals (monitoring) is required. For adaptive management to succeed, the knowledge gained must be translated into restoration policy and program redesign over time and must be shared across restoration programs at all levels of government. Also, given that policy and program design is based on more than technical information, surveillance and monitoring of restoration projects would include gathering and interpreting economic and organizational information.

An example of an adaptive management process is the Chesapeake Bay nutrient management strategy. The governments in the region initially established a goal to reduce nutrient loadings to the bay by 40 percent and agreed to an approach to achieve the goal. They also committed themselves to continuous study of the goal itself and of the cost and effectiveness of the means employed to attain it. As a result, both goals and approaches are subject to revision over time.

5. Evaluation and Ranking of Restoration Alternatives Should Be Based on an Assessment of Opportunity Cost Rather than on Traditional Benefit-Cost Analysis

The challenge in restoration management is to evaluate trade-offs not only between restoration and the current state of the aquatic ecosystem, but also between alternative approaches to restoration. No computational procedure by itself will establish either how far restoration should proceed or the relative priorities for funding alternative restoration efforts. Plans for a restoration and priorities for alternative restorations will depend on the current condition of a particular aquatic ecosystem, the knowledge base for restoration of that system, and the values gained and lost as a result of the restoration.

Because different structures and functions of an aquatic ecosystem yield different values, choosing whether and how to restore amounts to choosing one set of values over another. The appropriate basis for defining values is a central analytical question. The basis for value may be expressions of individual preferences (the basis for economic value) or expressions of collective preferences (social norms often expressed in multiple forms of collective action). Both these perspectives envision natural systems as having value in relation to human preferences for the environment, the economy, and the society. That is essential, because the decisions on the values that will be served by restoration are ultimately decisions made through human institutions.

Computation of values is often associated with benefit-cost analysis. For the benefit-cost analyst, value is appropriately expressed through individual preferences that are measured in terms of monetary equivalence. Success in such measurement efforts has been achieved in specific instances, but widespread application of the measurement approaches has not occurred. This limited use represents in part the experimental nature of the valuation approaches and in part a lack of agreement on the philosophical bases for assigning such values. Of even greater concern is that benefit-cost analysis requires a static view of human preferences (for the computations to be accomplished) that conflicts with the adaptive planning and management approach of Recommendation 4.

In lieu of benefit-cost analysis, the committee proposes a decision-making approach based on opportunity cost. Within the opportunity cost framework, the correct answer to the question, How much restoration is enough? emerges from legitimate social choice processes within governments that will determine the degree of restoration desirable. Confronting the decision process with cost information elicits "values" from that process. Continually questioning the value of restoration by asking whether an action is "worth" its cost is the most practical way to decide how much restoration is enough. Costs include both direct (life cycle) financial outlays by government and individuals, and benefits derived from the existing state of the aquatic ecosystem that would be forgone if the restoration were done. An example of an opportunity cost approach is the one taken by the U.S. Army Corps of Engineers (COE) in determining the extent of justifiable mitigation for environmental damages done by a water development project. The COE mitigation analysis prohibits the use of solely economic measures of environmental values. Instead the process requires the establishment of alternative mitigation goals and approaches to achieve these goals. Costs of the alternative strategies are identified, and a justifiable level of mitigation is chosen in recognition of the incremental costs of increasing mitigation levels and as part of a negotiation process with affected interests and other federal agencies.

Opportunity cost analysis is quite different from traditional benefit-cost analysis, and its implementation may be organizationally complex. Within the ecosystem to be considered are the simultaneous interdependencies among political decision points, social value systems, and the land-water resource complex. An opportunity cost analysis accepts a human-based determination of value but looks to collective action to define values achieved by restoration. It requires continual questioning of the value of a restoration by asking whether an action is "worth" its cost. This is the most practical way to decide how much restoration is enough. The interested parties who participate in these social choice processes must be informed in making the decisions—informed about the effectiveness of the technical options for restoration, the aquatic ecosystem functions that might be restored by the options, and the opportunity costs of different levels and strategies of restoration.

POLICY AND PROGRAM REDESIGN
FOR FEDERAL AGENCIES

6. The Definition of Restoration Used Throughout This Report Should Be Incorporated Into All Appropriate Federal Legislation

Restoration is defined as the return of an ecosystem to a close approximation of its condition prior to disturbance. The responsibility for attaining the goals of restoration falls on the multiple agencies of the federal government, as well as on nonfederal governments. Specific legislative language should define restoration uniformly in terms of the landscape and aquatic ecosystem process focus advocated in this report. Emphasis should then be put on how the particular historical mission of the appropriate agencies can serve this concept of restoration. For example, the Water Resources Development Act of 1986 and 1990 use terms such as *environmental improvement* and *environmental enhancement*, as well as *restoration*, to describe new missions and authorities for COE. Nowhere are these terms adequately defined to emphasize biological, physical, and chemical processes of aquatic ecosystems. As a result, in its own interpretations of restoration, COE has equated restoration with fish and wildlife habitat, rather than with hydrologic processes—COE's area of expertise. As another example, the Clean Water Act of 1977 includes in its statement of objectives the physical restoration of aquatic ecosystems. However, the act does not give EPA or any other agency the authority to develop a restoration program. The reauthorization of the Clean Water Act of 1977 in this Congress should request that a lead federal agency (see Recommendation 9) prepare a report on major aquatic ecosystems that should be restored, with a description of the national ecological and social benefits of such restoration and an identification of the federal programs that could be retargeted to provide federal financial and technical assistance for such restoration as part of a unified national program for aquatic ecosystem restoration (see Recommendation 8). These examples illustrate the necessity for the mandates of all appropriate federal agencies, including COE and EPA, as well as agencies in the Departments of the Interior and Agriculture, to be reviewed and rewritten as needed to match each agency's own programs to the broader national restoration strategy.

7. A National Restoration Strategy Should Allocate Leadership to the Federal Government for Landscape Restorations of National Significance and Should Rely on Nonfederal and Federal Units of Government to Coordinate Restoration Programs in Local Areas

Institutional arrangements should be appropriate to the scale of the restoration action. Successful restoration at the landscape level will depend on program coordination among those responsible for management decisions on the separate aquatic ecosystem components. Major federal leadership in aquatic ecosystem restoration projects is warranted in at least three instances: where such projects are the most cost-effective approach to achieving flood control or other traditional water development missions; where aquatic ecosystems to be restored have regional, national, or even international biological significance; and where federal water development projects have historically been a dominant factor in degradation of an aquatic ecosystem. Otherwise, the states are in the best position to coordinate between local (grass roots) restoration needs and federal programs that may serve those needs. Given the emerging role of the states in water management, it is logical that a significant organizational responsibility for integrating restoration programs at the landscape level fall to the states.

8. The Federal Government Should Initiate an Interagency and Intergovernmental Process to Develop a Unified National Strategy for Aquatic Ecosystem Restoration

Management responsibilities for the various components of an aquatic ecosystem—the lands, waters, and wetlands—are fragmented among multiple governmental organizations and between the public and private sectors. Individuals, groups, and government agencies with an interest in decisions about the use of resources have engaged in bargaining to resolve disagreement and reach decisions within the framework of law, executive order, and administrative regulations. Indeed, the history of river basin planning demonstrates that efforts to establish central control over the programs of independently funded agencies has not worked. Political allegiances, legislative mandates, and budgetary rules are all more powerful influences on the state agencies of government than are the directives of a central organization, such as a river basin authority.

State and federal governments can more effectively bring about coordination among their agencies by taking actions to improve interagency communication and agreement, a process supported by many of the recommendations made in this chapter. First, establishing common goals and precise legislative language should reduce conflict among agency missions. Second, requiring common decision criteria and analytical processes will help resolve analytical conflict. Third, organizational arrangements specifically developed for resolving agency and program conflicts can be defined in legislation, in executive order, in promulgation of administrative rules, or in mutually established memoranda of agreement or understanding.

Establishment of a "Unified National Program for Aquatic Ecosystem Restoration," modeled after the federal effort to develop a Unified National Program for Floodplain Management, is one approach that could be used to facilitate the needed communication process. Development of such a program for aquatic ecosystem restoration can become the focal point within the federal government for interagency discussion and program reform to take action on the goals and principles described in this chapter. It is particularly important to emphasize the planning and developing of programs that offer consistent, long-term support for restoration efforts.

9. The Development of a Unified National Program for Aquatic Ecosystem Restoration Should Be Facilitated, and Then Maintained, Under the Leadership of a Single Responsible Organizational Unit

Creation of a unified national program for aquatic ecosystem restoration requires improved interagency communication. Because the federal organizations necessary for an integrated approach to restoration currently are dispersed in numerous agencies, and will remain there, it is desirable to have an organizational unit that can develop a network for communication among agencies and develop guidelines for the identification, evaluation, and financing of restoration projects. Without showing favoritism toward any particular agency, the committee believes that a lead group must be designated with responsibility for achieving the improved communication necessary to the development of a unified national program.

Water resource policy coordinating bodies and authorities established in the past have had limited success. The history of river basin planning is traced in Chapter 2 of this report. In recent years, coordination was the basic purpose served by the Water Resources Council (WRC) and its reestablishment could be one option as a coordinating unit for restoration efforts. The WRC, which still exists under the Water Resources Planning Act of 1965 (P.L. 89-80), might be renewed and given the task of developing a unified program, but its weaknesses (described in Chapter 2) must be addressed. Another possibility is to reinvigorate and restructure the U.S. Council on Environmental Quality (CEQ), giving it the responsibility to develop a unified national program. The CEQ initially directed successfully the implementation of the National Environmental Policy Act and executive orders on floodplains and wetlands issued by President Carter. In the late 1970s, CEQ also served as the chair of a federal task force made up of COE, EPA, and the Department of the Interior that, together with the state of Louisiana, designed a nonchannelization water management plan for the Atchafalaya Basin. However, in the recent past, CEQ has not been as effective as envisioned at its creation.

The committee advocates correcting the weaknesses identified in past efforts at coordination. Uncertain budget support and limited authority are potential limitations. The history of the national floodplain management program makes clear that successful coordination can be achieved simply by agencies being willing to work together. However, the institutional and scientific complexity of restoration ecology demands formal designation of responsibility and adequate resources to execute that authority. There should also be a mechanism for nonfederal advisory groups to work with a federal restoration coordinating agency to assist in the overall direction and aid in coordination.

10. Current and Proposed Federal Programs Should Exploit Available Opportunities for Aquatic Ecosystem Restoration

Programs of many agencies of the federal government might be enlisted to achieve national aquatic ecosystem restoration goals. Indeed change is now occurring that will contribute to restoration. A comprehensive review of all federal programs, or the changes now under way, is not necessary here. Instead selected programs are described and recommendations for reform within these programs are offered, both as specific suggestions for those programs and as illustrations of the type of change that might be considered for programs that are not discussed.

CLEAN LAKES PROGRAM

The Clean Lakes Program has been a model of federal restoration efforts, especially with regard to its emphasis on causes of lake problems and on local and state financial participation. Although many lakes have been restored or improved as a result of the program, the need and demand for lake restoration far exceed its current resources, and its efforts have been seriously hampered by uneven agency and administration support. The Clean Lakes Program requires local and state sharing of the financial burden of lake restoration, but many agencies at this level do not have expertise in lake restoration.

The committee believes that the Clean Lakes Program should receive stable administrative support within EPA and increased funding from Congress. Its mandate should be broadened to include all aspects of lake ecosystems, including habitat restoration, elimination of undesirable species, and restoration of native species. The program should increase its research and development of effective tools for restoration and its role in guiding states in developing efficient and effective programs. Clean Lakes Program restorations should be coordinated with fisheries improvement programs, waterfowl enhancement programs, and restorations of wetland and stream ecosystems that interact with the target lake.

EPA CONSTRUCTION GRANTS PROGRAM

Current grant-in-aid programs, present throughout the federal government, should encourage aquatic ecosystem restoration. For example, the EPA construction grants program, although now modest by historical standards, might increase its flexibility to finance the restoration of wetlands as a means to accomplish wastewater treatment. Similarly, it may be desirable to allow use of such funds for control of nonpoint source pollution, if it can be demonstrated that nonpoint source controls are more effective and efficient than point source controls for meeting a water quality standard. The concept of making such trades in programs to help finance restoration is further discussed in Recommendation 16. Recent EPA decisions emphasizing storm water management should immediately prompt assessment of how storm water flows (both peak and volume) might be reduced by aquatic ecosystem restorations. Any federal funds (or regulations) directed to storm water management should allow for the possibility of using restoration as a method for managing storm water flow.

FLOOD INSURANCE PROGRAM

The disaster assistance programs that provide subsidized flood insurance premiums and disaster aid might also be used to enhance restoration and its benefits. These federal programs are constantly seeking ways to reduce repetitive flood insurance and disaster claims.

Revisions made to flood insurance and disaster aid programs that will evacuate from the floodplain those activities whose location is subsidized by these programs could help reduce future flood losses and help restore aquatic systems (NRC, 1990). Thus, it will be possible to move toward actuarially sound insurance premiums for the remaining properties and to minimize future disaster assistance payments. A modest increase in funds for such programs could, in many instances, allow the evacuated floodplain to be restored for environmental purposes.

Continued pressure to make flood insurance premiums meet the test of actuarial soundness is a start in this direction. This is a goal of the Federal Emergency Management Agency (FEMA), which will encourage evacuation from floodplain areas and then make these riparian areas available as sites for aquatic ecosystem restoration. A more proactive program under Section 1362 of the Flood Insurance Act of 1968 allows for acquiring structures with repeated damage and relocating them outside the floodplain. The criteria for this program might be expanded to permit relocations that restore flood storage and conveyance capacity, especially where such hydrologic restoration will contribute to aquatic ecosystem restoration on a larger scale. Indeed, such restoration may financially benefit a community by leading to reduced insurance premiums, because the community's risk rating for flood insurance is improved as high-risk properties are removed and the floodplain is restored. The FEMA "community rating system," which provides communities with lowered insurance premiums when structures are relocated out of the floodplain, is an example of a risk reduction strategy that could help support restoration.

Finally, disaster assistance legislation could be changed to finance restorations while also serving the primary intent of reducing disaster aid payments. At present, 15 percent of disaster funds can be used for public actions to reduce future losses, rather than simply compensate for past losses. Often modest structural or floodproofing actions or floodplain evacuations are undertaken. Authority should be given to allow restoration of upstream floodplain areas to modify peak flows and volumes, as one approach to reducing future losses, especially if such actions can also be shown to provide broader aquatic ecosystem restoration benefits.

FEDERAL WATER RESOURCE DEVELOPMENT

Reform of programs in the federal water project construction agencies might also serve aquatic ecosystem restoration. Expanded legislative authority and budget appropriations should be given to the federal water project construction agencies (Bureau of Reclamation, U.S. Army Corps of Engineers, Soil Conservation Service, and Tennessee Valley Authority) for restoration activities. Indeed, there has already been some movement in this direction. The Bureau of Reclamation has been positioning itself to manage natural resources, rather than project construction. The Water Resources Development Acts of 1986 and 1990 have directed COE to raise "environmental protection" to the same status as traditional agency missions. Recent COE budget guidance establishes "restoration" as a priority output. However, the legislative and administrative language is too focused on fish and wildlife. For example, the Great Lakes Fish and Wildlife Restoration Act of 1990 authorizes COE to undertake fish restoration activities. The role of water resource agencies in restoration needs to be expanded to the hydrologic structure of watersheds, with COE projects oriented to restoring hydrologic conditions, not fisheries or habitat alone.

In addition, the purposes of some federally funded and constructed water development projects may long since have been served, and they are aging. In cases where the economic and environmental costs of maintenance exceed the benefits, the planned phaseout or dismantling of project levees, dams, channel revetments, or other works— through active removal, reduced maintenance, or benign neglect— could provide major opportunities for aquatic ecosystem restoration. Congress should amend the Water Resources Development Act of 1986 and other acts, authorizing the federal water development agencies to identify such projects and plan for their deconstruction. Federal development funds that are saved, and other programs such as the Agricultural Wetland Reserve Program, could be used to acquire easements on private lands that could be affected hydrologically by such deconstruction. One possibility for deconstruction could be the locks and dams on the Missouri River. Based on their use for navigation, continued federal maintenance of those structures may not be justified economically, and the long-term environmental benefits of reestablishing major sections of the river to free-flowing status and restoring riparian wetland habitat could be great. The type of restoration under way for the Kissimmee River (restoring meanders and wetland habitat) is what is necessary, although additional effort is needed to extend the experience gained in that project. Both the construction agencies and other federal, or nonfederal entities should be aggressively seeking new restoration opportunities.

Grant-in-aid and water project construction agency programs are but two vehicles by which aquatic ecosystem restoration might be served within traditional agency missions. Other opportunities may be found within the landholding agencies of the U.S. Forest Service, the U.S. Fish and Wildlife Service, and the Bureau of Land Management. The list need not stop here; as part of the development of a unified national aquatic ecosystem restoration program, other opportunities will be identified.

NOAA WATER RESOURCES FORECASTING SERVICES (WARFS)

The National Oceanic and Atmospheric Administration (NOAA) Water Resources Forecasting Services (WARFS) initiative is being proposed to begin providing water resources and improved flood forecast services on a national basis. The WARFS initiative will provide urgent improvements in NOAA's hydrologic prediction services. The beginning infrastructure for WARFS is the current National Weather Service River Forecast System (NWSRFS). WARFS model and data improvements within NWSRFS will greatly benefit all scales of forecasting, bring badly needed improvements in flood warnings as well as longer-term forecast services.

WARFS will complement, provide support to, and benefit from the major NOAA thrusts in the climate and global change, coastal ocean, and environmental data areas. One of the earliest and most serious impacts from climate change will be on water resources. The operational focus of the WARFS initiative will provide information essential to assessing actual or potential climate change impacts on our nation's rivers and streams. Similarly, information from WARFS may be used to identify the hydrologic component to large-scale atmospheric models and to evaluate the validity of climate analyses and long-term experience and technology needed to enable the country to manage the consequences of global climate change. Equally strong relationships exist between the WARFS initiative and the Coast Ocean Program, especially as related to estuarine areas. Solutions to better management of estuarine ecosystems, including fisheries management, are critically dependent on accurate assessments of freshwater levels and inflows. In addition, WARFS extended forecasts will enable more accurate assessments of future water balances as they relate not only to quantities but to salinity and pollutant concentrations.

The advanced models, data integration techniques, and expanded historical and real-time hydrometeorological data bases, which will be implemented through WARFS, will provide a strong technological base for comprehensive water resources information to support the restoration of aquatic ecosystems. The historical and forecast information provided by WARFS will be essential to the planning, design, and operation of restoration projects. The NOAA Hydrologic Services Program, strengthened by WARFS and complemented by the activities of other agencies, is an ideal vehicle to provide national, authoritative, and reliable forecast information to support management of the nation's aquatic ecosystems.

INNOVATION IN FINANCING AND IN USE OF LAND AND WATER MARKETS

11. Congress Should Establish a National Aquatic Ecosystem Restoration Trust Fund

Perhaps the most significant obstacles to large-scale restoration are the absence of budget incentives for agency cooperation and the absence of adequate funds for executing restoration plans. One mechanism might be establishment of a national restoration trust fund which could reduce these obstacles. The trust fund would be a separate budget account, created by Congress and treated as other trust funds (for example, the Inland Waterways Trust Fund) are in the budget process. The design of a trust fund is complicated and needs to be monitored and reviewed to ensure that it is operating as originally intended. One way to ensure this is to require periodic reauthorization of the program and if the trust fund is not operating properly, it would cease to exist. Although the specific design of such a trust fund will be highly complex, a structural outline can be indicated.

The trust fund could be managed by a board appointed by the President and approved by Congress. This board would be responsible for disbursements from the fund for restoration projects that it approves as part of its restoration program. However, because expenditures for such restoration might be quite high, disbursements over a certain amount should be reviewed by the executive branch and approved by the Congress.

The trust fund, although national in scope, might be partitioned into 8 to 12 regions that cover the nation. The regional boundaries should have similar ecological properties. Projects would be selected for support within those regions. Applications for funding could be made by any group of federal and state agencies able to demonstrate that the funds would enhance their capacity to better integrate their existing missions and programs to reflect/incorporate the landscape perspective that is required for successful large-scale restoration. The trust fund's board would require a demonstration of the commitment of other resources to the plan before providing supplemental funds to the budgets of federal and state agencies. In this way the trust fund could add needed revenues to restoration efforts and create a budgetary incentive for the interagency cooperation necessary to achieve landscape-scale restoration.

In the current budget environment, the source of funds for a national restoration trust fund must be of concern. One possible source is general federal revenues that might be transferred to the fund. Initially it may be desirable to allocate to the fund an annual amount equal to 0.25 percent of the annual appropriation to all natural resource and environmental programs in the federal budget. This allocation could cease after 10 years as new revenue sources are phased in. These revenues would be distributed equally among the ecoregions.

Other funds could be collected and allocated to the regions in which the funds were raised. Electric power from federally financed hydroelectric dams could be auctioned to private power producers. The proceeds could be contributed to the trust fund to be used to pay for regional aquatic ecosystem restorations. A less drastic action might be to raise slightly the rates on federal power and dedicate the added revenues to restoration within the rate payer's region.

Awards granted as compensation for damage of natural resources are a particularly promising source of revenue for a national trust fund. Increasingly the courts are awarding damages to governments, which act as "trustees" for the natural resources that may have been damaged. At present, the sums awarded are expected to be sufficient to restore the resource to its predamage condition as well as to compensate users of the resource for the values forgone until the restoration becomes effective.

Finally, it might be possible over time to dedicate certain federal fees to a restoration trust fund. At present, user fees and user taxes, such as outer continental shelf leasing fees and taxes on sporting equipment, have been allocated to particular purposes. As these fees increase or as new fees are instituted (for example, fees for pollution permits might be instituted in the reauthorized Clean Water Act of 1977), it might be desirable to dedicate such fees to a restoration trust fund rather than to the general treasury.

12. There Should be Increased Reliance on Local Environmental Restoration Boards for Program Planning, Synthesis, and Leadership

At the substate level, environmental restoration boards should be formed and supported by trust fund financing under state statutes. These institutions would be created (or could be expansions of existing authorities) under state auspices and would be responsible for the planning process outlined earlier in this chapter, the alternative means of finance, and the use of resulting revenues. As a result, restoration strategies would at least be partially developed and implemented at the substate level. In the past, this special district-based approach has been applied across a range of functions from education to garbage collection to mosquito control to flood control. The federal government could facilitate the success of these institutions by making cash contributions to such districts, providing technical assistance if requested, and agreeing to comply as much as possible with each district's goals—recognizing the responsibilities of higher levels of government (following the example of the coastal zone management program).

13. Congress Should Allow States and Local Governments to Trade Federal Funds Designated for Development, Construction, Maintenance, and Major Repair of Water Projects and to Obtain Instead Funds for Aquatic Restoration Programs

The discussion of Recommendation 11 describes circumstances in which it makes economic sense for federal water development agencies not to maintain, repair, or rebuild a federal project. Yet Congress may be under political pressure to maintain or rebuild an aging project because such work creates jobs in the local community or because certain private lands would experience increased flooding. Without increasing total outlays, Congress could respond to these concerns by allowing states or local interests to receive some portion of the otherwise saved federal funds to pay for active deconstruction, aquatic ecosystem restoration, and where appropriate, purchase of flowage easements.

Consider the Arkansas River levee project. Local interests did not find it economically worthwhile to maintain these levees, and many levee sections were of questionable economic value. If federal water project funds could be dedicated for restoration, the political pressure to rebuild marginal structural projects such as the Arkansas River levees would be diminished. However, in Section 110 of the 1990 Water Resources Development Act, Congress instructed COE to rebuild the levees. Even under the new cost-sharing rules, the federal government will pay 75 percent of the cost of this rebuilding.

14. Congress Should Authorize Expansion of the Agricultural Wetland Reserve Program with Funds from Farm Program Cost Savings

Under the 1990 Farm Bill, the U.S. Department of Agriculture (USDA) is authorized to enter into long-term contracts with farmers to take former wetlands in agricultural use out of production and to allow those lands to be restored as wetlands. In exchange, the farmers receive annual payments. The 1990 Farm Bill limits the number of acres eligible for the program to 200,000 per year, with a maximum cap of 1 million acres. However, each acre of cropland taken out of production and restored as wetland will no longer be eligible for USDA program benefits. Congress should request that USDA investigate where and how an expansion of the agricultural Wetland Reserve Program (WRP) would result in savings in USDA farm program expenditures, and the saved funds should be reallocated to expanding the WRP beyond 1 million acres.

In addition, existing short-term agricultural set-aside programs, the Conservation Reserve Program (CRP), the Environmental Easement Program (EEP), and the Wetland Reserve Program of the Food, Agricultural, Conservation, and Trade Act of 1990 (P.L. 101-624) should be amended to ensure that riparian zones and floodplains of all kinds are eligible for inclusion in these programs along with wetlands.

15. The U.S. Government Should Encourage Water Pollution Credit Trading Programs to Finance Aquatic Ecosystem Restoration

Many lakes, rivers, wetlands, and estuaries are suffering from excessive loadings of nutrients. Prime sources of those nutrients may be discharges from municipal wastewater treatment plants and agricultural fertilizers. In some cases it may be cost-effective for municipalities operating sewage treatment plants or for adjacent landowners who would benefit directly from improved water quality to pay farmers to take land out of production, to pay for adoption of best management practices, or to pay to restore wetlands. Restored wetlands can provide important water quality benefits. Former wetlands adjacent to rivers, lakes, or estuaries that are restored as wetlands can contribute to reducing loadings of inorganic nutrients by recycling those nutrients as organic nutrients. The EPA and the states should allow contributors to point sources of pollution to substitute restoration of wetlands or other land use practices for point source control if it can be demonstrated that nonpoint source controls are less costly for meeting a water quality standard.

16. The Federal Government and State Governments Should Encourage the Trading of Water Rights to Promote Aquatic Ecosystem Restoration

Many aquatic ecosystems suffer from inadequate flows of water due to man-made diversions. This is true of both streams and lakes, such as Mono Lake, that depend on water inputs from tributary streams and also of wetland systems throughout the arid West. It is also true of large aquatic ecosystems, such as the Florida Everglades, that have experienced diversions of water for urban and agricultural use. In such areas, aquatic ecosystem restoration will be inhibited unless some of the water now in consumptive use is returned to the streams and lakes. Current users in these regions may be able to reduce their water consumption so that some of the "saved" water can then be rediverted to restoring affected aquatic ecosystems. The committee recommends that watershed planning efforts establish targets for returning water to the aquatic ecosystem, based on the planning principles described in Recommendations 3 to 5, and that these targets be used to stimulate total reductions in consumptive use.

Because of the wide variety of water law doctrines across the 50 states, it is not possible to make a recommendation that will apply in all instances. However, the committee does recommend that careful consideration be given to several points.

In many states, water rights transfers have become a tool used to reallocate water to more highly valued uses, thereby reducing the need to develop new sources (NRC, 1992). The committee encourages increased use of such water transfers as an effective means to avoid increases in diversions from aquatic ecosystems. The premise of such a transfer is that it benefits both the water purchaser and the seller. These combined gains represent a joint economic gain that encourages the transfer. States might consider either a monetary charge or a reservation of water for in-stream flow on these transfers as a means to restore some in-stream flow. For example, for each unit of water sold the state might require some percentage of the transferred water to be returned to in-stream use. Alternatively, an assessed monetary fee could be used to purchase water rights for restoring in-stream flow within the state. However, caution must be exercised to design a fee program that does not reduce the total or marginal gains from trade to the point that the water transfer is discouraged. Thus, the amount of the fee must be specific to each transfer and not be a uniform requirement.

In some instances it may be possible to obtain overall reductions in water use in a particular region or for a specific economic activity, with the amount saved (or some portion thereof) then returned to in-stream flow (NRC, 1992). In renewal of contracts for water delivery from federal or state water storage projects, it is possible to initiate a negotiation process in advance with all interested parties (see Recommendations 3 to 5) to establish some target for reduced deliveries. Following the model for sulfur oxide reductions now incorporated into the Clean Air Act, there may not need to be requirements for uniform reductions in water consumption among all users. Instead current rights holders could be allocated water consumption allowances that together equal the total water consumption target. These water use allowances would then be tradable to ensure that those who are able to most efficiently reduce water use should do so. In effect, such an approach encourages continued development of markets for water rights, but allows those markets to trade an amount of water that is constrained by a desired in-stream flow.

Where there is a general abundance of water relative to use, as in many humid eastern states, the creation of water markets would not be appropriate because trades would be too infrequent to justify the setup costs and administrative costs of such a market. With the exception of some parts of Florida, where markets may be more developed, this condition describes most eastern states. In these cases, the committee suggests establishment of special administrative procedures that facilitate negotiation over water rights conflicts (usually for urban water supply) and ensure protection of in-stream flows (Collins, 1990). However, the focus should be on protection of future flows.

17. Federal and State Agencies Should Cooperatively Design Landowner-Financed Regional Restoration Projects

Under the Section 404 program, EPA and COE follow a sequencing procedure that requires permit applicants first to avoid, through alternatives, then to minimize, and finally to compensate for damage permitted to wetlands within aquatic ecosystems. Compensation in this approach typically means that wetland acres and functions that are destroyed must be compensated for by "equivalent" creation or restoration of comparable wetland acres. At times the replacement ratio exceeds 1 to 1, to account for uncertainties involved in restoration or creation. However, in some cases it may be possible to modify the Section 404 permit program and constraints on development, such as density limits and zoning regulations, to achieve restoration of critical components with the aquatic ecosystem. This restoration could include large numbers of wetlands designated in accordance with a system restoration plan developed cooperatively by a municipality or state, COE, and EPA, and could be paid for through the permit processes.

Consider a case in which degraded wetlands, streams, or lakes are in or adjacent to an urban area. In such areas, the right to fill a wetland or otherwise intrude on a body of water may have enormous economic value to a developer. Under Section 404 sequencing procedures, such a developer may not be eligible to obtain a permit to fill that acreage. At the same time there may be other limits on the development, such as density limits in the zoning in the area. However, a public authority such as the Hackensack Meadowlands Development Commission (see Appendix A) or the state of New Jersey could develop a plan for restoring the large aquatic ecosystem of which the wetlands are a part. Under a revised permit approach the developer could receive the permit if there was a requirement that the developer make a substantial contribution to the achievement of the goals of the restoration plan. In one approach the public authority might ask developers to bid for the right to develop on degraded wetlands, for example. The highest monetary bid with the lowest environmental cost would be awarded development rights.

The lowest environmental cost means that the developer does not intrude on parts of the aquatic ecosystem that are not degraded and that minimum added damage to degraded parts of the system is achieved. As part of the bid process, increases in zoning density could be requested to minimize the wetlands affected by the development proposal. With funds from the bid, it would be possible to obtain private financing of a large-scale restoration program at a low cost to the environment. If no acceptable bid were received, no permit would be issued.

Where there are few potential bidders, an auction may not be possible. The public authority would then negotiate with single developers to establish a payment that might be offered in return for granting the permit. Negotiation over development proposals has a long history and is increasingly used in wetland management. The Hartz Mountain Project in Hackensack Meadowlands (see case study, Appendix A) was approved after a negotiation in which the developer's offer to restore a large area of wetlands to high-value marsh was accepted as part of allowing the development to proceed in a small area of low-value marsh.

However, in this and in other cases, the amount, type, and location of restoration initially proposed by the permit applicant are designed to replace only functions lost to development and are not part of the negotiation with the management agency or part of a regionwide restoration plan. Also, unlike the proposal offered in this example, the current permit process still requires avoidance of damage to wetlands as a first consideration and does not explore the possible revenues to be gained from a modified permit system, which can be spent to increase aquatic ecosystem functions. The current approach to planning permits may forgo both high development values and aquatic ecosystem restoration opportunities.

Conditions basic to operation of such a program (whether by auction or negotiation) are the presence of an aquatic ecosystem to be restored (not created), a scientifically sound restoration program approved by EPA and COE, values high enough so that restoration can reasonably be expected to be achieved at low environmental costs, and an organizational arrangement to facilitate the auction or, if needed, a negotiation process. Although this approach may not be applicable in all areas, it should be considered where very high development values are possible and where there are substantial opportunities for wetland restoration that may not be realized in any other way.

SUMMARY

The recommendations of this chapter were organized according to four elements that should comprise a national aquatic system restoration strategy: (1) goal setting, (2) priority setting and decision-making principles, (3) redesign of federal policies and programs, and (4) innovation in financing and use of markets. In addition, most of the recommendations suggest changes in federal programs. Even with the attention to federal programs, specific recommendations are not sufficiently comprehensive in detail or in number to define a complete national plan of action.

A central theme of this report is that there is much to learn about the physical, chemical, and biological restoration of aquatic ecosystems. The committee has continually stressed the need for adaptive management as particular aquatic ecosystems are targeted for restoration. Adaptation is as essential for the formulation of policy and programs as it is for the restoration of any specific aquatic ecosystem.

All governments must be willing to try new policy ideas, to assess their outcomes carefully, and to modify those ideas over time.

The recommendations in this chapter should be seen as the beginning of a search for policy innovation. As scientists and citizens learn more about how to put aquatic ecosystems back together—to some extent by trying new approaches—they must simultaneously learn how to make policies and programs to serve such ends.

REFERENCES

Clean Water Act of 1977. P.L. 95-217, Dec. 27, 1977, 91 Stat. 1566.

Collins, R. C. 1990. Sharing the pain: Mediating instream flow legislation in Virginia. Rivers 1(2):126-137.

Great Lakes Fish and Wildlife Restoration Act of 1990. P.L. 101-537, Nov. 8, 1990, 104 Stat. 2370.

National Research Council (NRC). 1990. Managing Coastal Erosion. National Academy Press, Washington, D.C. 182 pp.

National Research Council (NRC). 1992. Water Transfers in the West: Efficiency, Equity, and the Environment. National Academy Press, Washington, D.C.

Water Resources Development Act of 1986. P.L. 99-662, Nov. 17, 1985, 100 Stat. 4082.

Water Resources Development Act of 1990. P.L. 101-640, Nov. 28, 1990, 104 Stat. 4604.

Water Resources Planning Act of 1965. P.L. 89-80.

White, G. F. 1980. Environment. Science 209:183-190.

APPENDIXES

Appendix A

Restoration Case Studies

The following case studies were written by several members of the Committee on Restoration of Aquatic Ecosystems, a National Research Council (NRC) consultant, and NRC staff to give the reader more details of specific restoration efforts: Lake Michigan, Lake Apopka, the Atchafalaya Basin, the Upper Mississippi River, the Illinois River, the Willamette River, the Mattole River Watershed, the Merrimack River, the Blanco River, the Kissimmee Riverine-Floodplain System, the Bottomland Hardwood Wetland Restoration in the Mississippi Drainage, the Prairie Potholes, and the Hackensack River Meadowlands. The committee made site visits to the Kissimmee River Restoration Project, the Blanco River Restoration, the Prairie potholes regions in Minnesota, and the Bottomland hardwood wetlands in the Mississippi drainage.

Several case studies show that citizen participation (through either private citizen groups or public interest groups) in restoration activity was instrumental in beginning and continuing the restoration effort (i.e., Merrimack River, Upper Mississippi River, Hackensack Meadowlands, and Illinois River). Other case studies feature cooperative participation by citizens, industry, and the state, local, and federal governments working together to return an aquatic ecosystem to a superior condition, such as the Merrimack River, the Kissimmee River, and the Atchafalaya River. One case study (Lake Apopka) shows the problems that can occur over many years to render a restoration activity ineffective.

* * * * *
LAKES
* *' * * *

LAKE MICHIGAN

Claire L. Schelske and Stephen R. Carpenter

General Description

Restoration measures have been instituted as the result of a series of environmental problems that have occurred in Lake Michigan (Figure A.1) since the drainage basin was settled by Europeans. In the late nineteenth century, drinking water for the city of Chicago was contaminated with human and other wastes. In 1900, sewage was diverted from the lake to the Mississippi River drainage via the newly constructed Chicago Sanitary and Ship Canal. The diversion controlled waterborne vectors for diseases, including typhoid and cholera. More recently, water quality problems in the lake have resulted from accelerated nutrient enrichment. The fisheries of the lake have also been affected by changes that followed European settlement. Populations of commercially important fish have been eliminated sequentially from the combined effects of environmental degradation, overfishing, and eutrophication (Christie, 1974). In addition, the fish community has been altered by introductions and invasions of exotic species. Potentially toxic chlorinated hydrocarbons, which have been manufactured in the last four or five decades, have entered the food chain and now pose serious problems for the fish community.

Historical management strategies for Lake Michigan illustrate some of the consequences of attempts to restore degraded water quality and fishery resources. The main lesson is that management is imperfect and can remediate only some problems. Therefore, whenever possible, we should try to preserve natural systems and avoid having to restore them. Five examples can be cited. First, seriously contaminated water supplies were restored at great expense in 1900 by diverting sewage from Lake Michigan to a river basin (see Illinois River case study, Appendix A). (The cost of constructing the Chicago Sanitary and Ship Canal was $36 million; this was the largest channelization project prior to construction of the Panama Canal.)

Second, although problems of nutrient enrichment were alleviated initially by the Chicago sewage diversion, continued nutrient loading from sewage probably would have had severe environmental impacts because Chicago is located at the shallow end of the long cul-de-sac of Lake Michigan, where loading effects would have been magnified. Ironically, diversion would not have been needed if modern sewage treatment facilities had been available at the time. The diversion undoubtedly provided benefits for water quality long after there was a need to control waterborne diseases. Some estimate of the importance of diversion can be obtained by extrapolating the rapidly increasing loadings from human waste in the late nineteenth century. These would have continued if sewage from Chicago had not been diverted in 1900 (Figure A.2). However, these benefits of diversion caused serious water quality problems in the Illinois River (see Illinois River case study, Appendix A) and undoubtedly contributed to degraded water quality in the Mississippi River (Turner and Rabalais, 1991).

Lake Michigan

FIGURE A.1   Lake Michigan, the third largest of the Laurentian Great Lakes, is the only one to lie completely within the United States. The lake is bordered by four states: Illinois, Indiana, Michigan, and Wisconsin. Its length is 491 km, and its width is 190 km. The lake is divided into two distinct basins. The southern basin is gently sloping and has a maximum depth of 175 m. The northern basin has an irregular profile and a maximum depth of 288 m.

Third, problems of nutrient enrichment were controlled in the 1970s by strategies to reduce phosphorus loading, particularly from sewage treatment plants. These sources of nutrients had become especially important beginning in the 1940s. Improved water quality that resulted from better sewage treatment was obtained at a cost of $10 billion. Benefits other than reduced nutrient loading that may accrue in the future from improved sewage treatment include reduced loadings of potentially toxic materials and vectors for waterborne diseases.

Fourth, although water quality has improved, two examples can be cited to show that the chemical condition of the water in Lake Michigan has not been restored to pristine conditions. One example is that silica has been depleted as a result of phosphorus enrichment and consequent increased growth of diatoms, which require silica for growth (Figure A.2). With a shortage of this essential nutrient, the natural phytoplankton assemblages of the lake and the dependent

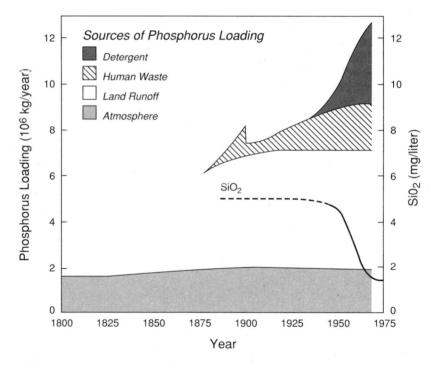

FIGURE A.2   Computer simulation of total phosphorus loads to Lake Michigan from 1800 to 1970 (adapted from Chapra, 1977). Source: Reprinted, by permission, from Schelske (1988). Copyright © by Akademie-Verlag Berlin, Leipziger Strasse Berlin, FRG.

trophic interactions cannot be restored. Because of the large volume of Lake Michigan, the reduction in silica concentration, a consequence of eutrophication, amounts to a loss of 15 million tons of silica from the lake. It is not likely that silica will be added to the lake because the cost of even partial restoration is prohibitive. The other example is that fish from the lake may not be safe to eat because they have accumulated high levels of potentially toxic chlorinated hydrocarbons. These materials are dispersed throughout the system and apparently are being renewed by atmospheric inputs.

Fifth, native fish stocks have been either decimated or severely depleted, and exotic species have invaded or have been introduced into the lake. Some of the ecosystem function attributed to fish has been restored by stocking and by other forms of management. The fish community (Figure A.3) is now largely dominated by exotic species, however, and this and other parts of the original biological community have been lost. The artificial fish community is susceptible to perturbations such as hatchery-transmitted diseases of exotic salmonids, the potential evolution of lampreys resistant to 3-trifluormethylnitrophenol (TFM), and invasions of exotic species. The Lake Michigan food web is a caricature of the ancestral one and lacks the stability of a self-sustaining natural community.

In summary, the case history for Lake Michigan provides important lessons about the limitations of restoration and other types of remedial action. Benefits resulting from restoration efforts include improved water quality and rehabilitation of fishery resources. An unanticipated benefit of remedial action may have been improvements in water quality that resulted from the diversion of sewage from Chicago to the Mississippi River drainage. However, this diversion of sewage undoubtedly contributed to degradation of water quality in downstream receiving waters, including the Mississippi River. Several examples show that corrective measures to restore ecosystem function were obtained only at very high costs, that some attributes can be maintained only with continuous management, and that certain losses in the ecosystem were irreversible.

Types of Disturbances

EUTROPHICATION AND NUTRIENT LIMITATION

Nutrient control in the Lake Michigan basin is devoted to phosphorus reduction because experimental studies of effects of nutrient limitation on phytoplankton have clearly established that Lake Michigan is a phosphorus-limited system (Schelske et al., 1986). In addi-

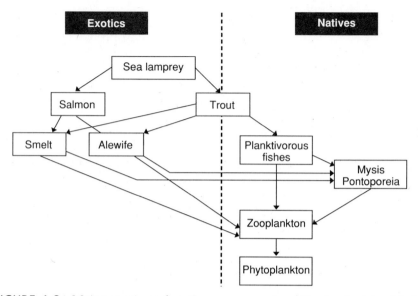

FIGURE A.3   Major exotic and native components of the food web of Lake
Michigan. Source: Reprinted, by permission, of Kitchell and Crowder (1986).
Copyright © 1986 by Kluwer Academic Publishers.

tion, these studies also have provided evidence that silica limitation
for diatom production can be induced with increased phosphorus
loading (Schelske and Stoermer, 1971). The demand for silica, an essential nutrient for diatom growth, increases as diatom production is
stimulated by increased phosphorus supplies. Some proportion of the
increased diatom production is sedimented, leading to silica depletion in the water column. Under these conditions, phosphorus supplies that would normally be used for diatom production can be used
for production of other types of phytoplankton, including blue-green
and green algae. Silica depletion and a shift in species composition
of phytoplankton, therefore, are expected consequences of eutrophication. Anthropogenic nutrient loadings of phosphorus have increased
rapidly, whereas loadings of silica have not increased proportion-
ately to meet the elevated silica demand for diatom production.

PHOSPHORUS LOADING

Profiles of historical phosphorus loading have been obtained by
computer simulation for 1800 to 1970 (Chapra, 1977). Until the 1970s,
these simulations provide the main source of information about phosphorus loading to the lake (see Figure A.2). Prior to the beginning of

European settlement in the mid-1800s, nearly 50 percent of the phosphorus loading was from atmospheric sources. After 1850, phosphorus loading increased, first as the result of forest clearance and associated soil erosion, and later as a result of added human waste from a rapidly increasing human population along the lake shore. The contribution from human waste increased rapidly until 1900, when sewage from Chicago was diverted to the Mississippi drainage by the Chicago Sanitary and Ship Canal. Without this diversion, phosphorus loading to and phosphorus concentration in Lake Michigan probably would have increased exponentially, following the pattern observed in Lake Erie and Lake Ontario. In Lake Michigan, the rapid, exponential increase did not occur until phosphate detergents were introduced after World War II. Phosphorus loads to Lake Michigan have decreased as a result of the 1972 Water Quality Agreement between Canada and the United States. (Although Lake Michigan is entirely within the United States, its water quality is pertinent in the international agreement because outflow from the lake enters Lake Huron, where it can affect the quality of international waters.) No detectable trend in total phosphorus loading occurred from 1974 to 1980, when loads ranged from 6,000 to 7,500 metric tons per year, with the exception of a load of 4,670 metric tons in 1977. Loads were much lower from 1981 to 1985, ranging from 3,500 to 4,500 metric tons annually. The effectiveness of phosphorus control programs is evident because loads from 1981 to 1985 were well below the target load of 5,600 metric tons per year established in the 1972 Water Quality Agreement (Rockwell et al., 1989).

Response to Phosphorus Loads

Long-term studies of phytoplankton standing crop, which have been restricted to nearshore sites, have shown that the annual standing crop of algae (measured either as counted cells or calculated biomass) increased from 1927 to 1965 and then decreased markedly until levels in 1976 to 1978 were essentially equal to those in the late 1920s (Makarewicz and Baybutt, 1981). Data on monthly average cell counts showed a slightly different pattern. The population density was 50 percent lower in the period 1972 to 1975 than in the preceding 4 years and lower than any 4-year period since 1953 to 1956 (Danforth and Ginsburg, 1980). The change in 1972 to 1975 resulted largely from decreases in the spring and fall maxima of diatoms. Both studies showed that the recent decreases in algal abundance were accompanied by increases in blue-green algae. Although decreases in algal standing crop occurred before the implementation of nutrient reduction programs in 1972, both studies suggest that the effect may be attributable at least partly to the reversal of eutrophication.

The trends in the data set for 1972 to 1984 are complicated by several factors. The first of these is the inherent problem of obtaining a representative sample from a lake the size of Lake Michigan (22,400 square miles). The second is that the trend apparently is confused by climatic factors, particularly the unusually cold winter of 1976-1977. As a result of the cold winter, when ice cover was 90 percent, compared to 20 to 50 percent under normal conditions, winter and spring resuspension of sedimented materials was minimal and the total concentration of phosphorus mean decreased from 8 to 5 µg per liter from 1976 to 1977. The third is that trends may be difficult to measure because concentrations of both total phosphorus and chlorophyll are relatively low. The final difficulty is variability introduced by biological factors. Water clarity increased dramatically in 1983, when the abundance of *Daphnia pulicaria* increased markedly (Scavia et al., 1986). The increase in water clarity was attributed to increased grazing pressure from this filter feeder and to the cascading trophic effect of decreased predation on *Daphnia* by alewife (Scavia and Fahnenstiel, 1988).

Evidence for cultural eutrophication has been obtained from the study of diatoms in a sediment core from the northern basin (Stoermer et al., 1990). These results indicate that the diatom community responded relatively little to nutrient enrichment from 1885 to 1925, with an accelerating trend between 1925 and 1954 and the most rapid change between 1954 and 1965. The reversal in trends in diatom species abundance after 1965 attributed to silica limitation was also inferred previously from trends in accumulation rates of biogenic silica in sediment cores (Schelske et al., 1983). This phasing of the effects of phosphorus loading on diatom production agrees well with historical changes in silica concentration in the water mass. Rapid silica depletion from 1955 to 1970 has been attributed to increased diatom production and sedimentation (Schelske, 1988). During these 15 years, silica concentrations decreased from approximately 4.5 to 1.5 mg per liter during the annual water maximum and from approximately 2.0 to 0.1 mg per liter in epilimnetic waters during summer stratification. Whether this rapid change in silica concentration (see Figure A.2) in the open waters of Lake Michigan can be substantiated from long-term data that were collected from nearshore waters at the Chicago Water Filtration Plant has been questioned by Shapiro and Swain (1983). In the case of Lake Michigan, several independent lines of evidence were available to document an historical decrease in silica concentration (see Schelske, 1988). The important reason here is that it may be impossible to establish the accuracy of historical data and that historical data, therefore, must be evaluated carefully. It has recently been documented that silica concentrations in the Mississippi River also decreased after 1950 (Turner and Rabalais, 1991).

All the responses to phosphorus loading that have been summarized above share a common temporal feature. Large responses that have been attributed to nutrient enrichment occurred after the introduction of phosphate detergents in the period from 1955 to 1970. Although these effects are correlated with increases in phosphate detergents, it should be pointed out that this source of phosphorus increased concomitantly with increased population growth and sewage.

EXOTIC SPECIES FOR LAKE MICHIGAN

In this century, the food web of Lake Michigan has been almost completely reconfigured by a combination of exotic species invasions and deliberate stocking of sport fishes (Christie, 1974). The ancestral offshore fish stocks were dominated by lake char and coregonines, although 114 native fish species representing 21 families were known from the lake. Waves of introductions of exotic species and collapses of native species began in the 1940s.

The collapse of lake char populations between 1946 and 1952 was correlated with an expansion of sea lamprey populations and an increase in harvest rates (Christie, 1974). Populations of the exotic, parasitic sea lamprey peaked between 1950 and 1957. Between 1950 and 1955, the gillnet fisheries of the lake converted from cotton and linen to nylon nets, which achieved at least a threefold increase in fishing efficiency. Selective harvesting of the largest lake char may have forced the lampreys to feed on smaller individuals, which are more likely to die from lamprey attacks (Kitchell, 1990). The relative importance of overfishing and of sea lamprey increases in the collapse of the lake char stock continues to be debated. Since the early 1960s, the sea lamprey has been successfully controlled by the regular additions of TFM, which kills the sedentary ammocoetes in the breeding streams.

The exotic rainbow smelt was first reported in Lake Michigan in 1923 and by the 1930s had attained sufficient numbers to support a fishery (Christie, 1974). Fishery yield peaked between 1953 and 1960. It is not certain when the exotic alewife entered Lake Michigan, but populations began to increase in 1949 and reached nuisance levels by 1957. The lake char was probably an important predator of both smelt and alewife, and collapse of the lake char likely contributed to the population growth of both of these forage fish. Expansion of rainbow smelt and alewife populations corresponded with the collapsing stocks of native lake herring, and the causal mechanisms of these changes in forage fish communities continue to be debated. After 1960, smelt populations declined, whereas alewife populations boomed, culminating in the infamous die-offs that littered Lake Michigan beaches in the late 1960s.

Control of the sea lamprey was followed by highly successful stocking of exotic coho and chinook salmon in Lake Michigan. By the 1980s, stocked salmonids formed the basis of a sport fishery valued in excess of a billion dollars per annum (Kitchell and Crowder, 1986). By 1978, careful analyses of salmonid diets and bioenergetic requirements indicated that heavy predation was likely to trigger a collapse of the alewife stock (Stewart et al., 1981). By 1983, it was evident that a severe decline in alewife abundance was under way (Kitchell and Crowder, 1986). It is ironic that "Save the Alewife" tee-shirts could be purchased in Milwaukee less than 20 years after massive die-offs fouled water intakes and beaches.

Lake Michigan cannot be viewed as a pristine, natural system. Ecosystem dynamics are determined mainly by nonnative species and decisions made by managers. At present, the food web's keystone species are exotic fish whose population dynamics are determined by management policies and are uncoupled from typical ecological feedbacks (Figure A.3). A substantial share of the variability in lower trophic levels is determined by the predatory effects of these fish (Kitchell and Crowder, 1986). Though the species composition of the community is dramatically different from the ancestral one, the extent to which ecosystem functions and trophic structure resemble those that existed prior to disturbance remains an open question. At present, fish biomass at all trophic levels is around twice as large as it was before collapse of the native stocks (J. F. Kitchell, Center for Limnology, University of Wisconsin, personal communication, June 1990).

The management of Lake Michigan's fish stocks must be judged a success by several criteria. An extremely successful sport fishery has created and sustained public interest in the resource while controlling the nuisance alewife. However, the ecosystem is an artificial one. The exotic salmonids are susceptible to outbreaks of disease, such as the current epidemic of bacterial kidney disease, exacerbated by complete dependence on hatcheries. Restoration of native species has not occurred and in many cases seems unlikely. It has proved very difficult to establish reproducing populations of lake char. Ironically, the native lake char is less respected as a game fish, and has far greater concentrations of organochlorine contaminants, than the other salmonids. High contaminant levels may contribute to low egg viability of the lake char.

Lake Michigan is also vulnerable to further invasions of exotic species, with consequences that are largely unpredictable. The recent invasion of the exotic zooplankter *Bythotrephes cederstroemii* has had a profound effect on the planktonic community structure of Lake Michigan (Lehman, 1988). Other potentially more significant invaders are already present in other Laurentian Great Lakes: zebra mussels (also in Lake Michigan), river ruff, and white perch. One prediction can be made with relative certainty: continual vigilance and management of Lake Michigan's food web will be essential to sustain the favorable conditions that currently prevail.

ORGANOCHLORINE CONTAMINANTS

In Lake Michigan, organochlorine contaminants have been a major environmental concern because of their potentially deleterious effects on wildlife and humans who eat fish. Because these lipophilic compounds tend to accumulate in higher concentrations at higher trophic levels, fish-eating organisms, such as ospreys, eagles, gulls, otter, mink, and humans can be exposed to chemicals at concentrations that far exceed those in the water (International Joint Commission, 1989). Dichlorodiphenyltrichloroethane (DDT) and polychlorinated biphenyls (PCBs) provide contrasting examples of efforts to reduce organochlorine contaminant levels in fish (U.S. FWS, 1989).

Until it was banned in 1970, DDT was used commonly as an insecticide. After the ban, concentrations in fish declined exponentially (Figure A.4). From a human health standpoint, concentrations reached acceptable levels by the mid-1970s and have continued to decline since then. In the case of DDT, point source reduction of inputs successfully restored concentrations to acceptable levels.

Polychlorinated biphenyls were used for a variety of industrial applications until a voluntary reduction was effected in 1972, followed by a ban on manufacture of the compounds in 1976. After the ban, concentrations in fish dropped significantly (Figure A.4) and have been stable since the early 1980s, although water column concentrations have continued to decline (Swackhamer and Armstrong, 1985; Anders Andren, Sea Grant Institute, University of Wisconsin, personal communication, June 1990). Concentrations in several exploited fish stocks remain above the Food and Drug Administration action level of 2 mg/kg (De Vault et al., 1985; Masnado, 1987). Why

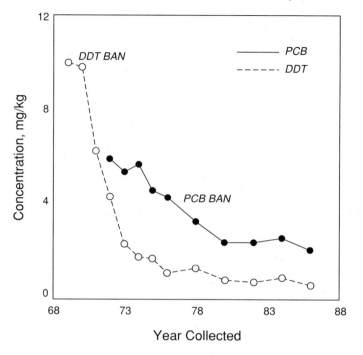

FIGURE A.4   Trends in levels of organochlorine contaminants DDT and PCB in bloater chubs (*Coregonus hoyi*) in Lake Michigan.   Source: U.S. Fish and Wildlife Service, 1989.

has the PCB ban failed to reduce concentrations in fish to acceptable levels?   First, nonpoint source inputs are relatively high.   The atmosphere is a major source of PCBs to the ecosystem, and atmospheric concentrations of PCBs in the Lake Michigan airshed did not decline during the 1980s (Manchester-Neesvig and Andren, 1989).   Also, poorly contained waste materials continue to add PCBs to ground water and surface water systems draining into Lake Michigan.   Annual inputs from atmospheric and fluvial sources exceed total losses, which include those to the atmosphere, outflow to Lake Huron, and sedimentation (Andren, 1983; Swackhamer and Armstrong, 1985).   Second, recycling of PCBs within the ecosystem appears to be relatively efficient and stabilizes concentrations in the biota.   Diet and growth rate are among the major factors that determine PCB concentrations in fish (Thomann, 1989a,b).   It may be possible to reduce PCB concentrations by restructuring the fishery around species that tend to have lower PCB concentrations.   Recycling from sediment may be another process that can be manipulated in some areas to reduce PCB flux into the biota. Because nonpoint inputs of PCBs are hard to control, research on internal cycling mechanisms has become increasingly important. Further efforts to reduce PCB concentrations in Lake Michigan fish will evidently require new scientific and institutional efforts. Research and management efforts must address not only point source reductions but also nonpoint and atmospheric inputs that now dominate the mass balance. New efforts are needed to understand the internal cycling of PCBs through the food web, and the interfaces between the food web and PCBs in water and sediments.

Institutional and Educational Issues

Restoration of Lake Michigan requires cooperation among water chemists, fisheries biologists, and wildlife ecologists—researchers in disciplines that traditionally have had very little interaction. No single agency is responsible for the Lake Michigan ecosystem. Responsibilities for water quality, fisheries, and wildlife are divided piecemeal among several state and federal agencies. Effective management of the ecosystem requires institutional collaborations that involve the U.S. Environmental Protection Agency, the U.S. Fish and Wildlife Service, and the water quality and fisheries management agencies of Illinois, Indiana, Michigan, and Wisconsin. The next decade will reveal whether the interdisciplinary and interagency efforts needed to reduce PCB levels in Lake Michigan fish can be implemented.

References

Andren, A. W. 1983. Processes determining the flux of PCBs across the air/water interface. Pp. 127-140 in D. MacKay, S. Paterson, S. J. Eisenreich, and M. Simmons, eds., Physical Behavior of PCBs in the Great Lakes. Ann Arbor Science, Ann Arbor, Mich.

Chapra, S. C. 1977. Total phosphorus model for the Great Lakes. J. Environ. Eng. Div., Am. Soc. Civ. Eng. 103:147-161.

Christie, W. J. 1974. Changes in the fish species compositions of the Great Lakes. J. Fish. Res. Bd. Can. 31:827-854.

Danforth, W. F., and W. Ginsburg. 1980. Recent changes in the phytoplankton of Lake Michigan near Chicago. J. Great Lakes Res. 6:307-314.

De Vault, D. S., W. A. Willford, and R. J. Hesselberg. 1985. Contaminant trends in lake trout (Salvelinus namavcush) of the Upper Great Lakes. U.S. Environmental Protection Agency Report 905/3-85-001.

International Joint Commission. 1989. Report on Great Lakes Water Quality. Windsor, Ontario, Canada N9A 6T3.

Kitchell, J. F. 1990. The scope for mortality caused by sea lamprey. Trans. Am. Fish. Soc. 119:642-648.

Kitchell, J. F., and L. B. Crowder. 1986. Predator-prey interactions in Lake Michigan: Model predictions and recent dynamics. Environ. Biol. Fish. 16:205-211.

Lehman, J. T. 1988. Algal biomass unaltered by food-web changes in Lake Michigan. Nature 332:537-538.

Makarewicz, J. C., and R. I. Baybutt. 1981. Long-term (1927-1978) changes in the phytoplankton community of Lake Michigan at Chicago. Bull. Torrey Bot. Club 108:240-254.

Manchester-Neesvig, J. B., and A. W. Andren. 1989. Seasonal variation in the atmospheric concentration of polychlorinated biphenyl congeners. Environ. Sci. Technol. 23:1138-1148.

Masnado, R. G. 1987. Polychlorinated biphenyl concentrations of eight salmonid species from the Wisconsin waters of Lake Michigan: 1985. Wis. Dep. Nat. Resour. Fish Manage. Rep. 132:55.

Rockwell, D.C., Salisbury, D.K., and Lesht, B.M. 1989. Water Quality in the Middle Great Lakes: Results of the 1985 U.S. EPA Survey of Lakes Erie, Huron and Michigan. U.S. Environmental Protection Agency, Great Lakes National Program Office, Chicago, IL. EPA-905/689-001, GLNPO Report No. 4. 207 p. and appendices.

Sandgren, C. D., and J. T. Lehman. 1990. Response of chlorophyll a, phytoplankton, and microzooplankton to the invasion of Lake Michigan by *Bythotrephes*. Ver. Int. Ver. Limnol. 24:386-392.

Scavia, D., and G. L. Fahnenstiel. 1988. From picoplankton to fish: Complex interactions in the Great Lakes. Pp. 85-97 in S. R. Carpenter, ed., Complex Interactions in Lake Communities. Springer-Verlag, New York.

Scavia, D., G. L. Fahnenstiel, M. S. Evans, D. J. Jude, and J. T. Lehman. 1986. Influence of salmonine predation and weather on long-term water quality in Lake Michigan. Can. J. Fish. Aquat. Sci. 43:435-443.

Schelske, C. L. 1988. Historic trends in Lake Michigan silica concentrations. Int. Rev. Ges. Hydrobiol. 73:559-591.

Schelske, C. L., and E. F. Stoermer. 1971. Eutrophication, silica depletion, and predicted changes in algal quality in Lake Michigan. Science 173:423-424.

Schelske, C. L., E. F. Stoermer, D. J. Conley, J. A. Robbins, and R. M. Glover. 1983. Early eutrophication in the lower Great Lakes: New evidence from biogenic silica in sediments. Science 222:320-322.

Schelske, C. L., E. F. Stoermer, G. L. Fahnenstiel, and M. Haibach. 1986. Phosphorus enrichment, silica utilization, and biogeochemical silica depletion in the Great Lakes. Can. J. Fish. Aquat. Sci. 43:407-415.

Shapiro, J., and E. B. Swain. 1983. Lessons from the Silica "decline" in Lake Michigan. Science 221:457-459.

Stewart, D. J., J. F. Kitchell, and L. B. Crowder. 1981. Forage fishes and their salmonid predators in Lake Michigan. Trans. Am. Fish. Soc. 110:751-763.

Stoermer, E. F., J. A. Wolin, C. L. Schelske, and D. J. Conley. 1990. Siliceous microfossil succession in Lake Michigan. Limnol. Oceanogr. 35:959-967.

Swackhamer, D. L., and D. E. Armstrong. 1985. Distribution and characterization of PCBs in Lake Michigan water. J. Great Lakes Res. 13:24-36.

Thomann, R. V. 1989a. Bioaccumulation model of organic chemical distribution in food chains. Environ. Sci. Technol. 23:699-707.

Thomann, R. V. 1989b. Deterministic and statistical models of chemical fate in aquatic systems. Pp. 245-277 in S. A. Levin, M. A. Harwell, J. R. Kelly, and K. D. Kimball, eds., Ecotoxicology: Problems and Approaches. Springer-Verlag, New York.

Turner, R. E., and N. N. Rabalais. 1991. Changes in Mississippi River water quality in this century. Bioscience 41:140-147.

U.S. Fish and Wildlife Service (FWS). 1989. National Fisheries Research Center—Great Lakes. Biannual Report 87/88. Ann Arbor, Mich. 10 pp.

CAN LAKE APOPKA BE RESTORED?

Claire Schelske and Patrick Brezonik

General Description and Type of Disturbance

Lake Apopka prior to the late 1940s was well known as an outstanding sport fishing lake with exceptionally clear water (Clugston, 1963). With an original surface area of 18,000 ha, it was the second largest lake in Florida (Schneider and Little, 1973). However, draining the marshlands along the northern shore in the 1940s reduced the surface area to 12,500 ha. This shallow lake with an average depth of less than 2.0 m is located in Orange and Lake counties, 20 km west of Orlando. The change in water quality in this lake was reported to have been dramatic in 1947 (Clugston, 1963). At that time, the lake was characterized as having abundant rooted aquatic vegetation and very clear water until rooted aquatic plants were uprooted by a hurricane. A week after the hurricane the first plankton bloom was reported. Aquatic plants have never been reestablished. The lake is now a classic hypereutrophic lake, with chlorophyll concentrations that exceed 100 µg per liter, and has changed from a highly regarded sport fishing lake to a lake with few desirable sport fish.

According to Schneider and Little (1973), human influence on the lake was evident by 1920, when citrus groves were being planted in central Florida. The well-drained southern shoreline was an excellent site for groves, but the marshland on the northern shore was not developed. In 1920, the town of Winter Garden constructed a sewerage system and two large septic tanks, permitting waste products to enter the lake directly. Nutrients from municipal waste and runoff from the citrus groves seemed at first beneficial to a popular sport fishery. Grass beds that covered the lake bottom provided cover for young fish and tied-up nutrients. The marshlands were used as spawning grounds. Sport fishing for largemouth bass, speckled perch, bluegill, and other pan fish provided record-size fish and a half-million-dollar annual income for 13 fishing resorts and camps. A thriving commercial fishery yielded more than 3 million kilograms (dressed weight) of catfish in one 8-month period.

The lake was altered severely in the 1940s (Schneider and Little, 1973). A plan for draining and farming part of the marshland was formulated in 1942. This resulted in the construction of a dike along the north shore and draining of about 6,000 ha of fertile lake bottom so it could be used for muck farms. Thus, fertile muck land and availability of lake water enabled several crops to be grown and harvested each year. During dry periods, lake water could be used for irrigation, in wet periods, agricultural water from the farms could be back-pumped from the drainage canals into the lake. The opening of Beauclair Canal in 1948 permitted better drainage of Lake Apopka to a downstream chain of lakes and lowered the lake level. This alleviated concerns of muck farm owners that future storms would weaken or destroy the dikes and reflood their croplands. In 1946 the lake waters were still clear. However, in the fall of 1947, soon after a severe hurricane had uprooted large quantities of aquatic plants, the first plankton bloom was observed. The dense beds of rooted aquatics were never reestablished, probably because they could not compete with planktonic algae in the nutrient-rich waters.

Clugston (1963) did not discuss the effects of the hurricane but stated that a combination of external factors probably increased the fertility of the lake that led to the first algal bloom in 1947. First, a water hyacinth control program resulted in large amounts of decaying vegetation. Second, a citrus-processing plant at Winter Garden increased its capacity considerably between 1946 and 1950 and its release of waste products. Third, muck farms at the north end of the lake were expanded greatly in the 1940s. Water pumped out of the farming areas may have added nutrients and contributed to siltation in the lake. Fourth, citrus groves located along the eastern and western shores may have contributed nutrients. Finally, a sewage treatment plant at Winter Garden was pumping effluent into the lake.

The game fish population comprised 35 percent of the species present, and gizzard shad made up 20 percent of the total fish population by weight in 1947 when the plankton bloom was first noted (Clugston, 1963). As the plankton bloom persisted, the gizzard shad probably increased greatly in numbers but were small in size, providing excellent forage for game species. As a result, game fish constituted 69 percent of the total population in 1950. From 1947 to 1950 the estimated weight of the total fish population increased tenfold. By 1956-1957, however, the game fish population had dropped to 18 percent of the total. Shad, which made up most of the remaining 82 percent, are thought to have become too large and numerous to be cropped by game species. In an effort to alleviate the shad problem, the Game and Fresh Water Fish Commission treated the lake with rotenone in three successive years; 1957, 1958, and 1959. An estimated 9 million kilograms of gizzard shad were killed with the three treatments. These fish were left in the lake to decompose and release nutrients. In May 1963, 1.4 million kilograms of fish were reported killed by gas (oxygen or nitrogen) embolism (Schneider and Little, 1973).

State Intervention

This series of environmental problems led the governor of Florida on April 4, 1967, to appoint a technical committee to evaluate the restoration of the lake. Sixteen agencies, including the Federal Water Pollution Control Administration (FWPCA), agreed to participate in the project (Schneider and Little, 1973). An FWPCA study begun in 1968 revealed that 90 percent of the bottom was covered with unconsolidated bottom sediment (muck) averaging 1.5 m thick. These sediments and peat sediments found along the shoreline were anaerobic and provided limited suitable substrate for desirable biota. Only 5 percent of the bottom was covered with sand, clay, and shell. The top meter of lake sediment contained 225 million kilograms of total nitrogen and 2 million to 4 million kilograms of total phosphorus. Chemical oxygen demand in the muck samples (dry weight) was 1,100 mg/g. The FWPCA also made a crude nutrient budget and emphasized that restoration of the lake must include reduction of nutrient input. Although direct rainfall on the lake and high nutrient input from citrus grove runoff were important, the principal controls on inputs emphasized by the FWPCA were point sources such as agricultural runoff pumped directly into the lake from muck farms, and municipal and industrial wastes.

In addition to control of external nutrient sources, several solutions for improving lake water quality were listed by Schneider and Little (1973). These include:

1. dredging to remove nutrient-rich unconsolidated bottom sediments to increase lake depth and reduce internal nutrient recycling;

2. using lake drawdown to expose and subsequently consolidate large areas of lake bottom by oxidation and compaction;

3. adding an inert sealing material to stabilize bottom sediments;

4. engaging in hydroponic farming to remove dissolved nutrients;

5. harvesting to remove algae by flotation, filtration, precipitation (not within the lake), or centrifugation (recovered algae could be used as a feed supplement); and

6. harvesting fish to remove nutrients "on a large scale." Harvested fish could be used as a protein supplement.

"Other bizarre schemes were considered but not seriously," according to Schneider and Little (1973).

The governor of Florida assigned complete responsibility for a 1970 restoration of Lake Apopka to the Florida Air and Water Pollution Control Commission. This agency decided to proceed with the lake drawdown approach by allowing gravity drainage to lower the lake level 60 cm beginning in December 1970. The effect of this lowering was to be evaluated, and the lake would then be drained further by pumping to 25 percent of its original area. This final drawdown would occur in the spring of 1971. It was anticipated that two beneficial effects would result from the drawdown. First, nutrient recycling would be reduced or eliminated from dried, compacted sediments. Second, suitable substrate for rooted aquatic vegetation would also be a result. This plan to lower the lake about 7 ft below normal water level was not implemented, however, because of the projected cost ($20 million) and because of concern about environmental and economic impacts (Lowe et al., 1985). For example, the loss of lake volume would minimize the freeze protection citrus growers received from the large heat capacity of the lake.

In the 1970s, additional studies were conducted on water quality problems and on restoration of Lake Apopka (Brezonik et al., 1978; Lowe et al., 1985). Studies of techniques that might be used to restore the lake have continued. Biomanipulation of algal standing crops with gizzard shad may actually increase standing crops of undesirable algae (Crisman and Kennedy, 1982). A multimillion-dollar feasibility study on growing and harvesting water hyacinths to remove nutrients from the lake was launched (Amasek, Inc., 1985). The field test of this project in Lake Apopka was abandoned when the enclosure that was to have been used for the experiment was destroyed by water movements in the lake.

A project on a smaller scale will be tested in a small lake. Many experts believe this approach will fail because of problems in harvesting water hyacinths, which have a very high water content; because nutrients, including nitrogen, would have to be added to lake water to obtain projected growth rates; and finally, because water hyacinths require high phosphorus concentrations for growth and do not reduce phosphorus to low levels under any conditions.

Currently, the Saint Johns Water Management District is beginning a feasibility study on using marsh restoration to improve water quality in the lake (Lowe et al., 1989; Lowe et al., in press). The water management district has purchased muck farmland that will be flooded to restore the wetland by using the wetland as a filter to remove nutrients. The hydrology of the wetland will be manipulated so that highly nutrient-enriched water will flow from the lake into the wetland and nutrient-depleted water from the wetland will be directed back to the lake. If successful, this project will result in both a restored wetland and a restored lake.

Overall Evaluation

There seem to be two divergent views about Lake Apopka. One group contends that the lake can be restored. This viewpoint is supported by the need to reduce nutrient inputs to prevent accelerated eutrophication. Schneider and Little (1973) commented that the history of Lake Apopka "is not atypical" because other lakes in Florida and reservoirs all over the South were being subjected to similar attacks. They stated that the lake could be restored, but only with great expense and difficult decisions (e.g., the extent to which a $10 million plus marginal muck farming operation could expend money for nutrient removal). "The technical capabilities to prevent accelerated eutrophication are and have been available for some time. The planning and foresight needed to prevent the early demise of our lakes, however, has come into being only lately. Today, we must consider the full ecological impact of all our resource development activities if we are to eliminate the Lake Apopka syndrome from our aquatic environment," they emphasized.

At the other extreme is the viewpoint that restoration should not be attempted because it will meet with failure or it is too expensive. This viewpoint can be supported to a certain extent with results of studies on Lake Tohopekaliga (Lake Toho), Florida. A number of restoration measures have been instituted on Lake Toho since 1971, with little evidence of improvement in water quality (Dierberg et al., 1988). In this lake, nutrient inputs have been reduced by sewage treatment and by storm water detention and filtration. In addition, drawdown has been used as a restoration measure. What is not known is whether water quality would have been degraded even more if remedial measures had not been instituted. Dierberg et al. (1988) point out that evaluation of restoration practices in Florida lakes has been hampered by the lack of long-term data and the consequent limitation on the use of robust statistical approaches in evaluating effectiveness.

References

Amasek, Inc. 1985. A Practical Concept for the Restoration of Lake Apopka. Report by Amasek, Inc., Cocoa, Fla. 38 pp.

Brezonik, P. L., C. D. Pollman, T. C. Crisman, J. N. Allenson, and J. L. Fox. 1978. Limnological and water quality studies on Lake Apopka and downstream lakes in the Upper Oklawaha River Basin. Historical trends and current status. Report to Florida Department of Environmental Regulation. Rept. No. Env-07-78-01. Department of Environmental Engineering Science, University of Florida, Gainesville. 283 pp.

Clugston, J. P. 1963. Lake Apopka, Florida; A changing lake and its vegetation. Q. J. Fla. Acad. Sci. 26:168-174.

Crisman, T. L., and H. M. Kennedy. 1982. The role of gizzard shad (Dorosoma cepedianum) in eutrophic Florida lakes. Publ. No. 64, Water Resources Research Center, University of Florida, Gainesville. 83 pp.

Dierberg, F. E., V. P. Williams, W. H. Schneider. 1988. Evaluating water quality effects of lake management in Florida. Lake Reservoir Manage. 4:101-111.

Lowe, E. F., J. Adams, J. Bateman, C. Fell, D. Graetz, J. Hulbert, W. Ingram, W. Johnson, P. Muller, L. Nall, R. Reddy, L. Snyder, and V. Williams. 1985. Proposed scope of work for the Lake Apopka pilot project. Unpublished Draft report. 55 pp.

Lowe, E. F., D. L. Stites, and L. E. Battoe. 1989. Potential role of marsh creation in restoration of hypereutrophic lakes. Pp. 710-717 in D. A. Hammer, ed., Constructed Wetlands for Wastewater Treatment: Municipal, Industrial and Agricultural. Lewis Publishers, Chelsea, Mich.

Lowe, E. F., L. E. Battoe, D. L. Stites, and M. F. Coveney. In press. Particulate phosphorus removal via wetland filtration: An examination of the potential for hypertrophic lake restoration. Environmental Management.

Schneider, R. F., and J. A. Little. 1973. Rise and fall of Lake Apopka: A case study in reservoir mismanagement. Pp. 690-694 in W. C. Ackerman, ed., Man-Made Lakes: Their Problems and Environmental Effects. American Geophysical Union, Washington, D.C.

* * * * * *

RIVERS

* * * * * *

THE ATCHAFALAYA BASIN

Richard E. Sparks[1]

Introduction

The 217-km Atchafalaya River and its basin comprise North America's largest river overflow swamp (5,700 km^2), excluding the fresh marshes south of the Gulf Intracoastal Waterway (Glasgow and Noble, 1974) (Figure A.5). The Atchafalaya begins at the confluence of the Red River with a distributary that receives about 30 percent of the flow from the Mississippi River and the Red River. Habitat types range from dry bottomland hardwoods in the Morganza and West Atcha-

---

[1]Material supplied by C. Frederick Bryan, Leader, Louisiana Cooperative Fish and Wildlife Research Unit, School of Forestry, Wildlife, and Fisheries, Louisiana State University, Baton Rouge, Louisiana 70803-6202; and Suzanne R. Hawes, U.S. Army Corps of Engineers, New Orleans District, P.O. Box 60267, New Orleans, Louisiana 70160.

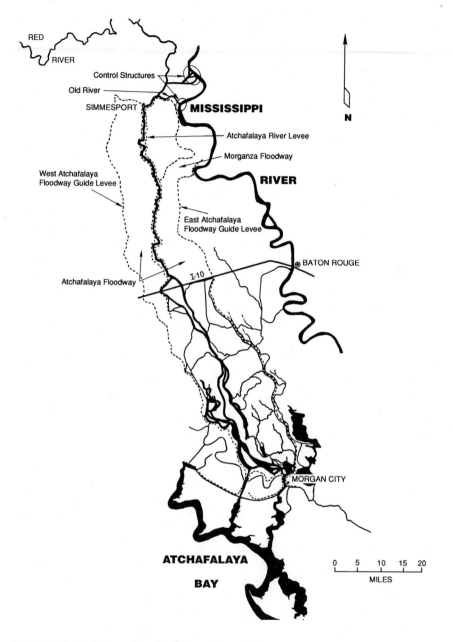

FIGURE A.5   Map of Atchafalaya River Basin.

falaya Basin floodways through cypress-tupelo swamps, to fresh and brackish marshes south of Morgan City and in the emerging delta in Atchafalaya Bay. The area is exceedingly rich in fish and wildlife resources. An acre of water can support up to 1,000 pounds of fish. The crawfish harvest averages 15 million pounds per year. Hunters, fishermen, and canoeists use the area extensively for recreational purposes.

The Delta Cycle

The problems of the Atchafalaya Basin can only be understood in the context of the dynamics of the much larger natural system of which it is a part, the Mississippi Delta. The Mississippi River builds, then abandons, deltaic lobes in an orderly cycle: six lobes in the last 8,000 years (Penland and Boyd, 1985). If left to itself, the Mississippi River would have switched most of its water and sediment flow from its present course to the Atchafalaya River, which is 307 km closer to the sea and therefore a much more hydraulically efficient channel (van Heerden and Roberts, 1980). The new Atchafalaya deltaic lobe would be building at a much faster rate than it is, and the old Plaquemines-Balize delta complex south of New Orleans would be regressing due to the combined effects of coastal erosion, sea level rise, and land subsidence as the delta sediments were compressed under their own weight (Penland and Boyd, 1985). Humans have intervened in this natural process of delta switching, thereby forestalling, or at least slowing, some of these events, in an attempt to protect an enormous investment in existing cities, port facilities, and waterways.

Problems

TOO MUCH AND TOO LITTLE WATER

From the point of view of New Orleans' citizens, many of whom live below the mean *low*-water elevation of the Mississippi River (and are sinking still farther), the dilemma is how to maintain enough flow to keep salt water from intruding on the city's water intakes and sediment from closing their deep-water connection to the sea. The engineering response has been to use both the Atchafalaya and the Bonnet Carre Spillway above New Orleans to divert high flows before they reach the city, and to prevent the switching of the Mississippi into the Atchafalaya by using several dams called the Old River Control Structures. The entire floodway on the Atchafalaya was enclosed within levees that were set well back from the river channel on the floodplain.

EROSION AND SEDIMENTATION

The ability of the floodway to convey floods was rapidly diminished by sedimentation. The U.S. Army Corps of Engineers (COE) responded by increasing the cross-sectional area of the main channel by dredging, although the geologically young river was also naturally enlarging and deepening its channel. Twenty-two distributaries were closed, which accelerated the natural erosive action in sections of the main channel. The dredge spoil was used on the upper 80 km to build mainstem levees set close to the river, which reduce inundation of adjacent floodplains except for record floods. Floodplain sedimentation, leveeing, and channel deepening together caused lakes, bayous, and seasonally flooded forests to dry out.

CONVERSION AND OCCUPATION OF FLOODPLAIN

Most of the bottomland hardwood forest in the lower basin has been cleared and much of the land converted to agriculture in the areas protected by the mainstem levees. Houses have been constructed within the West Atchafalaya Floodway because no easements were acquired to prevent construction, but such easements are now being purchased in the central and lower portions of the basin. These easements should forestall wholesale clear-cutting and agricultural development (Robert Campos, Project Manager, New Orleans District Office, U.S. Army Corps of Engineers, personal communication, May 7, 1990 to C. Frederick Bryan).

FLOOD PULSE ALTERATION AND BACKWATER STAGNATION

The ongoing natural and engineered enlargement of the main channel apparently will reduce the frequency of flooding in the adjacent swamps from approximately once every 1.5 years to once every 2.3 years (Wells and Demas, 1977), which could reduce fish production by reducing the frequency of access to backwater spawning and nursery areas (Bryan and Sabins, 1979; Fremling et al., 1989). Moderate floods function to purge the back swamps of accumulated organic debris but also bring inorganic sediments that fill back swamps and lakes (Bryan et al., 1974, 1975, 1976). Reduction of flood frequency may increase water stagnation and the potential for fish kills from anoxia (Bryan et al., 1974, 1975; Bryan and Sabins, 1979).

Impact of the National Environmental Policy Act

In 1969, the National Environmental Policy Act (NEPA) became law, and in the early 1970s, the COE started preparing an Environmental Impact Statement (EIS) on its plan to raise the levees and increase flow capacity by enlarging the main channel. The newly emerging environmental community was afraid that the deeper channel would hasten the drying out of the swamp. An agreement was reached between Thomas Kimball of the National Wildlife Federation and General Frederick Clark of the COE that levee raising could continue without an EIS, but that all other work would stop until the EIS was completed. The agreement stated that the COE would involve numerous agencies and individuals in this EIS effort.

The planning group was called the Steering Committee in the early 1970s and later evolved into the Agency Management Group (AMG). By 1981, the draft EIS, with a complex plan, was out for public review. The AMG plan, among other things, involved the purchase of greenbelts along numerous waterways in the basin for both forest preservation and public access.

A Public Relations Failure and a Political Reprieve

Just as the EIS was published, the U.S. Fish and Wildlife Service, also a member of the AMG, made public its own plan to purchase the entire lower basin for the Atchafalaya National Wildlife Refuge. The five public meetings, instead of focusing comments on all facets of the AMG plan, degenerated into protests against the "land grab." The public meetings were attended by 1,000 people and generated more than 4,000 written responses. Most complained about the proposed refuge or asked that the floodwaters be carried safely past Morgan City.

In 1982, Governor David Treen of Louisiana worked out a compromise plan that appeared in the Final EIS. The Treen plan called for purchase of 50,000 acres in fee for public access and purchase of flood control and environmental easements for over 367,000 acres of the lower basin to prevent construction of permanent houses and clearing for agriculture. The comprehensive plan, authorized in 1982, also included several features that restored wetlands or slowed their rate of degradation. These features are described below.

Inducing Flow, Reducing Sediment

Low "channel-training" banks will reduce the rate at which sediment spills over banks and fills back swamps, although these structures will also reduce the frequency and duration of flooding. The benefits of retaining the near-annual flood pulse have to be traded against the benefits of reducing the rate of overbank sedimentation in the floodplain and backwaters, because this system is in disequilibrium (due to both natural and man-made causes) and is rapidly aggrading to a new steady state. Another method of extending the life of the swamps is to reduce the amount of riverine bed-load sediment (the heavier sediment skidding along the bottom of the river, rather than in the water column) that enters the major distributaries by realigning the angle at which two major distributaries leave the main channel of the Atchafalaya River, thereby causing the main channel to retain and convey most of its bed load.

The banks will be lowered ("bank shaving") in selected areas to allow headwater flow, but not bed-load sediments, into the swamps and lakes. The openings also provide access to the backwaters for anglers. The combination of these actions does not restore the backwaters, but does slow the rate at which the aquatic habitat of lakes and swamps becomes willow thicket.

Habitat Management Units

If nothing further is done, 125,000 acres of swamp and bottomland hardwoods that flood annually today will no longer flood by the year 2030. The flooded wetlands are vital to the fishery in the basin because they provide feeding, spawning, and hiding areas for fish. As the basin dries out, the fishery will be reduced by 40 percent for both finfish and crawfish (U.S. Army Corps of Engineers, 1982).

In 1990, COE and the U.S. Fish and Wildlife Service updated earlier plans to modify management units in 11 hydrologically distinct areas within the Atchafalaya floodway. In the Buffalo Cove Management Unit, the existing canal system admits large amounts of sediment during floods, but during low summer flows the spoil banks of the canal prevent water circulation, thus causing anoxic conditions and excessive water temperature (more than 100°F). Plans are to plug one of the canal entrances to reduce sediment loading, but also to breach the spoil banks in several upstream locations to increase water flow through the entire unit, thereby reducing or eliminating stagnant areas. Downstream weirs will be set at levels to preserve the historical flooding regime and still keep water moving. Water may be kept on bottomland hardwoods about 2 weeks longer than now, which will probably slow the growth of ground cover, but increase timber growth and mast (the fruit of oak, pecan, and other trees used by wildlife) production. Similar concepts and techniques will be applied to Sherburne Wildlife Management Area and the Atchafalaya National Wildlife Refuge, where river levees keep floodwater from inundating and nourishing approximately 5,000 acres of bottomland forest. The management units can slow the rate of change, but not arrest or reverse it: the most realistic estimates are that the crawfish harvest, for example, will still drop by 28 percent by 2030 (U.S. Army Corps of Engineers, 1982).

Marsh Formation

The Corps of Engineers maintains a navigation channel through the new delta that is forming in Atchafalaya Bay. New procedures have been adopted to reduce the negative effects of the dredging that is required to maintain the channel. Instead of being piled into relatively high banks, the dredged spoil material is placed either in a distributary where currents carry it off to form new marsh, or at a height such that when it settles and compacts, it will be at the same height as the adjacent marshes. Marsh vegetation covers the spoil within one growing season, and the plant biomass, species diversity, and nursery function generally are the same as in the naturally formed marsh.

Summary

The Atchafalaya is an example of a fluvial system in disequilibrium from both natural and man-made causes. The Mississippi River is at a critical juncture in the natural delta switching cycle: if left alone, it would rapidly shift its flow of sediment and water to its Atchafalaya distributary, which provides a shorter and hydraulically more efficient route to the sea than does the present main channel past New Orleans. As a result, the Atchafalaya is enlarging its channel, filling its existing floodplain and backwaters with sediment, and building a new delta at the downstream end. The rate of sedimentation probably has been accelerated by floodway levees, which constrict the floodplain and tend to concentrate sedimentation on the remaining floodplain. Man-made canals can introduce sediments from the main channel into biologically productive backwaters, thereby filling them. The spoil banks associated with the canals prevent circulation of water and cause stagnation during low-flow periods. If left alone, the existing vegetative zones probably would migrate downstream over a period of decades to centuries due to the death of salt- and flood-tolerant species and their replacement by species adapted to fresh water or drier soil. Because the boundaries of refuges, nature preserves, and fish and wildlife management areas are fixed, a refuge that was originally an estuarine, intertidal wetland could in time become a forested wetland of the Atchafalaya River. The refuge would have to migrate downstream, keeping pace with the advancing delta, if it were to remain an intertidal wetland.

The existing and planned channel-training works, bank shaving or breaching, and wetland management units, where the inflow and outflow of water are controlled, are attempts to slow this process and are not really restoration. Instead of merely slowing the inevitable, additional attention ought to be directed to the natural reclamation or creation of wetlands from the sea that is now being performed by the river. For example, the COE is already employing a dredged material placement technique that augments, rather than disrupts, the natural formation of new marsh. Perhaps a decision could be made now that substantial portions of the new lands will be managed as natural areas or refuges, because virtually all of the existing Atchafalaya Delta is already a state wildlife management area, and the new land out to the 3-mile state limit will be too prone to flooding to be of use for development. The Atchafalaya Basin offers an opportunity to develop and test management concepts and techniques that are as dynamic as the system; habitats could be conserved as they are formed. In this case, restoration involves allowing spatial scope for a dynamic equilibrium to occur, where delta regression in one location is balanced by delta expansion in another.

References

Bryan, C. F., and D. S. Sabins. 1979. Management implications in water quality and fish standing stock information in the Atchafalaya River Basin, Louisiana. Pp. 293-316 in J. W. Day, Jr., et al., eds., Third Coastal Marsh and Estuary Management Symposium for Louisiana State University Press, Baton Rouge, La.

Bryan, C. F., F. M. Truesdale, D. S. Sabins, and C. R. Demas. 1974. A Limnological Survey of the Atchafalaya Basin. First Annual Report. U.S. Fish and Wildlife Service, U.S. Department of the Interior. 208 pp. and annotated bibliography.

Bryan, C. F., F. M. Truesdale, D. S. Sabins. 1975. A Limnological Survey of the Atchafalaya Basin. Second Annual Report. U.S. Fish and Wildlife Service, U.S. Department of the Interior. 203 pp. and appendix.

Bryan, C. F., D. J. DeMont, D. S. Sabins, and J. P. Newman, Jr. 1976. A Limnological Survey of the Atchafalaya Basin. Third Annual Report. Louisiana Cooperative Fisheries and Resources Unit, U.S. Fish and Wildlife Service. 285 pp.

Fremling, C. R., J. L. Rasmussen, R. E. Sparks, S. P. Cobb, C. F. Bryan, and T. O. Claflin. 1989. Mississippi River fisheries: A case history. Pp. 309-351 in D. P. Dodge, ed., Proceedings of the International Large River Symposium. Canadian Special Publication Fisheries and Aquatic Sciences 106.

Glasgow, L. L., and R. E. Noble. 1974. The Atchafalaya Basin. School of Forestry and
    Wildlife Management, Louisiana State University, Baton Rouge, La. 12 pp.
Penland, S., and R. Boyd, eds. 1985. Transgressive Depositional Environments of the
    Mississippi River Delta Plain: A Guide to the Barrier Islands, Beaches, and Shoals
    in Louisiana. Guidebook Series No. 3. Louisiana Geological Survey, B, Baton Rouge,
    La. 233 pp.
Sager, D. R., and C. F. Bryan. 1981. Temporal and spatial distribution of phytoplank-
    ton in the lower Atchafalaya River Basin, Louisiana. Pp. 91-101 in L. A. Krumholz,
    C. F. Bryan, G. E. Hall, and G. T. Pardue, eds., The Warmwater Streams Sympo-
    sium. American Fisheries Society, Bethesda, Md.
U.S. Army Corps of Engineers, New Orleans District. 1982. Atchafalaya Basin Flood-
    way System Louisiana, Environmental Impact Statement. Vol. 3, Appendix G. 250
    pp.
van Heerden, I. L., and H. H. Roberts. 1980. The Atchafalaya delta—Louisiana's new
    prograding coast. Trans. Gulf Coast Assoc. Geol. Soc. 30:497-506.
Wells, F. C., and C. R. Demas. 1977. Hydrology and water quality of the Atchafalaya
    River Basin. Technical Report No. 14. Louisiana Department of Transportation and
    Development, Office of Public Works. 53 pp.

THE UPPER MISSISSIPPI RIVER

Richard Sparks

The Upper Mississippi River (Figure A.6) is a 1,300-mile navigation system maintained by the U.S. Army Corps of Engineers (COE),
but it is also a national fish and wildlife refuge system, totaling
280,000 acres arranged like a corridor, maintained by the U.S. Fish
and Wildlife Service (Upper Mississippi River Basin Association, n.d).
The Izaak Walton League was largely responsible for persuading Congress to create the refuge in 1924 (Scarpino, 1985) and played a major
role again in 1974, when it joined with the Sierra Club and 21 western railroad companies to file a lawsuit to prevent COE from constructing a new dam and set of locks (Locks and Dam 26) near St.
Louis, Missouri (Upper Mississippi River Basin Commission, 1981).
The plaintiffs argued that the new locks had not been duly authorized by Congress and that more information was needed on the
effects on railroads and on the rivers of spending $1 billion (in 1991
dollars) to quadruple lock volume and thereby increase barge traffic
on the entire upper river. The U.S. District Court ordered COE to
obtain the consent of Congress, as well as more information on environmental and economic impacts. After considering the additional
reports, Congress authorized construction of a new dam and single
lock, imposed a fuel tax for the first time on commercial navigation,
and created a trust fund to use the revenues. Congress also ordered
that no further expansion of the navigation capacity of the system
occur until a Master Management Plan for the river was prepared by

the Upper Mississippi River Basin Commission (Inland Waterways Authorization Act of 1978, P.L. 95-502). The plan was submitted to Congress on January 1, 1982, and with some modifications ultimately became the Environmental Management Program for the Upper Mississippi River System that was authorized as Public Law 99-88 in 1985 and as part of the Water Resources Development Act of 1986 (P.L. 99-662). Most of the following description of the program was taken from the Fifth Annual Addendum to the program (U.S. Army Corps of Engineers, 1990), except where noted otherwise.

Funding

Approximately 97 percent of the $200 million authorized for the Environmental Management Program for 1986 to 1996 is for habitat restoration ($124.6 million) and long-term resource monitoring ($61.1 million). Although the program has yet to be funded at its fully authorized level of $20 million per year, the funding levels have increased steadily from $0.8 million in 1986 to approximately $15 million in 1990 and 1991, and authorization has been extended 2 years, to 1998. The five states bordering the Upper Mississippi River collectively have contributed $3 million during the first 4 years, or approximately 18.4 percent of the federal funding.

As of May 1990, 5 habitat projects had been completed, another 35 were under way, and a total of 54 had been scheduled out of several hundred that were submitted by the U.S. Fish and Wildlife Service and the five states bordering the Upper Mississippi River. Although the master plan recognized that sediment loading from tributaries created major problems in the mainstream rivers and recommended that the already well-established programs for soil erosion control be accelerated, the Environmental Management Program focuses on restoration projects located within the floodplain of the mainstem rivers where there was no preexisting program (Upper Mississippi Basin Commission, 1981; U.S. Army Corps of Engineers, 1990). Funding for the Environmental Management Program comes entirely from the general revenues of the United States, not from the fuel tax on commercial navigation that goes into the Inland Waterways Trust Fund (IWTF). The 11-member board which oversees the Trust Fund is selected by the Secretary of the Army from shippers and carriers who use the inland waterways. There is general agreement that the Fund can be used for new construction and rehabilitation of the inland navigation system but not for operation and maintenance, although there is no agreed-upon definition of what distinguishes rehabilitation from maintenance. The sections of the public law that

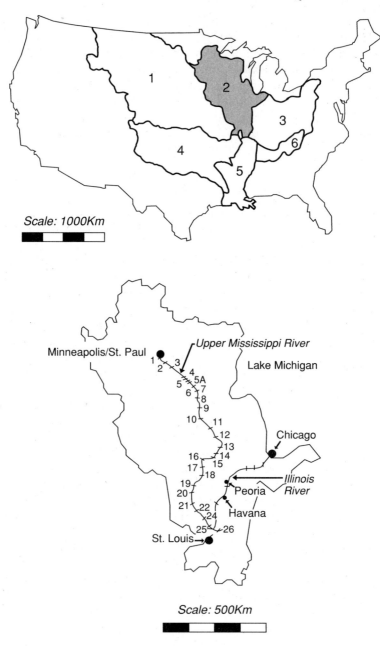

FIGURE A.6 Map of the six subbasins of the Mississippi River and an enlargement of the upper Mississippi subbasin (No. 2). The dams on the upper Mississippi River are numbered.

establish the Trust Fund (P.L. 99-662, the Water Resources Development Act of 1986) do not address environmental restoration. It is clear that rehabilitation does not include environmental restoration, because Public Law 99-662, the Water Resource Development Act of 1986, Sections 1404 and 1405, specifically excludes use of the Trust Fund for the Environmental Management Program.

As of March 15, 1991, 6 habitat projects had been completed, 6 were under construction, and 41 were in various stages of design and review, out of several hundred that were submitted by the U.S. Fish and Wildlife Service, and the five states bordering the Upper Mississippi River. Proposed projects are evaluated and prioritized by a panel of biologists from state and federal agencies, and are then screened by the Corps of Engineers for program eligibility and engineering feasibility. Projects on lands managed as federal refuges are 100 percent federally funded for construction and 75 percent federally funded for operation and maintenance. Lands managed by states or as part of federal navigation projects are 75 percent federally funded for both construction and maintenance. The majority of the projects redress sedimentation problems in side channels and backwaters, through a combination of dredging and alteration of flow patterns by channel structures, construction of enclosed levee systems with pumps for water level control, or construction of island breakwaters.

Monitoring

The projects include funds for monitoring for 1 to 2 years before construction and 2 to 5 years after completion. A longer time series of preconstruction data is often available from the management agency that originally proposed the project. The Upper Mississippi River includes three COE districts (St. Paul, Rock Island, and St. Louis), in two divisions (North Central and Lower Mississippi), and each district has its own monitoring plan, although a basic list of 17 physical-chemical factors is used by both St. Louis and Rock Island. Periodic hydrographic surveys will determine the longevity of the projects by measuring the rate of sedimentation. Populations of the animals or plants expected to benefit from the projects will also be monitored.

While site-specific effects of the restoration projects are being monitored, the separate Long-Term Resource Monitoring (LTRM) subprogram addresses data gaps identified in the original lawsuit: (1) lack of information on long-term trends in the Illinois and Upper Mississippi rivers, and (2) incomplete knowledge of the factors affecting fish and wildlife populations, including the impacts of navigation. The first of three elements in LTRM is long-term monitoring conducted on six study reaches spaced along the two rivers. Field stations on each reach are operated by the state natural resource agencies with federal funds and supervision provided by the U.S. Fish and Wildlife Service and the COE. Results from the study reaches are expected to document broad upstream-downstream and year-to-year trends in water quality, fish and invertebrate populations, and vegetation. The second program element uses short-term (1 to 3 years) experimental studies to determine what factors regulate or limit populations of key organisms. The third element is a Computerized River Information Center (CRIC) at Onalaska, Wisconsin, that maintains and analyzes long-term monitoring data as well as data provided by other agencies and programs, and supports data management at the field stations (U.S. FWS, 1988, 1989).

Evaluation

There have been problems in evaluating both the restoration proposals and the completed projects. One problem is quantification of expected benefits for purposes of selecting among competing proposals or among design alternatives within proposals. A habitat unit method for incremental analysis was first used in 1990 and will probably supplement, but not entirely replace, professional judgment. The method determines how many habitat units are gained by each increment in project cost. An example of a habitat unit might be the feeding area required to support a duck for a typical feeding period during its fall migration. Problems include the lack of quantitative data on habitat requirements of some species and the difficulty of defining optimum habitat for multiple species (although this limitation might be overcome by a guild approach; e.g., optimize for dabbling ducks rather than just mallards). The U.S. Army Corps of Engineers (1990) found that the cost of developing appropriate habitat models for diving ducks was $100,000, and application of the method to one proposal cost $4,000 but resulted in an estimated cost savings of $200,000 through elimination of a sediment-deflection dike that only marginally increased the number of habitat units. The validity of the habitat unit/incremental analysis method will be checked by the monitoring data collected after the projects are completed.

Criteria for Selecting Sites

A broader benefit/cost problem is whether restoration should be concentrated on areas that are least degraded or most degraded. More projects probably could be done, and the success rate might be higher, if resources were concentrated on the less degraded areas. However, this approach would favor the northernmost states on the Upper Mississippi River, where backwaters are relatively less degraded because sediment loading and boat traffic are not as heavy, and would not help the areas most in need of restoration (see Illinois River case study, below). The actual approach has been to strive for a reasonably even geographic distribution, although this approach has its difficulties because federal lands are concentrated along the northern portions of the river system, and COE lacks the authority to acquire lands for restoration, even if there are willing sellers (again see Illinois River case study).

Effectiveness

Although it is too early to judge whether the restorations undertaken under the Environmental Management Program are successful in ecological terms, the program has been successful in mobilizing federal and state resources in a coordinated approach to restoration. The program is important because it is among the first in the nation to address conflicting federal mandates for large interstate rivers and to redress habitat degradation caused by alterations within the rivers and their drainage basins. Programs based on this model are being proposed for other large rivers in the United States (Raymond Hubley, U.S. Fish and Wildlife Service, La Crosse, Wisconsin, personal communication, January 8, 1991). Finally, what may be most encouraging to citizens concerned about restoration of rivers is that these programs grew out of concerns first raised at the local level by chapters of national conservation organizations.

References

Scarpino, P. V. 1985. Great River. An Environmental History of the Upper Mississippi, 1890-1950. University of Missouri Press, Columbia, Mo. 219 pp.

U.S. Army Corps of Engineers. 1990. Upper Mississippi River System Environmental Management Program. Fifth Annual Addendum. U.S. Army Corps of Engineers, North Central Division. 221 pp.

U.S. Fish and Wildlife Service (FWS). 1988. 1985 National Survey of Fishing, Hunting, and Wildlife-Associated Recreation. Washington, D.C. 167 pp.

U.S. Fish and Wildlife Service (FWS). 1989. Long Term Resource Monitoring Program for the Upper Mississippi River System—First Annual Report. U.S. Fish and Wildlife Service, Environmental Management Technical Center, Onalaska, Wis. 125 pp.

Upper Mississippi River Basin Association. n.d. The Upper Mississippi River System Environmental Management Program. Upper Mississippi River Basin Association, St. Paul, Minn. 8 pp.

Upper Mississippi River Basin Commission. 1981. Comprehensive Master Plan for the
    Management of the Upper Mississippi River System. Upper Mississippi River Basin
    Commission, Minneapolis, Minn. 181 pp.
Water Resources Development Act of 1986. P.L. 99-662, Nov. 17, 1985, 100 Stat. 4082.

THE ILLINOIS RIVER-FLOODPLAIN ECOSYSTEM

Richard Sparks

Introduction

The history of change in the Illinois River is unusually well documented because scientific studies began in 1874 (Calkins, 1874;
Forbes, 1878; Hart, 1895; Kofoid, 1903; Baker, 1906; Forbes and Richardson, 1908, 1913, 1919; and, subsequently, many others). This history
provides examples of (1) the effects of an attempt to restore one ecosystem (Lake Michigan) by transferring a pollution problem to another ecosystem (the Illinois River), (2) partial success in subsequent
efforts to restore water quality in the river, (3) exceeding the limits of
ecosystem resistance to increased sediment loading from nonpoint
sources, (4) ongoing efforts to restore multiple functions of the river-
floodplain ecosystem, and (5) constraints on these efforts.

General Characteristics

HYDROLOGY

The Illinois River drains approximately 29,010 square miles (or
18.5 million acres) in three states (85.5 percent of the drainage is in
Illinois, 11.0 percent in Indiana, and 3.5 percent in Wisconsin) (Neely
and Heister, 1987) (see Figure A.6). The upper Illinois River was
connected to Lake Michigan as early as 1848 by the Illinois and Michigan Canal, but the major link, the Chicago Sanitary and Ship Canal,
opened in 1900, and the total distance from Chicago to the confluence
with the Mississippi just above St. Louis is now 327 miles (Injerd,
1987; Neely and Heister, 1987). The average discharge measured 71
miles upstream of the confluence is 21,895 ft^3/s for the 51-year period of record and includes approximately 3,200 ft^3/s released from
Lake Michigan (Sullivan et al., 1990). The maximum recorded discharge of 123,000 ft^3/s occurred in May 1943, and the minimum of
1,330 ft^3/s in September 1984 (Sullivan et al., 1990).

The lower 200 miles of river was unusually productive of fish and
wildlife because it is a river-floodplain ecosystem flanked by exten-

sive backwaters, floodplains (ranging from 2 to 12 miles in total width and totaling approximately 400,000 acres), and floodplain lakes, and lies on the Mississippi flyway for migratory waterfowl (Mills et al., 1966). The typical spring flood provides access to spawning areas for fish and to feeding areas for both fish and waterfowl. The flood is protracted because the capacity of the floodplain is large in relation to the flow of the river, and the gradient is very shallow: the rate of fall in the lower 371 km (223 miles) is only 2 cm/km (approximately 0.1 ft per mile) (Mills et al., 1966). Two natural mainstem lakes, the upper and lower Peoria lakes, occurred where the river was partially dammed by alluvial deposits from tributaries. Four dams downstream of the Chicago waterways maintain 9-ft minimum depths for navigation (Sparks, 1984).

BIOLOGICAL PRODUCTIVITY

In 1908, a 200-mile reach of the Illinois River produced 10 percent of the total U.S. catch of freshwater fish—more than any other river in North America (excluding rivers with anadromous fish, such as the Columbia). More than 2,000 commercial fishermen were employed on the river (U.S. Department of Commerce and Labor, 1911) and the commercial yield was 24 million pounds annually, or about 178 pounds per acre of permanent water (Lubinski et al., 1981). By the 1950s the yield had dropped to 38 pounds per acre; since the 1970s the yield has been a low 4 pounds per acre, totaling only 0.32 percent of the total U.S. freshwater harvest (Sparks, 1984). Similar downward trends were recorded over the same period for other indicators of biological productivity: waterfowl and sport fish populations (Bellrose et al., 1979; Sparks, 1977).

The declines are attributable to two major man-made changes: diversion of Chicago's pollution from Lake Michigan to the river, and intensification of agriculture in the upland drainage and floodplain.

Transferring Pollution from Lake Michigan to the Illinois River

In 1854 and 1885 major rainstorms caused untreated sewage to be carried into Lake Michigan, where it entered water intakes and caused outbreaks of cholera and typhoid (Injerd, 1987). In response to these epidemics, the flow of the Chicago and Calumet rivers was reversed, and sewage and Lake Michigan water were conveyed away from the lake and into the Illinois River via the Sanitary and Ship Canal, starting in 1900 (Injerd, 1987).

From 1900 to 1910 the organic loads from Chicago exerted their oxygen demand in the upper Illinois River and fertilized the middle and lower reaches (Palmer, 1903). The peak yield of commercial fish from the Illinois River occurred in 1908, not only because the nutrient loading and therefore the overall productivity of the river increased, but also because the diversion raised the minimum water levels in the river and its backwaters, thereby increasing the amount of aquatic habitat (Sparks, 1984). After 1910, however, the increasing pollution load from Chicago caused critically low oxygen levels in the water and putrescent conditions in the bottom sediments to progress further downriver each year. Forbes and Richardson (1919) and Richardson (1928) documented the alteration and destruction of the bottom fauna by this wave of pollution. They also reported that beds of aquatic plants, which once covered up to 50 percent of the total surface of the bottomland lakes, had disappeared by 1920. Fish populations declined because of the direct effects of low oxygen and loss of food, and habitat alteration produced by the die-off of aquatic plants (Sparks, 1984).

Restoration of Water Quality in the Illinois River

In the late 1920s and early 1930s, most of the larger cities along the Illinois River constructed sewage treatment plants; and so dissolved oxygen levels improved, and aquatic plants and fish populations recovered (Starrett, 1972). From the 1940s to the 1970s, municipal waste treatment capacity and technology did not always keep up with population growth: low oxygen levels occurred in both upper and middle reaches of the river in the mid-1960s, for example (Mills et al., 1966; Starrett, 1971).

IMPACT OF THE FEDERAL CLEAN WATER ACT

Since the enactment of the federal water pollution control acts, including the Clean Water Act of 1977 (P.L. 95-217), there has been an infusion of federal funds to help upgrade sewage plants. Between 1965 and 1975, approximately $4 billion in federal and local funds were spent on waste treatment by municipalities in the Illinois drainage basin (Briceland, 1976). In the Chicago area alone, a program to capture and treat combined sewer overflows (the Chicago Tunnel and Reservoir Plan, or TARP) will cost $3.8 billion if fully funded and completed in the 1990s (Lanyon and Lue-Hing, 1987). The combined sewer overflows constitute the largest remaining organic loading in the Chicago area that is untreated: the waterway absorbs the equivalent of raw waste from one million people per day (Lanyon and Lue-Hing, 1987). The first phase of TARP is partly complete and will reduce this load by approximately 85 percent, as well as reducing floodwater damage by 10 to 15 percent (Lanyon and Lue-Hing, 1987). Upon completion in the 1990s, TARP will reduce the organic loading by 99.8 percent and flood damage costs by 65 percent (Lanyon and Lue-Hing, 1987).

MONITORING AND EVALUATION OF SUCCESS

Water quality in the Illinois River has improved, according to the latest biennial report filed by the Illinois Environmental Protection Agency under the requirements of Section 305(b) of the federal Clean Water Act (Illinois Environmental Protection Agency, 1990). This assessment is based on a water quality index (WQI), a macroinvertebrate biotic index (MBI), and a trend analysis of physical-chemical factors. The WQI compares six factors (temperature, dissolved oxygen, pH, total phosphorus, turbidity in nephelometric turbidity units (NTUs), and conductivity) with Illinois water quality criteria for general use. If no state criteria are available the U.S. Environmental Protection Agency (EPA) criteria for warm-water fish are used. These subindices of departure from criteria are then summed to provide an overall index. The six factors were selected because they showed the greatest degree of correlation with the MBI of all combinations of factors monitored in the Illinois Ambient Water Quality Monitoring Network (Kelly and Hite, 1984). The MBI assesses the numbers of pollution-tolerant and pollution-intolerant macroinvertebrates that colonize artificial substrates (Hester-Dendy samplers) suspended in the water column. Seven of the eight water quality monitoring stations on the mainstem Illinois River showed improved WQIs when the period from 1985 to 1989 was compared with that from 1979 to 1984. The trend analysis was conducted in cooperation with the U.S. Geological Survey (USGS) on the most recent 12 years of data available from 10 stations on the mainstem Illinois River. The seasonal Kendall test, which is intended for monthly water quality time series with potentially large seasonal variability (Smith et al., 1987), was used both on flow-adjusted concentrations and on unadjusted concentrations. Factors that showed no trends at any of the stations included two used in the WQI: turbidity and total phosphorus. Conductivity, ammonia, total suspended solids, chemical oxygen demand (COD), and five other factors showed improvement at several stations, and only one factor, sodium, showed a worsening trend (higher concentrations in more recent samples). Another state agency, the

Illinois Department of Conservation, credits improved water quality for partial restoration of gamefish populations in the upper 100 miles of the river, including the appearance of a sauger (*Stizostedion canadense* (Smith)) population that has supported a nationally ranked annual fishing tournament (Conlin, 1987).

Although restoration of water quality is often a necessary component of river and stream restoration, improving water quality alone may not be sufficient to restore streams and rivers, as described next in the case of the Illinois River.

Stress Thresholds: Sedimentation in the Illinois River

Some ecosystems have a degree of biotic control or compensation that may mask a gradual, detrimental change in some physical-chemical factor until a threshold is reached. Once the threshold has been crossed, the ecosystem may degrade rapidly into a stable condition that is very difficult to restore to its previous condition. In the Illinois River, clear vegetated backwaters and lakes became excessively turbid, barren areas, and gamefish and duck populations declined drastically from 1958 to 1961 in what was formerly the most biologically productive reach of the river: the lower 321 km (200 miles) (Bellrose et al., 1979). These degraded conditions persist to this day and are the subject of restoration efforts ranging from the scale of the entire drainage basin to experimental plots a few meters square.

INCREASED SEDIMENT LOADING

The changes in biological productivity were associated with, and are probably attributable to, increases in sediment loading and sediment resuspension in the Illinois River (Bellrose et al., 1979). Sediment loading increased because of land use changes in the drainage basin and floodplain. Cropland accounts for 70.4 percent of the land area of Illinois, and so changes in farming practices have a major impact on streams and rivers (Herman, 1987). In the Illinois River basin, row cropland increased about 67 percent between 1945 and 1986, at the expense of pasture, forage, and small grains, which better protect the soil from erosion (Bellrose et al., 1979). As farms became larger and more specialized in row crops, fences and fence rows were taken out (Illinois Environmental Protection Agency, 1979). The size of farm machinery also increased, making it more difficult to do contour farming, and many contours and old terraces were removed (Illinois Environmental Protection Agency, 1979). The common practice of plowing fields soon after harvest in the fall leaves land susceptible to erosion for almost two-thirds of the year, including the period of heaviest rainfall in the early spring (Havera and Bellrose, 1985).

Stream modifications undertaken by farmers also contributed to sediment loading. Farmers channelize streams to improve drainage so that they can get on the fields earlier in the year and to straighten field borders to make it easier to use large equipment. Channelization shortens the stream length and thereby increases the slope. The water moves at greater velocity, gains erosive power, and tends to erode its bank and bed. Leedy (1979) estimated that more than 50 percent of the annual sediment yield of Illinois streams comes from bank and bed erosion. Because the rate of fall of the tributaries is approximately a foot per mile, they deliver most of their sediments to the Illinois River, where it settles out because the mainstem falls at only one-tenth of a foot per mile (Mills et al., 1966). Channelization extended to the marshy or forested deltas where tributaries enter the Illinois River, so these areas no longer trap sediments before they enter the river (Roseboom et al., 1989). Finally, approximately half the floodplain of the Illinois River was drained and leveed for agriculture (Bellrose et al., 1983; Thompson, 1989), so that sedimentation was concentrated in the remaining overflow areas, lakes, and backwaters.

The end result of these changes was an increasing sediment loading on the river, evidenced by sedimentation rates in a mainstem lake (Peoria Lake) that were twice as high in 1965 to 1985 (1.44 percent of the lake volume is lost per year) as in 1903 to 1965 (Demissie, 1989). Of an estimated 27.5 million tons of sediment delivered to the Illinois mainstem annually, approximately 12.1 million tons are delivered to the confluence with the Mississippi: the remaining 15.4 million tons are deposited in the remaining unleveed floodplain and backwaters (Lee, 1989). If this amount were spread evenly over the remaining floodplain, it would aggrade at the rate of 0.19 inch per year (Lee, 1989).

RESUSPENSION AND TRANSLOCATION

Once the sediment is in the river and backwaters it is resuspended by boat- and wind-driven waves and currents. The fine-grained sediments take 7 to 12 days to settle out, following a windstorm (Stall and Melsted, 1951). Because the average recurrence interval of moderate to strong winds in Illinois is less than 7 days, the river and its backwaters tend to remain turbid (Jackson and Starrett, 1959). Commercial and recreational boats not only cause bank erosion, but also resuspend and relocate sediments (Bhowmik and Schicht, 1980). By using infrared photography, Karaki and van Hoften (1974) showed that commercial tow boats on the Illinois River generate turbidity trails 1.6 km in length, some of which enter backwaters adjacent to the river. Sparks et al. (1980) reported that barge traffic in the Illinois River increased suspended solids by 30 to 40 percent. Simons et al. (1981) used simulation models to project that sediment volumes entering a backwater along the lower Illinois River will increase 46.8 percent by the year 2000, if the tow traffic increases as predicted.

THRESHOLD EFFECT

Although the sediment loading undoubtedly increased gradually, the major biotic changes occurred rather suddenly, probably because rooted aquatic vegetation helped control turbidity (Jackson and Starrett, 1959) until an effect threshold was exceeded. Rooted plants promote settling and reduce sediment resuspension in the following ways: the roots of aquatic plants anchor the bottom against disturbance by waves and bottom-feeding fish, the stems slow currents and cause sediment to drop out, and the leaves dampen wave action. Sparks et al. (1990) described the probable sequence of events in the Illinois River from 1955 to 1960: increasing turbidity reduced light penetration and photosynthesis that, in turn, weakened or killed submerged aquatic plants growing in the deepest parts of the shallow backwaters. With wave action undampened, larger waves resuspend more sediment and uproot more plants, further increasing turbidity. This positive feedback caused rapid degradation of the remaining plant beds. A vegetation survey by Havera et al. (1980) in 1978 showed that submergent plants and all but one species of floating aquatic plant had been virtually eliminated from the lower 200 miles of the Illinois River and its connecting backwaters.

SECONDARY EFFECTS

The secondary effects of the loss of aquatic plants have been dramatic. Waterfowl usage of the Illinois River has declined because of the loss of plants and plant-associated invertebrates that waterfowl fed upon (Havera and Bellrose, 1985). Fish populations have declined, and dominance has shifted from sight predators and nest builders to species that can locate their food by scent and scatter their eggs on silty substrates (Sparks, 1975). Rooted aquatic plants take up ammonia as a nutrient, and loss of the plants may explain the buildup of toxic ammonia in the sediments, which appears to be limiting benthic macroinvertebrates and possibly fish (Ross et al., 1989; Ruelle and Grettenberger, 1991).

Restoration of the submersed aquatic vegetation in the Illinois River is key to restoration of the other functional parts of the system, and control of sediment introduction and resuspension is necessary for revegetation. The institutional and technical approaches to these objectives are described below.

Public Recognition of the Need for a Comprehensive Restoration Program

In 1987, approximately 200 people representing private citizens, conservation and environmental organizations, elected officials, university faculty, and federal, state, and local governmental agencies met on the Illinois River at Peoria to define management problems relating to the river and to identify ways to solve the problems (Mathis and Stout, 1987). The meeting had been instigated by 34 organizations, most of which were concerned with the deterioration of the river and its lakes in the vicinity of Peoria. The meeting was coordinated by the Water Resources Center at the University of Illinois and organized in a format to invite exchange of information, discussion, and suggestions. Participants agreed that the river needed to be managed as a system and that soil erosion and sedimentation were major problems affecting functions of the river, including recreational use, fish and wildlife production, and flood conveyance. Secondary problems included (1) lack of a comprehensive management plan for the Illinois River system; (2) lack of coordination among local, state, and federal agencies; (3) loss of wetlands and wildlife habitat along the river; (4) lack of a central organization to deal with the entire watershed; and (5) a general apathy toward the Illinois River basin on the part of state officials and the general public.

Institutional Responsibilities and Actions

LOCAL

Several technical and lobbying groups organized during and after the conference to help promote the recommendations. The Illinois River Coalition/Father Marquette Compact was organized by citizens from five river counties to build a regional consensus and tap governmental resources for river restoration. The Heartland Water Resources Council focuses on managing and restoring the river and its lakes around Peoria. The Illinois River Soil and Water Conservation Task Force, which was formed in 1985, received a public relations boost for its programs as a result of the conference. The Soil and Water Conservation Task Force is made up of the elected directors of seven Soil and Water Conservation Districts along the Illinois River, as well as advisors from industry, state and federal agencies, and conservation organizations. The districts work together to accelerate and implement conservation practices in the Illinois River watershed. The task force supports traditional erosion control practices such as conservation tillage and terraces, as well as low-cost streambank and ravine stabilization practices. Many of the latter projects require extensive hand labor, which has been obtained through the Public Aid Program, Project Chance. As a side benefit, 20 out of the 50 Project Chance workers subsequently used the training they received and recommendations from their supervisors to obtain permanent jobs. The Soil and Water Conservation Task Force has received grants and equipment from the Caterpillar Tractor Co. and funding from the Illinois Department of Energy and Natural Resources. The Illinois State Water Survey monitors several of the projects.

STATE AND FEDERAL

At the request of the governor of Illinois, the Illinois State Water Plan Task Force reviewed the proceedings of the 1987 meeting and developed an action plan for state agency response. State agencies are now working with local groups and federal agencies on implementation of the plan, and progress is being monitored by the Water Plan Task Force (Vonnahme, 1989). Specific actions include (1) development of hydraulic simulation models and sediment transport models to evaluate the effects of management alternatives; (2) improved data collection, including hydrographic surveys, water quality monitoring, and sediment source identification; (3) installation of streambed and bank erosion controls on four of the tributaries that contribute the most sediment to the Illinois River above Peoria; (4) experimental rehabilitation of selected backwater areas; (5) nonstructural flood mitigation plans tailored for specific communities along the river (the plans are designed to reduce the $25 million average annual cost of Illinois River flooding [Wetmore, 1987] and include relocation of flood-prone structures, and public acquisition of flood-prone land and conversion to community parks and open areas); and (6) improvement of public access points and state parks along the river.

Approach to Restoration

The planned restoration is comprehensive: restoration techniques are being applied to the upland erosion sites and the tributaries to reduce sediment loading of the river; revegetation experiments have been conducted in selected backwaters of the main river; and plans are being made for large-scale projects involving dredging of deposited sediments and creation of island wave barriers. The techniques are described below, grouped according to where in the basin they are applied: uplands; tributary channels; or mainstem river and backwaters, or floodplain lakes.

TECHNIQUES: UPLANDS

Much of the restoration effort has been focused on the two mainstem lakes at Peoria. Together they total 4,784 surface hectares (8,033 acres) and have been characterized as the most important recreational lakes in central Illinois, because the Peoria-Pekin population center straddles them (Bellrose et al., 1983). Of the sediment in the lakes, 40 percent originates from just 3 percent of the upstream drainage area, and erosion control and sediment trapping techniques have focused on the problem tributaries (Semonin, 1989).

The Illinois River Soil Conservation Task Force accelerated existing state and federal programs that reduce erosion from row crop fields on a cost-share basis with farmers, using $85,000 per year provided by the Illinois Department of Conservation. These include well-established practices such as conservation tillage and grass waterways administered by the local Soil Conservation Service offices and the Agricultural Stabilization and Conservation Committees, and are not discussed further here.

TECHNIQUES: TRIBUTARY CHANNELS

Within problem tributaries, reaches contributing the greatest amounts of sediment are restored first. These are typically massive streambank or gully erosion sites (Condit, 1989). Selection of techniques was based on an experimental, long-term study conducted by the Illinois State Water Survey and the Illinois Department of Conservation on a tributary to the Illinois River, Spoon River, and one of its tributaries, Court Creek (Roseboom and White, 1990).

Tree cuttings are used at all of the sites on the three tributaries to establish vegetative cover and stabilize the banks (Condit, 1989). Most of the sites also use tree revetments cut from adjacent woods and anchored at the toe of the banks with Laconia earth anchors and cables. One of the sites uses a gabion basket structure filled with broken concrete (Condit, 1989).

Although all of the sites incorporate some experimentation (selection of tree species used for cuttings, alternative methods of inserting the cuttings in the bank), formal trials were designed by the Soil Conservation Service only for the most diverse site, where monitoring is being done by the Illinois State Water Survey (Condit, 1989).

TECHNIQUES: MAINSTEM RIVERS AND LAKES, BACKWATERS, AND FLOODPLAIN LAKES

Even if the sediment loading of the Illinois River is drastically reduced by the measures employed on the upland fields and in the tributary channels, the lakes and backwaters of the river will remain degraded by the sediments already deposited there, which are endlessly resuspended by wind- and boat-driven waves. The feasibility of restoring submersed aquatic plants under existing conditions was investigated in several field experiments. Based on these experiments, earlier projects on the Upper Mississippi River, and standard waterfowl management practices, nine major rehabilitation projects have been proposed for the lower 200 miles of the Illinois River. The field experiments are described below, and the design features of the major projects are discussed briefly.

Revegetation Experiments

Revegetation experiments were conducted with waterfowl food plants, duck potato or arrowhead (*Sagittaria latifolia*), and sago pondweed (*Potamogeton pectinatus*), in Peoria Lake from 1986 to 1990 (Roseboom et al., 1989) and with wild celery (eelgrass, tape grass), *Vallisneria americana*, in 1990 in backwaters about 50 miles downstream at Havana, Illinois (Korschgen, 1990). The following description is taken mostly from the two sources cited and from personal communication with Donald Roseboom, Illinois State Water Survey, on February 15, 1991. Two of the backwaters in the Havana area were isolated from the main channel during low flow, and one was contiguous with the river at all river stages. The plants survived and grew if they were rooted in a deep, cohesive soil layer or in buckets of soil. However, the plants in both sites attracted herbivores that ate them down to the roots. The herbivores were identified as Canada geese in Peoria Lake, with some consumption or removal attributed to muskrats. The herbivores were never seen at Havana but were presumed to have been waterfowl. The plants in Peoria Lake subsequently were protected from grazing by orchard netting, but were eventually uprooted by waves, except where a 40-ft tree had lodged and created a windbreak.

During the winter of 1986-1987, a 700-ft breakwater was constructed onshore with donated tires and labor, and towed out to the revegetation site, where it was tied to pilings made of pipes that had been water-jetted 11 ft into the sediment. The tires float because they trap air, and they are oriented to absorb the energy of waves coming from the prevailing wind direction. Survivors of the first planting and the new plantings thrived during the next three growing seasons, with the arrowhead producing flowers at the end of each growing season. Conditions were unusually favorable for revegetation because water levels remained relatively low and stable during the drought of 1988-1989. The air pockets trapped in the tires gradually diminish unless air is periodically blown into them. The breakwater was allowed to sink in the winter of 1989-1990, to test whether the plant beds could sustain themselves behind the sunken tires and the subsurface levee that had formed because of increased sedimentation induced by the breakwater and the plants. Also, the tire breakwater is not as esthetically pleasing to some as a natural island or vegetation bed.

Unfortunately, the water regime in 1990 was drastically different from that in 1988-1989; there were several low to moderate floods during the growing season, and it was impossible to determine whether the observed loss of the plant beds was attributable to the lack of a breakwater, the fluctuating water levels, or a combination of both. It would be desirable to continue the experiment with more than one breakwater, so that at least one could be left floating and one allowed to sink, although the logistics of building and putting 25-ton structures in place tax the limits of locally available volunteer help.

Proposed Side Channel, Backwater, and Bottomland Lake
Rehabilitation Projects

The partial success of the breakwater in Peoria Lake led to inclusion of breakwaters in the design of four of nine major habitat rehabilitation projects planned for the Illinois River (Donels, 1989). These projects are part of the Habitat Rehabilitation and Enhancement Subprogram of the Environmental Management Program for the Illinois and Upper Mississippi rivers. Advertisements for construction bids for the project in Peoria Lake are to be issued in December 1991 and include a barrier island (the material dredged for the island will also deepen a small portion of the lake), removal of a plug of downed trees and silt from a major side channel (the East River), and installation of water-control structures (low levees, gates, and pumps) for waterfowl management on an existing forested conservation area.

Lake Chautauqua, a federal wildlife refuge at Havana, will have improved water-control structures, island or levee breakwaters, and a reopened side channel. Wing dams are also proposed for another federal refuge, Swan Lake.

The rest of the projects include traditional methods for managing waterfowl: creation of impoundments, with low levees and gates or pumps, for water-level control. The impoundments are drawn down in the summer to encourage germination and growth of moist soil plants. Pumps or gates are used to flood the impoundments during the waterfowl migration in the fall, so ducks and geese can use the summer's seed production. The problem with these techniques is that low levees and water-control structures are barriers to fish movements. Although fish can often enter these areas during major floods in the spring (which overtop the low levees), they may not be able to escape through the shallow water or water-control structures when the impoundments are drawn down.

Constraints on Restoration

SCALE CONSIDERATIONS

A large-scale restoration usually implies that the restoration will also be long term. Because resources are limited and the problem of excessive sediment yield affects the entire predominantly agricultural basin, substantial reduction of sediment loading will probably take decades. In the meantime, the high turbidity and sediment concentrations must be factored into the design of restoration projects in the mainstem river and its backwaters. This usually means that not all functions of the river-floodplain system can be restored simultaneously. For example, in cases where former levee districts have been purchased for wildlife areas (or considered for purchase), plans are to retain the high levees, instead of opening them to the river (Roelle et al., 1988). Where natural backwaters were isolated from the river years ago, as at Spring Lake near Havana, the backwaters have retained the submersed aquatic macrophytes characteristic of the pre-1955 river. Opening the levees would admit sediment-laden water that would degrade the new areas. Most planners and managers believe that breaching of the levees in conservation areas must wait for reduction of sediment loading in the river; thus, restoration of these areas for flood storage and conveyance and for use by fish that migrate between channels and backwaters must be considered incomplete, although these areas do support both migratory and resident wildlife populations and outdoor recreation.

The revegetation experiments in the Illinois River also raise important research and management questions about the scale of restoration projects. What are the minimum density and area of aquatic plants that can sustain the anticipated grazing pressure and contribute to vertebrate production? Does the answer change if the herbivorous grass carp, *Ctenopharyngodon idella*, invades the Illinois River from the Upper Mississippi River and establishes reproducing populations? Is there some threshold surface area that has to be protected from waves or revegetated before plants begin to exert sufficient control over sediment resuspension and turbidity to maintain themselves or expand outward from planted areas (i.e., what levels of "treatment" will trigger rapid regeneration of plant beds)? These are questions that are likely to be answered only by continued field trials and careful monitoring.

CONFLICTS IN RESTORATION GOALS AND TECHNIQUES

As suggested in the previous section, there is a conflict between managing existing floodplains and backwaters for waterfowl versus managing for fish. The conflict is engendered by the present degraded condition of the river. The pristine river provided food and habitat for both waterfowl and fish. With the submersed aquatic plants and their invertebrate fauna gone, managers rely on moist soil plants for waterfowl food, so they build low levees and water-control structures to draw water levels down in the summer to create mud flats—practices that are often detrimental to migratory fish and other aquatic organisms. There has been a substantial investment in developing management techniques, installing water-control structures, and training managers so that there is a natural interest on the part of waterfowl hunters and refuge managers in continuing and even expanding moist soil units.

BIOTIC CONSTRAINTS

In some cases, reestablishment of vegetation was hindered, not by the herbivores for whose benefit the vegetation was planted (geese and muskrats), but by pests such as the willow leaf beetle that attacked the new leaves on willow posts installed along one of the tributaries. Mortality of the sprouts was held to 20 percent by spraying with insecticide (Condit, 1989).

ABIOTIC DISTURBANCES

Any biologically mediated restoration is most vulnerable to disturbance when newly planted. Environmental conditions that adversely affected some of the bioengineered bank stabilization projects in the Illinois River tributaries included the drought of 1988-1989 (some plantings dried up) (Condit, 1989).

Ice driven by winter or spring floods is a severe test for any bank stabilization method. In 1987, Palmiter revetments on the Court Creek tributary were heavily damaged or removed by such a flood less than a year after being installed (Condit and Roseboom, 1989). The revetments were rebuilt with Osage orange (*Maclura pomifera*), a more durable tree species; a better anchoring system (Laconia earth anchors); and willow whips planted through the revetments. The improved revetments withstood a severe flood the following year, when 3.5 inches of rain fell in an hour (Condit and Roseboom, 1989).

REDUCED WATER QUALITY

Aside from the beneficial effects on the mainstem river and lakes, the tributary restoration program was expected to restore fish and wildlife habitat along the streams. During the drought of 1988, feed-lot runoff killed fish in one of the demonstration areas, and pesticide and feedlot runoff is suspected of limiting fish populations in other reaches (Condit and Roseboom, 1989).

SOCIAL AND ADMINISTRATIVE CONSTRAINTS ON RESTORATION

In common with other large rivers that were used for economic development from the time of settlement, the drainage basin, the floodplain, and the bottom of the river itself (in many locations) represent a checkerboard of ownership parcels. The federally supported habitat rehabilitation projects can be installed only on public land, which is in short supply along the Illinois. Although the General Plan for the Environmental Management Program on the Upper Mississippi River System (U.S. Army Corps of Engineers, 1986, Appendix D: Guidance, Policy and Procedures Part III. Environmental Management Program Elements. Section A, Habitat Rehabilitation and Enhancement Projects) specifically stated that "Prime habitat areas for waterfowl and other wetland wildlife species can be acquired and restored," the Office of the Chief of Engineers directed that such projects not be pursued (memo of February 5, 1988 to the Commander, North Central Division). Lands are slowly being acquired, when there are willing sellers, by organizations such as the Nature Conservancy, local park districts, and the Illinois Department of Conservation. The Illinois-Michigan National Heritage Corridor, administered by the National Park Service, provides for the acquisition and management of some lands between the old Illinois and Michigan Canal and the Illinois River. The pending designation of the remainder of the river as a National Heritage Corridor may spur and focus local, state, and federal acquisition of the riverbank and conversion to open space.

Obtaining 30-ft riparian conservation easements along the tributaries for the bank-stabilization techniques has not been a problem because the landowners were losing cropland and crops to the streams (Condit and Roseboom, 1989). Private landowners were concerned about potential property damage and littering; thus, they retain the right to deny access to individuals. The conservation easements are given by the landowner to the local county Soil and Water Conservation District, which administers and maintains the area. The easement must remain in a natural state after reforestation with pin oak, green ash, red cedar, and gray dogwood (Condit and Roseboom, 1989).

Summary

The Illinois River was an unusually productive floodplain-river ecosystem, until impacted by municipal and industrial waste loading from Chicago and sediment loading resulting from land use changes associated with agriculture (drainage and leveeing of floodplains, channelization of tributaries, removal of riparian forests, and excessive soil erosion).

Expenditure of approximately $6 billion in federal and local funds on waste treatment by municipalities has resulted in improvement in water quality in the main channel, based on trend analysis of physical-chemical factors and biological indicators. Game fish have returned to the upper river, where they were formerly absent, and a sauger population supports a nationally ranked annual fishing tournament. The Illinois River provides 2.1 million angling days per year, valued at $25.2 million annually in 1983 dollars (Conlin, 1987). Hunters spend an additional $14 million per year (Conlin, 1987).

Although water quality has improved, the functions of the river-floodplain system remain impaired by excessive sediment loading. This loading probably increased gradually through the 1950s until a threshold was reached that caused rapid collapse of the submersed aquatic plant beds in the river and its associated backwaters. Since the plants acted as a biological mediator that made the ecosystem somewhat resistant to sediment loading, it is now difficult to restore the submersed aquatic vegetation in an environment where wind- and boat-generated waves continually resuspend bottom sediments.

Revegetation experiments indicate that submersed aquatic plants will grow if protected by wave barriers and if roots and tubers are planted more deeply in the sediments than they might naturally grow. As the plants gain a solid roothold in deeper, denser sediments, the breakwaters continue to protect the growing plants. Based on these successful results, artificial island breakwaters are being planned for four of nine restoration projects that are part of the federally funded Environmental Management Program for the Upper Mississippi and Illinois rivers. Material to build the islands will be dredged from the bottom, thereby re-creating deep areas that may be used as wintering areas by some fish.

The revegetation experiments raised important questions about the scale and effectiveness of restoration. Herbivores (Canada geese and muskrats) rapidly consumed the plants on the small test plots, which had to be protected with orchard netting. An introduced herbivore, the grass carp or white amur, may also constrain revegetation. There may be some threshold surface area that must be protected from waves and herbivory before plants can sustain the grazing pressure and begin to exert sufficient control over sediment resuspension and turbidity to maintain themselves and expand outward from planted areas. The threshold that might trigger rapid regeneration of the plant beds is not known.

In addition to treating the mainstem river and backwaters, efforts are being made to reduce sediment loading by reducing soil erosion in the drainage basin and along the tributaries. Funds have been provided by the state of Illinois to accelerate soil erosion control in the basins with the greatest sediment yields. Reaches with the highest rates of bank and bed erosion are being stabilized with bioengineering approaches, including willow whip and post plantings and anchored tree revetments. Techniques originally developed in Ohio had to be modified to work in the easily erodible soils of the Illinois River tributaries. New plantings were especially vulnerable to disturbances, such as the 1988-1989 drought, infestations of willow leaf beetles, and ice scour, and most of the tributary restorations require some repair and adjustment.

It may take decades before measures to control soil and bank erosion substantially reduce sediment loading of the river because of the large scale of the problem, limited resources, and lag effects (sediments already in channels may keep moving toward the main river). The present degraded condition of the river limits restoration options and brings about conflicts in restoration objectives and approaches. Breaching of levees in floodplain drainage districts acquired for conservation would admit sediment-laden river water that would degrade the restored wetlands. Hence, the functions of flood storage and conveyance and use by migratory fishes have not been restored in these areas. Because submersed aquatic plants will not grow in backwaters and lakes that are connected to or periodically overflowed by the river, waterfowl managers rely on low levees, gates, and pumps to lower water levels and produce moist soil plants on exposed mud flats. These techniques probably have been responsible for a recent upward trend in dabbling duck and goose populations but probably have been detrimental to fish that need access to backwaters.

The greatest uncertainties about the Illinois River restoration are whether thresholds exist for rapid regeneration of submersed aquatic plants, which are important biological mediators in the ecosystem, and if thresholds exist, how long it will take to reach them and what resources will be required. The most encouraging aspect is that the coalition of private landowners, advocacy groups, advisory groups, private businesses, and local, state, and federal agencies appears committed to a long-term, comprehensive restoration program that embraces accelerated soil erosion control on the uplands, bank stabilization and habitat improvement in tributaries, and rehabilitation of the mainstem river and its associated backwaters and bottomland lakes.

References

Baker, F. C. 1906. A catalogue of the mollusca of Illinois. Ill. Lab. Nat. Hist. Bull. 7(6):53-136.

Bellrose, F. C., F. L. Paveglio, Jr., and D. W. Steffeck. 1979. Waterfowl populations and the changing environment of the Illinois River valley. Ill. Nat. Hist. Surv. Bull. 32:1-54.

Bellrose, F. C., S. P. Havera, F. L. Paveglio, Jr., and D. W. Steffeck. 1983. The fate of lakes in the Illinois River valley. Illinois Natural History Survey Biological Notes 119.

Bhowmik, N. G., and R. J. Schicht. 1980. Bank erosion of the Illinois River. Illinois State Water Survey, Report 92.

Briceland, R. H. 1976. Statement of Dr. Richard H. Briceland, Director, Illinois Environmental Protection Agency, to the Task Force on Locks and Dam 26 of the Midwestern Conference, Council of State Governments, Clayton, Missouri, November 18.

Calkins, W. W. 1874. The land and fresh water shells of LaSalle County, Illinois. Ottawa Academy of Natural Science Proceedings. 48 pp.

Clean Water Act of 1977. P.L. 95-217, Dec. 27, 1977, 91 Stat. 1566.

Condit, D. 1989. Illinois River Soil Conservation Task Force. Pp. 106-109 in Management of the Illinois River System: The 1990's and Beyond. Illinois River Resource Management. A Governor's Conference held April 1-3, 1987, Peoria, Ill. 260 pp.

Condit, D., and D. Roseboom. 1989. Stream bank stabilization and the Illinois River Soil and Conservation Task Force. Pp. 110-132 in Management of the Illinois River System: The 1990's and Beyond. Illinois River Resource Management. A Governor's Conference held April 1-3, 1987, Peoria, Ill. 260 pp.

Conlin, M. 1987. Illinois River fish and wildlife considerations. Pp. 147-154 in Management of the Illinois River System: The 1990's and Beyond. Illinois River Resource Management. A Governor's Conference held April 1-3, 1987, Peoria, Ill. 260 pp.

Demissie, M. 1989. Peoria Lake sedimentation and proposed artificial islands. Pp. 46-57 in Management of the Illinois River System: The 1990's and Beyond. Illinois River Resource Management. A Governor's Conference held April 1-3, 1987, Peoria, Ill. 260 pp.

Donels, B. 1989. Environmental management program proposals—The Illinois basin. Pp. 77-80 in Proceedings of the Second Conference on the Management of the Illinois River System: The 1990's and Beyond. Illinois River Resource Management. A Governor's Conference held October 3-4. Peoria, Ill. 199 pp.

Forbes, S. A. 1878. The food of Illinois fishes. Ill. Lab. Nat. Hist. Bull. 1(2):71-89.

Forbes, S. A., and R. E. Richardson. 1908. The fishes of Illinois. Illinois Natural History Survey, Urbana, Ill. cxxxvi plus 357 pp.

Forbes, S. A., and R. E. Richardson. 1913. Studies on the biology of the upper Illinois River. Ill. Lab. Nat. Hist. Bull. 9(10):481-574.

Forbes, S. A., and R. E. Richardson. 1919. Some recent changes in Illinois River biology. Ill. Nat. Hist. Surv. Bull. 13(6):139-156.

Hart, C. A. 1895. On the entomology of the Illinois River and adjacent waters. Ill. Lab. Nat. Hist. Bull. 4(6):149-273.

Havera, S. P., and F. C. Bellrose. 1985. The Illinois River: A lesson to be learned. Wetlands 4:29-41.

Havera, S. P., F. C. Bellrose, H. K. Archer, F. Paveglio, Jr., D. W. Steffeck, K. S. Lubinski, R. E. Sparks, W. U. Brigham, L. Coutant, S. Waite, and D. McCormick. 1980. Projected effects of increased diversion of Lake Michigan water on the environment of the Illinois River Valley. Report prepared for the U.S. Army Corps of Engineers, Chicago District, Chicago, Ill.

Herman, R. J. 1987. National resources inventory and potential stream sediment reductions. Pp. 173-183 in Management of the Illinois River System: The 1990's and Beyond. Illinois River Resource Management. A Governor's Conference held April 1-3, 1987, Peoria, Ill. 260 pp.

Illinois Environmental Protection Agency. 1979. Water Quality Management Plan. Volume III. Nonpoint Sources of Pollution: Soil Erosion and Sedimentation, Livestock Wastes, Fertilizers, Pesticides, Forestry, and Fruit Production. Illinois Environmental Protection Agency, Springfield, Ill. 384 pp.

Illinois Environmental Protection Agency. 1990. Illinois Water Quality Report 1988-1989. IEPA/WPC/90-160. Illinois Environmental Protection Agency, Division of Water Pollution Control, Springfield, Ill. 352 pp.

Injerd, D. 1987. Illinois Lake Michigan water diversion. Pp. 56-64 in Management of the Illinois River System: The 1990's and Beyond. Proceedings of the Illinois River Resource Management, A Governor's Conference held April 1-3, 1987, Peoria, Ill. 260 pp.

Jackson, H. O., and W. C. Starrett. 1959. Turbidity and sedimentation at Lake Chautauqua, Illinois. J. Wildl. Manage. 23:157-168.

Karaki, S., and J. van Hoften. 1974. Resuspension of bed material and wave effects on the Illinois and Upper Mississippi rivers caused by boat traffic. Contract Report No. LMSSD 75-881, prepared for the U.S. Army Engineers District, St. Louis, Mo., by Engineering Research Center, Colorado State University, Fort Collins, Colo.

Kelly, M. H., and R. L. Hite. 1984. An evaluation of empirical correlations between the macroinvertebrate biotic index (MBI) and the STORET water quality index (WQI). Unpublished manuscript. Illinois Environmental Protection Agency, Springfield, Ill.

Kofoid, C. A. 1903. Plankton studies. IV. The plankton of the Illinois River, 1894-1899, with introductory notes upon the hydrography of the Illinois River and its basin. Part I. Quantitative investigations and general results. Ill. Lab. Nat. Hist. Bull. 6(2): 95-635.

Korschgen, C. E. 1990. Feasibility study impacts of turbidity on growth and production of submersed plants. U.S. Fish and Wildlife Service, Northern Prairie Wildlife Research Center, LaCrosse, Wis. 11 pp.

Lanyon, R., and C. Lue-Hing. 1987. MSDGC activities in the Upper Illinois basin. Pp. 103-130 in Management of the Illinois River System: The 1990's and Beyond. Illinois River Resource Management. A Governor's Conference held April 1-3, 1987, Peoria, Ill. 260 pp.

Lee, M. T. 1989. Soil erosion, sediment yield, and deposition in the Illinois River basin. Pp. 718-722 in Proceedings of the International Symposium on Sediment Transport Modeling, American Society of Civil Engineers, New Orleans, La., August 14-18.

Leedy, J. B. 1979. Observations on the sources of sediment in Illinois streams. Report of Investigations, No. 18. Illinois Water Information System Group, University of Illinois, Urbana, Ill.

Lubinski, K. S., M. J. Wallendorf, and M. C. Reese. 1981. Analysis of Upper Mississippi River system correlations between physical, biological and navigation variables. Technical Report in partial fulfillment of Contract No. 895-305. Upper Mississippi River Basin Commission, St. Paul, Minn.

Mathis, B. M., and G. E. Stout. 1987. Summary and Recommendations. Pp. 1-4 in Management of the Illinois River System: The 1990's and Beyond. Illinois River Resource Management. A Governor's Conference held April 1-3, 1987, Peoria, Ill. 260 pp.

Mills, H. B., W. C. Starrett, and F. C. Bellrose. 1966. Man's effect on the fish and wildlife of the Illinois River. Biological Notes, No. 57. Illinois Natural History Survey, Urbana, Ill. 24 pp.

Neely, R. D., and C. G. Heister, compilers. 1987. The Natural Resources of Illinois: Introduction and Guide. Special Publication 6. Illinois Natural History Survey, Havana, Ill. 224 pp.

Palmer, A. W. 1903. The pollution and self-purification of the waters of the Illinois River. Bull. Ill. State Water Surv. Bull. 2:62-240.

Richardson, R. E. 1928. The bottom fauna of the middle Illinois River, 1913-1925. Its distribution, abundance, valuation, and index value in the study of stream pollution. Ill. Nat. Hist. Surv. Bull. 17(12):387-475.

Roelle, J. E., D. B. Hamilton, and R. L. Johnson. 1988. Refuge management analyses: Restoration of Thompson Lake as an alternative to further development at Chautauqua National Wildlife Refuge. U.S. Department of the Interior, Fish and Wildlife Service, Research and Development, Washington, D.C. 65 pp.

Roseboom, D., and B. White. 1990. The Court Creek Restoration Project. Pp. 27-39 in Erosion Control: Technology in Transition. Proceedings of Conference XXI, International Erosion Control Association, Feb. 14-17, 1990. Washington, D.C.

Roseboom, D., R. Twait, and D. Sallee. 1989. Habitat restoration for fish and wildlife in backwater lakes of the Illinois River. Pp. 65-68 in Proceedings of the Second Conference on Management of the Illinois River System: The 1990's and Beyond. Illinois River Resource Management. A Governor's Conference held April 1-3, 1987, Peoria, Ill. 260 pp.

Ross, P. E., R. E. Sparks, and F. S. Dillon. 1989. Identification of toxic substances in the upper Illinois River. Annual Report. Illinois Department of Energy and Natural Resources, Contract No. WR36. 20 pp.

Ruelle, R., and J. Grettenberger. 1991. A preliminary contaminant and toxicological survey of Illinois River sediments. Special Project Report 90-1. U.S. Fish and Wildlife Service, Rock Island, Ill. 17 pp.

Semonin, R. G. 1989. Comments for Illinois River Conference. Pp. 41-45 in Proceedings of the Second Conference of the Management of the Illinois River System: The

1990's and Beyond. Illinois River Resource Management. A Governor's Conference held April 1-3, 1987, Peoria, Ill. 260 pp.

Simons, D. B., R. M. Li, Y. H. Chen, S. S. Ellis, and T. P. Chang. 1981. Investigation of effects of navigation traffic activities on hydrologic, hydraulic, and geomorphic characteristics. Working Paper 2 for Task D, submitted to Upper Mississippi River Basin Commission, Minneapolis, Minn.

Smith, R. A., R. B. Alexander, and M. G. Wolman. 1987. Water-quality trends in the nation's rivers. Science 235:1607-1614.

Sparks, R. E. 1975. Environmental inventory and assessment of navigation pools 24, 25, and 26, Upper Mississippi and Lower Illinois rivers. An electrofishing survey of the Illinois River. Contract Report No. Y-74-4 to U.S. Army Corps of Engineers District, St. Louis, Mo.

Sparks, R. E. 1977. Environmental inventory and assessment of navigation Pools 24, 25, and 26, Upper Mississippi and Lower Illinois rivers. An electrofishing survey of the Illinois River. Urbana-Champaign Water Resources Center Special Report No. 5. UILU-WRC-77-0005. University of Illinois, Urbana, Ill. 122 pp.

Sparks, R. E. 1984. The role of contaminants in the decline of the Illinois River: Implications for the Upper Mississippi. Pp. 25-66 in James G. Wiener, Richard V. Anderson, and David R. McConville, eds., Contaminants in the Upper Mississippi River. Proceedings of the 15th Annual Meeting of the Mississippi River Research Consortium. Butterworth Publishers, Stoneham, Mass. 368 pp.

Sparks, R. E., R. C. Thomas, and D. J. Schaeffer. 1980. The effects of barge traffic on suspended sediment and turbidity in the Illinois River. U.S. Fish and Wildlife Service, Rock Island Field Office, Rock Island, Ill.

Sparks, R. E., P. B. Bayley, S. L. Kohler, and L. L. Osborne. 1990. Disturbance and recovery of large floodplain rivers. Environ. Manage. 14(5):699-709.

Stall, J. B., and S. W. Melsted. 1951. The silting of Lake Chautauqua, Havana, Illinois. Illinois State Water Survey, in cooperation with Illinois Agriculture Experiment Station, Report of Investigations 8.

Starrett, W. C. 1971. A survey of the mussels (Unionacea) of the Illinois River. A polluted stream. Ill. Nat. Hist. Surv. Bull. 30:267-403.

Starrett, W. C. 1972. Man and the Illinois River. Pp. 131-169 in R. T. Oglesby, C. A. Carlson, and J. A. McCann, eds., River Ecology and the Impact of Man. Academic Press, New York.

Sullivan, D. J., P. D. Hayes, T. E. Richards, and J. C. Maurer. 1990. Water Resources Data. Illinois Water Year 1989. Volume 2. Illinois River Basin. U.S. Geological Survey Water Data Report IL-89-2. Urbana, Ill. 467 pp.

Thompson, J. 1989. Case Studies in Drainage and Levee District Formation and Development on the Floodplain of the Lower Illinois River, 1890s-1930s. Special Report 016. University of Illinois at Urbana-Champaign, Water Resources Center, Urbana, Ill. 152 pp.

U.S. Army Corps of Engineers. 1986. Upper Mississippi River System Environmental Management Program. General Plan. North Central Division, U.S. Army Corps of Engineers, Chicago, Illinois, 31 p. and 4 exhibits.

U.S. Department of Commerce and Labor. 1911. Special Reports. Fisheries of the United States 1908. U.S. Government Printing Office, Washington, D.C. 324 pp.

Vonnahme, D. R. 1989. Progress in the Illinois River watershed since the First Illinois River Conference. Pp. 8-14 in Management of the Illinois River System: The 1990's and Beyond. Illinois River Resource Management. A Governor's Conference held April 1-3, 1987, Peoria, Ill. 260 pp.

Wetmore, F. 1987. Flood damage protection programs. Pp. 89-102 in Management of the Illinois River System: The 1990's and Beyond. Illinois River Resource Management. A Governor's Conference held April 1-3, 1987, Peoria, Ill. 260 pp.

RESTORING ATTRIBUTES OF THE WILLAMETTE RIVER

John J. Berger

Introduction

The Willamette River cleanup (Figure A.7) has been called "the most successful river-rejuvenation program in the country" (Starbird, 1972). The U.S. Army Corps of Engineers (COE) has described the Willamette as "one of the cleanest streams of comparable size in the nation" (U.S. Army Corps of Engineers, 1989). According to Starbird (and to conventional wisdom about the Willamette), the river "has regained it unspoiled charm" from Eugene, Oregon to Corvallis, Oregon. A documentary videotape produced by the State of Oregon Department of Environmental Quality also reinforces this view with highly laudatory remarks about the river's condition (State of Oregon, Department of Environmental Quality, 1989). A department spokesperson recently stated that the river today is as clean and trouble free as in the 1970s. Gleeson (1972) characterized the river as "recovered" in terms of its water quality. The U.S. Environmental Protection Agency has described the Willamette as a water quality success story and pronounced the river clean (U.S. Environmental Protection Agency, n.d.). Yet various problems remain unsolved, and new problems have recently been discovered, specifically, the presence of dioxin in the river (State of Oregon, Department of Environmental Quality, 1989).

General Description

The Willamette River, the twelfth largest in the United States (Gleeson, 1972), drains 11,460 square miles (29,800 km^2) of northwestern Oregon (USGS, 1977). In its aboriginal condition, the Willamette flowed cleanly and freely through grassy meadows and shady woodlands until its confluence with the Columbia River. Thickets of alder, cottonwood, and willow grew along the river, whereas above the floodplain stood tall groves of cedar and fir, rooted in the valley's black, alluvial soil (State of Oregon, Department of Parks and Recreation, 1988). Two-thirds of the state's 2.8 million people live in the Willamette Basin, and two-thirds of its economic activity is conducted there (State of Oregon, Department of Environmental Quality, 1988a; Horner, 1989). About 95 percent of the watershed is still forest (in various stages of forest succession) or farmland (U.S. Environmental Protection Agency, n.d.), but the river is no longer pure and unfettered.

The Willamette River Basin is of roughly rectangular shape, trending north-south in length for 150 miles, at an average width of about 75 miles (State of Oregon, Department of Environmental Quality, 1988a). The basin is bounded on the north by the Columbia River, on the south by the Calapooya Mountains, and on the east and west by the Cascade and Coast Ranges (USGS, 1977). The middle third of the basin is occupied by the Willamette Valley (USGS, 1977). The river itself flows from south to north and is the basin's main waterway (State of Oregon, Department of Environmental Quality, 1988a).

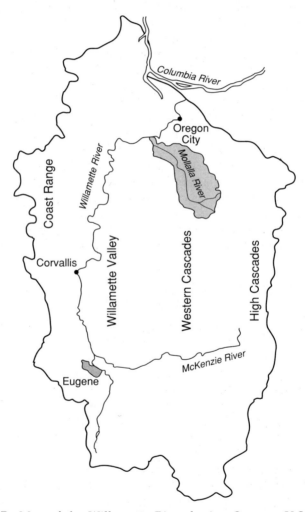

FIGURE A.7   Map of the Willamette River basin.   Source: U.S. Geological Survey, 1979.

Major subbasins include the Columbia, Tualatin, Molalla, Coast Range, Santiam, Long Tom, McKenzie, Middle Fork, and Coast Fork (State of Oregon, Department of Environmental Quality, 1988a). Major tributaries are the Middle Fork Willamette, McKenzie, Santiam, Molalla, Pudding, and Clackamas rivers (State of Oregon, Department of Environmental Quality, 1988a). The slopes and mountains of the basin tend to be forested, whereas the valleys are used for agriculture and urban settlement (State of Oregon, Department of Environmental Quality, 1988a). The river itself is used for municipal and industrial water supply, fish production, irrigation, electric power production, navigation, recreation (fishing, boating, and swimming), and receiving water for regulated wastewater discharges (U.S. Environmental Protection Agency, n.d.)

River Morphology

The mainstem of the Willamette River, formed near Eugene from the confluence of its Coast Fork and Middle Fork, flows for 187 miles (300 km) to its confluence with the Columbia River at Portland (USGS, 1977). The mainstem has three main reaches. The upper reach is the swiftest and shallowest; it extends for 135 miles (217 km) from Eugene almost to Newberg. This section is meandering, braided, and relatively turbulent, flowing over cobbles and gravel at velocities 10 to 20 times faster than the two downstream reaches (USGS, 1977). The next 25.5 miles, the "Newberg Pool," is a much slower, deeper depositional reach with fine sediments of sand and clay mixed with gravel and some cobbles (USGS, 1977). This is separated from the lower river by Willamette Falls near Oregon City. From there to the Columbia, the remaining 26.5 miles (42.6 km) of the river flows in a relatively stable incised channel subject to nonsaline tidal influences "transmitted from the Pacific via the Columbia River" (USGS, 1977). This portion of the riverbed is composed of "intermixed clay, sand, and gravel" and is dredged to maintain navigation (USGS, 1977).

Hydrological Data

Precipitation in the basin varies considerably with elevation and topography; the mean annual precipitation is 63 inches (1,600 mm) (USGS, 1977), with two thirds of it falling from November to March (State of Oregon, Department of Environmental Quality, n.d., unpublished material). Winters are generally wet in the basin; summers are dry. This accounts for the Willamette's annual pattern of Fall rising flows to a January peak, subsiding to annual low flows in August and September (State of Oregon, Department of Environmental Quality, n.d.). Natural mean average flow at Salem is about 23,000 cfs (Gleeson, 1972). The largest known flood was estimated at 500,000 cfs in December 1861 (Gleeson, 1972).

At Portland, the average discharge from 1972 to 1982 was 32,760 cfs or 23,730,000 acre-ft per yr. Maximum discharge was 283,000 cfs and minimum daily discharge was 4,200 cfs (USGS, 1984).

Augmentation of low summer flow rates by releases of water from reservoirs in the basin have significantly affected the river's natural flow and temperature patterns. Prior to regulation, which began having a major impact in 1952, average annual consecutive 30-day low flow was 3,670 cfs. Between 1953 and 1970 it averaged 6,010 cfs (USGS, 1977). Release of cool bottom waters from impoundments at Lookout Point-Dexter Reservoir complex on the Middle Fork often reduces the river's summer temperature above RM 120, but elsewhere has little effect (USGS, 1977).

History of Pollution Control Efforts

The river basin was first settled by people of European ancestry in 1812 (Weber, 1989). Population growth thereafter was extremely rapid. In less than 40 years, 6,000 people were making their homes in the basin. By 1900 the population had grown to 233,700. The basin had 691,204 residents by 1940 (Weber, 1989); and by 1970, 1.4 million (Gleeson, 1972); and the estimated 1990 population was 2.8 million. By the 1920s, with hundreds of thousands of people living in the Willamette region using the river, the Willamette was grossly polluted by sewage and industrial waste, most of it from the pulp and paper industry.

Although pollution control legislation had been passed as early as 1919, the laws were not well enforced. Citizens, private groups, and public agencies all endeavored to rectify the situation during the 1920s without success. Their concerns led to studies in the 1920s and 1930s of dissolved oxygen in the river; the studies revealed near-complete or complete oxygen depletion in the lower river below Portland (Gleeson, 1972).

At that time the river was an open sewer that stank and contained dangerously high concentrations of coliform bacteria; none of the cities or industries along the Willamette bothered to treat their waste materials. Wood-products industries, canneries, and slaughter-houses dumped their waste directly into the river. Decomposing wood fiber sludge from the river bottom rose to the surface, buoyed by the gases formed during its decomposition. These sludge rafts, swathed in sewage bacteria *(Sphaerotilus natans)*, floated downstream, consuming oxygen as they rotted (Gleeson, 1972). Aquatic life in the lower river suffocated and died. Zones of oxygen depletion in the lower river served as barriers to fish migration. Commercial and sport fishing as well as recreational uses of the river suffered. Slime coated the water's edge.

STATE, FEDERAL, AND PRIVATE RESTORATION EFFORTS

When efforts to get state antipollution legislation passed during the 1930s failed repeatedly, citizens sponsored a successful antipollution ballot initiative that was approved in 1938. The Water Purification and Prevention of Pollution Bill established a State Sanitary Authority with responsibility for cleaning up and protecting Oregon's public waters (U.S. Environmental Protection Agency, n.d.). The drafting of the legislation was preceded by a study of successful pollution control legislation elsewhere and an analysis of the principles that made controls work (Gleeson, 1972). World War II slowed pollution control efforts. Studies of river conditions resumed in 1944 and revealed that conditions were worse than in 1929.

Progress toward clean water came slowly. It took 8 years after the establishment of the State Sanitary Authority for the first municipal primary wastewater treatment plant to be built on the river, and it was not until 1957, 10 years later, that all municipalities on the river had primary treatment (U.S. Environmental Protection Agency, n.d.). Continued studies of the river in the 1950s and early 1960s revealed, however, that this was not sufficient to correct the problem of low dissolved oxygen. To do so, controlled release of water from reservoirs was begun in 1953 to increase river flow during low-flow months when pollutants were more concentrated and oxygen demand greatest. This was coupled with increasing regulation of paper and pulp mill discharges, beginning in 1950. Even these measures combined were insufficient to bring the Willamette up to standards for acceptable water quality. One indication of the river's condition was a fall chinook salmon run in 1965 of only 79 fish, counted at Oregon City Falls.

A turning point in the struggle to clean up the Willamette occurred in the late 1960s. Passage of the federal Clean Water Act of 1977 (P.L. 95-217) required all states to set water quality standards for their rivers and to prepare to enforce them. Hearings to set water quality standards were therefore held by the State Sanitary Authority in 1967, and that year, the state legislature rewrote and greatly strengthened the state's water quality laws. Citizens were very much involved in urging government to take decisive action (U.S. Environmental Protection Agency, n.d.).

The new state law of 1967 made it illegal to put waste materials in the river without obtaining a discharge permit from the State Sanitary Authority (U.S. Environmental Protection Agency, n.d.). All major point source discharges and some minor point sources were thus identified and subsequently controlled. The law set new water quality regulations and water pollution control standards for the state and a mid-1972 deadline for attainment of the new standards. Special water quality standards were established for the Willamette.

The broad state standards dealt with solids, microorganisms, oxygen content, pH, temperature, color, odor, turbidity, oils, aesthetics, and radiological properties of waste. The standards also prescribed minimum dissolved oxygen levels for different reaches of the mainstem, with a minimum of 5 mg per liter in the lower reach. To meet the new standards, the state gave grants to municipalities for sewage treatment plants and tax credits to industry for pollution control equipment.

Secondary treatment was accomplished by all municipalities by 1969, but even this was not sufficient because of the continuing release of waste from the pulp and paper industry (U.S. Environmental Protection Agency, n.d.). Reflecting heightened interest in and concern for the environment, the Oregon legislature in 1969 transferred the State Sanitary Authority to a new state Department of Environmental Quality under an Environmental Quality Commission (Gleeson, 1972; U.S. Environmental Protection Agency, n.d.).

Paper and pulp mills had been ordered in 1950 to stop releasing untreated sulfite waste liquors into the river during the low-flow months of June through October (Gleeson, 1972). They were ordered to provide year-round primary treatment in 1964, and sulfite mills were required to cut the biochemical oxygen demand in their effluents by 85 percent (Gleeson, 1972). Since 1973, under nondegradation provisions of the Federal Water Pollution Control Act Amendments of 1972, state discharge permits have allowed no further increase in total waste loadings to the river (U.S. Environmental Protection Agency, n.d.).

Low-Flow Augmentation

Despite the reduction in waste discharge to the river, low-flow augmentation is critical to the maintenance of water quality. "In spite of the remarkable reduction in pollutional loadings which have been accomplished, the quality of water is dependent upon natural water flow augmentation in the summer low flow periods" (Gleeson, 1972). The U.S. Geological Survey (USGS, 1976) concurred: "Even with the excellent pollution control program mounted over the years, [dissolved oxygen] standards would still be violated in certain sub-reaches during most summers without low-flow augmentation." The USGS scientists believe that flow augmentation is also critical to preventing algal blooms in the river, which might occur due to high nutrient levels, and USGS studies suggested that water detention time was the primary limitation on algal growth (USGS, 1976).

The Greenway

Water quality improvements on the Willamette have been accentuated by the development of a riverfront park system known as a greenway, authorized by the state in 1967. Millions of dollars in federal, state, and local funds have been combined in the endeavor, resulting in the creation of extensive shoreline preserves.

The greenway system now includes 255 river miles (J. Lilly, Rivers Program, Oregon Parks and Recreation Division, personal communication, 1990). These 255 miles are supposed to be protected from Lane County at the southern end of the Valley to the Columbia, including the lower reaches of the Coast Fork and Middle Fork under a state law passed in 1967 and updated in 1973. Other headwaters and Cascade Mountain reaches of the Willamette are not within the greenway system itself but are on National Forest land.

STATE ADMINISTRATIVE OBSTACLES

Two legal and administrative problems impede protection of the greenway. State law gives carte blanche to users of land for agricultural purposes in Oregon: farmers along the river can clear land all the way to the river's edge. Agriculture increased over the past 10 to 15 years along the river, then was on the decline, and is now stable (J. Lilly, Rivers Program, Oregon Parks and Recreation Division, personal communication, 1990). This provides an opportunity to have inactive agricultural lands put into the conservation reserve program.

The second problem affecting the greenway is timber harvesting. Until 4 years ago, local government could review and approve timber harvests along the greenway (J. Lilly, Rivers Program, Oregon Parks and Recreation Division, personal communication, 1990). However, as the result of a restructuring of the state forestry program and its rules, local government lost its right to approve timber harvests (J. Lilly, Rivers Program, Oregon Parks and Recreation Division, personal communication, 1990). Although forestry practice rules designed to protect fish and wildlife and water quality are in place, they do not address recreational impacts or scenic impacts of logging (J. Lilly, Rivers Program, Oregon Parks and Recreation Division, personal communication, 1990).

Another administrative problem affecting the greenway is lack of staff. The Oregon Department of Parks for the past 10 years has had only one person working only 20 percent time on the greenway acquisition process, although field staff and maintenance staff exist to manage the existing greenway sites.

In creating the greenway, not much thought has been given to restoration of natural ecosystems, and little or no actual ecological restoration has been done on the greenway lands, although urban park development has occurred along the river, and some riverfront land has been reclaimed from urban or highway use and converted to park land. The creation of Tom McCall Waterfront Park, for example, involved moving a highway away from the river. Uncharacteristically, a few acres of wetlands have been restored at the McCormick Pier condominium in Portland.

Greenway creation has been more reclamation than restoration. However, the Oregon Department of Parks and Recreation does have a joint management agreement with Fisheries and Wildlife to provide feed for game, especially grain for waterfowl.

Oregon Water Quality Programs

The State of Oregon has recently shifted the emphasis of its pollution control permit system from technology-based standards that focus on the point source facility to an emphasis on receiving-water quality (State of Oregon, Department of Environmental Quality, 1988c). In addition, the state has also begun to shift from its traditional emphasis on point source pollution to paying additional attention to nonpoint source pollution problems, prompted both by the federal Water Quality Act of 1987 and by a suit filed in 1986 by the Northwest Environmental Defense Center. Development and implementation of a nonpoint source management program is one of the state's six major water pollution control program goals (State of Oregon, Department of Environmental Quality, 1988c). The state has an overall water quality management plan as well as the Oregon Clean Water Strategy, a comprehensive geographic approach to meet clean water goals and fulfill requirements of the federal Clean Water Act of 1977 and the federal Water Quality Act of 1987 (State of Oregon, Department of Environmental Quality, 1988c).

Unresolved Problems

WATER QUALITY

The state in its 305B Report (State of Oregon, Department of Environmental Quality, 1988c) published the results of an extensive statewide water quality monitoring, revealing the location, type, severity, and causes of pollution. These are summarized briefly in Tables A.1 through A.4. Technical appendixes supporting these tables are available from the State of Oregon, Department of Environmental Quality.

As shown in Table A.1, the department found that 895 of the Willamette's 4,019 miles had severe water quality problems and that 1,696 miles had moderate pollution. The data in Table A.2 reveal that certain reaches of the river were contaminated with pesticides (e.g., dichlorodiphenyltrichlorethane (DDT)), toxic heavy metals (e.g., arsenic, lead, and zinc), polychlorinated biphenyls (PCBs), dioxin, phthalate, and anthracene. The data in Table A.3 showed that 24 percent of the fish sampled from river mile 51 had abnormalities, as did 17 percent of those from river mile 22. "Abnormalities included lesions, ulcers, deformed fins, missing eyes, or heavy mucous films" (State of Oregon, Department of Environmental Quality, 1988c). Abnormalities were not restricted to the heavily industrialized lower river. That reach, however, did show the highest levels of sediment contamination, including contamination by heavy metals, DDT, and PCBs (Table A.4). The 1988 Basin Status Summary attributes the pollution of the Middle Willamette to causes such as agriculture, industrial point sources, municipal point sources, leaky septic tanks, and urban and residential runoff, and attributes problems in the Coast Fork to septic systems, municipal waste, and agriculture. These plus urban and residential runoff were also cited as causes of pollution in the Lower Willamette.

As the data suggest, nonpoint source-related water quality problems remain in the Willamette Basin. Sources of these problems are land "surface erosion and disturbance of riparian vegetation and stream banks" caused by logging, farming, landslides, and surface runoff from roads (State of Oregon, Department of Environmental Quality, 1988a). "Waterbodies in which serious NPS pollution problems are known to exist or have been reported without challenge" include the Coast Fork of the Willamette and Willamette Harbor. Data indicate that moderate water quality problems exist in the river throughout the North Basin until the problems become severe north of the junction with the Clackamas River. Most of the river in the South Basin also has moderate water quality problems except for Lookout Point

TABLE A.1 Summary of Use Support for Rivers and Streams

Basin	Total Miles Assessed	Miles of Use Supported or Unknown	Miles of Use Partially Supported			Miles of Use Not Supported		
		Total	Monitored (based on DEQ Data)	NPS* Assessment Evaluation	Total	Monitored (based on DEQ Data)	NPS Assessment Evaluated	Total
North Coast/L. Columbia	905	165	0	490	490	59	191	250
Mid Coast	931	321	0	552	552	19	39	58
Umpqua	1,873	732	77	442	519	69	553	622
South Coast	1,368	496	0	582	582	39	251	290
Rogue	2,026	1,127	150	405	555	81	263	344
Willamette	4,019	1,428	374	1,322	1,696	257	638	895
Sandy	233	102	0	53	53	0	78	78
Mood	285	47	14	139	153	0	85	85
Deschutes	2,538	1,325	95	547	642	166	405	571
John Day	2,236	883	32	418	450	325	578	903
Umatilla/ Walla Walla	1,120	435	57	399	456	22	207	229
Grande Ronde	1,771	951	19	511	530	128	162	290
Powder Burnt	1,331	586	62	380	442	111	192	303
Malheur	1,613	902	0	353	353	110	248	358
Owyhee	1,659	1,254	10	161	171	18	216	234
Malheur Lake	1,902	1,035	0	301	301	0	566	566

Goose and Summer Lake	1,039	508	0	301	301	0	230	230
Klamath	889	249	14	237	251	220	169	389
Total	27,738	12,546	904	7,593	8,497	1,624	5,071	6,695
Percentage of total miles assessed	100	45	3	28	31	6	18	24

Note: Analysis was based on Department of Environmental Quality (DEQ) nonpoint source data base. Results should be treated as estimates. The assessment information is based on information provided by resource managers and others. That information has not been verified by DEQ. "Not supported" was interpreted as streams with "severe" water quality problems. "Partially supported" was interpreted as streams with "moderate" water quality problems. The nonpoint source assessment did not separate waters in which uses were supported and those waters where use support was quality problems. The nonpoint source assessment did not separate waters in which uses were supported and those waters where use support was unknown. Future update of the data base will provide this information. The nonpoint source assessment should be consulted for further information.

*National Park Service

SOURCE: State of Oregon, Department of Environmental Quality, 1988.

TABLE A.2 Documentation Ambient Monitoring Information

Basin-Waterbody	River Mile	Sample Type	Parameter of Concern	Number Samples	Median/ Mean Conc. (mg/kg)	Range Conc. (mg/kg)	Reason for Inclusion
COLUMBIA BASIN:							
Columbia River	41.0	ft	Dioxin	2	0.002	0.002-0.003	Dioxin in fish tissue.
Columbia River	141.0	wsr	As	12	1.0	<1.0-2.0	Exceeded 25% human health criteria (83%).
North Portland Harbor	102.5	sd	DDT	2	0.024	<.001-.047	Median exceeds threshold conc. for.
			DDD	2	0.014	<.001-.029	DDT, DDD, DDE in sediments.
			DDE	2	0.014	<.001-.009	
			As	2	4.05	3.6-4.05	Mod. polluted by GLG.
			Pb	2	18.4	17.4-19.4	Exceeds EPA median.
			Zn	2	145.5	140.0-151.0	Mod. polluted by GLG.
WILLAMETTE BASIN:							
Willamette River (St. John's Bridge)	6.0	sd	DDT	3	0.021	<.002-.041	Median exceeds threshold conc. for.
			DDD	3	0.027	<.002-.053	DDT, DDD, DDE in sediments.
			DDE	3	0.017	<.002-.033	
			As	3	2.4	.23-4.3	Range value exceeds EPA median.
			Cu	3	52.5	22.0-119.6	Highly polluted by GLC.
			Pb	3	56.4	20.4-116.4	Mod. polluted by GLC.
			Zn	3	123.0	109.0-210.0	Mod. polluted by Great Lakes Guidelines.
			enthracene	3	1.100	.059-4.2	Range value exceeds EPA median.
			phenanthrene	3	6.4	.490-28.0	Range value exceeds EPA median.
			PCB	3	.190	<.010-.380	Range value exceeds EPA median.
Willamette River (SP&S Bridge)	7.0	wsr	phthalate	3	2.3	0.3-5.6	Exceeds fresh H_2O chronic criteria.
		sd	DDT	3	0.003	<.001-.007	Median exceeds threshold conc. for.
			DDD	3	0.011	.006-.014	DDT, DDD, DDE in sediments.

Location	Value	Type	Analyte	n	Median	Range	Comment
Willamette River (Near Doanne Lake)	7.1	sd	DDE	3	0.006	.003-808	Mod. polluted by GLG.
			As	3	5.5	4.8-8.4	Range vlaue exceeds EPA median.
			Cd	3	0.63	0.1-1.15	Highly polluted by GLG.
			Cu	3	54.0	28.6-129.8	Range value exceeds EPA median.
			Pb	3	15.3	2.89-89.1	Range value exceeds EPA median.
			Zn	3	88.0	33.0-221.0	Range value exceeds EPA median.
			phthalate	3	0.172	<.290-580	Range value exceeds EPA median.
			PCB	3	0.076	<.015-103	Range value exceeds EPA median.
					2.7		
Willamette River (Near Doanne Lake)		sd	DDT	1	0.100		Median exceeds threshold conc. for. DDT, DDD, DDE in sediments.
			DDD	1	0.600		
			DDE	1	5.0		
			As	1	26.5		Mod. polluted by GLG.
			Cu	1	147.0		Mod. polluted by GLG.
			Zn	1	.314		Mod. polluted by GLG.
			PCB	1			Median exceeds EPA median.
Willamette River	22.0	fh		1			Fish health impairment.
Willamette River	51.0	fh		1			Fish health impairment.
Willamette River	184.6	ft	Dioxin	2	0.003	0.001-0.005	Dioxin in fish tissue.
Cottage Grove Reservoir	29.5	ft	Hg	4	0.685	.350-1.00	Exceeds 50% FDA action level (60%).
Columbia Slough (Below North Slough)	1.4	sd	As	2	9.3	6.5-12.0	Highly polluted by GLG.
			Cd	2	1.07	0.96-1.18	Median exceeds EPA median.
			Cu	2	51.9	42.7-61.0	Highly polluted by GLG.
			Pb	2	161.7	128.0-195.4	Highly polluted by GLG.
			Zn	2	295.5	260.0-331.0	Highly polluted by GLG.
Columbia Slough (At dumpsite)	2.7	ft / sd	PCB (+)	5	0.300	0.150-0.470	Not exceed 50% but need to note - (+).
			As	3	1.2	0.8-12.7	Range value exceeds EPA median.
			Cd	3	2.36	1.1-3.6	Range value exceeds EPA median.

TABLE A.2 (Continued)

Basin-Waterbody	River Mile	Sample Type	Parameter of Concern	Number Samples	Median/Mean Conc. (mg/kg)	Range Conc. (mg/kg)	Reason for Inclusion
WILLAMETTE BASIN: (continued)							
			Cu	3	44.7	41.0-95.8	Mod. polluted by GLG.
			Pb	3	113.9	75.5-396.8	Highly polluted by GLG.
			Zn	3	244.0	217.0-549.0	Highly polluted by GLG.
			phthalate	3	0.900	<.100-1.800	Range value exceeds EPA median.
			PCB (+)	3	0.420	.300-1.060	Range value exceeds EPA median.
Columbia Slough (B1 Landfill Bridge)	4.0-5.0	wsr	phthalate (+)	1	55.1		Exceeds Fr H_2O chronic criteria.
			PCB (+)	1	0.037		Exceeds Fr H_2O chronic criteria.
		sd	As	1	4.8		Mod. polluted by Great Lakes Guidelines.
			Cd	1	1.5		Median exceeds EPA median.
			Cu	1	27.5		Mod. polluted by Great Lakes Guidelines.
			phthalate (+)	1	0.266		Does not exceed but significant cause (+).
			PCB (+)	1	0.057		Does not exceed but significant cause (+).
Columbia Slough (B1 Denver Ave.)	5.7	wsr	PCE	1	18.0		Exceed human health criteria.
			PCB (+)	1	0.028		Exceed chronic and human health criteria.
		sd	As	1	9.0		Highly polluted by GLG.
			Cu	1	82.0		Highly polluted by GLG.
			phthalate	1	0.664		Medium exceeds EPA median.
			PCB (+)	1	0.115		Median exceeds EPA median.
			DDE	2	0.052	.042-.061	Median exceeds EPA median.

Location			Analyte	n	Value	Range	Comment
Beaverton Creek (B1 Tektronix)	5.5	wsr sd	PCE	1	39.0		Exceed human health criteria.
			DDT	1	0.140		DDT, DDD, DDE in sediments.
			DDE	1	0.024		
			As	1	4.8		Mod. polluted by GLG.
			Cu	1	115.0		Highly polluted by GLG.
			Pb	1	113.0		Highly polluted by GLG.
			phthalate	1	2.432		Median exceeds threshold conc.
Conser Slough	113.5	ft sd	PCB (+)	4	0.460	0.270-2.910	Exceeds FDA action levels.
			As	4	3.8	3.7-8.6	Mod. polluted by GLG.
			Cu	4	68.1	56.8-171.1	Highly polluted by GLG.
			Pb	4	30.2	22.2-165.6	Range value exceeds EPA medium.
			Hg	4	0.14	0.07-0.586	Do not know why it is here.
			Zn	4	158.0	136.0-224.0	Mod. polluted by GLG.
			phenanthrene	4	0.490	<.230-1.10	Range value exceeds EPA median.
			PCB (+)	4	2.280	1.52-7.63	Exceeds all sediment guidelines
ROGUE BASIN: Break Creek (At Medford)	9.0	sd	DDT	2	0.058	.009-107	Median exceeds threshold conc. for.
			DDD	2	0.017	.015-.018	DDT, DDD, DDE in sediments.
			Cu	2	26.7	26.0-27.4	Mod. polluted by GLG.
			Pb	2	30.9	29.0-32.8	Range value exceeds EPA median.
			Zn	2	65.5	65.0-66.0	Range value exceeds EPA median.
			phthalate	2	1.340	.7809-1.900	Median exceeds threshold conc.
Applegate River (Near Cooper)	45.7	wsr	As	6	1.3	<1.0-2.0	Exceeds 25% human health criteria (50%).
			Cd	6	1.3	<1.0-2.0	Exceeds 25% Fr H_2O chronic criteria (67%).
			Pb	6	7.2	1.0-20.0	Exceeds 25% Fr H_2O chronic criteria (83%).
			Hg	6	0.1	<0.1-0.4	Exceeds 25% Fr H_2O chronic criteria (63%).

TABLE A.2 (*Continued*)

Basin-Waterbody	River Mile	Sample Type	Parameter of Concern	Number Samples	Median/ Mean Conc. (mg/kg)	Range Conc. (mg/kg)	Reason for Inclusion
MALHEUR BASIN:							
Malheur River (At Mouth)	0.4	ft	DDT (+)	3	0.145	0.031-0.227	Not exceed but significant that detected. Since this is a (+).
			DDT	3	0.011	0.006-0.106	
			DDE	3	0.142	0.098-0.820	
		sd	DDT (+)	4	0.015	<.001-.019	Median exceeds threshold conc. for.
			DDD	4	0.009	<.001-.105	DDT, DDD, DDE in sediments.
			DDE	4	0.023	<.001-.053	
			As	4	4.1	0.5-4.3	Mod. polluted by GLG.
			Cu	4	34.3	21.7-37.8	Mod. polluted by GLG.
			Zn	4	61.5	35.0-69.1	Range value exceeds EPA median.
OWYHEE BASIN:							
Owyhee River (At Hwy. 201 Bridge)	2.9	ft	DDT (+)	1	0.480		Greater than 50% of FDA action level.
			DDD	1	0.239		
			DDE	1	2.060		
			Hg	1	0.840		>50% of FDA action level (84%).
		sd	DDT (+)	5	0.012	.006-.030	Median exceeds threshold conc. for.
Owyhee River (At Owyhee)			DDD	5	0.007	.005-.029	DDT, DDD, DDE in sediments.
			DDE	5	0.040	.032-.113	
			As	5	4.9	3.9-14.4	Mod. polluted by GLG.
			Zn	5	35.0	32.0-80.6	Range value exceeds EPA median.
	5.1	wsr	As	12	19.4	3.0-47.0	Exceeds 25% human health criteria (100%).
KLAMATH BASIN:							
Klamath River (At Hwy. 97)	248.3	ft	Hg	2	0.340	.20-1.10	Exceeds FDA action level of 1.

SOURCE: State of Oregon, Department of Environmental Quality, 1988.

TABLE A.3 Percent of River Fish Afflicted with Abnormalities.[a]

River Mile	No. of Individuals Collected	Percentage Abnormalities
185	105	6
176	99	2
160	152	2
93	128	2
77	119	3
70	122	1
58	74	3
51	45	24
39	50	6
29	85	6
25	125	7
22	35	17
19	25	8
17	44	5
3	98	5

[a]Fish abnormality is indicated by lesions/ulcers, deformed fins (genetic), eyes missing (genetic) or heavy mucus.

SOURCE: Reprinted, by permission, from Hughes and Gammon (1987). Copyright © 1987 by Transactions of the American Fisheries Society.

Reservoir, which has severe water quality problems, and river headwaters, where water quality is generally better. Moderate problems are defined as those that "interfere(s) with desired uses of the water body and with the normal life history or composition of aquatic populations" (State of Oregon, Department of Environmental Quality, 1987). Severe problems are those causing "substantial or nearly complete interference with the beneficial uses or opportunities to use the water" (State of Oregon, Department of Environmental Quality, 1987).

Among the specific water quality problems cited by the Department of Environmental Quality (DEQ) in the River are turbidity, low dissolved oxygen, bacteria/viruses, solids, erosion, low flow, sediment, pesticides, and plant growth. Probable causes of water quality cited by DEQ include changes in flow pattern and timing, pollution in runoff from roads and industrial sites, elimination of protective vegetation over streams, declines in alluvial water tables, flow alteration, water withdrawals, reservoir storage and releases, alteration of physical characteristics of the stream, channelization and wetland drainage, and diffuse waste disposal. (State of Oregon, Department of Environmental Quality, 1988b).

High levels of nutrients and bacteria have been found in certain parts of the river when precipitation is high (S. Kengla, Department of Environmental Quality, Portland, Oregon, personal communication, 1990). The most dangerous form of dioxin—2,3,7,8-tetrachlorodibenzo-p-dioxin (TCDD)—has recently been found in fish in the Willamette and Columbia rivers at levels of a few parts per trillion in whole fish and fillets. This persistent carcinogen is formed during the bleaching of paper pulp by pulp mills along the river. The DEQ

TABLE A.4 Areas Exhibiting Elevated Levels of Toxics in Sediment

Basin	Water Body of Concern	Location (river mile)	Elevated Parameters of Concern
Willamette Basin	Willamette River	6.0	As, Cu, Pb, Zn
		7.0	As, Cd, Cu, Pb, Zn
		7.1	As, Cu, Zn, DDT, PCB
		16.6	As, Cu, phthalates
		39.0	DDT
	North Portland Harbor	102.0	As, Pb, Zn, DDT
	Columbia Slough	1.4	As, Cd, Cu, Pb, Zn
		4.0	As, Cu
		5.7	As, Cu, phthalates, PCB
		10.0	phthalates
		12.0	As, Cd, Cu, Pb, Zn
		15.2	As, Cd, Cu, Pb, Zn, PCB, phthalates
	North Slough	2.8	As, Cu, Pb, Zn, PCB, phthalates
	Tuslatin River	8.7	Pb, Zn, phthalates
	Fanno Creek	1.2	As, DDT, phthalates
	Beaverton Creek	5.5	DDT, phthalates
	Yamhill River	5.0	As, Zn, phthalates
	Conser Slough	117.0	As, Pb, Zn
Malheur Basin	Malheur River	0.4	As, Cu, Zn, DDT
Owyhee Basin	Owyhee River	2.9	As, Zn, DDT
Klamath Basin	Klamath River	234.9	As, Pb, Zn
Umpqua Basin	South Umpqua River	46.6	As, Cu, Zn
Rogue Basin	Bear Creek	7.6	Cu, Pb, Zn, DDT, phthalates
Deschutes Basin	Crooked River	30.0	Zn
Umatilla Basin	Umatilla River	55.0	Zn

NOTE: Elevated levels of toxics determined by (1) sediment data medians exceed threshold concentrations, or (2) range values exceeded the national median.

SOURCE: State of Oregon, Department of Environmental Quality, 1988c.

is currently working to reduce dioxin discharges by adding dioxin discharge limits to National Pollutant Discharge Elimination System permits issued to pulp mills. The DEQ is also concerned about 3,000 absorbable organic halides formed during kraft pulp manufacture (State of Oregon, Department of Environmental Quality, 1989).

Some river miles today are not meeting water quality standards with respect to dissolved oxygen and dioxin (Lydia Taylor, Division Administrator, Oregon Department of Environmental Quality, personal communication, 1990). "Degradation has set in but the level of intensity is so small it would be difficult to measure," according to T. Morse (U.S. Army Corps of Engineers, Portland, Oregon, personal communication, 1990). "Dioxin is probably the most serious contaminant in the Willamette" (T. Morse, U.S. Army Corps of Engineers, Portland, Oregon, personal communication, 1990).

The consensus that some decline in river quality is being experienced was shared by James Monteith, director of the Oregon Natural Resources Council, who said, "In the last four to five years we've begun to experience some degradation as a result of agricultural practices and as a result of pulp mills discharges. The dioxin levels have become a very big issue recently. I'm not sure whether what we're seeing is an effect of increased awareness. The main issue today may be toxic loads from paper and pulp mills. This will be looked at very closely" (J. Monteith, Oregon Natural Resources Council, Portland, Oregon, personal communication, 1990).

FISHERIES

The U.S. Army Corps of Engineers has built 13 multipurpose dams in the basin since 1941. By COE calculations, these dams have, among other benefits, prevented $6.5 billion worth of flood damage (U.S. Army Corps of Engineers, 1989). Controlled water releases during low flow periods have maintained water quality by diluting pollution, but the price paid by the natural environment in lost spawning and rearing areas has been steep.

Today, high dams essentially block fish passage on the Willamette, and in response, the Oregon Department of Fish and Wildlife (DFW) has invested in replacing the natural spring chinook and winter steelhead runs with hatchery fish. To improve fisheries, a fish ladder was made fully operational at Willamette Falls in 1971, and nine salmon-steelhead hatcheries have been built in the basin (Gleeson, 1972). Even where fish ladders are constructed and operated properly, however, high head storage dams are drawn down by winter time; fish ladders do not operate well over such a wide range of water levels

(Max Smith, Oregon Department of Fish and Wildlife, personal communication, 1990). "We have not perfected fish passage technology," Smith noted. "Downstream migrant passage facilities don't work very well."

The wild components of the winter steelhead and spring chinook runs are very depleted. Juvenile salmonids are killed in passing through the power plant at Willamette Falls; adult salmon and steelhead are "delayed, stranded, injured, or killed" there (State of Oregon, Department of Fish and Wildlife, 1988).

An adequate water supply for fish is also a problem. According to the Department of Fish and Wildlife, "present water laws do not fully protect fish habitat. . . . Many tributaries do not have minimum flows established to protect fish and aquatic life" (State of Oregon, Department of Fish and Wildlife, 1988). Reservoir management practices resulting in turbidity, high dissolved nitrogen, and below-dam erosion are also problems.

Native winter steelhead and spring chinook share some spawning areas. Hatcheries, however, are being operated for chinook and winter steelhead, and for a summer steelhead and a fall chinook run that were never native (Max Smith, Oregon Department of Fish and Wildlife, personal communication, 1990). "Some of the anadromous fish stocks are in real trouble" due to competition from introduced exotic species, including bass, bluegills, catfish, pumkinseed, and strains of nonindigenous salmonids (Max Smith, Oregon Department of Fisheries and Wildlife, personal communication, 1990). Interbreeding between wild and hatchery fish and overharvesting of wild adults in mixed-stock fisheries are also problems (State of Oregon, Department of Fish and Wildlife, 1988).

The Department of Fish and Wildlife's philosophical orientation is reflected in its utilitarian goal of managing the basin's fish resource "to provide the greatest recreational, commercial, economic, and nonconsumptive benefits to . . . citizens." The restoration of natural ecological conditions and the management of ecosystems and habitats do not appear to be as strongly emphasized by the department as hatchery operations, although it does advocate cooperative efforts with other agencies to restore degraded habitat (State of Oregon, Department of Fish and Wildlife, 1988). Within the utilitarian framework mentioned, the department tends to prepare plans for the management of single fish species (e.g., the Coho Salmon Plan of 1981, the Trout Plan of 1987) or groups of (the Warmwater Fish Plan of 1987), rather than see the river as an aquatic ecosystem with valuable nongame components. The department manages some runs intensively, even sometimes trapping surplus fish that return to hatchery sites and trucking them "to downstream release sites for recycling through the fishery" (Max Smith, State of Oregon, Department of Fish and Wildlife, personal communication, 1990).

Significant fishery management actions in the basin have been taken with sparse knowledge of the resource base. Trout stocks in the basin have not been identified; recent estimates of natural spring chinook production are lacking; and "information on the habitat requirements of the native [winter steelhead] stock are inadequate to provide specific guidelines for habitat protection and enhancement . . ." (State of Oregon, Department of Fish and Wildlife, 1988).

Much fish habitat has been lost on the Willamette and has not been replaced. The river used to be braided throughout the Willamette Valley, and had side channels and slow places where fish could feed or spawn. About 400 river miles of habitat remain of the 1,400 river miles originally available (Sedell and Frogatt, 1984).

Probably less than half the fish production on the Willamette today is wild native fish. Dams more than any other single factor have the most detrimental effect on wild and native fish (Max Smith, Oregon Department of Fish and Wildlife, personal communication, 1990). There are 14 reservoirs in the Willamette Basin and 12 fish hatcheries or rearing ponds. Almost two-thirds of the Oregon Department of Fish and Wildlife budget is spent on fish propagation in hatcheries. Another 10 percent is spent on fish management activities, including stocking and the monitoring of harvest and population abundance. Only 8 percent of the budget is spent on habitat improvement (State of Oregon, Department of Fish and Wildlife, 1988).

Forestry also has had serious detrimental effects on the river and its fishery because of an emphasis on clearcutting on federal lands not covered by the Oregon Forest Practices Act.

THERMAL IMPACTS ON FISH

Referring to research efforts by the COE to devise ways to overcome thermal disturbances to the river resulting from reservoir releases, Morse (U.S. Army Corps of Engineers, Portland, Oregon, personal communication, 1990) noted, "[Our] thermal objective is to restore natural temperatures in the summer. The water is now 5-8 degrees C cooler than it was historically. In September, October, and November this has a warm water effect of about 4 degrees C which reduces growth rates and maturation times in the summer and accelerates hatching time in winter well before there are food resources for fish to take advantage of." The Pacific Northwest Power Planning and Conservation Act (P.L. 96-501) calls for the restoration and enhancement of salmonids in the Pacific Northwest.

Research Activities

The U.S. Army Corps of Engineers is currently conducting a $600,000 basinwide study of the operation and management of its 13 reservoirs in the Willamette Basin. Known as the River Basin Review, the study is to evaluate reservoir storage capacity and current operation on those reservoirs (Johnson, 1990).

The COE has also undertaken a major study of the temperature and flow effects of its dam releases in the basin. The Willamette System Temperature Control Study arose out of concerns expressed by resource management agencies since the 1960s about the effects on salmonids of releasing water in the spring and summer that is colder than normal and releasing water in the fall that is warmer than normal. A draft report was completed in 1990 stating that temperature control can be produced by construction and use of water withdrawal towers having selective multiple intake ports (R. Cassidy, U.S. Army Corps of Engineers, Portland, Oregon, personal communication, 1991).

The Oregon Department of Environmental Quality has recently received $25,000 from the Oregon legislature to form a technical steering committee to produce a work plan for a comprehensive River Water Quality Study. This study would reassess the current condition of the river, determine its capacity to accept waste, project the effects of further growth in the basin, and revise the current river management plan (State of Oregon, Department of Environmental Quality, 1990). The DEQ has outlined in detail the future research needs to achieve these goals (State of Oregon, Department of Environmental Quality, 1990). To further evaluate the risks from dioxin, a joint DEQ-Oregon State University study of fish tissue was conducted in 1991. Using current information, the DEQ is currently working to reduce the risks.

Overall Evaluation

The Willamette restoration has been directed primarily toward restoring attributes of water quality, protecting beneficial uses of the river water, and managing for certain species of gamefish. The restoration also includes a reservoir management and research effort to reduce temperature disturbances in the river caused by the release of water from reservoirs. Although attention has been given to land use planning in the basin and in some cases to stream-bank reclamation, there has been no holistic effort to re-create natural antecedent biological or ecological conditions on the Willamette.

Dams on the Willamette and its tributaries have altered normal temperature and flow regimes of the Willamette and its tributaries, and have damaged native wild salmonid populations. Much of the Willamette's water quality improvement has been accomplished by augmenting summer water flows with impounded water to dilute pollutants. Point source industrial discharges are also regulated in amount and concentration through a discharge permit system. As water treatment standards become more rigorous in the future to compensate for an increased human population in the Willamette Basin, more land treatment of wastewater may be employed, further reducing flow in certain Willamette tributaries. This may tend to lower water quality.

Little effort appears to have been made to restore native aquatic life other than anadromous game fish species, and much of the anadromous fish restoration has been the replacement of wild fish by hatchery stock. The river restoration effort has not yet been successful in maintaining natural fish migration routes or in re-creating the predisturbance structure of the native fish community, species by species, to its previous percentage composition. Dams serve not only as barriers to migration of organisms within the river, but also as sediment barriers and obstructions to flooding of riparian areas, which once returned nutrients and sediment to the land.

The Willamette River today is in an unnatural condition that requires constant management. Without flow management through augmentation of low river flow, water quality would be unacceptable. Without hatchery production and release of salmonids, the sport fishery would be severely limited, and without regulation of municipal and industrial waste discharges, high water quality could not be guaranteed. The 13 dams on the river, the past riprapping and channelization, and the dredging (in the lower river) are all indications of the inescapable major impacts that human activities have had on the river. Thus, to call the Willamette an example of river restoration is something of a misnomer. The Willamette is rather an example of river reclamation in which a severely polluted river was cleaned up so that some of its beneficial uses could again be enjoyed by the public. Just as clear-cutting a diverse, complex forest ecosystem and replacing it with a stand of Douglas fir produces a tree farm rather than a restored forest, neither does taking a highly disrupted and polluted river system and merely abating the pollution suffice to "restore" the river.

References

Clean Water Act of 1977. P.L. 95-217, Dec. 27, 1977, 91 Stat. 1566.
Federal Water Pollution Control Act Amendments. 1972. P.L. 92-500.

Gleeson, G. W. 1972. The Return of a River: The Willamette River, Oregon. WRRI13. Advisory Committee on Environmental Science and Technology and Water Resources Research Institute, Oregon State University, Corvallis, Ore. June.

Horner, E. R. 1989. The Almanac of the Fifty States. Information Systems, Palo Alto, Calif.

Johnson, K. 1990. Fact Sheet. River Basin Review, Oregon. U.S. Army Corps of Engineers. Portland, Ore. January 16.

Sedell, J. R., and J. L. Frogatt. 1984. Importance of streamside forests to large rivers: The isolation of the River, Oregon, USA from its floodplain by snagging and streamside forest removal. Int. Ver. Theoret. Angew. Limnol. Verh. 22:1828-1834.

Starbird, E. A. 1972. A river restored: Oregon's Willamette. Natl. Geogr. 141(6): 816-834.

State of Oregon. Department of Environmental Quality. 1987. Definitions. Pub. WH2399. Portland, Ore. October 14.

State of Oregon. Department of Environmental Quality. 1988a. Oregon Statewide Assessment of Nonpoint Sources of Water Pollution. Chapter 4: Assessment of NPS-Related Water Quality Problems; Appendix E: NPS Database Supporting Maps, Portland, Ore.

State of Oregon. Department of Environmental Quality. 1988b. Oregon Statewide Assessment of Water Pollution. Appendix E: NPS Database Supporting Maps, Portland, Ore.

State of Oregon, Department of Environmental Quality. 1988c. Oregon 1988 Water Quality Status Assessment Report. 305B Report. Portland, Ore.

State of Oregon, Department of Environmental Quality. 1989. A River Restored: Oregon's Willamette. Videotape of slide show. Portland, Ore.

State of Oregon, Department of Environmental Quality. 1990. River Water Quality Study. Draft study proposal. Portland, Ore. March 7.

State of Oregon, Department of Fish and Wildlife. 1988. Basin Fish Management Plan. Portland, Ore. March.

State of Oregon, Department of Parks and Recreation. 1988. Annotated Map. River Recreation Guide. Portland, Ore. March.

U.S. Army Corps of Engineers. 1989. River Basin Reservoir System Operation: Reservoir Regulation and Water Quality Section. Portland District Office. Portland, Ore. May. 29 pp.

U.S. Environmental Protection Agency. n.d. The River Lives Again. Water Quality Success Story Series. Office of Water Planning and Standards. Washington, D.C.

U.S. Geological Survey (USGS), U.S. Department of the Interior. 1976. Methodology for River Quality Assessment with Application to the Willamette River Basin, Oregon. Geological Circular 715 M by D. A. Rickert, W. G. Hines, and S. W. McKenzie.

U.S. Geological Survey, U.S. Department of the Interior. 1977. Dissolved-Oxygen Regimen of the River, Oregon, Under Conditions of Basinwide Secondary Treatment. Geological Circular 715 I by W. G. Hines, S. W. McKenzie, D. A. Rickert, and F. A. Rinella.

U.S. Geological Survey. 1979. Asynoptic Approach for Analyzing Erosion as a Guide to Land-Use Planning. USGS Circular 715-L. U.S. Geological Survey, Reston, Va.

U.S. Geological Survey. 1984. Statistical Summary of Streamflow Data in Oregon. Volume 2 Western Ore. Open File 1984-445-A, Department of the Interior, U.S. Geological Survey, Portland, Ore.

Weber, E. 1989. River Erosion Analysis Project: RM 56.0 to 121.0, 1989. Oregon Department of Agriculture, Soil and Water Division, Salem, Ore.

CITIZEN RESTORATION EFFORTS IN
THE MATTOLE RIVER WATERSHED

John J. Berger

General Description and Location

The Mattole River (Figure A.8) rises south of the town of Whitehorn in Northern California's Humboldt County and flows 62 miles northwest to the Pacific Ocean, which it meets 8 miles south of Cape Mendocino (Mattole Restoration Council [MRC], 1989). About 2,000 people inhabit the 306-square-mile watershed, which has a generally mild, Mediterranean climate due to its proximity to the ocean (MRC, 1989; House, 1990).

Mountain ridges and peaks on the western side of the Mattole watershed uplift winter storm clouds from the ocean and produce frequent heavy rains. Precipitation usually occurs in the winter months and ranges from an average of 50 inches (1,270 mm) of rain in the lower watershed to between 80 and 90 inches (2,032-2,286 mm) in the upper watershed (MRC, 1989).

Measurements of stream flow made since 1950 at Petrolia in the Lower Mattole watershed indicate an average annual flow rate of 1,340 ft^3/s and an average monthly winter flow of 1,710 to 4,170 ft^3/s (MRC, 1989). Summer and fall average flows are less than 60 ft^3/s (MRC, 1989). The minimum recorded flow was 20 ft^3/s, and the peak flood was 90,400 ft^3/s (MRC, 1989).

The watershed is in a seismically active area subject to rapid tectonic uplift and high rates of natural erosion and sedimentation (MRC, 1989).

The river bed drops an average of 22 ft per mile in elevation. The Upper Mattole has a relatively steep gradient and stable bedrock in much of the channel. The Middle Mattole is less steep, but the river and streams cut through highly erodible fractured sandstones and decomposing shales and clay (MRC, 1989). The Lower Mattole has a gentle gradient (less than 11 ft per mile) and meanders through a broad alluvial valley to end in a short estuary that becomes a lagoon when reduced summer river flows allow the river mouth to become plugged with sand (MRC, 1989). The gradual river slope results in deposition of gravel bars, islands, and sediment terraces (Focus, 1990).

The Mattole River was once much cooler, deeper, and narrower than it is today. In the 1940s, the river was still shaded by dense forests of Douglas fir, redwood, and native hardwoods (MRC, 1989). The watershed was rich in riverine fur-bearing wildlife, including fisher, mink, otter, and weasel, which were trapped commercially.

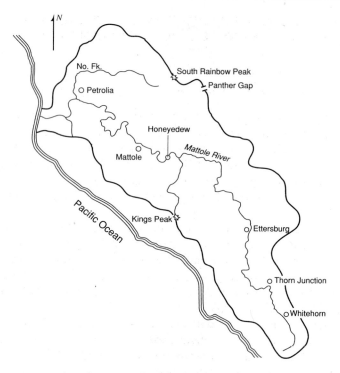

FIGURE A.8 Mattole River Basin. SOURCE: Reprinted, by permission, from MRC (1989). Copyright © by Mattole Restoration Council, Petrolia, Calif.

Immense runs of king (chinook) salmon *(Oncorhyncus tshawytscha)*, and silver (coho) salmon *(Oncorhyncus kisutch)*, and steelhead *(Salmo gairdneri)* the Mattole during the spawning season as the fish made their way to the clean, cold, well-aerated waters of the Upper Mattole and its numerous tributaries.

Dramatic changes in the watershed began in the nineteenth century. Settlers arrived in the 1850s and had destroyed the indigenous Mattole and Sinkyone Native American people by the 1860s (MRC, 1989). The raising of vegetables, fruits, nuts, and livestock and the pursuit of lumbering replaced the Native Americans' stable hunting, gathering, fishing, and agricultural economy. Trees were cut for homes and fences, and large numbers of tan-bark oaks were stripped of their bark to produce tannin for curing hides (MRC, 1989). Clear-cutting did not begin in the watershed until after World War II, when military tank technology led to creation of the steel-tracked bulldozer, giving timber companies access to steep, remote forestlands.

Type of Disturbance and Time Since Restoration

Today the Upper Mattole still has a closed forest canopy over the river and relatively intact riparian zone vegetation (MRC, 1989). However, between 1950 and 1970, more than three-quarters of the watershed's redwood and Douglas fir timber was cut. Severe damage occurred to the middle and lower sections of the river due to massive erosion from fires, logging, overgrazing, and road construction (MRC, 1989). A maze of thousands of skid trails and roads caused millions of cubic yards of sediment to enter the river. Newly eroded soil and rocks filled in the river channel and pools, silting in spawning gravels used by native king and silver salmon. The result was channel migration, flooding, bank erosion, loss of riparian vegetation, and disappearance of riverside farmland (MRC, 1989; House, 1990). Overflowing its old channel, the lower river reached a width of half a mile.

The heavy load of sediment transported by the river tends to settle in its lower reaches, especially near the mouth. Sometime in early summer, when river flows drop below the 120 ft^3/s needed to repel sand left at the river mouth by ocean waves, the river naturally closes and a lagoon forms behind the sandbar. In bygone days, smolts thrived in the slightly brackish waters here. However, since the river has been charged with sediment, the lagoon is much shallower and lacks the cool water necessary to king salmon. Thick silt on the river bottom discourages the benthic insects the salmon need for food. Overhanging vegetation that once cooled the water and provided habitat for insects has been scoured away.

Large nylon gillnets used by Oriental fishermen to catch squid cover many square miles of the Pacific and are also believed to be reducing the number of returning Mattole River salmon.

Restoration Methods and Techniques

Citizen restoration workers in the Mattole watershed intervene directly in the salmonids' spawning cycle in an effort to improve reproductive success. During salmon runs, fish are guided into wooden salmon traps by weirs in the river. Females are removed and stripped of their roe; the roe is fertilized in buckets, and the eggs are placed in clean gravel inside wooden hatch boxes. These handmade streamside salmon incubators are constructed so that clean, filtered, flowing river water can be provided to the fertilized eggs and fry, instead of forcing them to attempt survival in the muddy river. These simple, inexpensive hatching devices with no moving parts typically increase salmon hatch success rates from 15 to 80 percent (House, 1990).

Salmon are then raised in creek holding areas. When salmonid smolts were released in May for their downstream migration, mortality was high in the warm, shallow lagoon. Smolts are therefore now being held for release in the fall when high river flows reopen the plugged river mouth. Restoration workers hope that this will increase smolt survival. Citizens had to work in the river in May 1990, to keep the river mouth open long enough for salmon smolt to get to the sea. Because the river has been designated a Wilderness Study Area, mechanized equipment cannot be used for this task.

Public or Political Involvement

The watershed restoration was begun by a small group of about a dozen people residing in the Mattole watershed who called themselves the Mattole Watershed Salmon Support Group (MWSSG) (House, 1990) and who initiated erosion control, reforestation, salmonid habitat repair, and habitat enhancement. After meeting informally from 1979 to 1985, the Mattole Restoration Council was incorporated in 1985 to conduct active watershed restoration, to make long-range plans, and to oversee their implementation. The MRC is a consensual decisionmaking body representing 13 member organizations, including the MWSSG (House, 1990). The group's guiding principle in watershed restoration is to imitate natural processes as closely as possible (House, 1990). To date, the MRC and member groups have raised between $500,000 and $600,000 for Mattole watershed restoration. School children as well as local residents and landowners have been extensively involved in the restoration work, especially in the release of young salmon.

Early in the restoration process, a citizen salmonid habitat inventory was conducted in the watershed. Citizens also inventoried the remaining old-growth forests, and more recently, they systematically identified sources of erosion in the watershed and prescribed remedial actions.

Whenever the need and opportunity for work in the watershed arose, the MRC endeavored to train local citizens. For example, they were trained in techniques used for erosion mapping, for measuring siltation in the river channel, and for estimating reforestation success and the survival of remnant old-growth forest (House, 1990). With the assistance of professional geologists, 23 watershed residents performed erosion surveying and mapping. They found roads, including logging haul roads and skid trails, to be the source of 76 percent of all erosion problems mapped in the watershed (MRC, 1989).

In the Mattole estuary, the MWSSG has also been attaching driftwood structures to the riverbanks to provide shade and shelter to juvenile king salmon and steelhead. Other structures will be built to use winter high-water flows to rescour deep pools in the estuary. Currently, the MWSSG and the MRC are engaged in a 2-year effort to create a salmonid enhancement plan for the estuary.

Because of the silt deposition in the estuary area, the river is unstable and has moved its course away from the riparian areas. Alders and willows have been planted around the estuary to encourage the river to scour more deeply, to stabilize the banks and increase shading, and to create deeper channels and pools. The U.S. Bureau of Land Management and the California Conservation Corps planted 3,000 2-year-old alder seedlings in 1990. A few thousand willows cuttings 5 to 10 ft in length were augured into the banks in the winter of 1990 by the MWSSG in the south side of the floodplain a mile or two from the river's mouth. Since the restoration work began, more trees have been planted in the watershed by timberland owners than by restoration groups. However, much of the landowners' activity stems from community education work and activism by the MWSSG and the MRC. For example, the MRC published a map contrasting the distribution of old-growth forests in 1988 with that in 1947 and has also done a forest regeneration study.

Federal and Local Agency Roles

For the past 5 years, the MRC in cooperation with the State Coastal Conservancy and the Redwood Community Action Agency has been conducting annual cross-sectional surveys of river channel depth at 14 sites to monitor sediment movement in the river (Focus, 1990). During the same period, the MRC has been conducting a study of the estuary in cooperation with the Bureau of Land Management and, since 1983, has been studying a major landslide in cooperation with the California Department of Water Resources, the Redwood National Park, and the California State Coastal Conservancy (Focus, 1990).

The MWSSG and the MRC have coordinated closely with the California Conservation Corps (CCC), identifying target areas and tasks for CCC workers and, in some cases, working directly with them.

The Bureau of Land Management has been resistant to having local residents play a managerial role in the restoration (F. House, Mattole Restoration Council, personal communication, 1990). However, the bureau has proposed a plan for reintroduction of Roosevelt elk to the watershed.

Government agencies have, in general, provided expertise and recordkeeping for much of the restoration work. The Department of

Fish and Game and the State Coastal Conservancy have provided restoration grants, and the conservancy has provided planning funds.

Recovery Accomplished

The MRC states that 250,000 salmonids were released in the watershed between 1980 and 1990. But the MWSSG has scanty data on fish returns, because most of the fish released were not tagged. Thus, it is difficult in most cases to identify propagated returnees from native returnees. However, the MWSSG has also released fish into tributary streams where there were no king salmon or silver salmon before and has found salmon now occupying these habitats. For example, from 1986 to 1991 there have been silver salmon in Lower Mill Creek. Silver salmon have also reoccupied the North Fork of Honeydew Creek from 1989 to 1991. The population is thought to be self-sustaining in Lower Mill Creek, but introductions are continuing in Honeydew Creek, Squaw Creek, Thompson Creek, Bear Creek, and the mainstem of the river have also had introductions of king salmon.

The overall numbers of returning spawners have continued to decline in the river as a whole, despite the introductions. Whereas an estimated 20,000 king and silver salmon used the river in 1964, they had declined to 3,000 kings and 500 silvers in the winter of 1981-1982. In 1991, the king salmon count was only 200 fish, and no silver salmon were seen. The depletion has probably been intensified by a 4-year drought in California and by an offshore ocean temperature anomaly known as El Niño. Low water keeps the returning fish from reaching their preferred spawning areas and forces them to spawn in parts of the river where survival is poorer; low water also leads to early closure of the river mouth. Without the assistance of the MWSSG and the MRC over the past decade, the river's king salmon population might be extinct or at least closer to it.

Ecological Models Against Which to Measure Success

The MRC has not articulated a clear policy on what ecological model will be used to measure success, but the MRC is likely to consider its efforts highly successful if it is able to restore native salmonid populations to near-predisturbance levels, and if it can educate citizens to live in a sustainable relationship to their natural resources (House, 1990). According to House (F. House, Mattolle Restoration Council, personal communication, 1990), success will be achieved when hatchery operations can be halted and species can maintain themselves without human intervention.

Overall Evaluation

The work of the MWSSG and the MRC has prompted regulatory agencies to look more closely at the risks of extinction of old growth-linked species within hydrological and biological management units. Much public education has been conducted, and a cadre of local citizens has been created who have a deep concern and detailed knowledge of the watershed and its hydrological processes. The MRC sees environmental restoration as an opportunity to bring about social as well as ecological transformation by reeducating residents to restore and live in a harmonious, sustainable way within their watershed (House, 1990).

References

Focus on Mattole Estuary. 1990. Mattole Restoration Newsletter. Winter. 12 pp.

House, F. 1990. To Learn the Things We Need to Know. Whole Earth Review. Spring.

Mattole Restoration Council (MRC). 1989. Elements of Recovery: An Inventory of Upslope Sources of Sedimentation in the Mattole River Watershed with Rehabilitation Prescriptions and Additional Information for Erosion Control Prioritization. Mattole Restoration Council, Petrolia, Calif. December. 47 pp. (Note: This document is presented as a step toward a comprehensive watershed restoration plan.)

THE MERRIMACK RIVER

Sheila David

With good management and human commitment, nature often takes over and heals itself.

Rene Dubos

General Description and Type of Disturbance

The history of pollution of the Merrimack River reads like a horror story. Report after report describes the foul, polluted condition of the river, its headwaters, and its tributaries. In the 1930s, reports indicated that contamination along the length of the river made it too polluted for domestic water supply uses. Raw sewage, paper mill waste, tannery sludge, and other pollutants were dumped into the river untreated. By the end of World War II, the Merrimack was recognized as one of the 10 most polluted streams in the nation.

The Merrimack River is formed by the confluence of the Pemigewassett and Winnepesaukee Rivers in Franklin, New Hampshire (Figure A.9). It flows south through New Hampshire's capital, Concord, past its

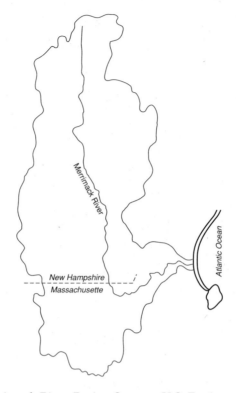

FIGURE A.9 Merrimack River Basin. SOURCE: U.S. Environmental Protection Agency, 1987.

most populated cities of Manchester and Nashua, and into Massachusetts. There it turns sharply east toward the Atlantic, flowing through the cities of Lowell, Lawrence, and Haverhill. Just below Haverhill, the river becomes tidal, widening into an important estuarine zone at Newburyport. The river then flows through a narrow channel between Plum Island and Salisbury Beach into the Atlantic Ocean. The Merrimack River Basin is the fourth largest in New England and has a maximum length of 134 miles and a maximum width of 68 miles. Of the 5,010-square-mile basin area, 3,810 square miles lie in New Hampshire and 1,200 square miles in Massachusetts.

Approximately 1,484,000 people lived in the basin in 1980. The economy of the basin depends primarily on manufacturing and service industries. Recreation and tourism are of great importance in the upper basin, where mountains and lakes have made the area a leading vacation spot in the Northeast.

The Cleanup Begins

Serious efforts to improve the quality of the Merrimack River began in 1972. The Clean Water Act of 1977 (P.L. 95-217) established standards for water quality, supplied funds to assist states in meeting those standards, and directed the states to maintain the standards through effective administration.

Since then there has been significant improvement in the quality of the river. More than 84 wastewater treatment plants have been constructed throughout New Hampshire at a cost exceeding $280 million. Initial emphasis was placed on cleaning up the tributaries. Treatment facilities came online one after the other during a period of generous federal support. Industries were brought into compliance, and they often made a substantial investment in pollution control equipment. In many cases, industries learned how to use water more efficiently, whereas others were unable to operate profitably under the clean water regulations and went out of business.

The river's recovery is not yet complete. Key municipalities along the river have not completed their wastewater treatment plants or are in need of interceptors and expansion. In its 1986 report to Congress, the New Hampshire Water Supply and Pollution Control Commission reported to the Nashua Regional Planning Commission that it assessed 488.3 miles of the Merrimack and its major tributaries in New Hampshire (Nashua Regional Planning Commission, 1986). A total of 420.9 of those miles met or exceeded the Class-B federal standards for fishable and swimmable waters, leaving 67.4 miles in need of water quality improvement.

The combination of reduced federal funds and impending deadlines for compliance has increased competition and friction between river communities. Whether the federal government continues to withdraw its support for wastewater treatment projects or not, greater cooperation between the states and communities is required.

Citizen Power

Local citizen groups have been very effective at helping to restore the Merrimack River. For example, the Merrimack River Watershed Council, begun in 1977, is an interstate organization dedicated to the protection of the Merrimack River. The council was formed by a small group of citizens who recognized the need for an independent organization to address the issues affecting river life and water quality. The council has been effective in defining resource values, increasing public awareness of water issues, and preserving open space in many towns.

The council's goal has been a revitalized Merrimack, and it has been working to achieve a balanced program of environmental protection and enlightened land use that is nonpolluting and in the public interest, and to build a citizenry alert to the issues and opportunities surrounding protection and restoration of the river. The group has been assisted by two regional planning agencies and the New England Rivers Center, and prepared a comprehensive greenway plan in its first 2 years.

Access points have been established, trails protected, and agricultural preservation restrictions enacted. Because of the council's efforts, the Merrimack has been designated a local scenic river in Massachusetts and may soon be granted National Wild and Scenic River status.

The council has worked with business and industry to stop pollution, has linked various groups with municipal and states agencies, and has created effective coalitions. *The River's Reach*, a first-class, issue-oriented publication circulated to 12,000 people twice annually, has focused attention on the river as a dominant and positive force in the valley. The council has helped people realize that the Merrimack, despite 100 years of neglect and abuse, is a natural resource of critical value to the economic and social well-being of the region. The Merrimack River Watershed Council has been key to managing in the public interest the increasing demands on the river and to achieving equitable resolution of upstream and downstream and out-of-basin conflicts.

Results

The Merrimack River is resilient and has responded favorably to the tremendous and successful efforts of government agencies and private citizens to make it fishable and swimmable. Pollution abatement activities on the Merrimack have resulted in at least partial achievement of water quality standards in 94.3 percent of the New Hampshire portion of the basin and 68 percent of the Massachusetts portion (Nashua Regional Planning Commission, 1986). The entire Massachusetts portion of the mainstem of the river is at least partially supporting designated uses. The river is the drinking water source for several Massachusetts and New Hampshire communities, supplying more than 237,000 people in Massachusetts alone. The use of the river as a source of drinking water intensifies the need to protect the integrity of the Merrimack's water quality.

The river now offers a good deal of enjoyable canoeing. In the marshlands, canoeists can enjoy a multitude of bird and other wildlife sightings. The state Division of Fisheries and Wildlife, Division

of Marine Fisheries, and the U.S. Fish and Wildlife Service are working on a major restoration effort for Atlantic salmon, American shad,
and other anadromous fish. The Lowell Heritage State Park (the first
urban national park of its kind) preserves examples of a textile mill
era that had a dramatic impact on the regional economy.

A stretch of 36.4 miles of the Merrimack River from its start in
Franklin to northern Manchester in New Hampshire meets water
quality standards for Class-B waters, with coliform concentrations of
240 ppm or less per 100 ml, drastically reduced total organic loadings, levels of suspended solids reduced by one-third, and a much
higher dissolved oxygen content (U.S. EPA, 1987). The river's appearance, and its biological and chemical makeup, have improved
markedly. For example, significant reductions in the input of pollutants have resulted in the replacement of sewage-laden sediments by
reestablished benthic fauna. Such results were obtained in numerous
improvement projects that included the Winnepesaukee basin plant
in Franklin, the Boscawen Wastewater Treatment Plant, a secondary
waste treatment facility in Hooksett, two facilities in Concord Penacook
and Hall Streets, a secondary facility at Allenstown-Pembroke, six
facilities along the Pemigewassett River and six on the Contoocook, a
secondary facility in Manchester, another secondary facility on the
Souhegan, and a primary treatment plant on the Piscataquog.

In Massachusetts, similar efforts have upgraded the river's water
quality. Secondary wastewater treatment facilities were built in six
communities along the Merrimack mainstem. Operation of three of
these plants—at Lowell, Lawrence, and Haverhill—reduced the amount
of Biological Oxygen Demands on the river from these three towns
by an estimated 80 percent (Nashua Regional Planning Commission,
1986). Also significant have been the increase in dissolved oxygen
and the decrease in coliform bacteria counts. Today, four drinking
water collection sites are in operation in Lowell, Andover, Methuen,
and Lawrence, Massachusetts, and one operates in Nashua, New Hampshire. A sixth withdrawal site is under construction in Tewkesbury,
Massachusetts.

Upgraded wastewater treatment has also led to a relaxation of the
prohibition against shellfishing in the estuary. A recent upgrade of
the Newburyport wastewater treatment facility is expected to yield
further improvement in the estuary.

Hydrology and Ground Water

Unlike surface water systems, which are hydrologically well understood, ground water systems and their intricacies are less well

understood. Understanding the ground water environment, with an assessment of the resource, is necessary before it can be adequately protected and efficiently utilized. This is critical because ground water serves as the principal source of domestic water supply throughout rural areas of the basin and also makes an important contribution to public systems. Past and present efforts within the basin have focused on ensuring the quality of ground water for public consumption as a prerequisite for managing ground water quality. Large-scale, general data on ground water quality have been compiled by New Hampshire. Information is slightly more refined in the Massachusetts portion of the basin because of hydrologic studies performed by the U.S. Geological Survey. Studies indicate that ground water quality conditions are generally the same in Massachusetts as in New Hampshire (Hanley, 1990).

The runoff of the Merrimack Basin flows through the interconnected system of surface water and ground water. Although surface water is the most visible manifestation of runoff, it is derived primarily from ground water via subsurface flow. The Merrimack mainstem is the axis in the basin, connecting five principal tributaries and ultimately channeling an average flow of 4.9 billion gallons per day to the Atlantic Ocean.

Overall Evaluation

The water quality of the Merrimack River has gradually improved, and the readily observable contamination has vanished. Although the projections of expected continued improvement in water quality for years 1990 through 2020 may be well founded for Class-B water characteristics, the increased population in the basin and subsequent activity along the river will inevitably affect the attainment of public drinking water standards. It is clear that improvements in the river have been achieved through a combination of local, state, and federal efforts. In addition, the cooperation and financial assistance of private industry have helped bring the Merrimack River to health.

According to the October 1988 bulletin of the Merrimack River Watershed Council, current policies affecting the water resources of the Merrimack include the following:

• The Clean Water Act of 1977 and its recent reauthorization, which reflect strong commitment to support water quality programs that improve or maintain water resources at fishable and swimmable quality.
• Implementation of the Anadromous Fish Restoration Program in the U.S. Fish and Wildlife Service.

• Mandate of October 1987 from the U.S. Environmental Protection Agency (EPA) for commencement of the Merrimack River Watershed Initiative, which recognized the need for managing water quality with funding provided to the states for implementation.

• Riparian rights in New Hampshire, meaning that ownership of riverfront land allows use of the water flowing by, provided the owner has a "reasonable use" and downstream riparian owners are not unreasonably affected.

• Congressional mandate that New Hampshire enforce the Clean Water Act of 1977. Although several miles of the Merrimack did not meet the July 1, 1988, fishable and swimmable criteria as required, state efforts have greatly improved the river's water quality.

• The Merrimack River Watershed Initiative (funded by a $50,000 grant from EPA) and the New Hampshire River Protection Program, which were implemented in 1988.

• The Water Supply Task Force, partially funded by major water users, which is studying the long-range water supply needs of 68 communities in southern New Hampshire.

• The fact that the state of New Hampshire has no specific policies that address development and use of public drinking water supplies. There has not been an act of legislature either establishing policy or establishing a program in the state administrative agencies.

• Lack of a clear mandate from the legislature, so that state agencies allow and encourage water users to rely more heavily on the Merrimack River. As the Department of Environmental Services Commissioner told the Merrimack River Watershed Council recently, "It is clear that there is sufficient minimum flow for usage of the Merrimack as a principal water supply source for south central New Hampshire in the future" (Hanley, 1990).

References and Recommended Reading

Brochures and bulletins from the Merrimack River Watershed Council, 694 Main Street, West Newbury, MA 01985.

Clean Water Act of 1977. P.L. 95-217, Dec. 27, 1977, 91 Stat. 1566.

Hanley, N. E. 1990. A Massachusetts Merrimack River Water Supply Protection Initiative 1990. Massachusetts Department of Environmental Protection, Technical Services Branch, Westborough, MA 01581.

Nashua Regional Planning Commission. 1986. The Merrimack River: Issues for Southern New Hampshire. Nashua Regional Planning Commission, 115 Main Street, Nashua, NH 03061.

U.S. Environmental Protection Agency (EPA). 1987. Merrimack River Watershed Protection Initiative: Past, Present, and Future. U.S. EPA, Water Management Division, Boston, Mass.

THE BLANCO RIVER

John J. Berger

Introduction

This case study discusses scientific, technological, and administrative aspects of the Blanco River reconstruction project in southwestern Colorado (Figure A.10). It focuses on the channel stabilization and fishery problems encountered and the processes used to solve them.

Before repair work began in 1987, target sites on both branches of the Blanco River were broad, shallow, and braided with no pools. In the course of the 3-year river reconstruction project directed by hydrologist D. L. Rosgen, the river's bank-full width was reduced from a 400-ft-wide braided channel to a stable, 65-ft channel with a high pool-to-riffle ratio (personal communication during site visit to

FIGURE A.10 Map of the Blanco River.

Blanco River, June 1990; D. L. Rosgen, telephone interview, January 1991). Even before project conclusion in 1990, major improvements had occurred in the fishery and in the appearance of the site.

General Description

The Blanco River is located 20 miles southeast of Pagosa Springs, Colorado, on Highway 84, 10 miles east of Blanco Basin Road. The project area is about 2.7 stream miles in length and drains a basin of approximately 56 square miles. The site has been used for grazing for about 50 years. Mean precipitation in the basin is approximately 42 inches per year; runoff is dominated by snowmelt. Major floods in recent years have been caused by late summer and fall high-intensity thunderstorms.

The river has a slope of about 1.5 percent. Bed materials are heterogeneous unconsolidated cohesive particles ranging from fine sand to very coarse cobbles. The mean river depth is 3.5 ft, and the river in the project vicinity is a fourth-order Horton stream. In the stream classification system of the hydrologist who repaired sections of the Blanco and San Juan Rivers (Rosgen, 1988), both project reaches of both rivers were designated as D1 streams and were reconstructed as C1 streams.

A D1 stream has a slope of 1.5 percent or greater; a braided channel; and a cobble bed with a mixture of coarse gravel, sand, and small boulders; it is slightly entrenched without valley confinement, and is found in coarse glacial outwash depositional material in a reach with an excess sediment supply of coarse-size material (Rosgen, 1988).

The C1 target stream has a gradient of 1.2 to 1.5 percent; a sinuous channel with a sinuosity ratio of 1.5 to 2.0; a width-to-depth ratio of 10 or higher (18 to 20 in the case of the reference streams used as models for the Blanco restoration); and a cobble bed with a mixture of small boulders and coarse gravel; it is moderately entrenched and moderately confined by its valley, and is found in predominantly coarse-textured, stable, high alluvial terraces (Rosgen, 1988; D. L. Rosgen, telephone interview, January 1991).

Origin of the Problem: Improper Flood Control

A major difference between the Blanco and the San Juan River is that the Blanco was channelized by the U.S. Army Corps of Engineers (COE) after a 1970 flood in an effort to protect adjacent land from flooding. The flood control effort resulted in channel instability and in the creation of a braided reach. By contrast, the channel instability in the San Juan River was caused by extirpation of stream-bank vegetation.

After the 1970 flood, COE straightened portions of the Blanco River, increased its slope, and entrenched the river within a levee system so that what once was the low-flow channel, terrace, and floodplain became a wide, flat-bottomed trapezoidal channel. The loss of meanders and steepening of the river caused the channel bed to degrade. This, in turn, resulted in stream-bank failure and erosion. This erosion typically travels upstream and eventually contributes to sedimentation and aggradation of downstream reaches.

Replacement of the natural river morphology by the wide artificial channel induced sediment deposition through a reduction in shear stress. The shear stress is a function of stream gradient, specific gravity of water, and the hydraulic radius. Enlarging the width-depth ratio by channelization reduces the shear stress or entrainment capacity of the stream at any flow. This can cause sedimentation and a braiding channel. Channel confinement also prevented the floodplain from functioning (the floodplain is necessary to dissipate energy). Another problem was COE's use of highly erodible riverbed material to build the levees.

After channelization, a broad range of hydrological problems began to appear. The river began to spread out from its channel, becoming broad and shallow, detaching riparian vegetation, and eroding banks as it migrated. The active bank erosion contributed high sediment loadings to the stream, which in turn led to bar building and other types of sediment deposition. Agricultural land along the river was made unusable and agricultural facilities, including a barn, were threatened. Because the shallow river experienced high summer temperatures and full freezing in the winter, and had lost its pools and other trout refugia, few fish could be found; those taken were generally small brook trout. As on the East Fork of the Blanco (another Rosgen channel stabilization site), much of the Blanco sediment was contributed by a relatively short stretch of the river.

Since COE's flood control intervention, there has been continual progressive erosion of the property owner's land near the Blanco, threatening portions of the remaining land and facilities. Since COE changed the natural riverine hydrology, the landowner has had to spend thousands of dollars over many years trying to stabilize the riverbanks, but to no avail. Until he learned of Rosgen's successful work on the nearby San Juan River, the landowner was uncertain how to proceed, because he did not want to use conventional COE engineering approaches that rely on unsightly concrete and riprap to imprison the river.

Restoration Goals and Objectives

The major goal of the Blanco reconstruction was to stabilize the river in a well-incised but natural looking permanent channel that would enable it to handle floods, without requiring creation of an artificial-looking concrete channel. The "soft engineering" approach used by Rosgen required rebuilding the river's width-to-depth ratio, and re-creating a natural channel geometry containing a low flow channel, floodplain, and terrace.

In selecting design criteria, Rosgen first located undisturbed similar streams in the vicinity of the Blanco and found that their dimensions and patterns were consistent with those of the C1 stream type in his stream classification system. He then sought to use as design criteria the values of this very stable stream type existing in the local area on the same gradient and within similar channel and bank materials.

In Rosgen's work on the East Fork of the Blanco, he modeled the reconstruction on a stable section of the river about a mile downstream from his project site. To verify that the candidate stream type selected should be stable, he studied a long time series of aerial photos taken from the 1940s until recent years. This historical record included a period that extended many years before and after major floods. Inspection revealed that the C1 stream type exhibited post-flood self-stabilization. Rosgen therefore concluded that the C1 stream type had held up and would hold up very well.

Another reconstruction goal was to increase bank storage. Previous to Rosgen's stabilization work, the Blanco River project sites were so wide that the floodplain had been eroded away, and no land was adjacent to the active channel.

Cost and Benefits

Restoration costs on the Blanco, about $30 per lineal foot of stream, were half those on the San Juan River project, because equipment operators had been trained during the earlier project and were able to work more efficiently (D. L. Rosgen, telephone interview, January 1991). Total costs to the private landowner on whose property the Blanco River work was done were about $400,000. Spawning channels and a new spring-fed, floodplain-level trout pond on the Blanco added about another 15 percent to total project costs (D. L. Rosgen, telephone interview, January 1991).

From his $400,000 investment, the landowner gained the 168 to 170 acres of agricultural land that was once again made available in the floodplain. The recreated floodplain varied from about 400 to 800 ft in width, and the cost of its creation per acre was about the going rate for purchase of land in the area.

Where fish were scarce and small, it is now not unusual to catch 16- to 18-inch brown and rainbow trout that have been stocked. Fishing has also been improved by construction of narrow, sinuous fish spawning channels ranging from 20 or 30 to 500 yards in length and connected to the main branch of the river, as well as by creation of an acre-and-a-half, spring-fed trout pond. Through revegetation with willows and cottonwoods, major aesthetic improvements were also made at the site.

Another way to assess the value of the project beyond landowner satisfaction with the fishery, land reacquisition, aesthetic improvements, and property protection values would be to estimate the avoided damage from stream sedimentation. Rosgen points out that Pacific Gas and Electric Company pays $4 per cubic yard of sediment kept out of Wolf Creek, which it manages in California. Sediment not kept out of the creek and accumulating in a company holding pond must be dredged and disposed of at a cost of about $6 per cubic yard. Because of the very active bank erosion in progress before the project began, use of this sediment-to-dollars conversion factor would result in a very large estimated project benefit based solely on avoidance of sediment damage.

Project Permits

The Blanco River project design was reviewed by the following agencies prior to the granting of a construction permit under Section 404 of the Clean Water Act of 1977 (P.L. 95-217): the Colorado Division of Wildlife, the U.S. Fish and Wildlife Service, the Environmental Protection Agency, and the U.S. Forest Service. The project almost failed to materialize when COE subjected the unique design to expert review and was told by its reviewers that the new system would not contain flood flows. The project design was then sent for review to Professor Luna Leopold at the University of California, Berkeley, Department of Geology and Geophysics; Leopold praised the project and expressed confidence that it would work. On the basis of his recommendation, COE withdrew its reservations, and the project was allowed to proceed.

No federal funding was applied for, although it is possible that matching funds could have been obtained through the Agricultural Conservation and Stabilization Service.

Restoration Process: Major Stages

Rosgen attempts in a mathematical way to match the observed stream morphology of stable streams to the reconstruction design criteria for his projects. He takes theoretical relationships regarding channel geometry and then observes the empirical relationships in the field for verification of the theory. Observations are then linked to his standardized empirical stream classification system. In effect, he matches stream data and other variables to the model. The principles employed work in any physiographic region, because sediment grain size and slope are physical characteristics that can be observed anywhere in the world, and the laws of physics are also universal. Over time, Rosgen continues expanding and refining his classification system by adding newly observed stream types.

Rosgen began the Blanco project with research to identify the causes of the river's problem. This entailed inventorying of hydrological conditions, locating stream flow records from the U.S. Geological Survey, and interpreting over time the behavior of stable and unstable channel forms and types.

Rosgen then created a design based on existing flow and other variables similar to the natural stable form for that flow. Dimensions for the channel were chosen based on flow data, and patterns for the channel were developed based on the dimensions and flow. The river's meander geometry—the radius of its curvature, curve amplitude, and meander length—was designed commensurate with its width in the same proportion as a natural river of the model type.

Next, Rosgen obtained necessary permits using calculations based on permanent stream cross sections to calculate the necessary amounts of cut and fill (i.e., the yardage of excavations). He was also required to do an environmental assessment for his project and mitigation. Clearance of the Section 404 permit took about 60 days.

Rosgen then field-staked the active channel and its proper alignment and other aspects of its meander geometry using a laser beam level. Before construction began, he diverted the stream into a constructed bypass channel so the stream work could be done dry. Construction was done during seasonal low-flow periods.

Downstream of the construction area, Rosgen had constructed a settling detention basin outside the active channel via a diversion. Thus any sediment from the project was flushed into the pond.

Rosgen then directed the shaping of the channel with bulldozers and scrapers, so that material from bars and channel was deposited in the floodplain. The effect of the entire project was to bring about a transformation of the river. The problem with braided rivers is that they generally do not recovery naturally; recoveries are known but are extremely rare; and braided channels typically get progressively worse and are not self-correcting (D. L. Rosgen, telephone interview, January 1991).

Once Rosgen establishes the river geometry including cross-sectional dimensions, he delineates the flow pattern and then performs bank revetment work utilizing native material, including logs, root wads, boulders, and live vegetation. On the Blanco River, he used cotton-woods and willows to reestablish streamwide tree cover, and he used fescue, bluegrass, and clover, as well as nonnative timothy and orchard grass, to cover bank areas. Willows in this project were trans-planted by front-end loaders from an adjacent terrace located about 150 yards from the river channel to the river banks. Cuttings were also taken from willows adjacent to the river, utilizing the same species of willows.

To reinforce banks, Rosgen used much the same procedure as on the San Juan River: he sank logs in the streambed, put boulders over them, and positioned logs on top of the boulders. After the logs had been covered with some soil, willows were planted in the newly created bank margins.

Project Indicators

Variables with which Rosgen was concerned on this project in-cluded river width, depth, velocity, discharge, slope, energy slope, roughness, sediment load, sediment size, sinuosity, width-to-depth ratio, dominant particle size of bed and bank materials, entrench-ment of channel, confinement of channel, landform confinement of channel, landform features, soil erodibility, and stability (Rosgen, 1988).

Rosgen measured sediment particle size, substrate, aggregation, degradation, slope, longitudinal profile, bed load, suspended sedi-ment, grading curves, and particle size on the very similar East Fork project reach on the Blanco and used that data in his design work on the main branch of the river.

In gathering this data, Rosgen's main concern was to verify that, given the sizes of sediment that moved through the new channel versus the old channel and the expected sediment input from the feeder channel, the reconstructed reach would be able to accommo-date the demands placed on it.

Conclusion

The Blanco River project site now has new meanders, deep pools, new flood terraces, rebuilt floodplains, riparian vegetation, verdant pasture grasses, and banks stabilized with locally obtained root wads, tree trunks, and boulders. The current is focused into the center of the channel by strategic placement of "vortex rocks" in the channel to aim the force of the water away from the banks. The new stable channel complex has a natural look, compared with cement trapezoidal channels, levees, and riprapped banks. The fishing is a delight to landowner and visitors alike.

References

Clean Water Act of 1977. P.L. 95-217, Dec. 27, 1977, 91 Stat. 1566.

Rosgen, D. L. 1988. A Stream Classification System. Pp. 163-179 in K. M. Mutz et al., eds., Restoration, Creation and Management of Wetland and Riparian Ecosystems in the American West, pages 163-179. A Symposium of the Rocky Mountain Chapter of the Society of Wetland Scientists, November 14-16. PIC Technologies, Inc./ CRS Sirrine, Inc., Denver, Colo.

THE KISSIMMEE RIVERINE-FLOODPLAIN SYSTEM

John J. Berger

I wondered . . . about this passion to make a place into something it isn't. We irrigate the desert and drain Florida. I suppose we'd bulldoze the Rockies if we could find a big enough bulldozer . . . What made south Florida unique was singled out for eradication.

G. Norman, 1984

Introduction

The restoration of the Kissimmee River needs to be understood in the larger context of the effort to restore the Florida Everglades. The Kissimmee River was once a broad, meandering 103-mile-long waterway that drained an upper basin consisting of a chain of lakes (Figure A.11). The river then flowed slowly through an expansive marshy floodplain into Lake Okeechobee, its southern terminus. The Kissimmee River basin, the enormous lake, and the Everglades together formed a 9,000-square-mile hydrological system. (The connection between the lake and the Everglades was intermittent, mainly during or after the rainy season, when water flowed over the lake

Kissimmee Basin

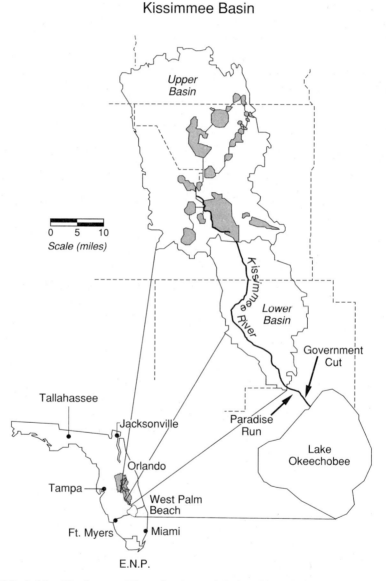

FIGURE A.11   Kissimmee River basin and Lake Okeechobee area. Source: Reprinted by permission of the South Florida Water Management District, n.d.

brim in a sheet. Although the connection was intermittent, the habitat that received periodic inundation was continuous.)

Then in 1961, the U.S. Army Corps of Engineers (COE), in response to a request from the state of Florida, began a far-reaching flood control program on the Kissimmee that brought major ecological changes to the river and its floodplain. The COE is currently conducting a feasibility study and design work for an effort to restore some of the natural hydrological functions that were lost in the course of that flood control effort. Restoration of the Kissimmee River would be a centerpiece of the effort to restore the Kissimmee-Okeechobee-Everglades ecosystem. This case study describes the Kissimmee River Restoration Demonstration Project, begun in the late 1980s, and assesses that experiment in the context of (1) ecological changes that have taken place in South Florida and (2) the state's comprehensive program to restore the Everglades.

Predisturbance Ecological and Hydrological Conditions

The Everglades to the south of Lake Okeechobee was once a vast, gently sloping area of wet marsh, tussocks, bayous, ponds, and sloughs inundated from the north by clean water that spilled gently over the brim of Lake Okeechobee (Brumbach, 1990). Water emptying out of Lake Okeechobee after heavy rains formed a shallow sheet 40 to 60 miles wide that flowed across much of the Everglades, which covered most of southeast Florida. From the inland freshwater marshes that compose most of the Everglades, the water continued south through mangrove swamps into coastal salt marshes and then into Florida Bay. As recently as 100 to 130 years ago, relatively little drainage of wetlands and, hence, little damage to the entire Kissimmee-Okeechobee-Everglades ecosystem had occurred.

Along the margins of the Kissimmee-Everglades floodplain marshes was wet prairie, home to diverse grasses, forbs, and rushes with their seeds, bulbs, and rhizomes. This prairie provided a valuable feeding, resting, nesting, and breeding ground for a great variety of wildlife, including species now extinct, rare, or endangered, such as the Florida panther.

Among the profusion of life in the Kissimmee-Everglades waterways and marshes, billions of shrimp grazed on algae and zooplankton. Above the shrimp on the complex trophic web, other billions of small fish fed, supporting the legions of multicolored wading and diving birds that foraged in the water for crustaceans and fish.

In the Kissimmee basin during the rainy season, river water would flow over adjacent lands, providing habitat and nourishment for the multitude of fish, invertebrates, and other aquatic life. During the dry season, the marshes slowly drained their nourishing contents back into the river, concentrating flood organisms there for large fish, waterfowl, and alligators.

FEDERAL FLOOD CONTROL EFFORTS

South Florida, a hurricane-prone land, was hit by several hurricanes in the 1920s. After a 1928 hurricane broke a low dike around Lake Okeechobee and caused a flood that took more than 2,000 lives, the federal government launched a major flood control program in South Florida, and COE impounded Lake Okeechobee with a 20-ft-high levee (Brumbach, 1990). More hurricane-induced flooding in the 1940s in the upper Kissimmee lakes basin and throughout South Florida led the state of Florida to request additional federal flood control help in the Kissimmee River basin (Dreher, 1986). The COE responded to the state's desire for flood control and developable land by proposing the excavation of a canal from Lake Kissimmee to Lake Okeechobee to replace the Kissimmee River (U.S. DOI, 1958). "The primary purpose of the Kissimmee River development [was] to permit improvement or more intensive use of grazing lands within the basin" (U.S. DOI, 1958).

Over the objections of the U.S. Fish and Wildlife Service, which proposed alternative flood control plans that did not require conversion of the Kissimmee River into a canal (U.S. DOI, 1958), COE in 1961 began the channelization effort that transformed the slow, winding, shady river, with its renowned largemouth bass fishery, into a straight, deep, unshaded 56-mile canal. The new channel was uniform in geometry, with water levels controlled by electrically operated steel and concrete locks that divided the river into five, reservoir-like, longitudinal pools. The results of this technological fix were far-reaching.

ECOLOGICAL IMPACTS OF WATER MANAGEMENT

Hydrology was the principal factor that made the Kissimmee-Okeechobee-Everglades ecosystem unique, and hydrological change brought ecological problems. The channelization of the Kissimmee River and the destruction of much of its associated wetlands and floodplains, the leveeing of Lake Okeechobee, the drainage of its wetlands, the digging of hundreds of canals, and, the construction of east-west roads, all interrupted the natural timing and flow of clean water into the Everglades (Brumbach, 1990). Even National Park status was not enough to prevent water control activities beyond park boundaries from bringing severe ecological degradation—in the form of periodic desiccation, flooding, and water pollution—to the Everglades and to much of its surroundings.

So scant did the water flow become at times that salt water seeped into freshwater streams and aquifers in parts of the Everglades. As water tables dropped, rapid land subsidence occurred in places such as the Everglades Agricultural Area just south of Lake Okeechobee. Oxidation ate away fragile peat soil. As the soil dried out and turned to dust, winds eroded it, sometimes down to the porous limestone bedrock. Elsewhere during the dry season, desiccated marshes caught fire and burned.

As hydrologic conditions changed, wildlife in vast numbers perished or departed. The wading bird population of South Florida has plummeted 90 percent since the 1930s (Lancaster, 1990) and is only 5 percent of what it was before drainage efforts began in the nineteenth century (Brumbach, 1990).

Ecological Effects of the Kissimmee River Channelization

The channelization of the Kissimmee River alone drained 34,000 acres of Kissimmee floodplain wetlands, wiping out 5 billion small fish and 6 billion shrimp (Loftin et al., 1990; Toth, 1990). In addition, 13,000 acres of natural Kissimmee wetlands were converted to "impounded wetlands," resulting in a loss of ecological values (Loftin et al., 1990). Another 7,000 acres of wetlands were obliterated along with about 35 miles of the original river channel when COE's new C-38 canal was excavated (Loftin et al., 1990), and the excavated spoil was piled along the canal banks to form levees. Six indigenous species of fish were extirpated from the river in the process (Toth, 1990).

Channelization caused profound alterations in the riverine-floodplain hydrologic system—changes in the hydroperiod, amounts of flow, rates of flow, flow distribution, and smoothness of the seasonal transition from high to low flows (Loftin et al., 1990). Natural hydroperiods were eliminated in favor of stable water levels. After channelization, stagnant sections of the old river channel remained as oxbows off the excavated canal but retained little habitat value because of low water flow, large in-channel accumulations of submerged organic matter, and consequent low dissolved oxygen levels (Loftin et al., 1990; Toth, 1990).

Water quality degradation in the Everglades, the Kissimmee, and Lake Okeechobee soon followed the flood control projects. Wastewater from sugarcane fields was regularly drained into the Everglades and Lake Okeechobee. Currently it drains southward of the lake into water conservation areas from which it enters the Everglades. The water contains large quantities of fertilizer-derived phosphorus and nitrogen that promote the proliferation of algae and cattails, the latter replacing native plants of higher ecological value to wildlife. Flood control activities in the Kissimmee basin also stimulated agricultural development on adjacent lands, sending contaminated runoff into the Kissimmee canal. As it flowed from the canal into Lake Okeechobee, the polluted water exacerbated the lake's grave eutrophication problem.

Agricultural enterprises in Florida are still allowed to pump polluted water into the Everglades pursuant to exemptions under the federal Clean Water Act of 1977 (P.L. 95-217). In 1988, U.S. Attorney Dexter W. Lehtinen sued both the South Florida Water Management District and the state's Department of Water Resources for allowing this. The suit was suspended for 2 months in February 1991 as part of a state-federal agreement to work together on a water cleanup plan (New York Times/AP, 1991; Schneider, 1991).

Although the canal replacing the Kissimmee River is by no means biologically dead, the Kissimmee River as a naturally flowing riverine-wetland system has ceased to exist. Everglades National Park, too, is in jeopardy: "Scientists, public officials and leaders in the National Park Service have known for years that the Everglades are dying" (Schneider, 1991). According to Park Superintendent Robert S. Chandler, the park is "at a stage of biological collapse" (Schneider, 1991).

Impetus to Restore Lost Ecological Values

An extensive, publicly funded program is now being pursued to save the Everglades. The program includes restoration of a more natural flow regime to the park, filling of drainage canals, expansion of the park by 107,000 acres, proposed treatment of agricultural wastewater by sending it through experimental cattail marshes, and proposed restoration of the Kissimmee River to much of its serpentine river channel. The estimated price tag was $700 million. Of all these measures, the Kissimmee restoration project at $300 million is by far the most expensive. Recent cost estimates for the Kissimmee restoration are $422 million.

The Kissimmee channelization was opposed from the start by conservation groups and by others, as previously noted. The Governor's Conference on Water Management in South Florida and the Central and South Florida Flood Control District began calling for reflooding the Kissimmee wetlands in 1971, even before COE was through with its work (Dreher, 1986). These objections soon led the Florida legislature to initiate a study to assess the impacts of channelization on the Kissimmee River and Lake Okeechobee (Dreher, 1986).

In 1976 the Florida legislature passed the Kissimmee River Restoration Act (Chapter 76-113, Florida Statutes) and established the Coordinating Council on the Restoration of the Kissimmee River Valley and Taylor Creek-Nubbins Slough Basin (KRCC) to guide the restoration of water quality in the Kissimmee River basin (Florida Statutes, 1976). The legislature directed the coordinating council to consider the merits of partial or total restoration of the Kissimmee River and to develop measures that would restore natural seasonal water-level fluctuations and make maximum use of the natural and free energies of the river (Florida Statutes, 1976). After 7 years of restoration studies, the coordinating council in 1983 issued a report that called for dechannelization of the Kissimmee River along with other measures (Dreher, 1986; Toth, 1991). The report, a milestone in the restoration effort, specifically recommended that the South Florida Water Management District (SFWMD) begin development of a program to dechannelize the Kissimmee River (McCaffrey, 1983). In complying with this recommendation, the district initiated its demonstration project (to be discussed later). The council then went out of existence in keeping with a "sunset" provision in the Kissimmee River Restoration Act of 1976 (Florida Statutes, 1976).

Florida Governor Bob Graham immediately replaced the council on November 4, 1983, by issuing an Executive Order creating the Kissimmee River-Lake Okeechobee-Everglades Coordinating Council (KOECC) to coordinate and promote restoration efforts in the Kissimmee River-Lake Okeechobee-Everglades ecosystem (Graham, 1983). The new council's mandate was broader than its predecessor's and included avoidance of further destruction of natural systems, reestablishment of ecological functions, improvements in overall management, and environmental preservation (Graham, 1983). The heads of six state agencies were appointed to the council by Governor Graham and were required to actively implement its restoration program (SFWMD, n.d.). That requirement greatly spurred the state's restoration efforts.

A major landmark in the effort to restore the Everglades and related ecosystems was Governor Graham's Save Our Everglades Program of 1983, which enlisted federal, state, and regional agencies in the effort (Brumbach, 1990) and included a call "to reestablish the values of the Kissimmee River" (State of Florida, 1983). It also contained a direct appeal to President Ronald Reagan for his support.

This well-publicized, comprehensive program captured the public's imagination and helped generate broad support for the goal of reestablishing the natural ecological functions of the Everglades, Lake Okeechobee, and the Kissimmee River (State of Florida, 1983). However, it did not go into much detail about the Kissimmee River. The SFWMD then, at Governor Graham's request, formulated a more detailed, seven-point Kissimmee River Restoration Strategy, which Governor Graham approved in 1985 (Loftin et al., 1990). It called for continuation of the SFWMD's demonstration restoration project, detailed monitoring of the project, expedited floodplain land acquisition, water quality improvement, and pursuit of restoration options that included dechannelizing the river by filling the C-38 canal (the designation given by COE to the channelized river).

Responding to the previously mentioned recommendation by the 1977 KRCC to the SFWMD to begin a dechannelization program (McCaffrey, 1983), the district in 1984 had begun a $1.4 million (excluding staff costs) experimental restoration demonstration project along a 12-mile stretch of the canal (between locks S-65A and S-65B) 14 miles due west of Sebring, Florida. This project's goal was to field test "methods of reestablishing a more natural water regime in the Kissimmee Valley" (SFWMD, n.d.) and was a highly significant step toward restoration of the Kissimmee.

Kissimmee River Restoration Demonstration Project

The physical restoration work was conducted in Pool B of the canal and began with construction of three notched weirs (steel walls) in the channelized river, which COE calls the C-38 canal. The plan was to reflood 1,300 acres of drained wetlands on the river floodplain and to increase water flow in remnant sections of the old river channel. The establishment of a 300-acre flow-through marsh was also attempted (Toth, 1991).

Other aspects of the demonstration project were adoption of a new water flow fluctuation schedule and conduct of hydrologic and hydraulic modeling studies. (Variations in stage and discharge regimes are important in creating a diverse mosaic of wetland plant communities [Toth, 1991].) The SFWMD, the Florida Game and Fresh Water Fish Commission, and the Florida Department of Environmental Regulation all agreed to share in monitoring and evaluating the demonstration project.

The focus of SFWMD monitoring was the effect of hydrologic changes on floodplain vegetation, floodplain fish, secondary productivity, benthic invertebrates, and river channel habitat characteristics (Toth,

1991). Data were collected from July 1984 to November 1988. The other Florida agencies agreed to count alligators, conduct bird surveys, sample fish populations, and monitor water quality and aquatic macroinvertebrate and periphyton responses (Toth, 1991).

Transects through representative postchannelization floodplain communities were set up in the Pool B floodplain, and aquatic invertebrates were sampled along the transects to measure changes in secondary productivity. Fish utilization was also measured. River channel cross sections were used to evaluate changes induced by increased flow in river channel habitat, bottom morphology, sediment characteristics, benthic invertebrate densities, and community structure.

DEMONSTRATION PROJECT RESULTS

The water-level manipulations and increased flow through remnant river channels produced encouraging results. Whereas the demonstration project experiment by no means "restored the Kissimmee River," it did demonstrate that wetland vegetation and other wildlife would readily recolonize the reflooded areas, and that riverine ecosystems would respond favorably to resumption of natural flow regimes. In response to seasonal and intermittent flooding, a diverse complement of wetland species became reestablished. Investigators found that inundation periods of 1 to 2 years may be the fastest way of reestablishing wetland species and that prolonged annual hydroperiods of about 250 days "shifted the competitive environment in favor of hydrophytic species" (Toth, 1991). The responses of the vegetation also proved that the reproductive potential and seedbank of many wetland plants were conserved over the more than two decades since drainage (Toth, 1991).

Although the frequency and distribution of wetland species increased in response to increased inundation, xerophytic and mesophytic species receded. In general, "plant community responses to Demonstration Project components showed that restoration of wetland communities on the Kissimmee River floodplain is feasible" (Toth, 1991). The experiment also provided evidence that some fish and invertebrate recolonization could be induced and that increased bird utilization of the floodplains could be expected. Restoration of water flow to the old river channels also helped reestablish more natural substrate characteristics, channel morphology, and benthic species diversity (Toth, 1991). The increased flow through the remnant channels swept away much organic debris and increased channel cross-sectional areas.

In many parts of the experimental area, the extent and depth of flooding and drying of the floodplain were not comparable to prechannelization conditions, and return drainage from floodplain to river channel occurred far more rapidly than in the natural prechannelization system, because the floodplain is very efficiently drained by the canal as soon as increased regulated water discharges from the upper Kissimmee lakes cease. Another departure from natural conditions was caused by the fact that water discharges from the upper Kissimmee basin depended on the Kissimmee lakes regulation schedules. This resulted in discharges during lake drawdown periods from January to April, rather than during the wet season months. In addition, the notched weirs installed in the C-38 canal to partially block it were inefficient in diverting water from the canal into the remnant river reaches and floodplains when discharges into the canal fell below 28 cubic meters per second.

In addition, the planned flow-through marsh could not be evaluated by demonstration project scientists because drainage of water from the site was impeded by a transverse ridge and spoil pile that impounded water instead of permitting it to flow over the site. River flow regimes also contrasted sharply in quantity and timing with prechannelization characteristics and included some no-flow periods (Toth, 1991). Before channelization, river flow had been continuous, with frequent overbank flow and much base flow (Toth, 1991). To summarize, "Because key hydrologic characteristics were not adequately reestablished, most structural and functional aspects of floodplain ecosystem integrity were affected temporarily and/or only partially restored" (Toth, 1991). Due to these constraints, the demonstration project achieved only partial success as an ecosystem restoration. It was, however, a very significant success as a demonstration that restoration of riverine-floodplain values and functions is possible. "It was not intended to restore the river or any section of the river," asserted project biologist L. A. Toth. "We demonstrated we can affect very positive changes to the biology of the system" (L. A. Toth, South Florida Water Management District, personal communication, 1991). Former project manager M. K. Loftin agreed: "The demonstration project was an overwhelming success, because it showed that when water conditions were correct, biological recovery was tremendous, and when water conditions were adverse, it showed catastrophic declines" (K. A. Loftin, former project manager, Kissimmee Alternative Plan Evaluation and Preliminary Design Report, South Florida Water Management District, West Palm Beach, Fla., personal communication, April 24, 1991). The demonstration project thus provided evidence that once more natural hydrological conditions have been restored to the lower Kissimmee basin by filling the C-38 canal, significant ecological recovery of the riverine-floodplain system is likely.

The end result of the demonstration project's scientific monitoring effort and the related hydrologic modeling studies was a recommendation by the SFWMD study team to backfill "long, continuous reaches of C-38" as the only way to restore the ecological integrity of the Kissimmee River and about 22,000 acres of its original floodplains. This backfilling option, known as the Level II Backfilling Plan, had previously been endorsed by Florida Governor Bob Martinez (L. A. Toth, South Florida Water Management District, personal communication, 1991).

Consistent with the Committee on Restoration of Aquatic Ecosystems' general procedural recommendations for restoration planning and evaluation, the SFWMD study team recommended that, in preparation for the major river restoration effort to come, baseline data should be collected immediately "on all components of the ecosystem, including wading birds, waterfowl, fisheries, fish communities, habitat, water quality, and ecosystem function . . ." (L. A. Toth, South Florida Water Management District, personal communication, 1991).

The fundamental challenge now remaining for Kissimmee restoration planners is to bring about the restoration of ecological integrity—species composition, physical structure, and ecological functions—of the Kissimmee's riverine-floodplain system, while avoiding unwanted future flood drainage.

Political Context

The pivotal role played by former Governor Bob Graham and his staff in establishing and providing firm guidance to the Kissimmee River-Lake Okeechobee-Everglades Coordinating Council, in launching Florida's Save Our Everglades Program in 1983, and in lobbying federal officials for their support in protecting the Everglades illustrates the tremendous political power that a governor committed to ecological restoration can wield. Not only can governors appoint ecologically minded individuals to regional management councils and water management districts, they can also propose and sponsor legislation supportive of restoration. Governor Graham, for example, strongly supported Florida's Save Our Rivers Act in 1981 to provide $300 million for the acquisition of river floodplains, wetlands, and recharge areas. In addition, Governor Graham backed a comprehensive statewide water policy that encouraged use of nonstructural water management methods, instead of ditches, dams, and levees. He also supported and approved the state's 1984 Wetlands Protection Act and proclaimed 1984 as the Year of the Wetlands. Governor Graham's successor, Governor Bob Martinez, also played a crucial role in the restoration program by galvanizing commitment to the most expensive of the Kissimmee restoration options, the dechannelization plan known as the Level II Backfilling Plan. Governor Martinez was also instrumental in persuading the SFWMD board to adopt that option (K. A. Loftin, former project manager, Kissimmee Alternative Plan Evaluation and Preliminary Design Report, South Florida Water Management District, West Palm Beach, Fla., personal communication, April 24, 1991).

Florida's current Governor, Lawton Chiles, has pledged to work with Senator Graham and the Florida delegation to obtain congressional appropriations for the East Everglades addition to the park and to continue restoration of water flows to the Shark River Slough area. In his first address after taking office in 1991, Governor Chiles, speaking to the Sixth Annual Everglades Coalition Conference, adopted the Save the Everglades Program and made restoration of the Kissimmee-Okeechobee-Everglades ecosystem his number one environmental policy.

By making an example of his or her own deep convictions about restoration, a governor can make restoration a high statewide priority. It is hard to imagine how a restoration program as challenging, controversial, and expensive as the program to restore the Kissimmee-Okeechobee-Everglades ecosystem could have gotten as far as it has without powerful leadership from the state's top executive. National and local environmental organizations have also been very active over the years on behalf of the restoration and protection program through a host of public education activities and lobbying.

Restoration Goals, Objectives, and Criteria

The overall mission of the restoration program was vividly articulated in the governor's Save Our Everglades Program of 1983: "Florida must take action to rejuvenate the Kissimmee-Okeechobee-Everglades ecological system and the environment of south Florida. Although the system can never be the same as it was before [drainage], many of its natural functions and values can be restored while providing water supplies and flood protection to south Florida." The standard against which the program was to be evaluated was to be its degree of success in ensuring "that the Everglades of the year 2000 looks and functions more like it did in 1900 than it does today" (State of Florida, 1983).

The Kissimmee River Restoration Symposium in October 1988 served a vital role in clarifying the approach to the Kissimmee restoration and underscored that the only way to realize the environmental goals identified in both the Kissimmee River Restoration Act of

1976 and Governor Graham's 1983 Executive Order would be by adopting a "holistic, ecosystem restoration perspective" (Loftin et al., 1990). (Further discussion of restoration in the Everglades took place in 1989 at another symposium [SFWMD, 1991].)

After years of studies and debates in Florida over what to do about the damage to the Kissimmee, a consensus emerged at the 1988 restoration symposium on what the specific goals and objectives for the Kissimmee River restoration ought to be. Reestablishment of the Kissimmee River ecosystem's ecological integrity emerged as the primary restoration goal. In addition, it was agreed that four broad goals needed to be accomplished and were to serve as guidelines for any major Kissimmee River restoration work:

1. Restoration should use the natural and free energies of the river system (not those of an impounded, highly managed system).

2. The natural ecological functions of the river system were to be restored.

3. The physical, chemical, and biological integrity of the river system was to be restored and maintained.

4. Lost environmental values were to be restored.

It was understood that in the restoration to come, these goals had to be met, subject to the retention of a specified level of flood control and without causing major adverse impacts to navigation, water supply, water chemistry, or sedimentation (Loftin et al., 1990).

Once these goals were agreed upon, detailed comparative studies of the Kissimmee riverine-wetlands system (Toth, n.d.; Toth, 1990) were performed, and five critical evaluation criteria were developed for use in appraising alternative methods of restoring the Kissimmee. The evaluation criteria were the following (from Loftin et al., 1990):

1. continuous flow with duration and variability characteristics comparable to prechannelization records;

2. average flow velocities between 0.8 and 1.8 feet per second (ft/s) when flows are contained within channel banks;

3. a stage-discharge relationship that results in overbank flows along most of the floodplain when discharges exceed 1,400 to 2,000 cubic feet per second;

4. stage recession rates on the floodplain that typically do not exceed 1 ft per month; and

5. stage hydrographs that result in floodplain inundation frequencies comparable to prechannelization hydroperiods, including seasonal and long-term variability characteristics.

Alternative Plan Evaluation

A major study was then conducted by the South Florida Water Management District, with hydrological modeling support from researchers of the University of California, Berkeley, to evaluate four principal alternative Kissimmee River restoration plans in terms of the above criteria (Loftin et al., 1990). The following four plans were evaluated:

1. Rely on weirs to restore river flow from the canal to remnant sections of the old river channel and, under some conditions, to the floodplain.
2. Block small portions of the canal with earthen plugs to produce the same effects as the first plan.
3. Extend the plugs from the same locations in the second plan so as to block longer sections of the canal, leaving portions of the canal at junctions, and to link portions of the original river channel.
4. Fill as much of the canal as possible without affecting flood control in the Upper Kissimmee River basin and at the outlet of the lower basin. (This option has become known as the Level II Backfilling Plan.)

The evaluation team concluded that only the fourth alternative plan, Level II Backfilling, would restore the ecological integrity of the riverine-floodplain system. All other options "would result in excessive river channel velocities, rapid stage recession rates, [and] inadequate floodplain inundation . . ." (Loftin et al., 1990). The analysts projected that Level II Backfilling could reestablish "prechannelization hydrologic characteristics along 52 contiguous miles of river channel and 24,000 acres of floodplain."

The cost of the Level II Backfilling Plan was estimated at $291,600,000, making it the most expensive of all the alternatives studied. Combined with ancillary related waterworks, the total cost would be $343,520,000. As noted earlier, recent estimates are $80 million higher. Because flood control would be abandoned over parts of the floodplain, all the plans studied provide for land acquisition in the floodplain and for the acquisition of flowage (flooding) rights (Loftin et al., 1990). It is obvious that the cost of restoration far exceeds the cost of the channelization of the river.

Anticipated Restoration Benefits

Restoration of the Kissimmee ecosystem will have favorable effects on the Okeechobee-Everglades system to which it is linked. Restored wetlands will retain water, improve water quality, reduce phosphorus inputs to Lake Okeechobee by 10 percent (Dreher, 1986), and aid in recharge of aquifers. The Kissimmee restoration will re-create the scenic beauty of a slowly meandering southern sub-tropical river with its floodplain marshes and abundant fish and wildlife.

Economists are not in agreement on the economic value of wet-lands and have proposed estimates that differ by more than a factor of four (Dreher, 1986); they do generally agree, however, that the value is considerable. In addition, the project would prevent further floodplain development that might lead to sizable flood damage claims later and would also avert spending by counties in the area on infra-structure to support development. Tax losses would be "insubstan-tial" (Dreher, 1986).

Areas of Critical State Concern: A Mechanism for Encouraging Restoration

The lower Kissimmee River basin along with the East Everglades area was one of the state's 12 designated resource planning and man-agement areas under an amendment to Florida's Areas of Critical State Concern Program (Graham, 1984; Brumbach, 1990). Each area was to have a resource and management committee under Chapter 380, Florida Statutes, to carry out tasks assigned to it. In the Kissimmee area, the governor charged the planning and management committee to work with local, regional, state, and federal agencies and private interest groups "to develop a Resource Management Plan for the lower Kissimmee River and Taylor Creek drainage basins" (Graham, 1984). The committee was directed to address issues of "land use manage-ment, land acquisition strategy, water quality protection and eco-nomic development," including a review of "local government com-prehensive plans and implementing regulations such as zoning and subdivision ordinances" (Graham, 1984). The committee was charged not only to provide policies for each issue identified but also to "as-sign implementation actions to appropriate federal, state, regional and local governments and a schedule for adoptions of these actions" along with measurable standards to ensure that these policies were carried out. Within 12 months of submitting its plan to the governor, the committee then had to evaluate its implementation by state, re-gional, and local governments.

In response to this firm guidance, the Kissimmee River Resource Planning and Management Committee chose to concentrate on water issues and wrote a model floodplain ordinance for all counties in the basin to adopt. The ordinance stated that low-intensity agriculture would be the highest-intensity land use allowed in the basin (K. A. Loftin, former project manager, Kissimmee Alternative Plan Evaluation and Preliminary Design Report, South Florida Water Management District, West Palm Beach, Fla., personal communication, April 24, 1991). One basin county adopted the ordinance; however, four have not. These counties will now have to submit their comprehensive growth management plans for review by the state's Department of Community Affairs. In general, if recommendations of a Resource Manangement Committee are not implemented by the local governing bodies, the state can then designate the locality as an Area of Critical State Concern and can set land use standards that the local government must meet. Having a resource planning and management committee is an alternative to being designated a state critical management area and is generally preferred by a locality to more explicit state control in which local areas must develop management plans that meet state standards. The Areas of Critical State Concern Program and the cogent directive given to the Kissimmee River committee are well worth study by resource managers interested in new institutional mechanisms for guiding complex restoration programs (Graham, 1984).

Current Status of the Kissimmee River Restoration

Congress in 1990 agreed to appropriate another $6 million in federal funds for the Kissimmee restoration, bringing the total federal contribution to $12.3 million in addition to the $20 million put up by the state of Florida (Woody, 1991). Federal spending was originally authorized under Section 1135 of the Water Resources Development Act (amended).

At the insistence of U.S. Senator and former Florida Governor Bob Graham, the 1990 version of the act directed the COE to do a feasibility study on the Level II Backfilling Plan recommended by the *Alternative Plan Evaluation and Preliminary Design Report* of the SFWMD (Loftin et al., 1990). The act also requires COE to submit the final feasibility report to Congress by April 1, 1992, and to complete a design memorandum, construction bidding plan, and all other preparatory work for the Kissimmee restoration by June 1, 1994 (Executive Office of the Governor, 1991). The COE in conjunction with the SFWMD is already doing design work on water control modifications that will be required in the upper Kissimmee lakes basin, and the SFWMD is acquiring flowage easements over floodplain lands so that more water can be stored in the headwaters region (R. Smith, government analyst,

State of Florida, Governor's Office of Planning and Budgeting, Environmental Policy Unit, personal communication, April 25, 1991).

The state of Florida has told the Office of Management and Budget (which opposed funding the whole Kissimmee restoration in 1990) that the full-scale Kissimmee restoration project, a 10- to 15-year effort, is the state's highest-priority project for inclusion in the 1992 Water Resources Development Act (T. Woody, personal communication, 1991). Meanwhile, the first phase of the project—modifications of the upper chain of lakes north of the Kissimmee by COE—can proceed in order to increase year-round water flow to the Kissimmee.

With the state of Florida solidly behind it, the biggest hurdle faced by the restoration program may be obtaining federal support in a recessionary period of large federal budget deficits—over the possible objections of the Office of Management and Budget and perhaps others in the Bush administration. If federal funds were appropriated by Congress as part of the 1992 Water Resources Development Act, Section 1135, the program could move forward decisively toward its final goal. Inclusion in the act would provide the federal cost sharing necessary for COE to conduct the dechannelization work. That role would be consistent with what some observers believe is COE's desire to be the environmental engineers of the 1990s.

Conclusion

The Kissimmee River Demonstration Project was the largest restoration project examined by this committee. (The much larger planned restoration of the Kissimmee River and the Everglades would probably be the largest restoration in the nation.) Several conclusions can be reached about the Kissimmee demonstration effort:

1. Although the general public eventually recognized the need to restore the river and its wetlands, and the citizens initially provided the impetus for restoring the river, many people believed this could be done merely by replacing the material removed during channelization, and that everything would then return to predisturbance ecological conditions. Scientists and engineers in the South Florida Water Management District had to make a major public education effort to acquaint people with the complexities of ecological restoration.

2. Because explicit goals for the Kissimmee restoration were set in advance, a number of alternative plans were excluded from consideration. For example, there could have been a minimal plan, such as regulating water levels differently in the canal or creating floodplain

impoundments. Both measures would have created more wetlands but would not have resulted in restoration of ecological integrity as defined in this report. They would instead have created a substitute ecosystem quite different from the one preceding channelization.

3. Although the words "self-maintaining" were not explicitly used, the requirement for using the natural and free energies of the river system and restoring and maintaining physical, chemical, and biological integrity has essentially the same effect. Nonetheless, all restoration options evaluated will require some regular maintenance.

4. The South Florida Water Management District wisely avoided establishing biological criteria in terms of numbers of fish or waterfowl to be restored. This would almost certainly have resulted in battles among different user groups, such as anglers, hunters, and bird watchers. More importantly, management for these particular species-oriented values would not have permitted natural, successional, and evolutionary ecosystem processes to operate. "No criteria specifying individual species requirements, whether alone or in combination, will reestablish the complex food webs, habitat heterogeneity, and physical, chemical and biological processes and interactions that determined the biological attributes of the former system" (Loftin et al., 1990).

5. The South Florida Water Management District prudently had a much more extensive scientific peer-review process than many restoration projects have, although others, such as the Des Plaines River restoration project, had scientific peer review. Intensive peer review lends credibility to restoration studies and makes more expertise and vision available to the restoration project.

6. The use of hydrologic models to estimate probable outcomes for some of the nonbiological aspects of alternative restoration plans reduced uncertainty about these outcomes.

7. Monitoring of the full-scale Kissimmee River restoration, should—like the restoration itself—be designed from an ecosystem perspective in order to "provide a thorough understanding of the ecosystem—with and without restoration; show direct cause and effect relationships between restoration measures and ecological responses; include quantified biological responses; and document changes that are of importance to society, as well as scientifically important" (Toth, 1991).

8. The Kissimmee River Demonstration Project and the Kissimmee River Alternative Plan Evaluation studies were particularly valuable in showing that the cost of restoring a riverine system is considerably greater than the cost of channelizing it, and that many important ecological values of the riverine-floodplain system can be restored if prompt and decisive action is taken by a competent, properly funded interdisciplinary team.

Acknowledgment

My thanks to John Cairns, Jr., for contributing to the preparation of this case study. The committee wishes to thank Louis A. Toth of the South Florida Water Management District for briefing it on ecological impacts to the Kissimmee River system and on restoration efforts, and for providing the committee with documents essential for understanding the situation. The committee also especially wishes to thank M. Kent Loftin, former project manager, Kissimmee River Alternative Plan Evaluation, SFWMD; Patricia Sculley of SFWMD; and Rick Smith, government analyst, State of Florida Governor's Office of Planning and Budgeting, Environmental Policy Unit, for their assistance.

References

Brumbach, B. 1990. Restoring Florida's Everglades: A strategic planning approach. In J. Berger, ed., Environmental Restoration: Science and Strategies for Restoring the Earth. Island Press, Washington, D.C.

Clean Water Act of 1977. P.L. 95-217, Dec. 27, 1977, 91 Stat. 1566.

Dreher, R. G. 1986. Comments Regarding Final Feasibility Report and Environmental Impact Statement, Kissimmee River, Florida. To Board of Engineers for Rivers and Harbors, U.S. Army Corps of Engineers. Printed material by Sierra Club Legal Defense Fund, Inc., San Francisco, Calif.

Executive Office of the Governor. 1991. Everglades Status. A report prepared by the Office of Environmental Affairs, Tallahassee, Fla. January 15.

Florida Statutes. 1976. Kissimmee River Restoration Act. Chapter 76-113, Laws of Florida. Section 373.1965 (1976 Supplement). Kissimmee River Valley and Taylor Creek-Nubbins Slough Basin: Coordinating Council on Restoration Project Implementation.

Graham, B. 1983. Executive Order Number 83-178. State of Florida, Office of the Governor, Tallahassee, Fla.

Graham, B. 1984. Letter to Mr. Timer E. Powers, Chairman, Kissimmee River Resource Planning and Management Committee. August 9.

Lancaster, J. 1990. Monumental salvage job is planned for Everglades. The Washington Post. February 20.

Loftin, K. A., L. A. Toth, and J. T. B. Obeysekera. 1990. Kissimmee River Restoration: Alternative Plan Evaluation and Preliminary Design Report. South Florida Water Management District, West Palm Beach, Fla. June.

McCaffrey, P. N. 1983. Memorandum of Findings and Recommendations Adopted at August 19, 1983 Meeting. Coordinating Council on the Restoration of the Kissimmee River Valley and Taylor Creek-Nubbin Slough Basin, Tallahassee, Fla. August 22.

New York Times/AP. 1991. U.S.-Florida deal raises Everglades cleanup hope. February 21.

Norman, G. 1984. Justice, just in time. Esquire. January.

Schneider, K. 1991. Returning part of Everglades to nature for $700 million. New York Times. March 11.

South Florida Water Management District (SFWMD). n.d. A Closer Look, Kissimmee River Restoration, Phase 1—A Demonstration Project. PIO 261 Rev 386 5M. West Palm Beach, Fla.

South Florida Water Management District. n.d. Florida's Kissimmee River—A Restoration Plan. West Palm Beach, Fla.

South Florida Water Management District. 1991. Spatial and Temporal Patterns as Guidelines for Ecosystem Restoration. Environmental Sciences Division, South Florida Water Management District. West Palm Beach, Fla.

State of Florida, Office of the Governor. 1983. Save Our Everglades Program. Press release and issue paper. August 9.

Toth, L. A. n.d. An Ecosystem Approach to Kissimmee River Restoration. Monograph. Environmental Sciences Division, South Florida Water Management District, West Palm Beach, Fla.

Toth, L. A. 1990. Impacts of Channelization on the Kissimmee River Ecosystem. Monograph. Kissimmee River Restoration Symposium (October 1988). Environmental Sciences Division, South Florida Water Management District, West Palm Beach, Fla.

Toth, L. A. 1991. Environmental Responses to the Kissimmee River Demonstration Project. Technical Publication 91-02. South Florida Water Management District, West Palm Beach, Fla.

U.S. Department of the Interior (DOI). Fish and Wildlife Service. 1958. Bureau of Sport Fisheries and Wildlife, Office of Regional Director, Region 4. Atlanta, Ga. A Detailed Report of the Fish and Wildlife Resources in Relation to the Corps of Engineers' Plan of Development, Kissimmee River Basin, Florida. Prepared by Branch of River Basins, Vero Beach, Fla. December 17.

Woody, T. 1991. News from the Kissimmee. Everglades Update. March-April.

* * * * * * * * * *
Wetlands
* * * * * * * * * *

BOTTOMLAND HARDWOOD WETLAND RESTORATION IN THE MISSISSIPPI DRAINAGE

Rebecca Sharitz

The lower Mississippi floodplain was selected as a case study because it provides an example of large-scale disturbance in which the physical condition of the wetland area has been altered and cumulative impacts have occurred. Both public and privately owned lands are involved. Restoration efforts have been limited, and most have focused on reestablishment of forest species for timber or wildlife habitat values. Actual site restoration, including recovery of original hydrologic conditions, is uncommon. Success is typically measured on the basis of early establishment of desirable woody species. Most such restoration activities have been undertaken within the last decade, and long-term evaluations of their success are not available.

General Description and Locations

HISTORY AND BACKGROUND

At the time of European settlement, approximately 80 million hectares of forested wetlands existed in the conterminous United States (Gosselink and Lee, 1989). Although substantial harvesting of timber resources began with the coming of pioneers, drainage and clearing for agriculture were extensive by the middle of this century. By the 1950s, forested wetland had been reduced to about 27 million hectares, and by the mid-1970s to 24.4 million hectares (Gosselink and Lee, 1989). The loss rate from 1954 to 1974 was about 0.51 percent per year (Harris and Gosselink, 1990). Conversion to agricultural use has accounted for 87 percent of these wetland losses (Tiner, 1984).

TYPE OF DISTURBANCE

One of the best-documented examples of conversion of bottomland hardwood wetlands has been on the 9.8-million hectare Mississippi alluvial plain (Figure A.12). In 1937, bottomland hardwood forests covered 4.9 million hectares of this alluvial plain, but by 1977 they had been reduced to 2.2 million hectares of natural wetlands (MacDonald et al., 1979). The greatest forest loss resulted from conversion to croplands. Currently, along the lower Mississippi River, areas of bottomland hardwood forest still are being cleared for agriculture in tracts up to 12,000 ha at a time (Gosselink and Lee, 1989). Disastrous floods of the Mississippi River in 1927 and 1929 led to massive government programs of levee construction and a myriad of other water control works. As a result of the reduced flood frequency and duration, agricultural development increased and bottomland forests were cleared for row crops such as soybeans. Other major factors include the continuing increase in urban areas and related uses.

REPRESENTATIVE SITES

Two of the best remaining examples of bottomland hardwood forest are the 24,000-ha Delta National Forest in western Mississippi and the 22,000-ha Tensas River National Wildlife Refuge in northeastern Louisiana (Newling, 1990). Sites representing the historical effects and current restoration and reforestation activities include (1) areas of the Yazoo National Wildlife Refuge in west central Mississippi, (2) part of the Tensas National Wildlife Refuge, (3) the Ouachita Wildlife Management Area in central Louisiana, and (4) research

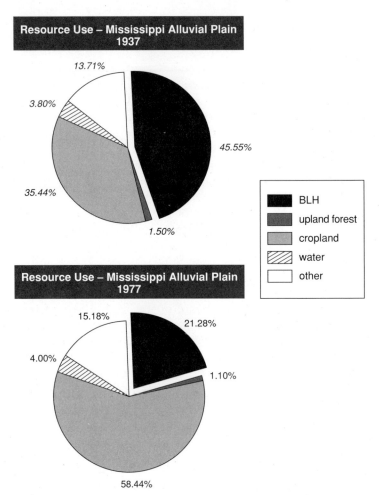

FIGURE A.12 Changes in resource use in the Mississippi alluvial plain from 1937 to 1977 (data from MacDonald et al., 1979). SOURCE: Sharitz and Mitsch, 1991.

plots in the Delta Experimental Forest near the Southern Hardwoods Laboratory at Stoneville, Mississippi.

The Yazoo National Wildlife Refuge complex consists of five national wildlife refuges located on the Mississippi-Yazoo rivers alluvial plain. The total area is approximately 24,000 ha. Reforestation to enhance wildlife usage began in the early 1980s. At the Tensas National Wildlife Refuge on the Tensas River, management of "wet soil" areas with seasonal flooding and crops encourages waterfowl and other wildlife use. Both sites are managed by the U.S. Fish and Wildlife Service, and have wildlife and waterfowl values of major importance. One of the largest bottomland forest restoration projects is currently under way near Monroe, Louisiana, where 1,821 ha purchased by the state in 1984 are being reforested to create a corridor between the existing Russell Sage and Ouachita wildlife management areas.

The Southern Hardwoods Research Laboratory is a research wing of the U.S. Forest Service's Southern Forest Experiment Station. Since the 1950s, some of the most complete and long-term research on regeneration of bottomland hardwood forests has been conducted on research plots established near this laboratory.

Political and Administrative Aspects

Public concern over losses of bottomland forests has increased in recent years with better awareness of the functions and values of wetlands and realization of the magnitude of past and continuing losses. However, most forested wetland restoration is driven by federal programs rather than by grassroots interests. Section 404 of the Clean Water Act of 1977 (P.L. 95-217), which requires that permits be issued by the U.S. Army Corps of Engineers for any discharge of dredged or fill material into the waters of the United States and adjacent wetlands, is intended to retard loss of wetlands, not restore them. Section 906 of the Water Resources Development Act of 1986 (P.L. 99-662) states that future mitigation plans for federal water projects should include specific plans to ensure that impacts to bottomland hardwood forests are mitigated in kind, to the extent possible (Haynes et al., 1988).

Opportunities for reestablishment occur when the initial loss or modification of the floodplain site, especially its hydrologic and geomorphologic condition, is not permanent and community reestablishment methods are technically feasible. These opportunities may include (1) reestablishment on abandoned, "high-risk" farmland in flood-prone areas; and (2) reestablishment in national forests, wildlife refuges and management areas, flood control projects, or public lands on which bottomland hardwood forest habitat serves management goals that are determined to be in the best public interest (Haynes et al., 1988).

The Food Security Act of 1985 (Farm Bill, P.L. 100-233), the Agricultural Credit Act of 1987, and Executive Orders 11990 and 11988 for protection of wetlands and floodplains provide for restoration of wetland habitat that is crucial to fish and wildlife resources and overall biological diversity. Under the Farm Bill, wetlands are restored on areas previously converted to agriculture through (1) easements on Farmers Home Administration (FmHA) lands and (2) enrollment of lands in the Conservation Reserve Program (CRP). Conservation easements established by FmHA are administered as a part of the National Wildlife Refuge System. The CRP program provides for cost-sharing of bottomland hardwood establishment on flood-prone croplands. Agreements with private landowners are for a minimum of 10 years.

Other support for bottomland forest restoration results from the Water Resources Development Act of 1986, the North American Waterfowl Management Plan, and the Environmental Protection Agency's policy of no net loss to wetlands.

Scientific Basis

Most bottomland forest restoration projects focus on techniques of planting and establishing forest species. Some of the most extensive research in this area has been conducted at the Southern Hardwoods Laboratory and in the Delta Experimental Forest (e.g., Johnson and Krinard, 1987; Krinard and Johnson, 1987; Krinard and Kennedy, 1987). A critical factor is to achieve adequate hydrological conditions for forest establishment and development. Other important factors may include substrate stability, availability of adequate soil rooting volume and fertility, and control of herbivores and competitive weeds (Clewell and Lea, 1989).

Restoration success is commonly judged, at least in the early phases, by the success of tree seedling establishment. For example, Allen (1990) reported densities ranging from 87 to 914 trees per acre in 10 stands of 4 to 8 years in age in the Yazoo National Wildlife Refuge. He found survival and growth of planted seedlings to be generally higher than those obtained from direct seeding.

ECOLOGICAL MODEL VERSUS WHAT MEASURES SUCCESS

The goal of duplicating an original forest stand in terms of species composition and age, structure, and function can only be approximated. Natural forests are themselves in constant flux. Also, land use activities may have modified soil or hydrologic conditions to the point that duplication of the original forest is impossible and an altered forest community is the only option.

Clewell and Lea (1989) recommend five criteria for judging success:

1. The watershed area within the same ownership shall be functioning in a manner that is consistent with project goals.

2. The substrate shall be stabilized and any erosion shall not greatly exceed that expected under normal circumstances in natural forests.

3. There shall be a density of at least 980 potential overstory trees per hectare (400 per acre) that are at least 2 meters tall. All trees shall be preferred species and shall occur in proper zonation (e.g., hydric trees in wet sites).

4. There shall be adequate representation of undergrowth species.

5. Streams and standing water bodies shall be of sufficient water quality so as not to inhibit reforestation or interfere with the attainment of other success criteria.

Universal acceptance of these criteria has not been achieved. For example, a lower stocking density of 150 to 200 trees per acre is generally preferred for wildlife habitat and enhanced mast production. Emphasis on evaluating success needs to be placed on presence of preferred species (indigenous trees and undergrowth characteristic of mature stands). Once a threshold density of trees 2 m tall has been attained, survival is virtually assured (Clewell and Lea, 1989). Natural regeneration relative to achieving a diversity of tree species is also an important consideration. Haynes and Moore (1988) suggest that bottom-land hardwood forests planted on abandoned farmland could become self-regenerating communities in 40 to 60 years.

RESEARCH BENEFITS AND NEEDS

Success criteria for evaluating wetland forest restoration projects in the Southeast are generally inadequately conceived. There are critical information gaps. For example, the silvicultural literature does not cover all aspects of wetland tree establishment, especially conditions conducive to natural regeneration and techniques for effective establishment of undergrowth species. Most important, research is needed to determine if successful forest replacement in terms of structure and species composition will provide the functions of the original wetland forest ecosystem.

Technical Basis

METHODS AND TECHNIQUES USED

Most research has been directed to methods of planting and establishing desired woody species. Much practical information comes from the Southern Hardwoods Research Laboratory. Allen and Kennedy (1989) have produced a useful booklet on reforestation techniques for the landowner. They provide guidelines on planting techniques, including seed sources, seed storage, site preparation, and planting depth and spacing; a list of nurseries from which seedlings can be obtained; and guidelines on flood tolerance of bottomland forest species and their suitability for wildlife use and timber production.

A further review of forest wetland restoration in the southeastern United States is given by Clewell and Lea (1989), along with a discussion of success criteria and research needs. Haynes et al. (1988) have produced an annotated bibliography for reestablishment of bottomland hardwood forests on previously disturbed sites.

Typical costs of direct seeding in 1989 were about $40 to $60 per acre (Allen and Kennedy, 1989), whereas planting seedlings costs two or three times as much. The species most often planted are Nuttall oak *(Quercus nuttallii)*, willow oak *(Q. phellos)*, cherrybark oak *(Q. falcata* var. *pagodaefolia)*, water oak *(Q. nigra)*, Shumard oak *(Q. shumardii)*, and pecan *(Carya illinoensis)*.

Direct seeding has several advantages: the cost is lower, and tree roots develop naturally without the disturbance caused by cutting roots and planting seedlings. A disadvantage of direct seeding is slower initial development of the forest and susceptibility of seeds to predation. Also, direct seeding is reliable only for oaks and, to a lesser degree, other large seeded species such as sweet pecan. Smaller seeds are more susceptible to heat and dry soil.

From a comparison of 4- to 8-year-old stands in the Yazoo National Wildlife Refuge, Allen (1990) recommended planting seedlings as a better method of establishing wildlife habitat quickly, even though direct seeding may cost only half as much. He reported extensive drought-caused mortality of newly germinated seedlings. However, he also reported effective invasion of light-seeded species, especially sweet gum *(Liquidambar styraciflua)*, green ash *(Fraxinus pennsylvanica)*, and American elm *(Ulmus americana)*, thus enhancing diversity.

The most successful planting technique to obtain mixtures of species involves planting of blocks or rows of a single species, interspersed with blocks or rows of other species (J. R. Toliver, U.S. Forest Service, Southern Forest Experiment Station, Stoneville, Miss, July 17, 1990). This approach enhances establishment of slower-growing or poorly competing species. Furthermore, this spacing arrangement allows placement of different species across a gradient of hydrologic and soil conditions within a site, according to their ecological tolerances. Such a planting approach is being used at the Ouachita Wildlife Management Area (L. Savage, Louisiana Department of Wildlife and Fisheries, Monroe, La., July 18, 1990).

Overall Evaluation

Reforestation and restoration efforts are proving successful in re-establishing bottomland hardwood forests for commercial and wild-life habitat values (Haynes et al., 1988; Allen and Kennedy, 1989; Clewell and Lea, 1989). A variety of forest establishment techniques have been employed, such as direct seeding of oaks and planting of seedlings of several bottom-land species. Most of these projects be-gan during the late 1980s. Although some may appear promising in terms of species composition and structure, it is too soon to assess the recovery of other wetland functions.

Many of the other functions of forested wetlands require full forest development before they can be evaluated. Thus, other ecological and societal values are seldom measured in evaluating the success of restoration in these wetland systems.

Conclusions and Recommendations

1. The wetland forests on these alluvial floodplain sites have been undergoing loss or conversion for several hundred years. Site alterations, such as diking, ditching, and channelization, have been extensive. It is not realistic to anticipate that major restoration to original geologic, hydrologic, and biological conditions is possible except in limited areas.

2. Smaller-scale areas (such as upland tributaries or watersheds) have greater possibilities of functional recovery than do larger-scale areas. Numerous small projects in such areas may be more effective in restoring floodplain forest values than one major project because restoration of the hydrologic regime is easier in smaller areas.

3. Most current bottomland forest restoration projects are restoring only a few of the functional values (e.g., timber values, wildlife or waterfowl habitat).

4. Most bottomland forest restoration projects are not restoring the original physical and hydrologic conditions.

5. Some management will remain necessary to maintain forested wetlands in some situations, especially where physical and hydrological alterations have been major.

6. Most bottomland forest restoration efforts are driven by agency activities (e.g., mitigation) and federal land programs, not by grassroots support.

7. Watershed effects must be considered in the restoration of forested wetlands. Activities in an upper portion of the drainage will affect downstream areas. Liability for downstream flooding or loss of water resources may limit possible restoration activities.

8. Cumulative effects of restoration on a landscape (watershed) scale must be assessed and incorporated into the planning process.

9. Longer-term leases under federal land programs (e.g., CRP) would increase the possibility of wetland forest restoration.

References

Allen, J. A. 1990. Stocking, growth, and natural invasion of bottomland oak plantations on the Yazoo National Wildlife Refuge Complex. Unpublished manuscript. U.S. Fish and Wildlife Services, Slidell, La.

Allen, J. A., and H. E. Kennedy, Jr. 1989. Bottomland Hardwood Restoration in the Lower Mississippi Valley. U.S. Fish and Wildlife Service, Slidell, La., and U.S. Forest Service, Southern Hardwoods Laboratory, Stoneville, Miss. 28 pp.

Clean Water Act of 1977. P.L. 95-217, Dec. 27, 1977, 91 Stat. 1566.

Clewell, A. F., and R. Lea. 1989. Creation and restoration of forested wetlands vegetation in the southeastern United States. Pp. 199-237 in J. A. Kusler and M. E. Kentula, eds., Wetland Creation and Restoration: The Status of the Science. Volume I: Regional Reviews. EPA 600/3-89/038. U.S. Environmental Protection Agency, Corvallis, Ore. 473 pp.

Gosselink, J. G., and L. C. Lee. 1989. Cumulative impact assessment in bottomland hardwood forests. Wetlands 9:89-174.

Harris, L., and J. G. Gosselink. 1990. Cumulative impacts of bottomland hardwood conversion on hydrology, water quality, and terrestrial wildlife. Pp. 259-322 in J. G. Gosselink, L. C. Lee, and T. A. Muir, eds., Ecological Processes and Cumulative Impacts: Illustrated by Bottomland Hardwood Wetland Ecosystems. Lewis Publishers, Chelsea, Mich.

Haynes, R. J., and L. Moore. 1988. Reestablishment of bottomland hardwoods within national wildlife refuges in the Southeast. Pp. 95-103 in Proceedings of a conference Increasing Our Wetland Resources. National Wildlife Federation, Washington, D.C.

Haynes, R. J., J. A. Allen, and E. C. Pendleton. 1988. Reestablishment of bottomland hardwood forests on disturbed sites: An annotated bibliography. U.S. Fish and Wildlife Service Biological Report 88(42). 104 pp.

Johnson, R. L., and R. M. Krinard. 1987. Direct seeding of southern oaks—A progress report. Pp. 10-16 in Proceedings of the Fifteenth Annual Hardwood Symposium. Hardwood Research Council, Memphis, Tenn.

Krinard, R. M., and R. L. Johnson. 1987. Growth of a 31-year-old bald cypress plantation. USDA Forest Service, Southern Forest Experiment Station Note SO-339. New Orleans, La. 4 pp.

Krinard, R. M., and H. E. Kennedy. 1987. Fifteen-year growth of six planted hardwood species on Sharkey clay soil. USDA Forest Service, Southern Forest Experiment Station Note SO-336. New Orleans, La. 4 pp.

MacDonald, P. O., W. E. Frayer, and J. K. Clauser. 1979. Documentation, Chronology, and Future Projections of Bottomland Hardwood Habitat Losses in the Lower Mississippi Alluvial Plain. 2 volumes. U. S. Department of the Interior, Fish and Wildlife Service, Washington, D.C.

Newling, C. J. 1990. Restoration of bottomland hardwood forests in the Lower Missis-
sippi Valley. Restoration Manage. Notes 8(1):23-28.
Sharitz, R. R., and W. J. Mitsch. 1991. Southern floodplain forest. In W. H. Martin, ed.,
Biodiversity of the Southeastern United States. Volume 1: Terrestrial Communities.
John Wiley & Sons, New York.
Tiner, R. W. 1984. Wetlands of the United States: Current Status and Recent Trends.
National Wetlands Inventory, U. S. Fish and Wildlife Service, Washington, D.C.,
59 pp.
Water Resources Development Act of 1986. P.L. 99-662, Nov. 17, 1985, 100 Stat. 4082.

PRAIRIE POTHOLES

Donald Hey

Introduction

The geographical region referred to as the prairie potholes com-
prises 192 million acres (Leitch, 1989). This area traverses the prov-
inces of Alberta, Saskatchewan, and Manitoba in Canada, and the
states of Montana, North Dakota, South Dakota, Minnesota, and Iowa
in the United States. Of this area, 40 percent falls in the United States
and 60 percent in Canada. The region is characterized by flat to
undulating glaciated topography with poorly defined natural drainage. Millions of potholes, remnant glacial depressions, are sprinkled
across the landscape.

Starting in the middle of the nineteenth century, the potholes and
their watersheds were altered by European settlers to facilitate farming. Engineered, agricultural drains converted the poorly defined
drainage to a well-defined system. Seasonal or perennial inundation
of potholes was eliminated by drain tiles and outlet ditches. From
the 1870s to the 1970s, 20 million acres of wetlands were reduced to
10 million acres. The effects on wildlife and water resources were
dramatic. Although early population estimates are not available, recent studies (Weller, 1982) show a direct relationship between wildlife populations and ponded areas. Given a 50 percent reduction in
ponded areas, wildlife populations were likely cut in half. At the
same time, flood storage may have been reduced by as much as 20
million to 30 million acre-feet. This loss, no doubt, contributed to
increased flooding along regional streams and rivers. Also, without
the long detention times provided by the storage, sediments and nutrients were flushed from and through these former wetlands to foul
receiving waters (Gilliam, 1986). However, no quantitative measure
of these effects has been made.

In recent years, a great deal of concern has been expressed about the loss of pothole wetlands. A number of studies have been undertaken to better understand the physical and biological functions of potholes (van der Valk, 1989), and some restoration programs have been initiated. Federal laws (e.g., the Clean Water Act of 1977 (P.L. 95-217), the Food and Security Act, and the Emergency Wetlands Resources Act) now discourage the destruction of these important aquatic ecosystems. In Minnesota, for example, there are 32 federal, state, local, and private assistance programs for preserving and restoring wetlands.

Background

The committee elected to review 18 restoration projects in pothole regions. They are located in Meeker and Rice counties, Minnesota. The work was accomplished under the direction of the U.S. Fish and Wildlife Service.

Meeker County lies 60 miles west of St. Paul, and Rice County is about 30 miles south. The population of Rice County is 46,000, approximately twice the population of Meeker County (21,000). On the other hand, Meeker has a larger area, approximately 396,000 acres. The area of Rice County is approximately 317,000 acres. The differences in land area and population result in the population density of Rice County being three times that of Meeker County. In Rice County, there is approximately 0.14 person per acre, whereas in Meeker County, the ratio is 0.05 person per acre.

Despite the large difference in population density, the land use of both counties is quite similar. Agricultural uses cover 84 percent of the land in Meeker County, whereas they cover 80 percent in Rice County. The remaining land is devoted to urban and transportation uses, and streams and lakes. Of the agricultural lands, 29 percent are drained in Meeker County and 25 percent in Rice County (Bureau of the Census, 1981).

The topography and surficial geology of both counties are also quite similar. Both were glaciated during the Wisconsinan stage; the surface material is till. Prior to settlement, both counties contained large areas of poorly drained soils and poorly defined drainage systems, as the extent of agricultural drainage implies.

The two counties are characterized by cool, subhumid conditions. Minimum temperatures range from –30°F in February to 54°F in July. Maximum temperatures range from 40°F in February to 98°F in August. Precipitation averages about 28 inches. Most of the precipitation occurs during the summer months, whereas the least accumulates in the winter. October is generally the driest month, and August the wettest. Soils are frozen 2- to 3-ft. for 4 to 5 months of the year.

Restoration Programs

The U.S. Fish and Wildlife Service is actively pursuing the restoration of wetlands on private, agricultural lands in both counties. The work is being done under a program developed by the service's Region 3. The program, called Stewardship 2000, was started in 1987 (U.S. Fish and Wildlife Service, 1990).

Twelve restoration projects were visited in Meeker County and six in Rice County. Except for their means of financing, the projects are all very similar. Following is a list of general topics covered in the initial planning work:

1. Land use
2. Location
3. Ownership
4. Easements, rights of way, and reservations
5. Wildlife use
6. Soils and topography
7. General habitat description
8. History of land use under private ownership (including uplands, wetlands, buildings, and so on)
9. Surrounding land use within 3 miles of the project (including private land, state land, federal waterfowl protection areas, and federal wetland easements)
10. Objectives
11. Past waterfowl protection activities

Financing played a significant role in defining the scope of the restoration activities. For example, the 160-acre Christenson project (named after the landowner) was financed with money from Ducks Unlimited as well as from the Conservation Reserve Program. Restoration of waterfowl habitat was the principal objective. Money from the Conservation Reserve Program was used to create a wildlife habitat buffer around the restoration area. On the other hand, money from the Luthens project was financed with money from Reinvest in Minnesota, a state program for habitat restoration. In this case, no buffer surrounds the restored pothole project.

In both cases, the restoration consisted of very simple changes to the drainage system. On the Christenson property, agricultural drainage structures in and around 10 farmed potholes were removed, blocked, or altered to emulate presettlement hydrologic conditions. The pot-holes ranged in size from 0.2 to 10 acres. The tributary watersheds ranged in size from 5 acres associated with the smallest wetland to 500 acres associated with the largest. The tiles draining the potholes were blocked. Drainage ditches were blocked by small earth fills or dikes, the longest of which was 125 ft. Each dike incorporated a spill-way.

On the Luthens farm, the drainage structures were modified for potholes of 1.5 and 0.7 acre. As in the Christenson case, earthen dikes were used to block the surface drainage and the tiles were removed to prevent subsurface drainage. No plant materials were introduced in the farmed wetlands being restored, and only a limited number of plant species, both warm and cool season grasses (no forbs), were planted in the buffer areas around the restored potholes. The costs for the restoration work were quite modest, in each case being less than $1,000.

Each property owner signed an agreement with the U.S. Fish and Wildlife Service. The service acquired the following rights:

1. to restore and maintain the wetlands described in the agree-ment by plugging drainage ditches or tiles and installing water con-trol structures;
2. to access the land for management purposes; and
3. to establish a vegetative cover on soils disturbed during con-struction.

In return, the property owner acquired the wildlife benefits (hunt-ing, fishing, and others) received from the restored potholes. The agreement could have been terminated within 30 days by a written notice from either party. If the property owner terminated the agree-ment within 4 years, the owner would reimburse the service for all improvements.

Conclusions

The overriding goal of these restoration projects was the devel-opment of waterfowl habitat. However, flood control and water quality were often mentioned as secondary goals. None of the goals had been quantified, nor had restoration criteria been estab-lished.

None of the 18 restoration projects inspected by the committee appeared to have a comprehensive plan concerning location, scale, or purpose. In fact, there is no comprehensive restoration plan for the counties, region, or state. The landowner, for one reason or another, had contracted the service for help in restoring wetlands. After the landowner agreed to participate in the service's program, a specific restoration plan was developed. After restoration, no monitoring or management of the restored areas was undertaken or planned.

The success of, and the degree to which, the restoration projects meet either site-specific objectives or regional objectives are unknown. The diversity of plant communities observed in the restored areas was extremely low. Even in the surrounding buffer areas the diversity of plants and, hence, wildlife habitat was extremely low. Because no monitoring has been undertaken, it will not be easy to ascertain success or failure or to improve future restoration projects. Providing a diverse habitat for animals other than waterfowl seems not to have been a consideration.

Despite the lack of well-thought-out restoration goals and criteria, this case study illustrates an extremely important aspect of any restoration strategy. The U.S. Fish and Wildlife Service is creating and responding to individual, local interest in and support for restoration. With only a meager staff commitment, the service is having considerable success. Potholes are being taken out of agricultural production and returned to their natural functions of water storage, nutrient cycling, and wildlife habitat. The other lesson involves the innovative financial program that weaves together a variety of funding sources. The ingenuity of the service's project officers and the creative dedication of its administration in Region 5 should serve as an excellent example to other states and other Fish and Wildlife Service regions. If better design criteria and management programs were available and used, the chances of success would be improved and a wider range of aquatic functions achieved.

References

Bureau of the Census. 1981. 1978 Census of Agriculture. U.S. Department of Commerce, Washington, D.C.

Clean Water Act of 1977. P.L. 95-217, Dec. 27, 1977, 91 Stat. 1566.

Leitch, J. A. 1989. Politicoeconomic Overview of Prairie Potholes. Northern Prairie Wetlands. Iowa State University Press, Ames, Iowa.

U.S. Fish and Wildlife Service. Region III, North Central Region. 1990. Stewardship 2000. Fort Snelling, Minn.

van der Valk, A. 1989. Northern Prairie Wetlands. Iowa State University Press, Ames, Iowa.

Weller, M. W. 1981. Wetlands of Canada. Polyscience Publications, Montreal, Quebec.

Weller, M. W. 1982. Ecology and Wildlife Management. Freshwater Marshes. University of Minnesota Press, Minneapolis, Minn.

THE HACKENSACK RIVER MEADOWLANDS

John Berger

Introduction

The Hackensack Meadowlands is a 21,000-acre estuarine area of freshwater and saltwater marshes and meadows situated in the lower Hackensack River basin amidst the New York-northeastern New Jersey metropolitan area (Figure A.13). Almost 18,000 acres of the Hackensack Meadowlands was originally wetland (M. Thiesing, U.S. Environmental Protection Agency, personal communication, 1991),

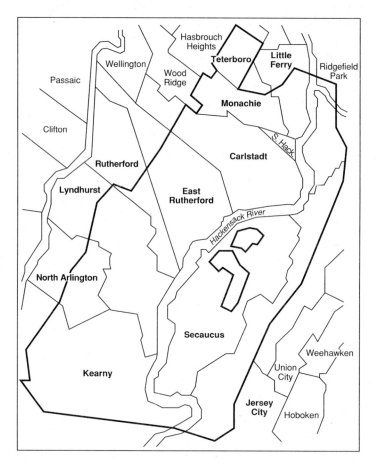

FIGURE A.13 Hackensack Meadowlands district. SOURCE: Reprinted, by permission, of the Hackensack Meadowlands Development Commission, 1990.

but extensive development, drainage, diking, filling, garbage dumping, and sewage pumping have occurred in the Meadowlands, disturbing many of the area's natural ecological processes.

Thirty-two square miles of the Meadowlands have been administered since 1969 by the Hackensack Meadowlands Development Commission (HMDC). At the time the HMDC was established, the Hackensack River reportedly was "nearly dead," and the Meadowlands' wetlands were being used as a disposal site for 30 to 40 percent of New Jersey's garbage (Scardino, 1990). Illegal waste dumping was also common, and development was proceeding in a haphazard manner (Scardino, 1990).

However, commission documents report that during the commission's tenure, "[t]he district has seen drastic improvement in its environment; the Hackensack River has returned to a state of health; wildlife is returning to the district in abundance [and] water quality has greatly improved . . ." (HMDC, 1989a). Former New Jersey Governor Thomas H. Kean in 1989 commended the HMDC on "the restoration of the environment of this once blighted landscape" (HMDC, 1989a). Responding to similar accounts of environmental restoration, the Committee on Restoration of Aquatic Ecosystems visited the Meadowlands in 1990 to gather evidence of environmental restoration.

The Hackensack Meadowlands Development Commission

Established by an act of the New Jersey legislature, the HMDC was set up to provide for the reclamation, planned development, and redevelopment of the Hackensack Meadowlands within Bergen and Hudson counties, a zone including 14 municipalities (HMRDA, 1968). The commission was also charged with providing garbage disposal sites for 116 communities (HMDC, 1989b).

Currently operating with a $5.5 million annual budget (A. Galli, Hackensack Meadowlands Development Commission, personal communication, 1991), the commission usually monitors 500 to 600 development projects in the district at a time (HMDC, 1989b) and by 1989 had overseen privately funded development worth more than $1 billion (HMDC, 1989a). Another $450 million in "publicly backed funds" have been spent on a 750-acre sports complex in the Meadowlands.

Some of this growth has impinged on natural areas. From the commission's inception until 1984, more than 863 acres of wetlands were filled in accordance with the HMDC's master plan. Little filling has occurred since then (D. Smith, Hackensack Meadowlands Development Commission, personal communication, 1991). Wetland habitat enhancement work has been performed on only 190 acres in mitigation for the wetland filling or drainage. All but a few acres of this mitigation work was paid for by developers. (Although the HMDC controls all construction in the district, permits to fill wetlands are principally the responsibility of the U.S. Army Corps of Engineers (COE) under Section 404 of the Clean Water Act of 1977, subject to concurrence by the U.S. Environmental Protection Agency.) Commission literature states that "the Meadowlands wetlands are in need of restoration, not simple protection," but that neither the state nor the federal government would pay for the restoration, "so the HMDC is left on its own to solve the problem" (HMDC, 1989b). The HMDC's solution is not to use any substantial part of its operating revenue to restore the Meadowlands, but to allow certain Hackensack wetlands to be filled in exchange for developer-sponsored mitigation (HMDC, 1989b).

The commission's emphasis on development was consistent with its original 1968 mandate. The Hackensack Meadowland Reclamation and Development Act (HMRDA), which established the commission, noted that, whereas extensive portions of the Meadowlands "have so far resisted development . . . the orderly, comprehensive development of these [Meadowlands] areas can no longer be deferred. . . ."

The commission has pursued this goal while also taking action to improve environmental conditions in the Meadowlands by exercise of its zoning powers and advisory role on discharge permit applications. In general, the commission sought the upgrading of sewage treatment plants, and the closure and cleanup of chemical manufacturing plants and toxic waste sites. It also oversaw the closure of 23 of 24 operating landfills in the Meadowlands; it blocked the use of wetlands for new garbage dumps; and it has generally served as a "watchdog" on environmental conditions for the New Jersey Department of Environmental Protection and the U.S. Environmental Protection Agency (R. Smith, Hackensack Meadowlands Development Commission, personal communication, 1991).

This environmentally oriented activity was in keeping with the declaration, in the Act establishing the HMDC, that "the ecological factors constituting the environment of the meadowlands and the need to preserve the delicate balance of nature must be recognized to avoid any artificially imposed development that would adversely affect not only this area but the entire state . . ." (HMRDA, 1968).

Exactly what constitutes the "delicate balance of nature" in a highly disrupted area has been a matter of some controversy in the years following the establishment of the commission. The HMDC's 1972 master plan and zoning regulations, for example, were approved by the Office of Coastal Zone Management (of the National Oceanic and

Atmospheric Administration) over the objections of the COE, the U.S. Fish and Wildlife Service, the Environmental Protection Agency, and the National Marine Fisheries Service (U.S. Army Corps of Engineers, 1982). (The HMDC is now engaged in preparation of a programmatic Environmental Impact Statement for a new master plan.)

The 190 acres of mitigation work performed in the Meadowlands to date has been conducted mainly by the New Jersey Turnpike Authority, the Bellemead Development Corporation, the Hartz Mountain Development Corporation, and to a minor extent, the HMDC itself (on a small wetland and swale around a landfill). The most detailed mitigation information available to this committee deals with the Hartz Mountain project and its mitigation. To assess its merits as a restoration, one must compare the conditions produced by the project with the ecological conditions prior to disturbance.

History

The Hackensack Meadowlands rest in the ancient basin of a lake formed during the retreat of the Wisconsin glaciation, when glacial meltwater was trapped behind a terminal moraine of rock and earth (HMDC, 1984). Over long periods of time, sediments were deposited in the lake bed, and vegetation took root in the lake's shallow reaches. Eventually, thousands of years ago, the moraine was breached, the lake drained, and tidal flows mingled with fresh waters in the resulting estuary (HMDC, 1984).

Much time passed, and a succession of plant communities came and went, competing with each other and struggling to adapt to environmental fluctuations, including the sea level changes that altered salinities in the estuary. Ecological studies dating back to the late nineteenth century indicate that in the last phase of its natural succession, the Hackensack Meadowlands was a boggy area dominated by Atlantic white cedar (*Chamacyparis thyoides*) in a region of black ash (*Fraxinus nigra*) and tamarack (*Larix laricina*) (Kraus and Smith, n.d.). The Hartz Mountain project site may have been highly brackish marsh dominated by salt hay (*Spartina patens*) and salt grass (*Distichlis spicata*) with a white Atlantic cedar bog at its upland edge, before the whole area was ditched and then diked for mosquito control between 1914 and 1950 (Kraus and Smith, n.d.; HMDC, 1984; D. Smith, Hackensack Meadowlands Development Commission, personal communication, 1991). (This issue was not independently verified by the committee.)

The altered hydrology quickly led to major changes in vegetation. With tidal flow excluded and water salinity reduced, the common reed (*Phragmites australis*) invaded the area and became the dominant vegetation. Another major change in the area's hydrology occurred in 1922 when the Oradell Dam was built across the Hackensack River upstream from the project site by the Hackensack Water Company. Reduction in freshwater flow further altered ecological conditions in the estuary by allowing greater saltwater intrusion upstream into the Hackensack basin below the dam.

The original Hartz Mountain project—a mall, office complex, and condominiums—was proposed for the Cromakill Creek and Mill Creek basins of the Meadowlands in the Township of North Bergen and the Town of Secaucus. Subsequently, the condominium component was dropped or delayed, and the company received permission to fill 127 acres of wetlands to build the mall at Mill Creek and the office complex. The COE reviewed the Hartz Mountain proposal and issued a finding in 1982 that it would have "no significant adverse environmental impacts" (U.S. Army Corps of Engineers, 1982). Thus COE did not require an environmental impact statement.

Based on an interagency wetland evaluation conducted by several federal agencies, the site—still dominated by common reed—was deemed to be of only average value as a wetland. Water quality was found to be poor; vegetation diversity was low; and benthic invertebrates and fish were rated "low to medium" (U.S. Army Corps of Engineers, 1982). The COE noted, however, that filling the wetland would "essentially destroy all wildlife values within the fill area" and that loss of the wetland would reduce the highly desirable isolation of other remaining wetlands, increasing noise levels and the probability of further human encroachment on the remaining wetlands (U.S. Army Corps of Engineers, 1982).

The COE also observed that the land would have much greater potential wildlife value if the water quality were improved and the diversity of wetland vegetation increased. "With improved water quality, loss of wetlands would be of much greater concern" (U.S. Army Corps of Engineers, 1982). The Hartz Mountain project was allowed to proceed with the stipulation that the company would have to mitigate its impacts by construction of a 63-acre brackish marsh. The mitigation site was slightly less than half the size of the filled wetlands, but the new marsh was intended to be of much higher ecological value.

The brackish marsh ecosystem is in Secaucus, N.J., approximately south of Hackensack River mile 10.5, adjacent to the eastern shore of Mill Creek, and west of the eastern branch of the New Jersey Turnpike. As noted, the site before 1985 was a degraded tidal marsh with poor water quality, dominated by tall, dense stands of common reed

(*Phragmites australis*) (TAMS, 1990). Tidal inundation was limited by site elevations ranging from +9.6 ft National Geodectic Vertical Datum (NGVD) to 0.0 NGVD.

The mitigation goals were to enhance wildlife diversity and abundance by converting the site from a common reed-dominated community to a cordgrass (*Spartina alterniflora*) intertidal marsh. The plan adopted was to remove the common reed in the process of lowering the site's elevation by excavation to increase tidal inundation. An effort was also made to construct a more heterogeneous habitat, including open water and raised areas of woody vegetation in order to increase vegetative diversity and wildlife use.

Replacement of the common reed by cordgrass offers several ecological benefits. Cordgrass detritus regularly enters marsh waters and breaks down relatively quickly, releasing nutrients. The detritus from common reed, which grows on higher ground, is only washed into the water on an irregular basis and decomposes relatively slowly (HMDC, 1984). Very dense stands of common reeds are not considered to be of high value to waterfowl, marsh mammals, and wading shorebirds (TAMS, 1990). In addition, the reed is very persistent, invasive, and robust, contributing to drying of marsh soil, reduction of water flow, and increases in site elevations through growth and accumulation of organic matter and ensuing entrapment of sediment.

However, among the common reed's ecological services are provision of habitat for large populations of aphids that in turn support large numbers of ladybugs, which provide food for praying mantises, birds (HMDC, 1984), and occasionally for fish.

Methods

The mitigation site was sprayed with the herbicide RODEO by helicopter and later by hand-sprayer to eliminate the common reed. The site was then shaped and graded with Priestman variable counterbalanced excavators imported from England for the marsh work, because of their low ground pressure and ability to accomplish the very fine gradations in elevations necessary to successfully establish the elevation-sensitive cordgrass. The horticultural contractor was Environmental Concern, Inc., of St. Michaels, Md., a firm well known for its pioneering work in salt marsh restoration.

The high marsh was sculpted into channels and open water, lower-elevation intertidal zones, and raised areas (berms) from +5.73 to +10.33 NGVD, built up of excavated materials. The earthwork was done from March 1985 to July 1987. Cordgrass seed was planted each spring from 1986 through 1988. Detailed biological and other monitoring has been done by TAMS Consultants, Inc., on the site and at an untreated 131-acre control site, also dominated by common reed and similar to the pretreatment mitigation site.

Results

Initial plantings of trees, shrub root stocks, and herbaceous vegetation experienced high mortality due to high soil salt content that was allowed to leach out with rainwater during the next 2 to 3 years. Plant survival on the berms improved in 1988 as leaching continued (TAMS, 1990). The plantings are in an early stage of establishment and are fenced to discourage muskrat depredation.

More than 80 percent of the site is now inundated during part of the mean tide cycle, and a vigorous growth of cordgrass has become established on more than 75 percent of the intertidal zones between +2.0 and +3.0 NGVD. Common reeds have not reappeared in the cordgrass zones. Where they reemerged on berms, they have apparently been controlled by hand-spraying with RODEO. Some native marsh species, such as fleabane, rushes, and sedges, have reappeared naturally on the site. Channels appear to be stable throughout the site (TAMS, 1990).

Although it is too early for the mitigation site to have fully recovered from earth-moving operations, fish, benthic organisms, and zooplankton already appear similar to those at the ecologically impaired control site, whereas bird life has become much more abundant and diverse (TAMS, 1990). Because of the creation of more channels, greater water surface area for oxygen exchange, and greater tidal flushing, water quality on-site seems to approximate values in the adjacent Hackensack River. High levels of coliform bacteria are still found in water samples from the site, and benthic organism samples contained a large proportion of a few pollution-tolerant species, indicative of a stressed ecosystem. Almost all the fish found at the control and mitigation site were mummichogs (*Fundulus heteroclitus*), an important secondary consumer in the eastern salt marsh food web, but their physical distribution was wider in the mitigation site, as was the case for zooplankton (TAMS, 1990).

Bird species diversity was markedly greater on the mitigation site (46 species) versus the control (32 species), and the distribution among species was also more equitable on the mitigation site, probably in response to its greater habitat diversity and secondary productivity (TAMS, 1990).

Conclusion

The intertidal cordgrass marsh created out of high marsh at the mitigation site appears to have met the goals of enhancing habitat heterogeneity, vegetational diversity, and wildlife utilization, principally by birds. In this sense the project has been a success, and the engineering and biological science used appears to be of a high caliber. However, the project should be viewed as habitat enhancement and conversion rather than ecosystem restoration for the following reasons:

1. The mitigation did not endeavor to re-create the particular estuarine ecosystem that existed on the site prior to the damming of the Hackensack River and prior to other significant environmental modifications that have occurred in the Meadowlands. By altering the hydrology of the area and the salinity of its water and soil, the Oradell Dam made restoration of vegetation adapted to less saline conditions impossible without the reintroduction of additional fresh water to the project site.

2. Because of the limited areal scope of the mitigation work and the limited goals, the mitigation project had virtually no impact on the regionwide ecological degradation of the Meadowlands—exemplified by the damming and ditching of Meadowland marshes, the blockage of the Hackensack River, the presence of sewage and toxic substances in soil, and the extirpation of certain species. Therefore the resulting ecosystem cannot be considered "restored" because of the influence of these intractable conditions on the mitigation project site.

3. Where once there was probably a high marsh of *Spartina patens*, *Distichlis spicata*, and other species, the contractors produced an intertidal marsh with mud flats and raised inlands of woody vegetation. There is no evidence that the ecosystem created on the mitigation site has existed there within human memory.

The regulated development of the HMDC is far better than the indiscriminate dumping and haphazard development that preceded the HMDC in the 1950s and 1960s. Water quality in the Hackensack River appears to be far better than the sewer-like conditions reported 20 years ago. Evidence is undeniable that certain aquatic organisms, such as grass shrimp and mummichog, are now thriving in vast numbers and that certain species of waterfowl and fish have returned.

However, as the Committee on Restoration of Aquatic Ecosystems has pointed out elsewhere, river restoration involves more than water quality improvement and increased wildlife use. Also required are a return of ecological integrity, structure, function, and ecosystem processes, beginning with natural hydrological conditions and including restoration of communities of organisms and their interactions. An increase in the presence of a wildlife species is generally a promising indication that ecological health is returning but is insufficient cause for proclaiming that restoration has occurred.

The HMDC has in the past sanctioned the development of substantial wetland acreage rather than protecting all wetlands. The commission thereby set a precedent of trading wetland development for wetland enhancement, with a resulting net loss of wetland acreage in a quest for increased wetland functional values. There are alternatives to that strategy. The commission might instead gradually begin to invest some of its own not inconsiderable revenues directly in wetland restoration year by year (and solicit federal, state, local, and private funds to augment its contribution), without choosing to sacrifice additional wetland acreage to subsidize wetland improvement.

In the future, too, the commission may wish to consider developing a systematic mitigation or ecological restoration program for the Meadowlands in which individual mitigations are conducted as part of a broader overall restoration strategy.

References

Axelrod, H. R., C. W. Emmens, D. Sculthorpe, and W. Vorderwinkler. 1962. Exotic Tropical Fishes. T. F. H. Publications, Jersey City, N.J.

Clean Water Act of 1977. P.L. 95-217, Dec. 27, 1977, 99 Stat. 1566.

Hackensack Meadowlands Development Commission (HMDC). 1984. Wetland Bio-Zones of the Hackensack Meadowlands: An Inventory. Lyndhurst, N.J. June.

Hackensack Meadowlands Development Commission. 1989a. Annual Report. Lyndhurst, N.J.

Hackensack Meadowlands Development Commission. 1989b. Fact Sheet. Lyndhurst, N.J. October.

Hackensack Meadowlands District. 1990. Special Area Management Plan. Lyndhurst, N.J.

Hackensack Meadowland Reclamation and Development Act (HMRDA). 1968. State of New Jersey Statutes. Chapter 17, Sections 13:17-1 to 13-17-86.

Kraus, M. L., and D. J. Smith. n.d. Competition and Succession in a Perturbed Urban Estuary: The Effects of Hydrology. Monograph. Hackensack Meadowlands Development Commission, Lyndhurst, N.J.

Scardino, A. 1990. Executive Director, Hackensack Meadowlands Development Commission. Briefing to Committee on Restoration of Aquatic Ecosystems. Lyndhurst, N.J.

TAMS Consultants, Inc. 1990. Comprehensive Baseline Studies: IR-2 Site and Off-Site Mitigation Areas—Evaluation of Harmon Meadow Western Brackish Marsh Mitigation Area. January. New York, N.Y.

U.S. Army Corps of Engineers. 1982. Statement of Findings for Application No. 81-391-J2 by the Hartz Mountain Development Corporation. New York District. Regulatory Branch. December 16.

Appendix B

Glossary

If you wish to converse with me, define your terms.

Voltaire

ARTIFICIAL MARSH CREATION—Simulation of natural wetland features and functions by topographic and hydraulic modification of nonwetland landscapes. Typical objectives of artificial marsh creation include ecosystem replacement or storm water management.

BANK STABILIZATION—Methods of supporting the structural integrity of earthen stream channel banks with structural supports to prevent bank slumping and undercutting of riparian tress, as well as overall erosion. Recommended bank stabilizing techniques include the use of willow stakes, overlapping riprap, or brush bundles.

BIOGEOGRAPHIC REGION—Any region delineated by its biological and geographic characteristics.

BIOLOGICAL MONITORING—Surveillance to ensure that previously established quality control conditions are being met. Biological monitoring surveys can span the trophic spectrum from algae and other aquatic plants, to macroinvertebrates, to fish species.

CHANNELIZATION—Straightening of the meanders in a river system to create more navigable waterways, or when accompanied by channel deepening to provide flood control.

COMPENSATION—Provision for creation or restoration of "equivalent" wetland acres comparable to wetland acres and functions that have been destroyed.

CREATION—Bringing into being a new ecosystem that previously did not exist on the site.

DIATOMS—Microscopic plants are abundant in plankton. They sometimes produce a water "bloom" and give a yellowish or brownish tint to the water. Diatoms are notable for their shells of silica and the siliceous character imparted to bottom deposits by their remains.

ECOLOGICAL INTEGRITY—Maintenance of the structure and functional attributes characteristic of a particular locale, including normal variability.

ECOREGIONS—Ecological regions that have broad similarities with respect to soil, relief, and dominant vegetation.

ECOSYSTEM—A biological community together with the physical and chemical environment with which it interacts.

ECOSYSTEM CREATION—Construction of an ecosystem with new species, soil, and vegetation on a site that had a different type of system before destruction or damage occurred. (See creation.)

ECOSYSTEM FUNCTION[1]—Any performance attribute or rate function at some level of biological organization (e.g., energy flow, detritus processing, nutrient spiraling).

EMERGENT PLANTS—Aquatic plants that are rooted in the sediment, but whose leaves are at or above the water surface. These wetland plants provide habitat for wildlife and waterfowl in addition to removing urban pollutants.

ENHANCEMENT—In the context of restoration ecology, any improvement of a structural or functional attribute. Odum et al. (1979) have defined whatever contributes to enhancement as a subsidy.

EPILIMNION—In a thermally stratified lake, the turbulent layer of water that extends from the surface to the metalimnion.

EUTROPHICATION—Enrichment of lakes with nutrients. Increase in nutrients required for the growth of organisms may come about by natural processes, or rapid enrichment may take place due to some cause such as introduction of sewage effluent.

EUTROPHIC LAKES—"Rich" lakes; those well provided with the basic nutrients required for plant and animal production. In some lakes this enrichment becomes harmful, and light penetration and oxygen production are insufficient to maintain productivity. Oxygen is then consumed at a rate equal to that at which it is produced.

FLOODPLAIN—Defined by hydrologists as the area flooded at a recurrence interval of once in 100 years (Bhowmik and Stall, 1979). Ecologists define floodplains as areas that are periodically inundated (usually annually) by the lateral overflow of rivers or lakes, or by direct precipitation or ground water; the resulting physicochemical environment causes the biota to respond by morphological, anatomical, physiological, phenological, and/or ethological adaptations, and to produce characteristic community structures (Junk et al., 1989).

FLOOD PULSE CONCEPT—The idea that the pulsing of river discharge, the flood pulse, is the major force controlling biota in river-floodplain systems (Junk et al., 1989). Lateral exchange between the floodplain and river channel, and nutrient cycling within the floodplain, are postulated to have a more direct impact on biota than does nutrient spiraling from upstream to downstream in the river channel; the bulk of the animal biomass in such a system is believed to be derived from production in the floodplain, not from downstream transport of organic matter produced elsewhere in the basin.

FUNCTIONAL EQUIVALENCY[1]—Ability of a restored or created ecosystem to perform ecosystem services, such as floodwater storage or pollutant detoxification, that are indistinguishable in effects from corresponding services performed by natural ecosystems.

FUNCTIONAL IMPROVEMENT[1]—Improving a functional attribute by restoring an ecosystem, preferably without impairing other attributes.

FUNCTIONAL VALUES[1]—With regard to ecosystems, any attributes perceived to have societal value (e.g., waste transformation, carbon dioxide removal).

GREENWAY—A protected linear open-space area that is either landscaped or left in its natural condition. It may follow a natural feature of the landscape, such as a river or stream, or it may occur along an unused railway line or some other right of way.

HYPOLIMNION—In a thermally stratified lake, the layer of water below the thermocline and extending to the bottom of the lake; water temperature in the hypolimnion is virtually uniform.

HYPORHEIC—Pertaining to the hyporheos.

HYPORHEOS—The saturated zone beneath a river or stream consisting of substrate, such as sand, gravel, and rock, with water-filled interstitial pore. The zone often extends beyond the width of the stream channel and is typically used by certain aquatic organisms during their normal life cycle and as a refuge.

LANDSCAPE ECOLOGY—An emerging specialty that deals with the patterns and processes of biological systems in spatially and temporally heterogeneous environments.

LANDSCAPE PERSPECTIVE—A way to view interactive parts of a watershed not necessarily all within one watershed.

LENTIC WATERS—Still waters such as lakes or ponds; any body of standing water.

LOTIC WATERS—Any body of moving or flowing water such as a river or other stream.

MESOTROPHIC LAKE—A lake that is intermediate in fertility, neither notably high nor notably low in its total productivity. Intermediate between oligotrophic and eutrophic.

METALIMNION—In a thermally stratified lake, a layer of water between the epilimnion and hypolimnion that is characterized by a sharp change in temperature or density with depth.

MITIGATION—Actions taken to avoid, reduce, or compensate for the effects of environmental damage. Among the broad spectrum of possible actions are those that restore, enhance, create, or replace damaged ecosystems.

NATIVE FISH REINTRODUCTION—The final phase of a stream restoration effort, in which fish are reintroduced into a physically restored stream system to reestablish the original fish community.

OLIGOTROPHIC LAKES—Lakes poorly provided with the basic nutrients required for plant and animal production.

OPPORTUNITY-COST ANALYSIS—Accepts a human-based determination of value but looks to collective action to define values achieved by restoration. This type of analysis of a restoration effort differs from traditional benefit-cost analysis. Continually questioning the value of a restoration by asking whether an action is "worth" its cost is the most practical way to decide how much restoration is enough.

PATCH OF AN ECOSYSTEM—The term *patch dynamics* is used regularly in the newly developing fields of landscape ecology and conservation biology, but it has been used previously as a standard ecological term. Use of the term *patch* recognizes that most ecosystems are not homogeneous, but rather a group of patches or ecological islands that are recognizably different from the parts of the ecosystem that surround them but nevertheless interact with them.

RECLAMATION—A process designed to adapt a wild or natural resource to serve a utilitarian human purpose. Putting a natural resource to a new or altered use. Often used to refer to processes that destroy native ecosystems and convert them to agricultural or urban uses.

REHABILITATION—Used primarily to indicate improvements of a visual nature to a natural resource; putting back into good condition or working order.

RESOURCE SPIRALING—Combined process of resource cycling and downslope or downstream transport (Elwood et al., 1983). A nutrient atom or organic molecule may pass through the same trophic level or chemical state (cycling) during its residence in a stream, but completion of the cycle involves some downstream displacement before the cycle is closed (spiraling).

RESTORATION—Return of an ecosystem to a close approximation of its condition prior to disturbance.

RIPARIAN REFORESTATION—Replanting of the banks and floodplains of a stream with native forest and shrub species to stabilize erodible soil, improve both surface and ground water quality, increase stream shading, and enhance wildlife habitat.

RIPARIAN VEGETATION—Hydrophytic vegetation growing in the immediate vicinity of a lake or river close enough so that its annual evapotranspiration represents a factor in the lake or river regimen.

RIPARIAN ZONE—The border or banks of a stream. Although this term is sometimes used interchangeably with floodplain, the riparian zone is generally regarded as relatively narrow compared to a floodplain. The duration of flooding is generally much shorter, and the timing less predictable, in a riparian zone than in a river floodplain.

RIVER CONTINUUM CONCEPT—The idea that a continuous gradient of physical conditions exists from headwaters to mouths of rivers, and that structural and functional characteristics of biological communities are adapted to conform to the most probable position or mean state of the physical system (Vannote et al., 1980). Producer and consumer communities establish themselves in harmony with the dynamic physical conditions of a given reach, and downstream communities are fashioned to capitalize on the inefficiencies of upstream procession of organic matter. Both upstream inefficiency (leakage) and downstream adjustment seem predictable.

RIVERINE WETLANDS—Wetland systems of less than 0.5 ppt ocean salts, exposed to channelized flow regimes. Riverine wetlands are categorized according to flow regimes such as tidal waters, slow-moving waters with well-developed floodplains, fast-moving waters with little floodplain, and intermittent systems.

ROUGH FISH—Species, such as carp and sucker, considered undesirable by anglers.

SELF-MAINTAINING SYSTEM—An ecosystem that can perform all of its natural ecological functions without human intervention or dependence on engineered structures.

SERIAL DISCONTINUITY CONCEPT—The idea that dams shift the physical and biological characteristics of streams and rivers away from the pattern predicted by the river continuum concept (Ward and Stanford, 1983). A dam may make conditions more like those of the headwaters (an upstream shift) or more like those downstream, or it may have a negligible effect. Multiple dams create multiple discontinuities in the expected or natural pattern of streams and rivers.

SPATIAL HETEROGENEITY—Variation in the attributes of an environment over space.

SPIRALING—See resource spiraling.

STREAM BUFFER—A variable-width strip of vegetated land adjacent to a stream that is preserved from development activity to protect water quality and aquatic and terrestrial habitats.

STREAM ORDER—Rivers and streams are classified by order. The order of a river or stream is a dimensionless number that indicates how many tributaries it has. The smallest unbranched tributary in a watershed is designated order 1. A channel formed by the confluence of two such tributaries is designated order 2. Where 2 order tributaries join, a channel segment of order 3 is formed, and so on. In general, the higher the order number, the larger is the watershed, and the greater are the channel dimensions and discharge (Chow, 1964).

STREAM RESTORATION—Various techniques used to replicate the hydrological, morphological, and ecological features that have been lost in a stream due to urbanization, farming, or other disturbance.

SURVEILLANCE—Periodic and systematic surveys undertaken to determine the condition of an ecosystem.

SYSTEMS PERSPECTIVE—A synthesis of all the components that collectively interact in such a way that effects on one component affect the entire system.

THERMOCLINE—In thermally stratified lakes, the layer below the epilimnion. It is the stratum in which there is a rapid rate of decrease in temperature with depth; a minimum of one degree centigrade per meter in depth.

TIDAL MARSH—Salt or brackish wetland systems subject to tidal flow patterns. Marsh estuaries are subdivided into subtidal (permanently flooded) or intertidal (temporarily flooded) wetland systems.

WATERSHED—The entire surface drainage area that contributes water to a lake or river.

WATERSHED-SCALE APPROACH—A consideration of the entire watershed, including the land mass that drains into the aquatic ecosystem.

WETLAND MITIGATION—Replacing wetland areas destroyed or impacted by proposed land disturbances with artificially created wetland areas.

WETLANDS—Term for a broad group of wet habitats. Wetlands are lands transitional between terrestrial and aquatic systems where the water table is usually at or near the surface or the land is covered by shallow water (Cowardin et al., 1979). Wetlands include features that are permanently wet, or intermittently water covered, such as swamps, marshes, bogs, muskegs, potholes, swales, glades, slashes, and overflow land of river valleys. According to the 1989 federal wetlands delineation manual, wetlands include lands saturated for at least 7 days to a depth of 12 inches. A newly proposed definition by the Bush Administration would be lands that have 15 days of standing water and 21 days of surface saturation.

ENDNOTE

1. For extended discussion of functional attributes, see Cairns, J., Jr., and J. R. Pratt, eds. 1989. Functional Testing of Aquatic Biota for Estimating Hazards of Chemicals. Special Technical Publication 988. American Society for Testing and Materials, Philadelphia, Pa. 242 pp. For contrast with structural values, see Cairns, J., Jr., and J. R. Pratt. 1986. On the relation between structural and functional analyses of ecosystems. Environ. Toxicol. Chem. 5:785-786.

REFERENCES

Bhowmik, N. G., and J. B. Stall. 1979. Hydraulic Geometry and Carrying Capacity of Floodplains. Water Research Center, Research Report No. 145, University of Illinois, Urbana, Ill.

Chow, V. T. 1964. Handbook of Applied Hydrology: A Compendium of Water-resources Technology. McGraw-Hill Book Co., New York.

Cowardin, L. M., V. Carter, F. C. Golet, and E. T. LaRoe. 1979. Classification of Wetlands and Deepwater Habitats of the United States. Office of Biological Services, U.S. Fish and Wildlife Service, U.S. Department of the Interior, Washington, D.C.

Elwood, J. W., J. D. Newbold, R. V. O'Neill, and W. Van Winkle. 1983. Resource spiraling: An operational paradigm for analyzing lotic ecosystems. Pp. 3-27 in Thomas D. Fontaine III and Steven M. Bartell, eds., Dynamics of Lotic Ecosystems. Ann Arbor Science Publishers, Ann Arbor, Mich. 494 pp.

Junk, W., P. B. Bayley, and R. E. Sparks. 1989. The flood pulse concept in river-floodplain systems. Proceedings of the International Large River Symposium (LARS). Canadian Special Publication of Fisheries and Aquatic Sciences 106:110-127.

Odum, E. P., J. T. Finn, and E. H. Franz. 1979. Perturbation theory and the subsidy-stress gradient. BioScience 34:558-562.

Vannote, R. L., G. W. Minshall, K. W. Cummins, J. R. Sedell, and C. E. C. Cushing. 1980. The river continuum concept. Can. J. Fish. Aquat. Sci. 37:130-137.

Ward, J. V., and J. A. Stanford. 1983. The serial discontinuity concept of lotic ecosystems. Pp. 29-42 in T. D. Fontaine III and S. M. Bartell, eds., Dynamics of Lotic Ecosystems. Ann Arbor Science Publishers, Ann Arbor, Mich. 494 pp.

Appendix C

Biographical Sketches of Committee Members and Professional Staff

JOHN CAIRNS, JR., received his Ph.D. in zoology from the University of Pennsylvania in 1953. His expertise lies in the ecology of freshwater protozoa and the response of aquatic organisms to toxic substances; water management; rapid biological information systems; the ecology of polluted water; regional environmental analysis; hazard evaluation, including evaluation of toxic chemicals; and restoration of damaged ecosystems. Since 1968, he has held the positions of University Distinguished Professor of Environmental Biology and director, University Center for Environmental and Hazardous Materials Studies, Virginia Polytechnic Institute and State University. He was a founding member of the Water Science and Technology Board and was elected to the National Academy of Sciences in 1991.

G. RONNIE BEST obtained his M.S. (1971) and Ph.D. (1976) at the University of Georgia, Athens, in ecology and botany. Since 1979, Dr. Best has held the positions of associate director, Center for Wetlands, and associate research scientist at the University of Florida. His areas of expertise include plant community ecology, ecosystems, mineral cycling, wetlands and wetlands management, mycorrhizae, and reconstruction of disturbed ecosystems.

PATRICK L. BREZONIK obtained his Ph.D. (water chemistry) in 1968 from the University of Wisconsin. His professional experience includes positions as assistant professor of water chemistry and environmental engineering, University of Florida (1966-1970); associate professor of water chemistry (1970-1976), professor of environmental science (1976-1981), and professor of environmental engineering, University of Minnesota since 1981; and director, Water Resources Research Center since 1985. His research interests are the eutrophication of lakes, nitrogen dynamics in natural water, acid rain, and organic matter in water.

STEPHEN R. CARPENTER received his Ph.D. in botany (1979) from the University of Wisconsin, Madison. He served from 1979 to 1989 on the faculty at the University of Notre Dame, where he taught courses in biostatistics and aquatic ecology and developed limnological research programs at the Environmental Research Center near Land O' Lakes, Wisconsin. Dr. Carpenter has been Bassett Research Professor at the Center for Limnology, University of Wisconsin, Madison, since 1989. His research interests include ecosystem experiments, ecological modeling, plankton, and macrophytes. He has served on several committees and editorial boards for the Ecological Society of America and the American Society of Limnology and Oceanography.

G. DENNIS COOKE received his B.S. in biology (1959) from Kent State University, and his M.S. (1963) and Ph.D. (1965) in zoology from the University of Iowa. He has been involved in basic and applied research in zoology since the late 1960s, has served as a consultant to the U.S. Environmental Protection Agency's Clean Lakes Program, was one of the originators and the first president (1980-1981) of the North American Lake Management Society, and is associate editor of both the *Water Resources Bulletin* and *Lake and Reservoir Management*. Dr. Cooke has been a professor of biological sciences at Kent State University since 1976.

DONALD L. HEY received a B.S. in civil engineering (1963) from the University of Missouri at Rolla, an M.S. in water resources engineering (1968) from Kansas University, and a Ph.D. in environmental engineering (1974) from Northwestern University. He is currently director of Wetlands Research, Inc., in Chicago, Illinois, and president of Hey and Associates, Inc., in Chicago. A few of his past positions include adjunct professor, Northeastern Illinois University, Chicago; vice president, Hydrocomp, Inc., Palo Alto, California and Chicago, Illinois; and research analyst and engineer, Public Works Department, City of Chicago. His research interests include producing criteria necessary for restoring river systems through the use of wetlands and developing management programs for the continued operation of the new structure. Dr. Hey is a member of the American Geophysical Union, the Society of Sigma Xi, Lambda Alpha International, and the American Society of Civil Engineers.

JON A. KUSLER obtained a B.S. in English and geology from the University of Wisconsin (with work at the University of Minnesota and the University of San Francisco); an M.S. in water resources management; his Ph.D. in land and water use management (an interdisciplinary degree involving law, geology, and economics); and a J.D. in law. Dr. Kusler is a lawyer, writer, and educator with 26 years of experience working with legal, science, and policy issues. He has served on the staffs of the Universities of Wisconsin and Massachusetts, and is currently the executive director of the Association of State Wetland Managers and legal counsel to the Association of State Floodplain Managers. He has served as a policy advisor to the U.S. Environmental Protection Agency's Office of Wetlands Protection, the former U.S. Water Resources Council, the U.S. Army Corps of Engineers, and the U.S. Fish and Wildlife Service. He has also acted as an advisor to the National Wetland Policy Forum and many state governments.

CLAIRE L. SCHELSKE, who has a Ph.D. in zoology, is currently at the University of Florida. His professional experience has included work at the University of Georgia Marine Institute and as a fishery biologist, Radiobiological Laboratory, Bureau of Commercial Fisheries. At the University of Michigan he was a research scientist in the Great Lakes Division from 1967 to 1987 and acting director from 1973 to 1976. He also served as an associate professor of limnology at the University of Michigan. Dr. Schelske's research interests include eutrophication, biogeochemistry, and paleolimnology of the Great Lakes and freshwater ecosystem ecology.

LEONARD SHABMAN received a Ph.D. in agricultural economics in 1972 from Cornell University. He has been with Virginia Polytechnic Institute and State University since 1972, first as an assistant professor and currently as a professor in the Department of Agricultural Economics. His responsibilities include the conduct and management of a research program in resource and environmental policy analysis; classroom teaching; and undergraduate and graduate student advising. Dr. Shabman has conducted economic research since 1972 over a wide range of topics in natural resource and environmental policy. He was an economic advisor to the Water Resources

Council in 1977-1978 and scientific advisor to the Assistant Secretary of the Army, Civil Works in 1984-1985.

REBECCA R. SHARITZ acquired her Ph.D. in botany in 1970 from the University of North Carolina, Chapel Hill. She is currently a professor of botany at the University of Georgia and a senior ecologist at the Savannah River Ecology Laboratory, where she is also head of the Division of Wetlands Ecology. Her research interests are plant population biology, structure and diversity of plant communities, and wetlands ecology.

SOROOSH SOROOSHIAN received his B.S. in mechanical engineering from California Polytechnic State University, San Luis Obispo, in 1971; an M.S. in systems engineering from the University of California-Los Angeles (UCLA) in 1973; and a Ph.D. from UCLA in 1978 in systems engineering, with specializations in water resources and hydrologic systems analysis. He is currently a professor and head of the Department of Hydrology and Water Resources, and a professor in the Department of Systems and Industrial Engineering at the University of Arizona, Tucson. He is also serving a four-year term (1988-1991) as editor of *Water Resources Research*, published by the American Geophysical Union. His research interests are surface hydrology, including rainfall-runoff modeling; flood forecasting; application of remote sensing in hydrology; and climate studies.

RICHARD E. SPARKS acquired his Ph.D. (biology) in 1971 from Virginia Polytechnic Institute and State University. He is currently an aquatic biologist, Illinois Natural History Survey, and director of the River Research Laboratory of the Forbes Biological Station on the Illinois River at Havana, Illinois, where he has worked since 1972. Dr. Sparks's interests include biological monitoring for pollution control, by using organisms as sensors; restoration of degraded aquatic ecosystems; ecology of large floodplain rivers; bioassays using aquatic organisms; and river surveys. He is a member of the American Fisheries Society, the Ecological Society of America, and Sigma Xi.

JAMES T. B. TRIPP obtained his L.L.B. from Yale Law School in 1966 along with an M.S. in philosophy from Yale's graduate school. He was Assistant U.S. Attorney for Southern New York State from 1968 to 1973. Since 1973, he has been with the Environmental Defense Fund as head of its Eastern Water Resources and Land Use Program. His interests include wetlands protection and ground water quality protection. He was formerly a member of the National

Research Council Committee on Ground Water Quality Protection: State and Local Strategies.

DANIEL E. WILLARD is an ecologist and wetlands biologist. He received an A.B. in biology (1959) from Stanford University and a Ph.D. in zoology (1966) from the University of California, Davis. He taught zoology at the University of Texas from 1966 to 1970, and at the University of Wisconsin through 1977. He has taught at the University of Oregon's Institute of Marine Biology and Cornell University's Shoals Marine Laboratory. He has served on the Office of Technology Assessment's Wetland Committee, on committees of the National Research Council's Water Science and Technology Board, on the U.S. Environmental Protection Agency's wetland committees, on the National Wetland Policy Forum, and on Indiana's Water and Minerals Advisory Board. He is a professor and the director of environmental science and policy programs in the School of Public and Environmental Affairs, as well as a professor of biology, at Indiana University, Bloomington.

JOY B. ZEDLER holds a Ph.D. in botany (plant ecology) from the University of Wisconsin. Since 1969, she has been at San Diego State University (SDSU) and is currently a professor of biology at SDSU and director of the Pacific Estuarine Research Laboratory. Her research interests include salt marsh ecology; structure and functioning of coastal wetlands; restoration and construction of wetland ecosystems; effects of rare, extreme events on estuarine ecosystems; dynamics of nutrients and algae in coastal wetlands; and use of scientific information in the management of coastal habitats. She recently worked on a compilation of literature on the creation and restoration of wetlands for the U.S. Environmental Protection Agency. Dr. Zedler was appointed as a member of the Water Science and Technology Board in July 1991.

CONSULTANT

JOHN J. BERGER received a B.A. in political science from Stanford University in 1966; an M.A. in energy and resources from the University of California, Berkeley, in 1980; and a Ph.D. in ecology from the University of California, Davis, in 1990. Dr. Berger is currently a visiting associate professor in the environmental policy branch of the University of Maryland's School of Public Affairs. He is also the founder and director of a private, nonprofit organization, Restoring the Earth, that fosters the repair of ecological damage through research, consulting, public policy development, model restoration work, and public education. He has authored and edited books on energy and environmental restoration, and is working to produce a public television documentary on restoration ecology.

PROFESSIONAL STAFF

SHEILA D. DAVID, a senior program officer at the Water Science and Technology Board, served as study director for the Committee on Restoration of Aquatic Ecosystems. On the staff of the National Research Council (NRC) since 1976, she has served as staff director for various NRC study projects, including studies on coastal erosion, ground water protection, water quality and water reuse, and natural resource protection in the Grand Canyon.

JEANNE AQUILINO is the administrative specialist for the Water Science and Technology Board and for the Committee on Restoration of Aquatic Ecosystems. She has been on the staff of the National Research Council since 1979.